Instruments and Techniques of Atmospheric Measurement

Instruments and Techniques of Atmospheric Measurement

Edited by **Bruce Mullan**

SYRAWOOD
PUBLISHING HOUSE
New York

Published by Syrawood Publishing House,
750 Third Avenue, 9th Floor,
New York, NY 10017, USA
www.syrawoodpublishinghouse.com

Instruments and Techniques of Atmospheric Measurement
Edited by Bruce Mullan

International Standard Book Number: 978-1-68286-361-9 (Hardback)

Printed in China.

Contents

Preface

Atmospheric measurement techniques have constantly progressed over the years, owing to the technological advancements. This has fueled more accurate predictions and studies. This book provides comprehensive insights into the techniques of atmospheric measurement. It covers topics like remote sensing, development of measurement instruments, techniques of data processing, etc. This book explores all the important aspects of data processing techniques in the present day scenario. From theories to research to practical applications, case studies related to all contemporary topics of relevance to this field have been included in this text.

This book has been the outcome of endless efforts put in by authors and researchers on various issues and topics within the field. The book is a comprehensive collection of significant researches that are addressed in a variety of chapters. It will surely enhance the knowledge of the field among readers across the globe.

It gives us an immense pleasure to thank our researchers and authors for their efforts to submit their piece of writing before the deadlines. Finally in the end, I would like to thank my family and colleagues who have been a great source of inspiration and support.

Editor

Application of high-resolution time-of-flight chemical ionization mass spectrometry measurements to estimate volatility distributions of α-pinene and naphthalene oxidation products

P. S. Chhabra[1,*], A. T. Lambe[1,2], M. R. Canagaratna[1], H. Stark[1,3], J. T. Jayne[1], T. B. Onasch[1,2], P. Davidovits[2], J. R. Kimmel[1,3,4], and D. R. Worsnop[1]

[1]Aerodyne Research, Inc. Billerica, Massachusetts, USA
[2]Chemistry Department, Boston College, Chestnut Hill, Massachusetts, USA
[3]Cooperative Institute for Research in Environmental Sciences (CIRES), University of Colorado, Boulder, Colorado, USA
[4]TOFWERK AG, Thun, Switzerland
[*]now at: Department of Chemical Engineering, University of Texas, Austin, Texas, USA

Correspondence to: P. S. Chhabra (pschhabra@utexas.edu)

Abstract. Recent developments in high-resolution time-of-flight chemical ionization mass spectrometry (HR-ToF-CIMS) have made it possible to directly detect atmospheric organic compounds in real time with high sensitivity and with little or no fragmentation, including low-volatility, highly oxygenated organic vapors that are precursors to secondary organic aerosol formation. Here, using ions identified by high-resolution spectra from an HR-ToF-CIMS with acetate reagent ion chemistry, we develop an algorithm to estimate the vapor pressures of measured organic acids. The algorithm uses identified ion formulas and calculated double bond equivalencies, information unavailable in quadrupole CIMS technology, as constraints for the number of possible oxygen-containing functional groups. The algorithm is tested with acetate chemical ionization mass spectrometry (acetate-CIMS) spectra of O_3 and OH oxidation products of α-pinene and naphthalene formed in a flow reactor with integrated OH exposures ranged from 1.2×10^{11} to 9.7×10^{11} molec s cm^{-3}, corresponding to approximately 1.0 to 7.5 days of equivalent atmospheric oxidation. Measured gas-phase organic acids are similar to those previously observed in environmental chamber studies. For both precursors, we find that acetate-CIMS spectra capture both functionalization (oxygen addition) and fragmentation (carbon loss) as a function of OH exposure. The level of fragmentation is observed to increase with increased oxidation. The predicted condensed-phase secondary organic aerosol (SOA) average acid yields and O / C and H / C ratios agree within uncertainties with previous chamber and flow reactor measurements and ambient CIMS results. While acetate reagent ion chemistry is used to selectively measure organic acids, in principle this method can be applied to additional reagent ion chemistries depending on the application.

1 Introduction

Oxygenated organics are an abundant class of compounds in the atmosphere, representing significant fractions of the total organic mass in the gas, particle, and cloud droplet phases (Goldstein and Galbally, 2007; Zhang et al., 2007; Jimenez et al., 2009). Much of the oxygenated organic mass is secondary in origin, generated from the gas-phase oxidation (O_3, OH / HO_2 / NO_x, NO_3) of anthropogenic and biogenic volatile organic compounds (VOCs) (Kroll and Seinfeld, 2008; Seinfeld and Pandis, 2006; Goldstein and Galbally, 2007). Recent studies have shown that secondary organic aerosol oxygen content is correlated with the photochemical age of an air mass (Ng et al., 2011; Heald et al., 2010; Jimenez et al., 2009). With increasing oxidation, gas-phase organic compounds can partition into the particle phase to contribute to secondary organic aerosol (SOA) mass or into the aqueous phase to contribute to water-soluble organic carbon (WSOC) (Hallquist et al., 2009; Ervens et al., 2011; Lim et al., 2010). As a result, their potential impact on climate by increasing and changing the properties of total aerosol mass and cloud condensation nuclei is substantial. However, sources, processing, and sinks of atmospheric organics are poorly characterized owing to their vast chemical complexity (Goldstein and Galbally, 2007). As a result, there is significant uncertainty in modeled secondary organic aerosol loadings and chemistries (Spracklen et al., 2011; Hallquist et al., 2009; Dzepina et al., 2011), necessitating more comprehensive measurements of gas- and particle-phase organics and better predictions of their SOA forming potentials.

Mass spectrometry has become a ubiquitous tool in identifying and quantifying atmospherically relevant organic species. Further, high-resolution mass spectrometry has increasingly become an important technique in elucidating the composition of organics because of its ability to resolve high molecular weight compounds with the same nominal mass with high sensitivity. Offline methods like electrospray ionization coupled with ultrahigh-resolution mass spectrometry are useful in identifying the many different compounds extracted from aerosol filter collection (Laskin et al., 2012; Kundu et al., 2012; Putman et al., 2012) as well as fog samples (Mazzoleni et al., 2010). Such methods, however, are highly selective with variable sensitivities and are not typically suited for online or field measurements. Conversely, the widely used Aerodyne high-resolution time-of-flight aerosol mass spectrometer (HR-ToF-AMS) couples electron ionization with the TOFWERK (Thun, Switzerland) high-resolution time-of-flight mass spectrometer (HTOF) to provide sensitive, quantitative online measurements of organic aerosol elemental composition (Jayne et al., 2000; Canagaratna et al., 2007; DeCarlo et al., 2006; Aiken et al., 2007). The harsh electron ionization of the HR-ToF-AMS limits its speciation capability and introduces uncertainty in its elemental ratio calculation (Chhabra et al., 2011; Farmer et al., 2010; Aiken et al., 2008). Thus, improvements to high-resolution mass spectrometry of atmospherically relevant high molecular weight compounds have sought to combine fast online detection with soft ionization sources (Zahardis et al., 2011).

Recently, the TOFWERK HTOF has been paired with various chemical ionization sources to allow for the sensitive detection of organic compounds at high time resolutions (≤ 1 s) with little to no molecular fragmentation. Chemical ionization sources employ specific reagent ions to initiate reactions that ionize analyte species; different reagent ions tend to be selective to different classes of compounds. The use of acetate chemical ionization mass spectrometry (acetate-CIMS) as a way to quantitatively measure organic acids was first demonstrated by Veres et al. (2008) using a quadrupole mass spectrometer. Veres et al. (2010) used the acetate-CIMS technique to measure acids of carbon number (n_C) 1 to 4, benzenediols, and inorganic acids from biomass burning; furthermore, Veres et al. (2011) used acetate-CIMS to measure small organic acids produced in urban air. Bertram et al. (2011) first described the use of a low-mass resolution ($R = 900$) time-of-flight CIMS which, unlike the quadrupole mass spectrometer, could acquire whole mass spectra at high time resolutions. Yatavelli et al. (2012) demonstrated the potential of a high-mass resolution (HTOF, $R = 4000$) acetate-CIMS to measure a large range ($1 \geq n_C \geq 30$) of organic acids in both particle and gas phases from α-pinene ozonolysis and subsequently used it in a remote forest atmosphere (Yatavelli et al., 2014). Aside from traditional organic acids, the technique has also been used to measure WSOC generated from α-pinene ozonolysis (Aljawhary et al., 2013) and nitrophenols from biomass burning (Mohr et al., 2013). The ability of the HTOF to acquire whole mass spectra at high time and mass resolutions represents a substantial improvement over quadrupole technology.

As soft ionization time-of-flight mass spectrometry techniques become more widely used to study gas- and aerosol-phase organics, methods will be needed to relate the many different species depicted in complex mass spectral data to the physical properties of the detected species. Volatility, expressed as vapor pressure, p^0, or saturated mass concentration, c^*, is an important property that governs whether a species partitions into the condensed phase. There has been much effort to relate HR-ToF-AMS data to volatility; thermal denuder measurements and dilution experiments with the AMS have been used to constrain the volatility of organic aerosol (Cappa and Jimenez, 2010; Cappa, 2010; Huffman et al., 2009). The lack of carbon number information from AMS data, a key input in volatility prediction models, may produce uncertainty. Soft ionization high-resolution mass spectrometry retains full molecular information in its spectra, i.e., the chemical formula, $C_xH_yO_z$, of unfragmented parent molecules can be obtained. Thus, it has the potential to provide the inputs for more accurate volatility estimations.

In this study, we explore the high-resolution acetate-CIMS spectra of photochemically produced organic acids and predict the volatility of the species detected. Acids are of particular interest because carboxylic acid functionality dramatically reduces the vapor pressure of its parent molecule and also represents the oxidative endpoint for a terminal carbon before fragmentation. Carboxylic acids also contribute a significant portion of total SOA mass (Ng et al., 2011; Vogel et al., 2013). We choose to examine the α-pinene, and naphthalene systems because their gas-phase compositions and mechanisms have been studied in detail, and various acids have been identified in each system (Yu et al., 1999; Claeys et al., 2013; Kautzman et al., 2009). Using the Van Krevelen diagram (hydrogen-to-carbon ratio, H / C, plotted against the oxygen-to-carbon ratio, O / C) and Kroll diagram (oxidation state, OS_C, plotted against carbon number, n_C), we identify small organic acids and tracer compounds and examine the changing distributions of carbon as a function of OH-exposure for the first time using the potential aerosol mass (PAM) flow reactor. We note that the high-resolution time-of-flight chemical ionization mass spectrometry (HR-ToF-CIMS) spectra contain many different mass-to-charge ions – on the order of 100 or more unknown ions. Since calibration of each individual measured species may not be feasible due to unknown identity or unavailability of standards, we adopt a simplifying approach in which the measured species are quantified using an average calibration value for organic acids. Acetate ionization may have an advantage in this respect since it is primarily selective for acids. While this assumption can lead to larger errors for any given individual species, it allows for a practical way of calculating average (ensemble) properties and mass concentrations for the large number of organic ions in the HR-ToF-CIMS spectra. This will be an on going developmental issue with the HR-ToF-CIMS systems and the uncertainty it introduces will be discussed. Additionally, we develop an algorithm to estimate the volatility of the species detected in each system and discuss the implications of the results.

2 Experimental method and data analysis

2.1 Acetate-CIMS

The Aerodyne HR-ToF-CIMS, hereby referred to as the CIMS, using acetate reagent ion chemistry has been described in detail in previous publications (Bertram et al., 2011; Yatavelli et al., 2012). A sample from the PAM reactor is drawn through a critical orifice at $2.0\,\mathrm{L\,min^{-1}}$ into the ion-molecular reaction (IMR) chamber. Acetate reagent ions are generated by bubbling approximately 200 sccm of N_2 through a reservoir of acetic anhydride, diluting the flow to $2\,\mathrm{L\,min^{-1}}$ with N_2 and passing it through a commercial ^{210}Po alpha emitter (P-2021, NRD) before introducing it into the IMR orthogonally to the sample flow (Veres et al., 2008). Within the IMR, acetate ions abstract protons from acids having gas-phase acidities greater than that of acetic acid or cluster with gas-phase species to form adduct ions, as shown in Reactions (R1) and (R2).

$$CH_3CO(O)O^- + HX \rightarrow CH_3C(O)OH + X^- \quad \text{(R1)}$$

$$CH_3CO(O)O^- + HX \rightarrow CH_3C(O)O^-(HX) \quad \text{(R2)}$$

The IMR (100 mbar) is coupled to the HTOF mass analyzer (1×10^{-7} mbar) by a series of differentially pumped stages that includes DC and RF focusing optics (AP interface). The first pumping stage contains a segmented, RF-only quadrupole operated at 2 mbar. Voltages in this quadrupole can be tuned to transmit (weak field) or dissociate (strong field) non-covalent clusters. In these experiments, voltages were adjusted in order to minimize the clustering, as indicated by the by the signal at m/z 119 (acetic acid acetate cluster, $C_4H_7O_4^-$). As a result, during normal operation, the ratio between m/z 59 (acetate reagent ion, $C_2H_3O_2^-$) and m/z 119 was approximately 5 to 1.

The ion source is coupled to a TOFWERK HTOF, identical to the mass spectrometer of the Aerodyne HR-ToF-AMS (DeCarlo et al., 2006; Canagaratna et al., 2007). The HTOF is capable of running in both positive and negative polarities but only the negative polarity was operated due to the reagent ion scheme. As with the Aerodyne HR-ToF-AMS (henceforth referred to as the AMS), the CIMS can be operated in a higher sensitivity, lower-resolution V-mode, or a lower sensitivity, higher-resolution W-mode. All experiments presented here were performed in V-mode. Although the HTOF is identical, the effective resolution of the CIMS is better than the AMS since the ion source is cooler in the former. In these experiments, the achieved resolution was approximately 4000.

A formic acid sensitivity calibration of the acetate-CIMS was conducted by flowing N_2 over a constant temperature permeation tube which itself was calibrated using an Aerodyne Quantum Cascade Laser Tracer Gas Monitor (Herndon et al., 2007). We determined a sensitivity of $5.5 \pm 0.9\,\mathrm{Hz\,ppt^{-1}}$ at an extraction frequency of 16.7 kHz, similar to the sensitivity $3\,\mathrm{Hz\,ppt^{-1}}$ at an extraction frequency of 25 kHz found by Yatavelli et al. (2012). Based on previous acetate-CIMS studies (listed in Table S1 in the Supplement), we assume a constant relative ionization efficiency of 1.3 ± 0.8 for all other species.

2.2 Organic vapor and aerosol production

Figure 1 shows a schematic of the experimental setup. Oxidized organic vapors were generated in a PAM flow reactor (Kang et al., 2007; Lambe et al., 2011a). The PAM reactor is a horizontal 13.3 L glass cylindrical chamber through which a carrier gas of N_2 and O_2 flowed at rates of 8.5 and $0.5\,\mathrm{L\,min^{-1}}$, respectively. Four mercury lamps (BHK Inc.) with peak emission intensity at $\lambda = 254$ nm are mounted in teflon-coated quartz sleeves inside the chamber and are flushed continuously with N_2. The CIMS was connected to the outlet of the PAM reactor with 0.635 cm (1/4 in.) OD, 0.476 cm (3/16 in.) ID PFA tubing, approximately 0.46 m (1.5 ft) long and was heated to 200 °C. These conditions were chosen to maintain laminar flow and to minimize condensation of organic vapors within the tube. Thus, collected spectra represent the combined contribution of gas-phase and particle-phase organics

OH radicals were produced via the reaction $O_3 + h\nu \rightarrow O_2 + O(^1D)$ followed by the reaction $O(^1D) + H_2O \rightarrow 2OH$. O_3 was generated by irradiating O_2 in the carrier gas flow with a mercury Pen-Ray™ lamp ($\lambda = 185$ nm) before it entered the PAM reactor. Ozone concentrations were measured using an O_3 monitor (2B Technologies); O_3 ranged between 5 ppm at the highest OH exposures to 6 ppm for ozonolysis experiments. Water vapor ($\sim 30\,\%$) was introduced by humidifying the carrier N_2 flow using a Nafion™ membrane humidifier. OH exposure was varied by changing the volt-

Table 1. Experimental conditions and results.

Expt no.	VOC system	VOC conc. (ppb)	OH exposure (molec s cm^{-3})	OH days[a]	$\overline{n_C}$[b]	$\overline{O/C}$[c]	$\overline{H/C}$[d]	$\overline{OS_C}$[e]
1	α-pinene + O_3	15	–	–	3.35	0.84	1.46	0.23
2	α-pinene + O_3	30	–	–	4.08	0.73	1.47	−0.02
3	α-pinene + OH	15	3.7×10^{11}	2.8	3.41	0.90	1.39	0.41
4	α-pinene + OH	15	7.0×10^{11}	5.4	2.97	1.02	1.34	0.71
5	α-pinene + OH	30	7.0×10^{11}	5.4	3.17	0.99	1.33	0.65
6	α-pinene + OH	30	9.7×10^{11}	7.5	3.09	1.01	1.32	0.71
7	α-pinene + OH	15	9.7×10^{11}	7.5	3.03	1.01	1.32	0.70
8	naphthalene + OH	23[f]	1.2×10^{11}	0.96	3.55	0.99	1.29	0.69
9	naphthalene + OH	46[f]	1.2×10^{11}	0.96	4.92	0.75	1.18	0.32
10	naphthalene + OH	46[f]	1.2×10^{11}	0.96	4.56	0.84	1.28	0.40
11	naphthalene + OH	23[f]	1.9×10^{11}	1.5	4.42	0.85	1.29	0.41
12	naphthalene + OH	23[f]	2.8×10^{11}	2.1	4.51	0.79	1.16	0.42
13	naphthalene + OH	23[f]	3.7×10^{11}	2.8	3.40	0.94	1.16	0.72
14	naphthalene + OH	23[f]	5.3×10^{11}	4.1	2.85	1.08	1.21	0.94
15	naphthalene + OH	23[f]	9.7×10^{11}	7.5	2.80	1.08	1.22	0.94

[a] Based on a diurnally averaged OH concentration of 1.5×10^6 molec cm^{-3}. [b] Average carbon number. [c] Average oxygen-to-carbon ratio. [d] Average hydrogen-to-carbon ratio. [e] Average carbon oxidation state. [f] Concentration estimated from equilibrium vapor pressure at 25 °C.

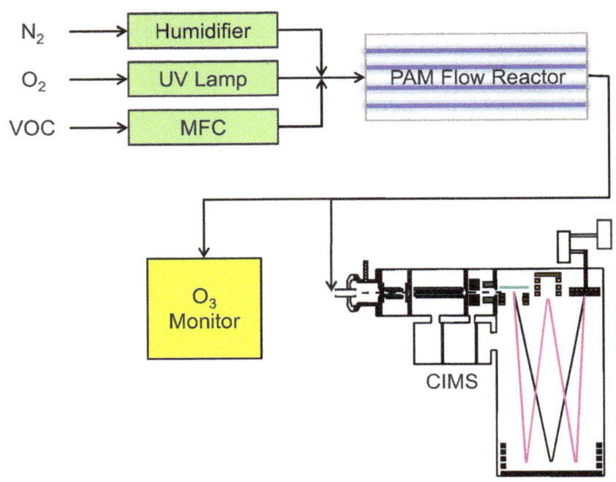

Figure 1. Schematic of the experimental setup. CIMS diagram adapted from Fig. 2 of Yatavelli et al. (2012).

age applied to the PAM reactor lamps between 0 and 110 V. The OH exposure was calculated by measuring the decay of (^{2}H)formic acid (measured in high-resolution spectra at $m/z = 46.00393$, as deuterated formate ion $DCOO^-$) introduced into the PAM reactor with a permeation tube assuming $k_{OH} = 4.62 \times 10^{-13}$ cm^3 molec^{-1} s^{-1} (Wine et al., 1985).

Although ozone concentrations were similar between ozonolysis and photooxidation experiments, OH concentrations were high enough that α-pinene + OH ($k_{OH} = 5.4 \times 10^{-11}$ cm^3 molecule^{-1} s^{-1}) reactions are 5 to 40 times faster than ozonolysis reactions ($k_{O_3} = 8.7 \times 10^{-17}$ cm^3 molecule^{-1}s^{-1}), and naphthalene + OH ($k_{OH} = 2.59 \times 10^{-11}$ cm^3 molecule^{-1} s^{-1}) reactions are 700 to 5700 times faster than ozonolysis ($k_{O_3} = 3 \times 10^{-19}$ cm^3 molecule^{-1} s^{-1}) reactions (Atkinson and Aschmann, 1986; Seinfeld and Pandis, 2006). These calculations suggest that photooxidation products measured with the CIMS are mostly from OH reactions, although we cannot completely rule out minor contributions from O_3 reactions particularly at the lowest OH exposures that were used.

Typical OH exposures ranged from 1.2×10^{11} to 9.7×10^{11} molec s cm^{-3} which is equivalent to 1.0–7.5 days of atmospheric oxidation assuming an average atmospheric OH concentration of 1.5×10^6 molec cm^{-3} (Mao et al., 2009). Ozonolysis experiments were performed by turning the UV lamps off. OH exposures in units of molec s cm^{-3} and equivalent atmospheric exposure times in OH Days are listed in Table 1. Before each experiment, the PAM flow reactor was conditioned with OH radicals until a particle background of less than 10 particles cm^{-3} was achieved.

The VOC precursors investigated in this study were α-pinene and naphthalene. α-pinene was prepared in compressed gas cylinders at known concentrations and introduced into the PAM reactor at controlled rates using a mass-flow controller. Naphthalene vapor was introduced by flowing N_2 over solid naphthalene placed in a Teflon tube (Chan et al., 2009). Naphthalene concentrations listed in Table 1 are estimated from equilibrium vapor pressures at 25 °C. We estimate an error of 38 % in the naphthalene concentrations based on the variability of literature vapor pressure values. Additionally, we estimate a 20 % error in the α-pinene concentrations. The VOC concentrations that were used resulted in particle mass loadings $\leq 7\,\mu$g m^{-3} (assuming SOA density of 1.4 g cm^{-3}), measured after the heated sample line, or approximately 3 % of the total organic carbon. Thus, the influence of slowly evaporating organic aerosols in the heated sample line after the PAM chamber on total gas-phase CIMS measurements is likely minimal under these conditions. Additional uncertainties in the measurements due to PAM-specific issues, such as direct photochemistry, wall effects, and high OH concentrations, have been discussed in detail in Lambe et al. (2011a, 2012).

2.3 Data analysis

Spectral data were analyzed using the Tofware (TOFWERK and Aerodyne) toolkit, developed for the IGOR Pro 6.x analysis software (WaveMetrics, Lake Oswego, OR, USA). Similar to the Squirrel and PIKA toolkits developed for Aerodyne AMS data, the Tofware software can index, preprocess (m/z calibration, baseline correction, and peak interpretation), and integrate raw mass spectra to produce unit mass resolution (UMR) data or take advantage of the high-resolution capability of the HTOF to assign signal contributions to individual ions with different elemental compositions. The latter method allows for the calculation of bulk elemental ratios ($\overline{O/C}, \overline{H/C}$), carbon numbers ($\overline{n_C}$), and carbon oxidation states ($\overline{OS_c}$) (Kroll et al., 2009). Here, we calculated O / C, H / C, OS_c, and n_C values assuming all measured ions are (1) produced from reactions with the acetate reagent ion (Reactions R1 and R2) and (2) have the same ionization efficiency.

All HR mass spectra were blank corrected using measurements of the PAM reactor outflow in the absence of VOC precursors. Spectra were normalized to the acetate reagent ion signal at m/z 59 to account for variations in the source and analyte loadings. We assumed that all identified peaks were molecular anions formed from Reaction (R1) (M^-) or cluster ions formed from Reaction (R2) ($CH_3C(O)O^-(MH)$); the latter were corrected using the method described below. Since no chemically labile nitrogen was added to the experiments, it was assumed no nitrogen-containing molecules were generated in the PAM. Thus, ions with even nominal masses were not included in the fitted peak list except for even-mass isotopes of odd-mass parent ions, constrained by their natural abundances. Signals at nominally even mass to charges were dominated by NO_2^- and NO_3^-, originating as contaminate signal from metal tubing and the ionizer. By summing high-resolution signals from even-massed ions that are not inorganic or isotopes of odd-massed ions, we estimate that less than 7 % of the signal at even masses is from organic ions. Using these criteria, approximately 350 molecular ions were fitted in each system. Of the total number of fitted ions, only those which had a blank subtracted signal greater than 1 standard deviation (1σ) of the time-averaged signal were included in the analysis, approximately 100–200 ions.

Adduct ion signals in the CIMS spectra have the potential to bias the measured composition toward higher molecular weight species. However, because of the elemental composition of acetate, distinguishing adduct ion signal from true molecular ion signal is not straightforward. To estimate adduct ion contributions, we first define that a non-clustered base ion has a molecular formula i, and when the acetate mass is added, its formula is $i + C_2H_4O_2$. We denote the adduct ion as i' and a non-clustered ion at the same mass as j. The signal at the clustered mass, $I_{i+C_2H_4O_2}$ is equal to the sum of the adduct ion signal $I_{i'}$ and non-adduct ion signal I_j:

$$I_{i+C_2H_4O_2} = I_{i'} + I_j. \tag{1}$$

We assume that across all experiments of a given precursor, the ratio of $I_{i'}$ to I_i is constant and is no more than the ratio of signal at m/z 119 to signal at m/z 59, or 0.2. The ratio of $I_{i+C_2H_4O_2}$ to I_i represents an upper bound to this value. Thus, across all experiments of a given precursor, the minimum ratio of $I_{i+C_2H_4O_2}$ to I_i or 0.20, whichever is smaller, is used to approximate the ratio of $I_{i'}$ to I_i. In other words,

$$I_{i'} \approx \min\left\{\frac{I_{i+C_2H_4O_2}}{I_i}, 0.2\right\} \times I_i. \tag{2}$$

To correct for clustering, the estimated value of $I_{i'}$ calculated from Eq. (2) is added back to I_i and subtracted from $I_{i+C_2H_4O_2}$. The effect of this correction on bulk composition values is minimal, on the order of a couple percent. The true clustering correction (and associated uncertainty) is likely significantly larger for individual molecules due to compound-specific acetate proton transfer reaction rates that are not known. Figures S3 and S4 in the Supplement illustrate how much signal in example spectra is estimated to be from clustering and redistributed to base ion signals.

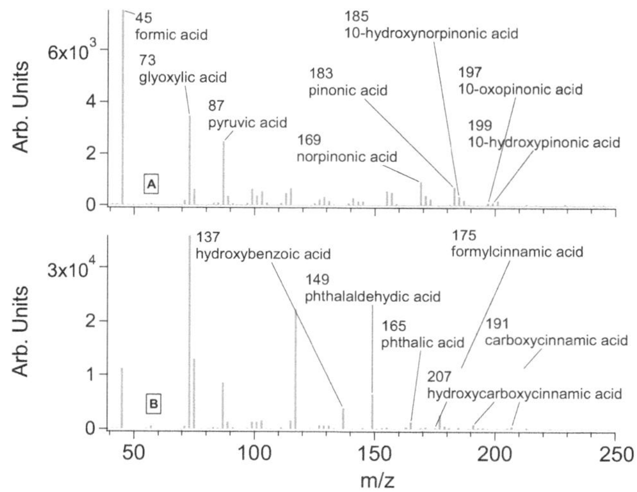

Figure 2. Acetate-CIMS unit mass resolution spectra of α-pinene ozonolysis (**a**, Expt 1), and naphthalene photooxidation (**b**, Expt 12). Even-massed ions, reagent ions, and dominant background ions are removed, and blank spectra are subtracted. Select ions identified in previous studies are labeled (Yu et al., 1999; Kautzman et al., 2009).

CIMS spectra of oxidized organic vapors produced from naphthalene contained contaminant ion signals from the α-pinene products. To remove ions overlapping both systems, an α-pinene spectra corresponding to the nearest exposure and normalized to the $C_9H_{14}O_3^-$ ion was subtracted from the naphthalene spectra (possibly introducing some additional uncertainty in small acid signals). Subsequently, because of the highly unsaturated nature of naphthalene tracer products ($H/C < 1$) some ions could be exclusively attributed to α-pinene ($n_C \geq 6$ and $H/C > 1$, supported by the spectral separation in Van Krevelen space illustrated in Fig. 3) and were removed.

2.4 Volatility estimation method

In this work, we use SIMPOL.1 (Pankow and Asher, 2008) to predict the vapor pressures of the species detected by the acetate-CIMS. The challenge in applying SIMPOL to conventional electron ionization mass spectra, e.g., AMS, is that the average carbon number ($\overline{n_C}$) and functional group composition needed as inputs are both typically unknown. This makes it necessary to calculate $\overline{n_C}$ indirectly, which introduces additional uncertainty (Daumit et al., 2013; Donahue et al., 2011; Kessler et al., 2010, 2012). Simplifying assumptions must also be made about the functional group composition and their effect on vapor pressure (Donahue et al., 2011; Cappa and Wilson, 2012). However, soft ionization techniques, such as acetate-CIMS, generate mass spectra that retain the unfragmented parent ions; thus, molecular formulas are known. This makes determination of the functional group composition more straightforward.

Here, we expand upon the work of Daumit et al. (2013) to determine functional group composition with additional constraints provided by high-resolution acetate-CIMS. Because acetate ionization is sensitive to acids, we assume at least one carboxylic acid functionality is present provided there are at least two oxygens in the molecule and at least one site of unsaturation. The number of sites of unsaturation is determined by the double bond equivalency (DBE, i.e., degrees of unsaturation) formula:

$$DBE = 1 + n_C - \frac{1}{2}n_H. \quad (3)$$

The remaining sites of unsaturation are referred to as DBE_r (which is one less than DBE if there is an acid group present). We assume that the remaining oxygen, n_{rO} (which is two less than the total if there is an acid group present), are either contained in hydroxyl (-OH) or keto (=O) groups and that any extra acid groups are a combination of one of each:

$$n_{rO} = n_{\text{-OH}} + n_{=O}. \quad (4)$$

We assume that other oxygen-containing functional groups that may be present in the spectra are well represented by ketone and/or hydroxyl surrogates. This assumption is reasonable for epoxide groups that have similar bonding as carbonyls, but not for other functional groups such as acyclic

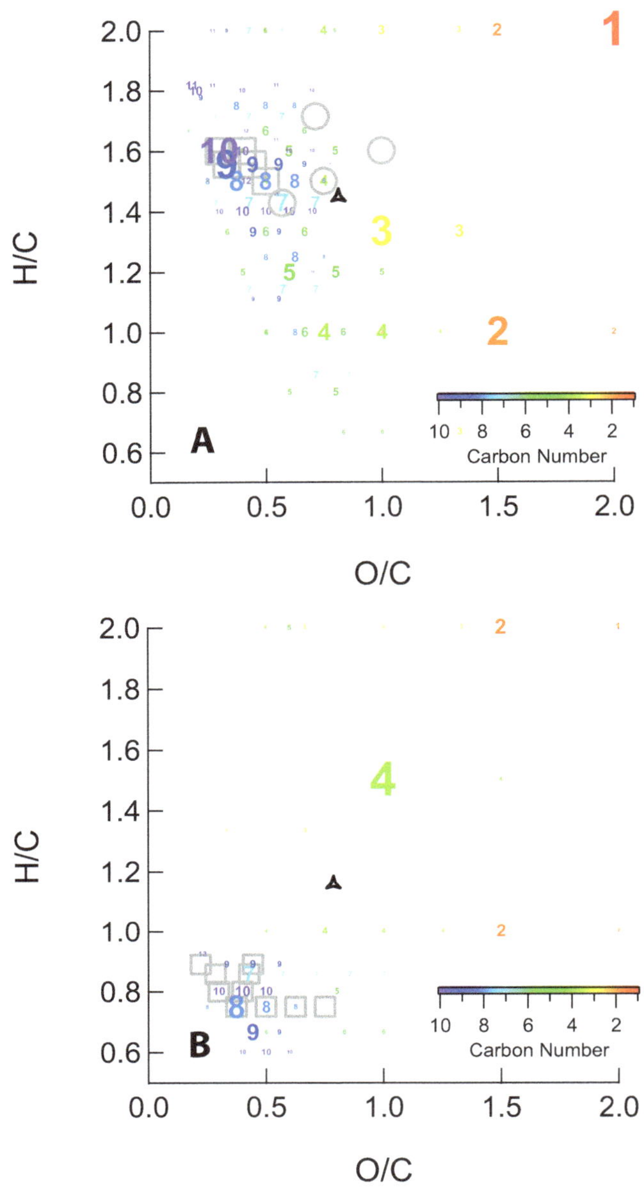

Figure 3. Modified Van Krevelen diagram of gas-phase acetate-CIMS high-resolution spectra of **(a)** α-pinene ozonolysis (Expt 1) and **(b)** naphthalene photooxidation (Expt 12). Each integer marker represents a fitted ion with the number representing the carbon number of the ion. The color bar is representative of the carbon number range of the high-resolution spectrum. The size of each number is proportional to the carbon number-weighted signal of its corresponding ion; the largest markers in **(a)** and **(b)** represent 11 % and 38 % of the carbon-weighted signal, respectively. Gray square markers in **(a)** indicate tracer acids identified by Yu et al. (1999) and gray circle markers indicate SOA tracer acids noted in Claeys et al. (2013). Gray square markers in **(b)** indicate tracer acids identified by Kautzman et al. (2009). Names and locations of acids on the VK are shown in the Supplement, Figs. S5 and S6. The bulk O / C and H / C values are marked by black triangles.

ethers, esters, or hydroperoxides that might be present (Daumit et al., 2013).

If $n_{rO} \leq DBE_r$ then all oxygens are assigned as carbonyls and the remaining sites of unsaturation are considered C = C double bonds:

$$n_{rO} = n_{=O}, \quad (5)$$

$$n_{-OH} = 0, \quad (6)$$

$$DBE_r - n_{=O} = n_{C=C}. \quad (7)$$

If $n_{rO} > DBE_r$ then all remaining sites of unsaturation are considered to be carbonyl groups and the remainder oxygens are hydroxyl groups:

$$n_{=O} = DBE_r, \quad (8)$$

$$n_{-OH} = n_{rO} - n_{=O}. \quad (9)$$

Pairs of carbonyls and hydroxyl groups are combined to form additional carboxylic acid groups, with the total number of acid groups constrained to no more than two assuming linear carbon chains.

Additional constraints can be applied as needed if more information is known. For example α-pinene oxidation products often retain a four-carbon ring; thus, we impose the constraint that if $n_C > 8$ and $DBE_r \geq 1$, a non-aromatic ring functionality (n_r) is added to the SIMPOL calculation. Likewise, for naphthalene spectra, if $n_C > 6$ and $DBE_r \geq 4$, then at most one aromatic ring functionality is added (n_{Ar}, equivalent to four sites of unsaturation). In both cases these steps are done after the initial acid is assigned and the DBE_r is updated.

Algorithm outputs from Eqs. (3) to (9) are used as inputs to the SIMPOL model, which parameterizes the vapor pressure of a pure compound as a linear combination of contributions from its functional groups:

$$\log_{10}\left(p^0_{L,i}\right) = b(T)_0 + \sum_k n_{k,i} b(T)_k, \quad (10)$$

where $b(T)_k$ is the temperature-dependent contribution term of functional group k and $n_{k,i}$ is the number of functional groups of type k in species i; $b(T)_0$ is a constant term. For example, the addition of a hydroxyl group lowers the vapor pressure by a factor of $10^{2.23}$ at 293 K; this is represented as $b(293\,K)_{OH} = -2.23$. To convert p^0 to c^*, an additional term is added to the above equation:

$$\log_{10}\left(c^*_i\right) = \log_{10}\alpha + b(T)_0 + \sum_k n_{k,i} b(T)_k, \quad (11)$$

where $\alpha = \frac{10^6 \times MW_i}{RT}$ which equals 8.314×10^9 at 293 K. The factor MW_i refers to the molecular weight of species i, R is the ideal gas constant (8.21×10^{-5} atm m^3 mol^{-1} K^{-1}), and T is the temperature (293 K). Technically, c^* is the product of the saturation concentration of the pure sub-cooled liquid, c^0, and its activity coefficient in solution γ, but the activity coefficient is assumed to be unity. Derivation of the estimation in uncertainty in SIMPOL vapor pressure prediction is detailed in the Supplement.

Table 2 lists several example oxidation products from each experimental system along with the number of each functional group assigned to it. Across the top of the table are the values of each group contribution term used in this work. The calculated c^* for product is listed at the right of the table.

3 Results

3.1 CIMS Spectra, $\overline{O/C}$ and $\overline{H/C}$ ratios, $\overline{n_C}$

Figure 2 displays representative UMR spectra obtained from α-pinene ozonolysis and naphthalene photooxidation conditions in the PAM reactor. In all cases the largest signals are observed at m/z 45, 73, 75, and 87, which correspond to formic acid, glyoxylic acid, glycolic acid, and pyruvic acid, respectively. The prominence of these small acids in acetate-CIMS spectra has been previously observed by Yatavelli et al. (2012). Production of small acids in urban atmospheres has been correlated with photochemical activity (Veres et al., 2011) and although their exact formation mechanism is unknown, it is likely they are generated from the oxidative fragmentation of the parent VOCs and subsequent products.

Analyzing the mass spectra shown in Fig. 2 in high resolution allows for calculation of average oxygen-to-carbon and hydrogen-to-carbon ratios ($\overline{O/C}$ and $\overline{H/C}$), average carbon oxidation states ($\overline{OS_C}$), and average carbon numbers ($\overline{n_C}$). Table 1 presents $\overline{O/C}, \overline{H/C}, \overline{OS_C}$, and $\overline{n_C}$ for all of the studied conditions. In general, these parameters follow the expected trends as a function of precursor type and OH exposure. First the $\overline{n_C}$ values of α-pinene ($\overline{n_C} = 2.97$–4.08) and naphthalene ($\overline{n_C} = 2.80$–4.92) oxidation products span similar ranges with the $\overline{OS_C}$ values of naphthalene being generally higher than those of α-pinene, especially at low OH exposures. This is consistent with naphthalene itself having a higher $\overline{OS_C}$ value than α-pinene. Second, the average $\overline{OS_C}$ and $\overline{O/C}$ for α-pinene and naphthalene oxidation products increase with increasing OH exposure while the average $\overline{n_C}$ decreases with OH exposure. Third, α-pinene oxidation products have higher $\overline{H/C}$ ratios than naphthalene, consistent with previous studies (Chhabra et al., 2011; Lambe et al., 2011b).

3.2 Van Krevelen diagram

A Van Krevelen diagram can be used to plot the corresponding H / C and O / C ratios of ions presented in the mass spectra. Figure 3a shows a Van Krevelen diagram corresponding to the α-pinene ozonolysis spectrum presented in Fig. 2a. Markers are numbered and colored by n_C, sized by the ion signal intensity, and weighted by n_C to emphasize the distribution of carbon across the spectra. Identified ions with known chemical formulae are marked with gray cir-

Table 2. Examples of detected species, their double bond equivalencies ($DBE = 1 + n_C - \frac{1}{2} n_H$), and their calculated saturation concentrations based on the number of each type of functional group, n_k. Note that $DBE = n_{=O} + n_{C(O)OH} + n_{C=C} + n_r + 4 \times n_{Ar}$. Values in parentheses are the SIMPOL group contribution coefficients for the corresponding functional group (Pankow and Asher, 2008). Underneath each type of functional group, the associated additivity constant is shown. Errors are calculated based on the method detailed in the Supplement.

Species Formula	DBE	Constant ($b_0 = 1.79$)	n_C ($b_1 = -0.44$)	$n_{=O}$ ($b_9 = -0.94$)	n_{-OH} ($b_7 = -2.23$)	$n_{C(O)OH}$ ($b_{10} - 3.58$)	$n_{C=C}$ ($b_5 = -0.10$)	n_r ($b_4 = -0.01$)	n_{Ar} ($b_3 = -0.68$)	$\log c^*$	c^* ($\mu g\,m^{-3}$)
α-pinene											
$C_9H_{14}O_4$	3		9			2		1		0.60 ± 1.0	3.68 ± 8.47
$C_{10}H_{14}O_7$	4		10	1	2	2		1		-5.14 ± 3.38	$7.15 \times 10^{-6} \pm 5.56 \times 10^{-5}$
$C_8H_{12}O_6$	3		8	1	1	2				-2.12 ± 1.70	$7.53 \times 10^{-3} \pm 3.04 \times 10^{-2}$
naphthalene											
$C_8H_6O_3$	6		8	1		1			1	2.89 ± 0.61	$7.77 \times 10^2 \pm 1.08 \times 10^3$
$C_9H_6O_4$	7		9	2		1			1	1.59 ± 0.74	$4.28 \times 10^1 \pm 67$

cles and squares in Fig. 3a (squares: Yu et al., 1999; circles: Claeys et al., 2013; proposed identities are given in Fig. S5 in the Supplement). Figure 3a illustrates those ions with $n_C = 1$ to 12 are measured by the acetate-CIMS for α-pinene ozonolysis. Much of the carbon (~ 45 %) is characterized by molecules with $O/C = 0.3$ to 0.7 and $n_C = 7$ to 10, including ions corresponding to well-characterized gas-phase oxidation products, such as $C_{10}H_{16}O_3$ (pinic acid) and norpinonic acid ($C_{10}H_{14}O_3$), which by themselves contribute ~ 14 % of the measured carbon. Other $n_C = 10$ ions are also detected, with O/C values as high as 0.7 such as $C_{10}H_{16}O_7$ and $C_{10}H_{14}O_7$. Highly oxidized ions with these formulas have been previously measured by Ehn et al. (2012) in ambient and chamber experiments. The acids identified by Yu et al. (1999) contribute approximately 20 % to the measured carbon of acidic α-pinene ozonolysis products. Acids identified by Claeys et al. (2013), such as 3-methyl-1,2,3-butanetricarboxylic acid (MBTCA, $C_8H_{12}O_6$) and terebic acid ($C_7H_{10}O_4$), have smaller contributions on the order of ~ 3 %.

Approximately 33 % of the carbon is contained in small highly oxidized acids with $n_C = 1$ to 3 and $O/C \geq 1$. Because the relative abundance of these acids increases as a function of OH exposure they are presumably formed following fragmentation of early generation oxidation products with larger n_C. Small-n_C ions with large signals include CH_2O_2 (formic acid), $C_2H_2O_3$ (glyoxylic acid), and $C_3H_4O_3$ (pyruvic acid) which represent 26 % of the total carbon measured for α-pinene ozonolysis products.

The formation of formic acid from α-pinene ozonolysis and photooxidation has been previously observed in several studies (Lee et al., 2006a, b; Librando and Tringali, 2005; Orlando et al., 2000). We measured a formic acid molar yield from α-pinene ozonolysis of 10 %, which is similar to the molar yield of 7.5 % previously observed by Lee et al. (2006a). Given that our experiments were performed under low-NO_x conditions, it is possible that formic acid was formed via the reaction of formaldehyde with HO_2 (Su et al., 1979). Orlando et al. (2000) also proposed formic acid as a first generation product from terpene-OH reactions. Formation of glyoxylic acid and glycolic acid from the gas-phase oxidation of α-pinene or naphthalene has not previously been reported. We note that glyoxal can be produced from the photooxidation of terpenes (Fu et al., 2008) and naphthalene (Nishino et al., 2009), and that glyoxal may be a source of glyoxylic and glycolic acid (Galloway et al., 2009; Lim et al., 2010). Similarly, methylglyoxal is also a product of terpene oxidation (Fu et al., 2008) and is a possible source of pyruvic acid (Lim et al., 2010).

The Van Krevelen diagram of naphthalene is shown in Fig. 3b. Measured naphthalene oxidation products that have been observed in previous chamber studies (depicted by gray square boxes, proposed identities given in Fig. S6 in the Supplement) represent at most 33 % of the carbon and include $C_8H_6O_3$ (phthalaldehydic acid), $C_8H_6O_4$ (phthalic acid), and $C_{10}H_8O_3$ (formylcinnamic acid) (Kautzman et al., 2009). We measured several oxidation products with $n_C = 9$ and 10, including $C_9H_6O_4$ and $C_{10}H_8O_5$ that to our knowledge have not been previously identified. In naphthalene spectra, a strong $C_4H_5O_4$ ion signal was detected that corresponds to a previously unidentified compound. Two possible structures are (1) a diacid with two saturated, unoxygenated carbons, or (2) a mono-acid with one double bond and two hydroxyl groups, which may be more plausible.

3.3 Oxidation state vs. carbon number

Figure 4 displays oxidation state as a function of carbon number for α-pinene and naphthalene oxidation products as a function of OH exposure. The utility of plotting OS_C as a function of n_C (i.e., Kroll Diagram) was introduced by Kroll et al. (2011) to provide a visualization of the chemical complexity of atmospheric organics and their corresponding oxidation trajectories. The top inset panels show the fraction of carbon signal as a function of n_C. A multimodal n_C distribution is observed for α-pinene and naphthalene oxidation products (Fig. 4a and b top insets). The mode defined by $n_C = 8$ to 10 species can be viewed as functionalized products where a net addition of oxygen occurs while the carbon backbone of the precursor is mostly retained (Lambe et al., 2012; Kroll et al., 2009). Figure 4a and b show that the fraction of acids with $n_C = 8$ to 10 decrease as a function of OH

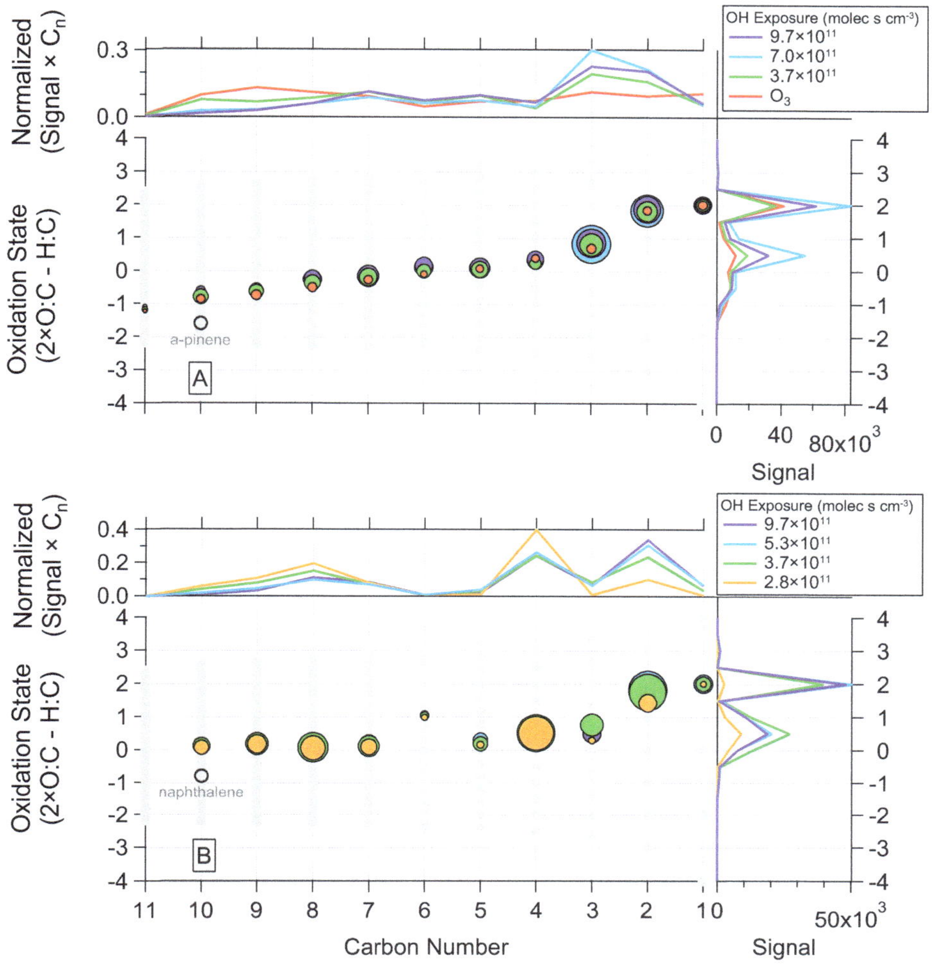

Figure 4. Kroll diagram for α-pinene **(a)** and naphthalene **(b)** oxidation experiments. The main panel displays the average OS_C per carbon number for each oxidation experiment. The area of the marker is proportional to the signal times the carbon number. The top panel plots the fraction of carbon-weighted signal at each carbon number. The right panel displays the signal distribution across oxidation state.

exposure. The fact that this decrease is directly correlated with an increase in the fraction of acids with $n_C < 4$ suggests that the $n_C < 4$ species are largely produced by fragmentation processes in which carbon–carbon bond cleavage occurs during oxidation of the $n_C = 8$ to 10 species.

Several other precursor-specific features are evident from Fig. 4. First, as is illustrated by the marker size in the main panel and curves in the right panel, CIMS signals peak at OH exposures of 3.7×10^{11} and 7.0×10^{11} molec s cm^{-3} for naphthalene and α-pinene oxidation products, respectively. Second, the carbon distribution of naphthalene oxidation products has negligible contributions from molecules with $n_C = 3, 5$, and 6. This is because naphthalene is unlikely to form pyruvic acid, an $n_C = 3$ acid with a methyl group, given naphthalene's aromatic structure. Additionally, naphthalene oxidation is unlikely to form acids with $n_C = 5$ or 6 because fragmentation of three or more carbon–carbon bonds is required.

3.4 Calculation of SOA mass from CIMS spectra

Spectra represented in Figs. 2–4 are for organic acids in the combined gas and particle phases. To estimate their relative contributions in each phase, we first calculated effective saturation concentrations (c^*) for each species shown in Fig. 3 using the algorithm described in Sect. 2.4. Table 2 lists examples of detected species from each precursor system and their corresponding c^* values. As shown, SIMPOL.1 estimates pinic acid ($C_9H_{14}O_4$) to have a $c^* = 3.68 \pm 8.47\,\mu g\,m^{-3}$ which is in good agreement with the $c^* = 5.34\,\mu g\,m^{-3}$ measured by Bilde and Pandis (2001). As shown in the Supplement and in Table 2, uncertainty in SIMPOL estimated vapor pressures increases with decreasing volatility.

To estimate the total mass concentration, M_i of each species i in $\mu g\,m^{-3}$ in a spectra, the formic acid sensitivity is applied to all ions using the following equation:

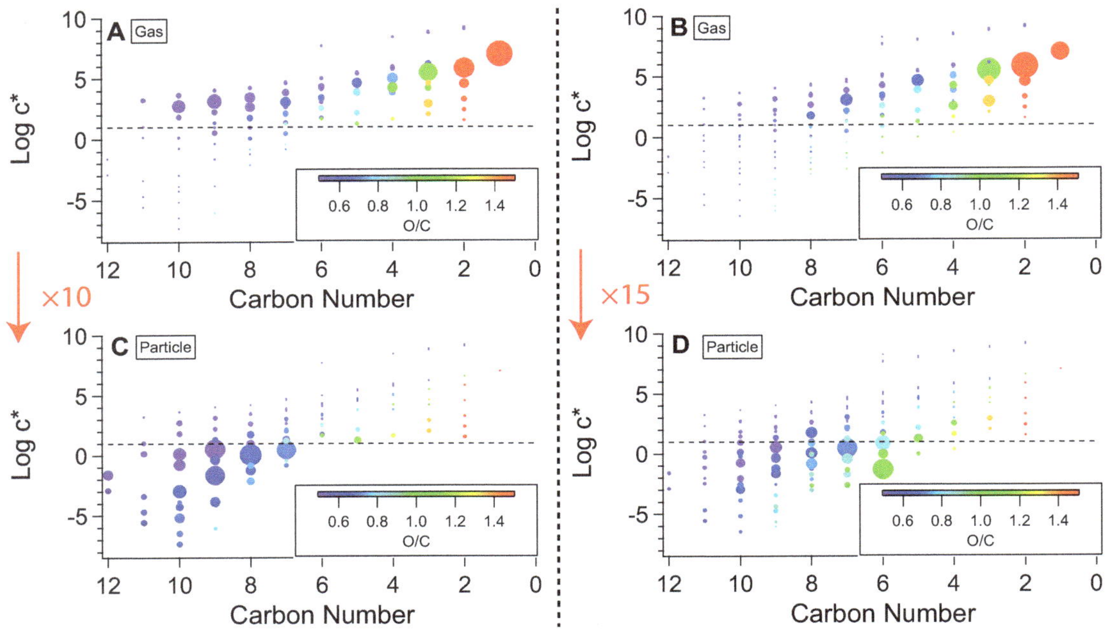

Figure 5. Estimated $\log c^*$ as a function of carbon number and O / C. **(a)** and **(c)** represent extracted gas and particle spectra of α-pinene ozonolysis (Expt 1), respectively, and **(b)** and **(d)** represent the same of α-pinene photooxidation at high OH exposures (Expt 7). Marker area is proportional to the fraction of mass in the depicted spectra. The factor and arrow in red represents the relative mass scale between gas and particle spectra. For example $\times 10$ between **(a)** and **(c)** indicate that markers of equal area represent the same fraction in each phase and 10 times more mass in the gas-phase spectrum.

$$M_i = \frac{\mathrm{MW}_i I_i}{RT S_{\mathrm{FA}}} \times 10^{-6}, \tag{12}$$

where I_i is the signal of corresponding ion i in Hz, S_{FA} is the formic acid sensitivity, MW_i is the molecular weight of species i, R is the ideal gas constant (8.21×10^{-5} atm m^3 mol^{-1} K^{-1}), and T is the temperature in Kelvin. The mass fraction of species i that partitions into the condensed phase is determined using Eq. (13) (Donahue et al., 2006):

$$\xi_i = \frac{1}{1 + \frac{c_i^*}{C_{\mathrm{OA}}}}, \tag{13}$$

where C_{OA} is the organic aerosol concentration. Here, instead of being a quantity that is solved for iteratively, we assume C_{OA} to take on decadal values of 1, 10 or 100 μg m^{-3} to determine how much of the measured spectra partitions into the condensed phase. The total mass estimated to be in the particle phase, C_{CIMS}, is then calculated as

$$C_{\mathrm{CIMS}} = \sum M_i \xi_i. \tag{14}$$

Application of Eqs. (12)–(14) allows for isolation of gas- and condensed-phase components of the CIMS spectra at a specific C_{OA}. Figure 5 plots c^* as a function of n_{C} for α-pinene ozonolysis and α-pinene photooxidation products (OH exposure $= 9.7 \times 10^{11}$ molec s cm^{-3}). Signals are apportioned into separate gas and particle-phase spectra assuming $C_{\mathrm{OA}} = 10\,\mu$g m^{-3} which is representative of urban conditions (Zhang et al., 2007). Under these conditions, compounds with $c^* = 10\,\mu$g m^{-3} will partition equally between the gas and particle phases. The factor and arrow in red represents the relative mass scale between gas and particle spectra. For example $\times 10$ between panels A and C indicate that markers of equal area represent 10 times more mass in the gas-phase spectrum. Figure 5 indicates that calculated c^* values decrease with increasing n_{C} and O / C ratio. Specifically, low-volatility $n_{\mathrm{C}} = 7$ to 10 compounds ($c^* = 1$ to $10^{-8}\,\mu$g m^{-3}) dominate the particle-phase spectra while higher-volatility $n_{\mathrm{C}} = 1$ to 3 compounds ($c^* = 100$ to $10^6\,\mu$g m^{-3}) dominate the gas-phase spectra. Signals in both phases shift to lower n_{C} and higher O / C ratio with increasing OH exposure. An acid of significant importance in the particle phase at high OH exposures is $C_6H_8O_6$. Though modeled as a diacid, this species may be a $-CH_2$ homologue of the triacid MBTCA. In total, the modeled particle phase represents 10 and 7.0 % of the measured mass and 5 and 3.5 % of the measured ion signal for ozonolysis and high OH conditions, respectively. Figure S7 in the Supplement shows c^* as a function of n_{C} for naphthalene oxidation products; in general, similar trends are observed as in the α-pinene system.

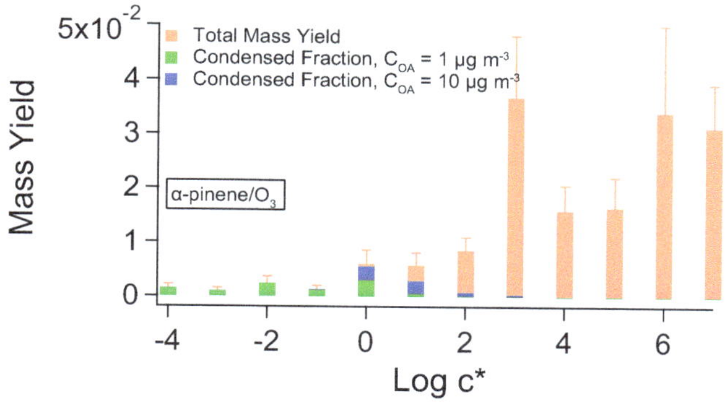

Figure 6. Estimated mass yield of compounds measured by the acetate-CIMS for α-pinene ozonolysis (Expt 1), binned by $\log c^*$. Colored in green is the fraction that would condense with $C_{OA} = 1\,\mu g\,m^{-3}$ and colored in blue is the additional mass that would condense with $C_{OA} = 10\,\mu g\,m^{-3}$. Error bars represent 1σ uncertainty from spectral variation, precursor concentration, and sensitivity

3.4.1 Derivation of acetate-CIMS volatility basis set and yields

Figure 6 shows the corresponding volatility basis set plot for α-pinene ozonolysis products measured with acetate-CIMS. Here, the calculated CIMS mass is normalized to the mass of α-pinene injected and plotted as a function of c^* (calculated c^* values are binned to the nearest c^* decade). The sum of the blue or green bars represents the aerosol mass fraction at $C_{OA} = 10$ and $C_{OA} = 1\,\mu g\,m^{-3}$, respectively. The c^* bins at the far right of the plot ($\log c^* = 6$ to 7) are dominated by formic, glyoxylic, and pyruvic acids which contribute 40 % of the total mass measured by the CIMS. Additionally, pinonic acid ($C_{10}H_{16}O_3$) and norpinonic acid ($C_9H_{14}O_3$) contribute significantly to the $\log c^* = 3$ bin. Error bars in Fig. 6 represent the combined 1σ uncertainty from spectra variation, sensitivity uncertainty, and precursor concentration uncertainty

Aerosol mass fractions (i.e., yields, ξ) were calculated for α-pinene experiments listed in Table 1 and are shown in Fig. 7 as a function of OH exposure. The ξ were calculated using particle-phase acetate-CIMS signals for decadal C_{OA} values of 1 to $100\,\mu g\,m^{-3}$:

$$\xi_{CIMS} = \frac{C_{CIMS}}{M_{pre}}, \tag{15}$$

where M_{pre} is the mass concentration of the injected aerosol precursor. The CIMS yields are compared to previously published yields determined from PAM experiments (Lambe et al., 2011a; Chen et al., 2013) and smog chamber experiments (Ng et al., 2006; Shilling et al., 2008; Eddingsaas et al., 2012). PAM yields as a function of OH exposure are represented by the gray shaded region. For α-pinene ozonolysis experiments conducted in the PAM, SOA yields, ξ_{SOA}, spanned from 11 to 24 %. These yields, applied to the α-pinene precursor concentrations used in this study, results in predicted SOA concentrations of 9 to $40\,\mu g\,m^{-3}$, corresponding to the CIMS yield curves with $C_{OA} = 10$ and $100\,\mu g\,m^{-3}$. For ozonolysis conditions, our estimate of SOA yields determined by the acetate-CIMS spectra, ξ_{CIMS} for 10 to $100\,\mu g\,m^{-3}$ of partitioning mass, range from about 2 ± 0.3 to 3 ± 0.7 %. Since acids are a subset of the entire SOA mass created, we would expect $\xi_{CIMS} \le \xi_{SOA}$. Using a Micro-Orifice Volatilization Impactor (MOVI) inlet with an acetate-CIMS, Yatavelli et al. (2012) estimated 11–34 % of α-pinene ozonolysis SOA mass was acidic, in agreement with measurements by Yu et al. (1999). Using the acid yields for 10 to $100\,\mu g\,m^{-3}$ of partitioning mass and the range of SOA yields from PAM experiments, our calculations suggest that acids comprise 8–27 % of the measured mass of α-pinene ozonolysis SOA and are thereby consistent with the results of Yu et al. (1999) and Yatavelli et al. (2012). We also find that ξ_{CIMS}/ξ_{SOA} increases systematically with increasing exposure. This observation is consistent with previous studies demonstrating an increase in acid content of SOA, indicated by the increase in the fraction of m/z 44 (f_{44}), as a function of OH exposure (Lambe et al., 2011b) suggesting that acids are an important class of compounds in aged air masses (Ng et al., 2011). ξ_{CIMS} and ξ_{SOA} increase and subsequently decrease as a function of OH exposure. This trend is consistent with a transition from functionalization- to fragmentation-dominated reaction pathways (Kroll et al., 2009; Lambe et al., 2012) and supports our interpretation of $n_C = 7$ to 10 and $n_C < 4$ as tracers for functionalization and fragmentation processes, respectively.

Implicit in the calculation of the CIMS yield curves shown in Fig. 7 is the assumption that the total (sum of gas and particle phases) distribution of acids generated is only a function of OH exposure and is independent of the amount of hydrocarbon injected. In reality, variations in the amount of injected aerosol precursor can introduce non-linearity in the formation of product species, for example, through competition for oxidants or bimolecular reactions of radical products. We ignore such non-linearities here but note that they should be studied in the future.

Naphthalene CIMS yields range from 1 to 11 % with the maximum yield occurring at an OH exposure of 3.7×10^{11} molec s cm^{-3}, thus also exhibiting a transition from functionalization to fragmentation processes. Previous smog chamber studies have estimated that the acid yield of naphthalene under low-NO_x conditions is approximately 24 % of the SOA mass acidic in nature (Kautzman et al., 2009). Reasons for this discrepancy are unclear at present.

Error bars in Fig. 7 represent the combined uncertainty in sensitivity, spectral variation, estimated vapor pressure, and precursor concentration which varied from 11 to 37 % of the estimated yield. As was stated earlier, we use an average relative sensitivity for all acids based on the measured sensitivity of formic acid. While this may introduce significant error for individual species, it provides a practical way to estimate en-

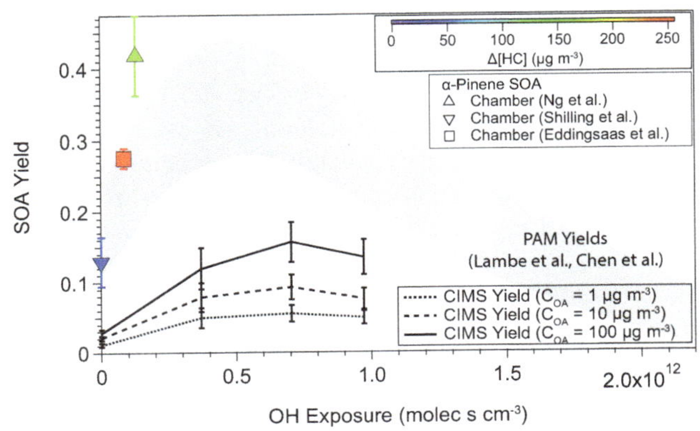

Figure 7. SOA yields estimated by the acetate-CIMS in this study and previously reported SOA yields from scanning mobility particle sizer (SMPS) measurements for α-pinene oxidation systems. The dotted black lines depict CIMS yields calculated for $C_{OA} = 1$, 10, and $100\,\mu g\,m^{-3}$. The gray shaded region represents the domain of yields determined from PAM experiments across OH exposure where the amount of α-pinene reacted ranged from 227 to $556\,\mu g\,m^{-3}$. Zero OH exposure corresponds to ozonolysis. Square and triangle points indicate yields determined from chamber experiments; their colors are indicative of the amount of α-pinene reacted. Error bars represent 1σ uncertainty from spectral variation, precursor concentration, sensitivity, and vapor pressure estimation.

semble properties. Its use for the acetate-CIMS may be particularly applicable since the acetate reagent scheme is primarily sensitive to acids. Based on previously reported sensitivities, we estimate an error of 60 % in the relative sensitivity of acids to formic acid. Aljawhary et al. (2013) measured the sensitivity of several different acids to acetate ionization, and found they were within a factor of 4. Of the acids measured, pinonic acid is the most relevant to our system and has a sensitivity approximately double that of formic acid. If this sensitivity is more representative of highly oxidized, lower-volatility compounds extracted from acetate-CIMS spectra, then the calculated ξ_{CIMS} is lower by roughly a factor of 1.5.

Another uncertainty involves the SIMPOL vapor pressure estimation method. Pankow and Asher (2008) report that calculated vapor pressures were typically within a factor of 2 of measured values. However, this variability is larger at low volatilities due to the lack of experimental vapor pressure measurements to constrain the parameterization. Compernolle et al. (2010) applied seven different vapor pressure estimation methods to a modeling study of SOA formation from α-pinene ozonolysis. They found that while there was general agreement for semi-volatile compounds, the vapor pressures estimated for low-volatility α-pinene acid tracers varied 3 orders of magnitude, leading to large differences in SOA yields. SIMPOL in particular tended to underestimate vapor pressures more than a similar method developed by (Capouet and Müller, 2006) due to the former's equal treatment of primary, secondary, and tertiary alcohols and lower volatilities given to carbon backbones and thus overestimating α-pinene ozonolysis yields in their model. At best, model simulations of α-pinene oxidation and SOA formation have agreed with experimental yields within a factor of 2 (Capouet et al., 2008). Although uncertainty in vapor pressure may be largest for the lowest volatility species, these species will completely partition into the condensed phase. It is the uncertainty in vapor pressure of compounds with c^* close to C_{OA} that will have the largest contribution to the error in the yield.

Recently, hydroperoxide moieties have been implicated in the composition of highly oxidized, extremely low-volatility organic compounds (ELVOC) (Ehn et al., 2014). In the functional group attribution method presented in Sect. 2.4, hydroperoxide moieties are indistinguishable from hydroxyl groups and one $-OOH$ group would be attributed as two $-OH$ groups. According to the SIMPOL model, hydroxy and hydroperoxide groups produce similar reductions in volatility ($b_{OH} = -2.23$ and $b_{OOH} = -2.49$ in Eq. 11) and thus the existence of hydroperoxide groups in detected molecules would lead to an underestimate in volatility and overestimate in acid yield.

Lastly, a potentially significant uncertainty is the estimation of the contribution of clustering in the high-resolution spectra. Here we set constraints on the ratio of cluster to molecular ion signals based on the assumption that the efficiency of clustering is constant for a given elemental composition. However, isomers may have different clustering efficiencies.

3.4.2 Comparison of particle-phase CIMS and Aerodyne AMS measurements

Figure 8 shows a Van Krevelen diagram of SOA species directly measured by the AMS and estimated from the acetate-CIMS for the oxidation of α-pinene and naphthalene. To facilitate this comparison, we calculated average O / C and H / C ratios of the particle-phase CIMS species (solid colored markers) for all experiments listed in Table 1 with exposure values similar to the published AMS data from Lambe et al. (2011b). Figure 8 shows that the O / C and H / C ratios measured by CIMS and AMS fall in the same region of the Van Krevelen diagram for α-pinene and naphthalene oxidation products. Additionally, the range of elemental ratios measured for the α-pinene system are consistent with the range observed by Aljawhary et al. (2013) in their experiments of WSOC formed from aqueous-phase α-pinene oxidation. This supports our interpretation that the low-volatility molecules measured by acetate-CIMS represent species contributing to SOA formation. However, we note that CIMS elemental ratios of α-pinene and naphthalene oxidation products span a smaller range of O / C than the AMS measurements conducted at similar exposures. One possible explanation is CIMS inlet losses of extremely low-volatility compounds; specifically, thermodenuder measurements have

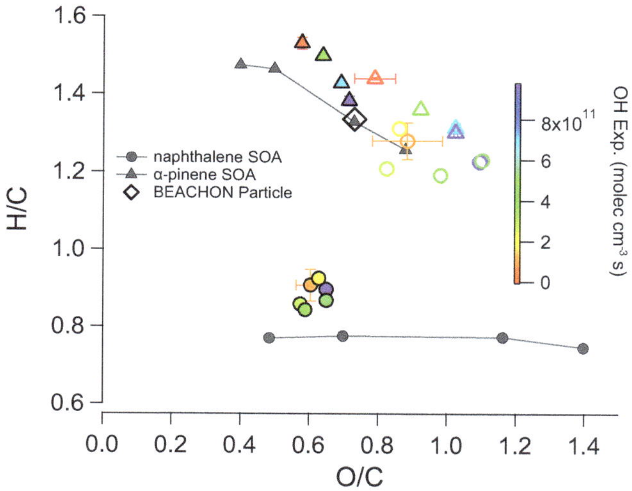

Figure 8. Average O / C and H / C ratios for extracted particle-phase (solid colored markers) and gas-phase (outlined colored markers) acetate-CIMS spectra (at $C_{OA} = 10\,\mu g\,m^{-3}$) as a function of OH exposure. Gray points represent AMS elemental ratios from Lambe et al. (2011a). Average particle-phase elemental ratios from the BEACHON-RoMBAS field campaign obtained from MOVI-HR-ToF-CIMS measurements are depicted by a black diamond (Yatavelli et al., 2014; Ortega et al., 2014). Error bars represent 1 standard deviation for multiple experiments.

shown that up to 30 % of particle-phase organic acid mass is not volatilized at 200 °C (Huffman et al., 2009). Another possibility is that highly functionalized organic acids may decompose in the inlet into products that are neither detected by the CIMS nor converted into smaller n_C acids (Yatavelli et al., 2014; Moldoveanu, 2010). Additionally, in the case of naphthalene, it may be difficult for oxidation products with acid groups to reach O / C values obtained in aerosol phase generated in the PAM at the highest OH exposures without extensive fragmentation.

Also shown in Fig. 8 are the average O / C and H / C ratios of the gas-phase CIMS species, illustrated by outlined colored markers. We observe much higher average O / C ratios for the estimated gas phase than the particle phase consistent with the production of small acids formed from the fragmentation of a larger carbon backbone. The H / C ratios of α-pinene gas-phase products are lower than those of the particle phase suggesting that the gas-phase acids might have more carbonyl groups and thus carbon with a higher oxidation state than carbon in the particle phase. In contrast, the H / C ratios of naphthalene gas-phase products are higher than those of the particle phase indicating that oxidation and fragmentation increase the oxidation state of carbon relative to carbon in aromatic rings. Particularly, the difference suggests that the particle phase is dominated by aromatic ring-retaining products while the gas phase is dominated by non-aromatic acid fragmentation products. The elemental ratios of the gas-phases products from both precursors display a convergence with increased OH exposure indicating the predominance of the same $n_C = 1$ to 4 acids, especially glyoxylic and glycolic acids.

Figure 8 also shows average O / C and H / C ratios of ambient organic aerosols measured with an acetate-MOVI-HR-ToF-CIMS during the Bio-hydro-atmosphere interactions of Energy, Aerosols, Carbon, H_2O, Organics and Nitrogen – Rocky Mountain Biogenic Aerosol Study (BEACHON-RoMBAS) field campaign (Yatavelli et al., 2014; Ortega et al., 2014). The BEACHON-RoMBAS measurements were obtained in a ponderosa pine forest and thus were influenced by emissions of biogenic compounds such as α-pinene. CIMS elemental ratios of highly oxidized α-pinene products produced in the PAM reactor agree with ambient measurements within 10 %. Interestingly, Yatavelli et al. (2014) were able to successfully model the partitioning of organic acids in bulk as alkanoic acids with excess oxygen in the form of hydroxyl groups, each lowering the vapor pressure by SIMPOL estimated factors. Our method uses the measured ion formulas to constrain the double bond equivalency and limit additional oxygen functional groups. These similarities suggest that the volatility estimation algorithm successfully captures atmospheric photochemistry of atmospherically relevant SOA precursors, and that our results can be used to interpret ambient measurements in different source regions.

4 Discussion and conclusions

The gas-phase SOA precursors for O_3 and OH oxidation of α-pinene and naphthalene were measured for the first time using a PAM flow reactor and an HR-ToF-CIMS with acetate reagent ion chemistry. The measured organic acid distributions exhibited similarities to gas-phase measurements from previous environmental chamber studies (Yu et al., 1999; Claeys et al., 2013; Kautzman et al., 2009). These measurements extend the previous measurements to higher OH exposures and include observations of several previously unidentified organic acids, particularly for naphthalene. Particularly, they illustrate the concurrent functionalization and fragmentation processes occurring in the gas-phase yielding high and low carbon number acids, the latter dominating the spectra.

In addition, we have presented an approach to estimate the vapor pressures and gas-to-particle partitioning of select organic compounds using HR-ToF-CIMS gas-phase measurements in combination with an algorithm based on the SIMPOL group contribution parameterization method introduced by Pankow and Asher (2008). Previous applications of SIMPOL attempted to use Aerodyne AMS measurements to calculate vapor pressures of organic compounds from their measured elemental ratios (O / C and H / C). However, since functional group information is not available from AMS, spectra simplifying assumptions are required which can introduce additional uncertainty (e.g., Donahue et al., 2011). Most recently Yatavelli et al. (2014) applied SIMPOL to

MOVI-HR-ToF-CIMS spectra using the acetate reagent ion. They were able to capture the bulk partitioning of species by carbon number, by modeling detected compounds such as alkanoic acids, and adding the remainder of oxygens in the form of different functional groups with hydroxyl groups giving the best agreement. Our approach builds on these studies by incorporating the DBE content of the measured species towards the goal of explicit characterization of oxygen-containing functional groups. This analysis is made possible by the high-mass resolution of the TOFWERK HTOF which can resolve the chemical formulas of detected ions.

Our model performs well in predicting the acid contribution to SOA formed from α-pinene ozonolysis compared to previous measurements (Yu et al., 1999; Yatavelli et al., 2012) and to our knowledge, this is the first attempt to estimate the mass of a class of compounds from CIMS spectra using a group contribution model. Semi-explicit model simulations of α-pinene oxidation and SOA formation have agreed with experimental yields within a factor of 2 (Capouet et al., 2008) while models with more degrees of freedom can be tuned to have even greater accuracies (Cappa and Wilson, 2012). Although we use SIMPOL parameterizations as a case study, our method can be adapted for the use of different vapor pressure formulation methods (Nannoolal et al., 2004, 2008; Hilal et al., 1994; Moller et al., 2008). The advantage of SIMPOL is that only functional group composition and carbon number information are needed in the parameterization. Various studies have compared vapor pressure models and have shown that volatility estimation, and thus aerosol formation, is highly sensitive to model choice (Compernolle et al., 2010; Barley and McFiggans, 2010; Clegg et al., 2008). Additionally, different reagent chemistries (e.g., $(H_2O)_nH^+$, I^-, NO_3^-) can and should be tested with our model to validate its effectiveness on different classes of compounds. For instance, iodide spectra may be more easily analyzed since its clusters do not contain carbon

With recent advances in CIMS measurement techniques that allow for separation of gas- and particle-phase analytes (Yatavelli and Thornton, 2010; Lopez-Hilfiker et al., 2014), uncertainties in the quantification of unknown organics will have to be addressed. For instance, epoxides and peroxides have been implicated in SOA formation in low-NO_x isoprene systems. Thus, our model needs to be tested against additional SOA systems and subsequently refined. In our study we assume that all detected species have the same ionization efficiency, relative to formic acid. While recent studies suggest that this approximation may be acceptable (Aljawhary et al., 2013) more data are needed to constrain acetate ionization efficiencies across carbon number and oxidation state. Additionally, the variables that control clustering, including operational voltages, IMR pressure, and reagent ion and neutral concentrations in the IMR, need to be explored. Use of isotopically labeled reagent ions may aid in this these investigations.

Acknowledgements. This work was supported by the Atmospheric Chemistry Program of the National Science Foundation, grants no. AGS-1244918 and no. ATM-0854916 to Boston College and no. AGS-1244999 and no. AGS-0904292 to Aerodyne Research, Inc. We would also like to acknowledge DOE SBIR grant no. DE-SC0004577 and Office of Science (BER), Department of Energy (Atmospheric Systems Research) grants no. DE-SC0006980 and DE-FG02-05ER63995. The authors would like to thank William Brune for canisters of α-pinene and Erika Hale at the University of Texas for assistance in error calculations.

Edited by: P. Herckes

References

Aiken, A. C., DeCarlo, P. F., and Jimenez, J. L.: Elemental Analysis of Organic Species with Electron Ionization High-Resolution Mass Spectrometry, Anal. Chem., 79, 8350–8358, doi:10.1021/ac071150w, 2007.

Aiken, A. C., DeCarlo, P. F., Kroll, J. H., Worsnop, D. R., Huffman, J. A., Docherty, K. S., Ulbrich, I. M., Mohr, C., Kimmel, J. R., Sueper, D., Sun, Y., Zhang, Q., Trimborn, A., Northway, M., Ziemann, P. J., Canagaratna, M. R., Onasch, T. B., Alfarra, M. R., Prevot, A. S. H., Dommen, J., Duplissy, J., Metzger, A., Baltensperger, U., and Jimenez, J. L.: O/C and OM/OC Ratios of Primary, Secondary, and Ambient Organic Aerosols with High-Resolution Time-of-Flight Aerosol Mass Spectrometry, Environ. Sci. Technol., 42, 4478–4485, doi:10.1021/es703009q, 2008.

Aljawhary, D., Lee, A. K. Y., and Abbatt, J. P. D.: High-resolution chemical ionization mass spectrometry (ToF-CIMS): application to study SOA composition and processing, Atmos. Meas. Tech., 6, 3211–3224, doi:10.5194/amt-6-3211-2013, 2013.

Atkinson, R. and Aschmann, S. M.: Kinetics of the reactions of naphthalene, 2-methylnaphthalene, and 2,3-dimethylnaphthalene with OH radicals and with O_3 at 295 ± 1 K, Int. J. Chem. Kinet., 18, 569–573, doi:10.1002/kin.550180507, 1986.

Barley, M. H. and McFiggans, G.: The critical assessment of vapour pressure estimation methods for use in modelling the formation of atmospheric organic aerosol, Atmos. Chem. Phys., 10, 749–767, doi:10.5194/acp-10-749-2010, 2010.

Bertram, T. H., Kimmel, J. R., Crisp, T. A., Ryder, O. S., Yatavelli, R. L. N., Thornton, J. A., Cubison, M. J., Gonin, M., and Worsnop, D. R.: A field-deployable, chemical ionization time-of-flight mass spectrometer, Atmos. Meas. Tech., 4, 1471–1479, doi:10.5194/amt-4-1471-2011, 2011.

Bilde, M. and Pandis, S. N.: Evaporation Rates and Vapor Pressures of Individual Aerosol Species Formed in the Atmospheric Oxidation of alpha- and beta-Pinene, Environ. Sci. Technol., 35, 3344–3349, doi:10.1021/es001946b, 2001.

Canagaratna, M. R., Jayne, J. T., Jimenez, J. L., Allan, J. D., Alfarra, M. R., Zhang, Q., Onasch, T. B., Drewnick, F., Coe, H., Middlebrook, A., Delia, A., Williams, L. R., Trimborn, A. M., Northway, M. J., DeCarlo, P. F., Kolb, C. E., Davidovits, P., and

Worsnop, D. R.: Chemical and microphysical characterization of ambient aerosols with the aerodyne aerosol mass spectrometer, Mass Spectrom. Rev., 26, 185–222, doi:10.1002/mas.20115, 2007.

Capouet, M. and Müller, J.-F.: A group contribution method for estimating the vapour pressures of α-pinene oxidation products, Atmos. Chem. Phys., 6, 1455–1467, doi:10.5194/acp-6-1455-2006, 2006.

Capouet, M., Müller, J. F., Ceulemans, K., Compernolle, S., Vereecken, L., and Peeters, J.: Modeling aerosol formation in alpha-pinene photo-oxidation experiments, J. Geophys. Res.-Atmos., 113, D02308, doi:10.1029/2007JD008995, 2008.

Cappa, C. D.: A model of aerosol evaporation kinetics in a thermodenuder, Atmos. Meas. Tech., 3, 579–592, doi:10.5194/amt-3-579-2010, 2010.

Cappa, C. D. and Jimenez, J. L.: Quantitative estimates of the volatility of ambient organic aerosol, Atmos. Chem. Phys., 10, 5409–5424, doi:10.5194/acp-10-5409-2010, 2010.

Cappa, C. D. and Wilson, K. R.: Multi-generation gas-phase oxidation, equilibrium partitioning, and the formation and evolution of secondary organic aerosol, Atmos. Chem. Phys., 12, 9505–9528, doi:10.5194/acp-12-9505-2012, 2012.

Chan, A. W. H., Kautzman, K. E., Chhabra, P. S., Surratt, J. D., Chan, M. N., Crounse, J. D., Kürten, A., Wennberg, P. O., Flagan, R. C., and Seinfeld, J. H.: Secondary organic aerosol formation from photooxidation of naphthalene and alkylnaphthalenes: implications for oxidation of intermediate volatility organic compounds (IVOCs), Atmos. Chem. Phys., 9, 3049–3060, doi:10.5194/acp-9-3049-2009, 2009.

Chen, S., Brune, W. H., Lambe, A. T., Davidovits, P., and Onasch, T. B.: Modeling organic aerosol from the oxidation of α-pinene in a Potential Aerosol Mass (PAM) chamber, Atmos. Chem. Phys., 13, 5017–5031, doi:10.5194/acp-13-5017-2013, 2013.

Chhabra, P. S., Ng, N. L., Canagaratna, M. R., Corrigan, A. L., Russell, L. M., Worsnop, D. R., Flagan, R. C., and Seinfeld, J. H.: Elemental composition and oxidation of chamber organic aerosol, Atmos. Chem. Phys., 11, 8827–8845, doi:10.5194/acp-11-8827-2011, 2011.

Claeys, M., Szmigielski, R., Vermeylen, R., Wang, W., Shalamzari, M., and Maenhaut, W.: Tracers for Biogenic Secondary Organic Aerosol from alpha-Pinene and Related Monoterpenes: An Overview, book section 18, 227–238, NATO Science for Peace and Security Series C: Environmental Security, Springer Netherlands, doi:10.1007/978-94-007-5034-0_18, 2013.

Clegg, S. L., Kleeman, M. J., Griffin, R. J., and Seinfeld, J. H.: Effects of uncertainties in the thermodynamic properties of aerosol components in an air quality model – Part 2: Predictions of the vapour pressures of organic compounds, Atmos. Chem. Phys., 8, 1087–1103, doi:10.5194/acp-8-1087-2008, 2008.

Compernolle, S., Ceulemans, K., and Müller, J.-F.: Technical Note: Vapor pressure estimation methods applied to secondary organic aerosol constituents from α-pinene oxidation: an intercomparison study, Atmos. Chem. Phys., 10, 6271–6282, doi:10.5194/acp-10-6271-2010, 2010.

Daumit, K. E., Kessler, S. H., and Kroll, J. H.: Average chemical properties and potential formation pathways of highly oxidized organic aerosol, Faraday Discuss., 165, 181–202, doi:10.1039/C3FD00045A, 2013.

DeCarlo, P. F., Kimmel, J. R., Trimborn, A., Northway, M. J., Jayne, J. T., Aiken, A. C., Gonin, M., Fuhrer, K., Horvath, T., Docherty, K. S., Worsnop, D. R., and Jimenez, J. L.: Field-Deployable, High-Resolution, Time-of-Flight Aerosol Mass Spectrometer, Anal. Chem., 78, 8281–8289, doi:10.1021/ac061249n, 2006.

Donahue, N. M., Robinson, A. L., Stanier, C. O., and Pandis, S. N.: Coupled Partitioning, Dilution, and Chemical Aging of Semivolatile Organics, Environ. Sci. Technol., 40, 2635–2643, doi:10.1021/es052297c, 2006.

Donahue, N. M., Epstein, S. A., Pandis, S. N., and Robinson, A. L.: A two-dimensional volatility basis set: 1. organic-aerosol mixing thermodynamics, Atmos. Chem. Phys., 11, 3303–3318, doi:10.5194/acp-11-3303-2011, 2011.

Dzepina, K., Cappa, C. D., Volkamer, R. M., Madronich, S., DeCarlo, P. F., Zaveri, R. A., and Jimenez, J. L.: Modeling the Multiday Evolution and Aging of Secondary Organic Aerosol During MILAGRO 2006, Environ. Sci. Technol., 45, 3496–3503, doi:10.1021/es103186f, 2011.

Eddingsaas, N. C., Loza, C. L., Yee, L. D., Chan, M., Schilling, K. A., Chhabra, P. S., Seinfeld, J. H., and Wennberg, P. O.: α-pinene photooxidation under controlled chemical conditions – Part 2: SOA yield and composition in low- and high-NO_x environments, Atmos. Chem. Phys., 12, 7413–7427, doi:10.5194/acp-12-7413-2012, 2012.

Ehn, M., Kleist, E., Junninen, H., Petäjä, T., Lönn, G., Schobesberger, S., Dal Maso, M., Trimborn, A., Kulmala, M., Worsnop, D. R., Wahner, A., Wildt, J., and Mentel, Th. F.: Gas phase formation of extremely oxidized pinene reaction products in chamber and ambient air, Atmos. Chem. Phys., 12, 5113–5127, doi:10.5194/acp-12-5113-2012, 2012.

Ehn, M., Thornton, J. A., Kleist, E., Sipila, M., Junninen, H., Pullinen, I., Springer, M., Rubach, F., Tillmann, R., Lee, B., Lopez-Hilfiker, F., Andres, S., Acir, I.-H., Rissanen, M., Jokinen, T., Schobesberger, S., Kangasluoma, J., Kontkanen, J., Nieminen, T., Kurten, T., Nielsen, L. B., Jorgensen, S., Kjaergaard, H. G., Canagaratna, M., Maso, M. D., Berndt, T., Petaja, T., Wahner, A., Kerminen, V.-M., Kulmala, M., Worsnop, D. R., Wildt, J., and Mentel, T. F.: A large source of low-volatility secondary organic aerosol, Nature, 506, 476–479, doi:10.1038/nature13032, 2014.

Ervens, B., Turpin, B. J., and Weber, R. J.: Secondary organic aerosol formation in cloud droplets and aqueous particles (aqSOA): a review of laboratory, field and model studies, Atmos. Chem. Phys., 11, 11069–11102, doi:10.5194/acp-11-11069-2011, 2011.

Farmer, D. K., Matsunaga, A., Docherty, K. S., Surratt, J. D., Seinfeld, J. H., Ziemann, P. J., and Jimenez, J. L.: Response of an aerosol mass spectrometer to organonitrates and organosulfates and implications for atmospheric chemistry, P. Natl. Acad. Sci., 107, 6670–6675, doi:10.1073/pnas.0912340107, 2010.

Fu, T.-M., Jacob, D. J., Wittrock, F., Burrows, J. P., Vrekoussis, M., and Henze, D. K.: Global budgets of atmospheric glyoxal and methylglyoxal, and implications for formation of secondary organic aerosols, J. Geophys. Res.-Atmos., 113, D15303, doi:10.1029/2007JD009505, 2008.

Galloway, M. M., Chhabra, P. S., Chan, A. W. H., Surratt, J. D., Flagan, R. C., Seinfeld, J. H., and Keutsch, F. N.: Glyoxal uptake on ammonium sulphate seed aerosol: reaction products and reversibility of uptake under dark and irradiated conditions, Atmos. Chem. Phys., 9, 3331–3345, doi:10.5194/acp-9-3331-2009, 2009.

Goldstein, A. H. and Galbally, I. E.: Known and Unexplored Organic Constituents in the Earth's Atmosphere, Environ. Sci. Technol., 41, 1514–1521, doi:10.1021/es072476p, 2007.

Hallquist, M., Wenger, J. C., Baltensperger, U., Rudich, Y., Simpson, D., Claeys, M., Dommen, J., Donahue, N. M., George, C., Goldstein, A. H., Hamilton, J. F., Herrmann, H., Hoffmann, T., Iinuma, Y., Jang, M., Jenkin, M. E., Jimenez, J. L., Kiendler-Scharr, A., Maenhaut, W., McFiggans, G., Mentel, Th. F., Monod, A., Prévôt, A. S. H., Seinfeld, J. H., Surratt, J. D., Szmigielski, R., and Wildt, J.: The formation, properties and impact of secondary organic aerosol: current and emerging issues, Atmos. Chem. Phys., 9, 5155–5236, doi:10.5194/acp-9-5155-2009, 2009.

Heald, C. L., Kroll, J. H., Jimenez, J. L., Docherty, K. S., DeCarlo, P. F., Aiken, A. C., Chen, Q., Martin, S. T., Farmer, D. K., and Artaxo, P.: A simplified description of the evolution of organic aerosol composition in the atmosphere, Geophys. Res. Lett., 37, L08803, doi:10.1029/2010GL042737, 2010.

Herndon, S. C., Zahniser, M. S., Nelson, D. D., Shorter, J., McManus, J. B., Jiménez, R., Warneke, C., and de Gouw, J. A.: Airborne measurements of HCHO and HCOOH during the New England Air Quality Study 2004 using a pulsed quantum cascade laser spectrometer, J. Geophys. Res.-Atmos., 112, D10S03, doi:10.1029/2006JD007600, 2007.

Hilal, S. H., Carreira, A., and Karikhoff, S. W.: Estimation of chemical reactivity parameters and physical properties of organic molecules using SPARC, book section 9, 291–353, Elsevier, 1st Edn., 1994.

Huffman, J. A., Docherty, K. S., Aiken, A. C., Cubison, M. J., Ulbrich, I. M., DeCarlo, P. F., Sueper, D., Jayne, J. T., Worsnop, D. R., Ziemann, P. J., and Jimenez, J. L.: Chemically-resolved aerosol volatility measurements from two megacity field studies, Atmos. Chem. Phys., 9, 7161–7182, doi:10.5194/acp-9-7161-2009, 2009.

Jayne, J. T., Leard, D. C., Zhang, X., Davidovits, P., Smith, K. A., Kolb, C. E., and Worsnop, D. R.: Development of an Aerosol Mass Spectrometer for Size and Composition Analysis of Submicron Particles, Aerosol Sci. Technol., 33, 49–70, doi:10.1080/027868200410840, 2000.

Jimenez, J. L., Canagaratna, M. R., Donahue, N. M., Prevot, A. S. H., Zhang, Q., Kroll, J. H., DeCarlo, P. F., Allan, J. D., Coe, H., Ng, N. L., Aiken, A. C., Docherty, K. S., Ulbrich, I. M., Grieshop, A. P., Robinson, A. L., Duplissy, J., Smith, J. D., Wilson, K. R., Lanz, V. A., Hueglin, C., Sun, Y. L., Tian, J., Laaksonen, A., Raatikainen, T., Rautiainen, J., Vaattovaara, P., Ehn, M., Kulmala, M., Tomlinson, J. M., Collins, D. R., Cubison, M. J., E., Dunlea, J., Huffman, J. A., Onasch, T. B., Alfarra, M. R., Williams, P. I., Bower, K., Kondo, Y., Schneider, J., Drewnick, F., Borrmann, S., Weimer, S., Demerjian, K., Salcedo, D., Cottrell, L., Griffin, R., Takami, A., Miyoshi, T., Hatakeyama, S., Shimono, A., Sun, J. Y., Zhang, Y. M., Dzepina, K., Kimmel, J. R., Sueper, D., Jayne, J. T., Herndon, S. C., Trimborn, A. M., Williams, L. R., Wood, E. C., Middlebrook, A. M., Kolb, C. E., Baltensperger, U., and Worsnop, D. R.: Evolution of Organic Aerosols in the Atmosphere, Science, 326, 1525–1529, doi:10.1126/science.1180353, 2009.

Kang, E., Root, M. J., Toohey, D. W., and Brune, W. H.: Introducing the concept of Potential Aerosol Mass (PAM), Atmos. Chem. Phys., 7, 5727–5744, doi:10.5194/acp-7-5727-2007, 2007.

Kautzman, K. E., Surratt, J. D., Chan, M. N., Chan, A. W. H., Hersey, S. P., Chhabra, P. S., Dalleska, N. F., Wennberg, P. O., Flagan, R. C., and Seinfeld, J. H.: Chemical Composition of Gas- and Aerosol-Phase Products from the Photooxidation of Naphthalene, J. Phys. Chem. A, 114, 913–934, doi:10.1021/jp908530s, 2009.

Kessler, S. H., Smith, J. D., Che, D. L., Worsnop, D. R., Wilson, K. R., and Kroll, J. H.: Chemical Sinks of Organic Aerosol: Kinetics and Products of the Heterogeneous Oxidation of Erythritol and Levoglucosan, Environ. Sci. Technol., 44, 7005–7010, doi:10.1021/es101465m, 2010.

Kessler, S. H., Nah, T., Daumit, K. E., Smith, J. D., Leone, S. R., Kolb, C. E., Worsnop, D. R., Wilson, K. R., and Kroll, J. H.: OH-Initiated Heterogeneous Aging of Highly Oxidized Organic Aerosol, J. Phys. Chem. A, 116, 6358–6365, doi:10.1021/jp212131m, 2012.

Kroll, J. H. and Seinfeld, J. H.: Chemistry of secondary organic aerosol: Formation and evolution of low-volatility organics in the atmosphere, Atmos. Environ., 42, 3593–3624, doi:10.1016/j.atmosenv.2008.01.003, 2008.

Kroll, J. H., Smith, J. D., Che, D. L., Kessler, S. H., Worsnop, D. R., and Wilson, K. R.: Measurement of fragmentation and functionalization pathways in the heterogeneous oxidation of oxidized organic aerosol, Phys. Chem. Chem. Phys., 11, 8005–8014, doi:10.1039/B905289E, 2009.

Kroll, J. H., Donahue, N. M., Jimenez, J. L., Kessler, S. H., Canagaratna, M. R., Wilson, K. R., Altieri, K. E., Mazzoleni, L. R., Wozniak, A. S., Bluhm, H., Mysak, E. R., Smith, J. D., Kolb, C. E., and Worsnop, D. R.: Carbon oxidation state as a metric for describing the chemistry of atmospheric organic aerosol, Nat. Chem., 3, 133–139, doi:10.1038/nchem.948, 2011.

Kundu, S., Fisseha, R., Putman, A. L., Rahn, T. A., and Mazzoleni, L. R.: High molecular weight SOA formation during limonene ozonolysis: insights from ultrahigh-resolution FT-ICR mass spectrometry characterization, Atmos. Chem. Phys., 12, 5523–5536, doi:10.5194/acp-12-5523-2012, 2012.

Lambe, A. T., Ahern, A. T., Williams, L. R., Slowik, J. G., Wong, J. P. S., Abbatt, J. P. D., Brune, W. H., Ng, N. L., Wright, J. P., Croasdale, D. R., Worsnop, D. R., Davidovits, P., and Onasch, T. B.: Characterization of aerosol photooxidation flow reactors: heterogeneous oxidation, secondary organic aerosol formation and cloud condensation nuclei activity measurements, Atmos. Meas. Tech., 4, 445–461, doi:10.5194/amt-4-445-2011, 2011a.

Lambe, A. T., Onasch, T. B., Massoli, P., Croasdale, D. R., Wright, J. P., Ahern, A. T., Williams, L. R., Worsnop, D. R., Brune, W. H., and Davidovits, P.: Laboratory studies of the chemical composition and cloud condensation nuclei (CCN) activity of secondary organic aerosol (SOA) and oxidized primary organic aerosol (OPOA), Atmos. Chem. Phys., 11, 8913–8928, doi:10.5194/acp-11-8913-2011, 2011b.

Lambe, A. T., Onasch, T. B., Croasdale, D. R., Wright, J. P., Martin, A. T., Franklin, J. P., Massoli, P., Kroll, J. H., Canagaratna, M. R., Brune, W. H., Worsnop, D. R., and Davidovits, P.: Transitions from Functionalization to Fragmentation Reactions of Laboratory Secondary Organic Aerosol (SOA) Generated from the

OH Oxidation of Alkane Precursors, Environ. Sci. Technol., 46, 5430–5437, doi:10.1021/es300274t, 2012.

Laskin, A., Laskin, J., and Nizkorodov, S. A.: Mass spectrometric approaches for chemical characterisation of atmospheric aerosols: critical review of the most recent advances, Environ. Chem., 9, 163–189, doi:10.1071/EN12052, 2012.

Lee, A., Goldstein, A. H., Keywood, M. D., Gao, S., Varutbangkul, V., Bahreini, R., Ng, N. L., Flagan, R. C., and Seinfeld, J. H.: Gas-phase products and secondary aerosol yields from the ozonolysis of ten different terpenes, J. Geophys. Res.-Atmos., 111, D07302, doi:10.1029/2005JD006437, 2006a.

Lee, A., Goldstein, A. H., Kroll, J. H., Ng, N. L., Varutbangkul, V., Flagan, R. C., and Seinfeld, J. H.: Gas-phase products and secondary aerosol yields from the photooxidation of 16 different terpenes, J. Geophys. Res.-Atmos., 111, D17305, doi:10.1029/2006JD007050, 2006b.

Librando, V. and Tringali, G.: Atmospheric fate of OH initiated oxidation of terpenes. Reaction mechanism of alpha-pinene degradation and secondary organic aerosol formation, J. Environ. Manage., 75, 275–282, doi:10.1016/j.jenvman.2005.01.001, 2005.

Lim, Y. B., Tan, Y., Perri, M. J., Seitzinger, S. P., and Turpin, B. J.: Aqueous chemistry and its role in secondary organic aerosol (SOA) formation, Atmos. Chem. Phys., 10, 10521–10539, doi:10.5194/acp-10-10521-2010, 2010.

Lopez-Hilfiker, F. D., Mohr, C., Ehn, M., Rubach, F., Kleist, E., Wildt, J., Mentel, Th. F., Lutz, A., Hallquist, M., Worsnop, D., and Thornton, J. A.: A novel method for online analysis of gas and particle composition: description and evaluation of a Filter Inlet for Gases and AEROsols (FIGAERO), Atmos. Meas. Tech., 7, 983–1001, doi:10.5194/amt-7-983-2014, 2014.

Mao, J., Ren, X., Brune, W. H., Olson, J. R., Crawford, J. H., Fried, A., Huey, L. G., Cohen, R. C., Heikes, B., Singh, H. B., Blake, D. R., Sachse, G. W., Diskin, G. S., Hall, S. R., and Shetter, R. E.: Airborne measurement of OH reactivity during INTEX-B, Atmos. Chem. Phys., 9, 163–173, doi:10.5194/acp-9-163-2009, 2009.

Mazzoleni, L. R., Ehrmann, B. M., Shen, X., Marshall, A. G., and Collett, J. L.: Water-Soluble Atmospheric Organic Matter in Fog: Exact Masses and Chemical Formula Identification by Ultrahigh-Resolution Fourier Transform Ion Cyclotron Resonance Mass Spectrometry, Environ. Sci. Technol., 44, 3690–3697, doi:10.1021/es903409k, 2010.

Mohr, C., Lopez-Hilfiker, F. D., Zotter, P., Prévôt, A. S. H., Xu, L., Ng, N. L., Herndon, S. C., Williams, L. R., Franklin, J. P., Zahniser, M. S., Worsnop, D. R., Knighton, W. B., Aiken, A. C., Gorkowski, K. J., Dubey, M. K., Allan, J. D., and Thornton, J. A.: Contribution of Nitrated Phenols to Wood Burning Brown Carbon Light Absorption in Detling, United Kingdom during Winter Time, Environ. Sci. Technol., 47, 6316–6324, doi:10.1021/es400683v, 2013.

Moldoveanu, S. C.: Pyrolysis of Carboxylic Acids, Vol. 28 of Techniques and Instrumentation in Analytical Chemistry, book section 17, 471–526, Elsevier, doi:10.1016/S0167-9244(09)02817-0, 2010.

Moller, B., Rarey, J., and Ramjugernath, D.: Estimation of the vapour pressure of non-electrolyte organic compounds via group contributions and group interactions, J. Mol. Liq., 143, 52–63, doi:10.1016/j.molliq.2008.04.020, 2008.

Nannoolal, Y., Rarey, J., Ramjugernath, D., and Cordes, W.: Estimation of pure component properties: Part 1. Estimation of the normal boiling point of non-electrolyte organic compounds via group contributions and group interactions, Fluid Phase Equilibr., 226, 45–63, doi:10.1016/j.fluid.2004.09.001, 2004.

Nannoolal, Y., Rarey, J., and Ramjugernath, D.: Estimation of pure component properties: Part 3. Estimation of the vapor pressure of non-electrolyte organic compounds via group contributions and group interactions, Fluid Phase Equilibr., 269, 117–133, doi:10.1016/j.fluid.2008.04.020, 2008.

Ng, N. L., Kroll, J. H., Keywood, M. D., Bahreini, R., Varutbangkul, V., Flagan, R. C., Seinfeld, J. H., Lee, A., and Goldstein, A. H.: Contribution of First- versus Second-Generation Products to Secondary Organic Aerosols Formed in the Oxidation of Biogenic Hydrocarbons, Environ. Sci. Technol., 40, 2283–2297, doi:10.1021/es052269u, 2006.

Ng, N. L., Canagaratna, M. R., Jimenez, J. L., Chhabra, P. S., Seinfeld, J. H., and Worsnop, D. R.: Changes in organic aerosol composition with aging inferred from aerosol mass spectra, Atmos. Chem. Phys., 11, 6465–6474, doi:10.5194/acp-11-6465-2011, 2011.

Nishino, N., Arey, J., and Atkinson, R.: Yields of Glyoxal and Ring-Cleavage Co-Products from the OH Radical-Initiated Reactions of Naphthalene and Selected Alkylnaphthalenes, Environ. Sci. Technol., 43, 8554–8560, doi:10.1021/es902018v, 2009.

Orlando, J. J., Nozière, B., Tyndall, G. S., Orzechowska, G. E., Paulson, S. E., and Rudich, Y.: Product studies of the OH- and ozone-initiated oxidation of some monoterpenes, J. Geophys. Res.-Atmos., 105, 11561–11572, doi:10.1029/2000JD900005, 2000.

Ortega, J., Turnipseed, A., Guenther, A. B., Karl, T. G., Day, D. A., Gochis, D., Huffman, J. A., Prenni, A. J., Levin, E. J. T., Kreidenweis, S. M., DeMott, P. J., Tobo, Y., Patton, E. G., Hodzic, A., Cui, Y. Y., Harley, P. C., Hornbrook, R. S., Apel, E. C., Monson, R. K., Eller, A. S. D., Greenberg, J. P., Barth, M. C., Campuzano-Jost, P., Palm, B. B., Jimenez, J. L., Aiken, A. C., Dubey, M. K., Geron, C., Offenberg, J., Ryan, M. G., Fornwalt, P. J., Pryor, S. C., Keutsch, F. N., DiGangi, J. P., Chan, A. W. H., Goldstein, A. H., Wolfe, G. M., Kim, S., Kaser, L., Schnitzhofer, R., Hansel, A., Cantrell, C. A., Mauldin, R. L., and Smith, J. N.: Overview of the Manitou Experimental Forest Observatory: site description and selected science results from 2008 to 2013, Atmos. Chem. Phys., 14, 6345–6367, doi:10.5194/acp-14-6345-2014, 2014.

Pankow, J. F. and Asher, W. E.: SIMPOL.1: a simple group contribution method for predicting vapor pressures and enthalpies of vaporization of multifunctional organic compounds, Atmos. Chem. Phys., 8, 2773–2796, doi:10.5194/acp-8-2773-2008, 2008.

Putman, A. L., Offenberg, J. H., Fisseha, R., Kundu, S., Rahn, T. A., and Mazzoleni, L. R.: Ultrahigh-resolution FT-ICR mass spectrometry characterization of alpha-pinene ozonolysis SOA, Atmos. Environ., 46, 164–172, doi:10.1016/j.atmosenv.2011.10.003, 2012.

Seinfeld, J. and Pandis, S.: Atmospheric Chemistry and Physics: From Air Pollution to Climate Change, Wiley, Hoboken, 2006.

Shilling, J. E., Chen, Q., King, S. M., Rosenoern, T., Kroll, J. H., Worsnop, D. R., McKinney, K. A., and Martin, S. T.: Particle mass yield in secondary organic aerosol formed by the dark ozonolysis of α-pinene, Atmos. Chem. Phys., 8, 2073–2088, doi:10.5194/acp-8-2073-2008, 2008.

Spracklen, D. V., Jimenez, J. L., Carslaw, K. S., Worsnop, D. R., Evans, M. J., Mann, G. W., Zhang, Q., Canagaratna, M. R., Allan, J., Coe, H., McFiggans, G., Rap, A., and Forster, P.: Aerosol mass spectrometer constraint on the global secondary organic aerosol budget, Atmos. Chem. Phys., 11, 12109–12136, doi:10.5194/acp-11-12109-2011, 2011.

Su, F., Calvert, J. G., and Shaw, J. H.: Mechanism of the photooxidation of gaseous formaldehyde, J. Phys. Chem., 83, 3185–3191, doi:10.1021/j100488a001, 1979.

Veres, P., Roberts, J. M., Warneke, C., Welsh-Bon, D., Zahniser, M., Herndon, S., Fall, R., and de Gouw, J.: Development of negative-ion proton-transfer chemical-ionization mass spectrometry (NI-PT-CIMS) for the measurement of gas-phase organic acids in the atmosphere, Int. J. Mass Spectrom., 274, 48–55, doi:10.1016/j.ijms.2008.04.032, 2008.

Veres, P., Roberts, J. M., Burling, I. R., Warneke, C., de Gouw, J., and Yokelson, R. J.: Measurements of gas-phase inorganic and organic acids from biomass fires by negative-ion proton-transfer chemical-ionization mass spectrometry, J. Geophys. Rese.-Atmos., 115, D23302, doi:10.1029/2010JD014033, 2010.

Veres, P. R., Roberts, J. M., Cochran, A. K., Gilman, J. B., Kuster, W. C., Holloway, J. S., Graus, M., Flynn, J., Lefer, B., Warneke, C., and de Gouw, J.: Evidence of rapid production of organic acids in an urban air mass, Geophys. Res. Lett., 38, L17807, doi:10.1029/2011GL048420, 2011.

Vogel, A. L., Äijälä, M., Brüggemann, M., Ehn, M., Junninen, H., Petäjä, T., Worsnop, D. R., Kulmala, M., Williams, J., and Hoffmann, T.: Online atmospheric pressure chemical ionization ion trap mass spectrometry (APCI-IT-MSn) for measuring organic acids in concentrated bulk aerosol – a laboratory and field study, Atmos. Meas. Tech., 6, 431–443, doi:10.5194/amt-6-431-2013, 2013.

Wine, P. H., Astalos, R. J., and Mauldin, R. L.: Kinetic and mechanistic study of the hydroxyl + formic acid reaction, J. Phys. Chem., 89, 2620–2624, doi:10.1021/j100258a037, 1985.

Yatavelli, R. L. N. and Thornton, J. A.: Particulate Organic Matter Detection Using a Micro-Orifice Volatilization Impactor Coupled to a Chemical Ionization Mass Spectrometer (MOVI-CIMS), Aerosol Sci. Technol., 44, 61–74, doi:10.1080/02786820903380233, 2010.

Yatavelli, R. L. N., Lopez-Hilfiker, F., Wargo, J. D., Kimmel, J. R., Cubison, M. J., Bertram, T. H., Jimenez, J. L., Gonin, M., Worsnop, D. R., and Thornton, J. A.: A Chemical Ionization High-Resolution Time-of-Flight Mass Spectrometer Coupled to a Micro Orifice Volatilization Impactor (MOVI-HRToF-CIMS) for Analysis of Gas and Particle-Phase Organic Species, Aerosol Sci. Technol., 46, 1313–1327, doi:10.1080/02786826.2012.712236, 2012.

Yatavelli, R. L. N., Stark, H., Thompson, S. L., Kimmel, J. R., Cubison, M. J., Day, D. A., Campuzano-Jost, P., Palm, B. B., Hodzic, A., Thornton, J. A., Jayne, J. T., Worsnop, D. R., and Jimenez, J. L.: Semicontinuous measurements of gas-particle partitioning of organic acids in a ponderosa pine forest using a MOVI-HRToF-CIMS, Atmos. Chem. Phys., 14, 1527–1546, doi:10.5194/acp-14-1527-2014, 2014.

Yu, J., Cocker, DavidR, I., Griffin, R., Flagan, R., and Seinfeld, J.: Gas-Phase Ozone Oxidation of Monoterpenes: Gaseous and Particulate Products, J.Atmos. Chem., 34, 207–258, doi:10.1023/A:1006254930583, 1999.

Zahardis, J., Geddes, S., and Petrucci, G. A.: Improved Understanding of Atmospheric Organic Aerosols via Innovations in Soft Ionization Aerosol Mass Spectrometry, Anal. Chem., 83, 2409–2415, doi:10.1021/ac102737k, 2011.

Zhang, Q., Jimenez, J. L., Canagaratna, M. R., Allan, J. D., Coe, H., Ulbrich, I., Alfarra, M. R., Takami, A., Middlebrook, A. M., Sun, Y. L., Dzepina, K., Dunlea, E., Docherty, K., DeCarlo, P. F., Salcedo, D., Onasch, T., Jayne, J. T., Miyoshi, T., Shimono, A., Hatakeyama, S., Takegawa, N., Kondo, Y., Schneider, J., Drewnick, F., Borrmann, S., Weimer, S., Demerjian, K., Williams, P., Bower, K., Bahreini, R., Cottrell, L., Griffin, R. J., Rautiainen, J., Sun, J. Y., Zhang, Y. M., and Worsnop, D. R.: Ubiquity and dominance of oxygenated species in organic aerosols in anthropogenically-influenced Northern Hemisphere midlatitudes, Geophys. Res. Lett., 34, L13801, doi:10.1029/2007GL029979, 2007.

2

Comparison of continuous in situ CO_2 observations at Jungfraujoch using two different measurement techniques

M. F. Schibig[1], M. Steinbacher[2], B. Buchmann[2], I. T. van der Laan-Luijkx[1,*], S. van der Laan[1,], S. Ranjan[***], and M. C. Leuenberger[1]**

[1]Climate and Environmental Physics, Physics Institute and Oeschger Centre for Climate Change Research, University of Bern, Bern, Switzerland
[2]Swiss Federal Laboratories for Materials Science and Technology (Empa), Dübendorf, Switzerland
[*]now at: Meteorology and Air Quality, Wageningen University, Wageningen, the Netherlands
[**]now at: Centre for Ocean and Atmospheric Sciences, School of Environmental Sciences, University of East Anglia, Norwich, UK
[***]formerly at: Climate and Environmental Physics, Physics Institute and Oeschger Centre for Climate Change Research, University of Bern, Bern, Switzerland

Correspondence to: M. C. Leuenberger (leuenberger@climate.unibe.ch)

Abstract. Since 2004, atmospheric carbon dioxide (CO_2) is being measured at the High Altitude Research Station Jungfraujoch by the division of Climate and Environmental Physics at the University of Bern (KUP) using a nondispersive infrared gas analyzer (NDIR) in combination with a paramagnetic O_2 analyzer. In January 2010, CO_2 measurements based on cavity ring-down spectroscopy (CRDS) as part of the Swiss National Air Pollution Monitoring Network were added by the Swiss Federal Laboratories for Materials Science and Technology (Empa). To ensure a smooth transition – a prerequisite when merging two data sets, e.g., for trend determinations – the two measurement systems run in parallel for several years. Such a long-term intercomparison also allows the identification of potential offsets between the two data sets and the collection of information about the compatibility of the two systems on different time scales.

A good agreement of the seasonality, short-term variations and, to a lesser extent mainly due to the short common period, trend calculations is observed. However, the comparison reveals some issues related to the stability of the calibration gases of the KUP system and their assigned CO_2 mole fraction. It is possible to adapt an improved calibration strategy based on standard gas determinations, which leads to better agreement between the two data sets. By excluding periods with technical problems and bad calibration gas cylinders, the average hourly difference (CRDS – NDIR) of the two systems is -0.03 ppm ± 0.25 ppm. Although the difference of the two data sets is in line with the compatibility goal of ± 0.1 ppm of the World Meteorological Organization (WMO), the standard deviation is still too high. A significant part of this uncertainty originates from the necessity to switch the KUP system frequently (every 12 min) for 6 min from ambient air to a working gas in order to correct short-term variations of the O_2 measurement system. Allowing additional time for signal stabilization after switching the sample, an effective data coverage of only one-sixth for the KUP system is achieved while the Empa system has a nearly complete data coverage. Additionally, different internal volumes and flow rates may affect observed differences.

1 Introduction

Carbon dioxide (CO_2) is an important greenhouse gas which plays a major role for the radiative forcing of the atmosphere and therefore contributes significantly to the greenhouse effect (Arrhenius, 1896). Since natural as well as anthropogenic processes alter the mole fraction of CO_2 in the atmosphere, it is important to survey the mole fraction changes over a long time period to understand these processes and draw conclusions about the involved CO_2 fluxes (Ramonet et al., 2010; Schimel et al., 2001). Additionally, in situ CO_2 measurements can be used in modeling studies as input parameters or to verify model output (Uglietti et al., 2011; Chevallier et al., 2010; Peters et al., 2010). Therefore it is crucial that the data sets are well calibrated and traced back to a common reference scale and that the measurement systems demonstrate high precision and accuracy. This guarantees minimal biases between different measurement sites and laboratories and improves the compatibility of the data sets. For CO_2 measurements, the World Meteorological Organization (WMO) recommends a compatibility of ± 0.1 ppm for the northern hemisphere (WMO, 2011).

High-altitude observatories such as the High Alpine Research Station Jungfraujoch (JFJ) are predestined to monitor the background mole fraction of trace gases like CO_2 because they are mostly in the free troposphere.

That is why the division of Climate and Environmental Physics of the Physics Institute, University of Bern (KUP) started measuring CO_2 with a non-dispersive infrared (NDIR) gas analyzer (Sick Maihak, Germany, model S710) embedded in a combined CO_2 / O_2 analyzing system at JFJ in late 2004. At the beginning of 2010, a cavity ring-down spectrometer (until July 2011 a Picarro Inc., USA, model G1301, afterwards a Picarro Inc., USA, model G2401) was installed by the Swiss Federal Laboratories for Materials Science and Technology (Empa) as part of the Swiss National Air Pollution Monitoring Network (NABEL), enabling a long-term performance of CO_2 measurements at JFJ by Empa. To ensure that both records could be merged smoothly, both systems ran in parallel for a significant period, not only to detect potential biases but also to provide information about the compatibility of the two different measurement systems as had been done at other stations, e.g., at Mauna Loa (Komhyr et al., 1989; Peterson et al., 1977). Because of the year-round manned infrastructure and excellent accessibility by train, it was possible to test the compatibility of the two systems in situ under real conditions at a background site featuring only limited atmospheric variation.

In this study we report the data of the KUP and the Empa in situ CO_2 measurement system from January 2010 until December 2012.

2 Methods

2.1 Sampling site

The High Altitude Research Station Jungfraujoch (JFJ) is located 7°59′20″ E, 46°32′53″ N on the northern ridge of the Swiss Alps. The station itself is on a mountain saddle between the mountains Mönch (4099 m a.s.l.) and Jungfrau (4158 m a.s.l), at an altitude of 3580 m a.s.l. and is accessible by train the whole year round. Because of the high elevation, the station is above the planetary boundary layer most of the time and receives predominantly air from the free troposphere. It is therefore an ideal location to measure the atmospheric background air of continental Europe (Henne et al., 2010; Zellweger et al., 2003; Baltensperger et al., 1997). Nevertheless, the station is sometimes influenced by polluted boundary layer air, especially during Föhn events (Zellweger et al., 2003) or during hot summer days, when air from the surrounding valley is thermally uplifted to JFJ (Zellweger et al., 2000; Baltensperger et al., 1997). A comprehensive in situ measurement program with more than 70 trace gases and a large suite of aerosol properties is run continuously at JFJ by an international consortium of research institutions. JFJ is also one of the currently 29 Global Atmosphere Watch (GAW) stations.

2.2 NDIR measurements (KUP)

The KUP CO_2 measurements are based on a combined system to monitor CO_2 and O_2 changes in the atmosphere. The ambient air enters through a strongly ventilated (600 $m^3\,h^{-1}$) common inlet on the observatory's roof to a manifold, which serves many air detectors, where an aliquot is drawn to the KUP system. The air is cryogenically dried to a dew point of -90 °C (FC-100D21, FTS systems, USA). CO_2 is measured by a NDIR spectrometer (Maihak S710) with a working range of 350 to 450 ppm and a frequency of 1 Hz; O_2 is measured by a paramagnetic cell. To avoid influences caused by ambient air temperature and density fluctuations, the temperature as well as the pressure is stabilized. The box with the pressure and flow control system is embedded in a temperature-controlled box (45 ± 0.05 °C); the precision of the pressure regulation is ± 0.05 mbar and the measurement cell of the NDIR analyzer is heated to 55 ± 0.05 °C. Measurements are done in a cyclic sequence of 18 h: each gas is measured for 6 min, with only the last 115 s of a 6 min period used for mole fraction determination to allow for signal stabilization after changing the sample source. At the beginning of each sequence, the system is calibrated with two reference gases (high and low span). A working gas is measured between two ambient air measurements to correct for short-term variations. Therefore, an 18 h measurement sequence looks as follows:

$$B-G-K-B-H-B-A-A-B-A-A-B- \\ \ldots -A-A-B,$$

where B represents the working gas, G the high span CO_2/low span O_2, K the low span CO_2/high span O_2, H the target cylinder and A represents ambient air.

All measurements ending at a particular hour are used for the calculation of hourly mean CO_2 observations, which in our case includes six ambient observation values per hour. Cylinder measurements (see Table 1) with a known mole fraction show a precision better than 0.04 ppm for 1 h analysis. The precision is calculated as the square root of the sum

Table 1. KUP calibration cylinders and their measured CO_2 values on the NDIR and the CRDS system. These values were used to calculate the second degree polynomial to correct the nonlinearity of the NDIR system (See also Fig. 9).

Cylinder number	Used before as	CO_2 Picarro [ppm]	CO_2 Maihak [ppm]	ΔCO_2 (Picarro – Maihak) [ppm]	ΔCO_2 (Picarro – Maihak corrected) [ppm]	Pressure [bar]
LK542510	G1	431.99 ± 0.02	431.74 ± 0.04	0.25	0.05	22.1
LK94521	K5	360.57 ± 0.07	359.96 ± 0.05	0.61	−0.06	23.8
LK570931	B20	438.01 ± 0.02	437.73 ± 0.04	0.28	−0.01	21.9
LK542204	B21	385.09 ± 0.02	384.87 ± 0.04	0.22	0.04	26.8
LK3137	B22	384.99 ± 0.02	387.71 ± 0.05	0.28	0.10	16.1
LK3077	B24	418.68 ± 0.08	418.78 ± 0.03	−0.06	−0.12	14.1
LK542469	B26	432.06 ± 0.03	431.87 ± 0.02	0.19	−0.01	23.7
LK564848	B27	448.40 ± 0.03	447.83 ± 0.02	0.57	0.08	22.7
LK570916	B29	437.66 ± 0.07	437.43 ± 0.05	0.23	−0.06	185

of the squared standard deviations of the individual measurements. The CO_2 values are reported on the WMO X2007 scale. In February 2011, the Maihak analyzer was replaced with an identical Maihak device because of technical problems.

2.3 Cavity ring-down spectroscopy measurements (Empa)

From late December 2009 until July 2011, Empa was measuring CO_2 with a commercially available wavelength-scanned cavity ring-down spectrometer (Picarro Inc., USA; G1301) coupled to a custom-built calibration/drying unit. Initially, the sample air from the manifold was dried prior to analysis by means of a Nafion dryer to a dew point of $< -30\,°C$. Along with CO_2, the instrument also measured CH_4 and H_2O at the same sample frequency of approximately 0.5 Hz. The H_2O measurements allowed for correction of the CO_2 mole fraction in case of interferences (i.e., dilution and pressure broadening) due to potential water moisture in the system. From August 2010 on, the Nafion dryer was short cut, no water vapor removal was used and CO_2 dry air mole fractions were determined after application of an empirical humidity correction. The empirically determined water vapor corrections for CRDS analyzers do not significantly change over time and are very similar for different analyzers. According to Rella et al. (2013), the error in the reported dry air CO_2 mole fractions when using different correction coefficients is < 0.1 ppm, up to a H_2O volume mixing ratio of 2.5 Vol-%. 2.5 Vol-% of H_2O at JFJ refers to a dew point of more than 14 °C, a value that was never reached at JFJ. The error can be further reduced when using the dedicated correction coefficients for the respective analyzer and doing repeated water vapor interference measurements, as is the case here.

After a system breakdown due to a faulty electronic board in July 2011, the G1301 analyzer was replaced in September 2011 by the newer G2401 model (Picarro Inc., USA) that is also capable of monitoring CO mole fractions. Since then, the sample gas has been dried again for the beneficial effects with respect to the precision of the CO analysis.

Calibrations are performed every 46 h with two calibration gases (high and low span). In addition, a target gas is analyzed every 15 h to detect potential shorter-term instrument sensitivity changes. The mole fractions of the calibration gases are determined by the World Calibration Center for CH_4, CO_2, CO and O_3 at Empa (Table 3). Measurements of used KUP standard cylinders at JFJ show a precision better than 0.04 ppm for the 15 min analysis (Table 1). Similar to the KUP, CO_2 mole fractions are also reported on the WMO X2007 scale.

2.4 Calibration gases

The gas used to calibrate the KUP measurements is compressed outside air filled in steel cylinders delivered by Carbagas (Switzerland). Normally, each gas cylinder is first delivered to our laboratory at the University of Bern and measured for its CO_2 mole fraction and $\delta O_2 / N_2$ and $\delta Ar / N_2$ values by mass spectrometry and a combined Licor 7000 NDIR and Oxzilla (Sable Systems, USA) system (CO_2 and O_2). The measured values are calibrated with a subset of three standards from the WMO/GAW Central Calibration Laboratory (CCL) run by the National Oceanic and Atmosphere Administration (NOAA, Boulder, USA) and are used as assigned values of the KUP system at JFJ (see Table 2). The cylinders are shipped to JFJ and stored in the basement until their usage as calibration gases. A maintained stock of calibration gases at JFJ is necessary because the KUP system has to be calibrated frequently for O_2 measurements (Uglietti et al., 2008) and has a rather high gas consumption of up to 150 mL per minute.

The calibration gases used by Empa consist of compressed dry natural air in aluminium cylinders filled by themselves with a modified oil-free compressor (Rix Industries, USA). The mole fractions of the gases are determined by

Table 2. List of calibration gas cylinders used by the NDIR system in the period from 2010 to 2012: K, G, H and B correspond to the usage as low span, high span, target and working gas, respectively. The gases were measured at the University of Bern and calibrated with a suite of NOAA standard cylinders.

Cylinder number	Used as	From	To	Assigned CO_2[ppm] WMO X2007
LK94521	K5	10.03.2009	31.12.2010	360.31 ± 0.50
LK541816	K6	31.12.2010	06.11.2012	380.77 ± 0.25
LK542332	K7	06.11.2012	now	384.41 ± 0.07
LK542217	G2	23.07.2009	now	429.02 ± 0.25
LK542119	B23	11.12.2009	27.04.2010	379.90 ± 0.25
LK542171	H1	08.09.2010	now	440.16 ± 0.25
LK3077	B24	27.04.2010	07.09.2010	416.60 ± 0.10
LK541869	B25	07.09.2010	09.02.2011	416.79 ± 0.10
LK542469	B26	09.02.2011	17.06.2011	431.76 ± 0.25
LK564848	B27	17.06.2011	19.10.2011	448.04 ± 0.25
LK557112	B28	19.10.2011	23.2.2012	425.35 ± 0.05
LK570916	B29	23.02.2012	03.07.2012	437.55 ± 0.10
LK564850	B30	03.07.2012	05.10.2012	395.62 ± 0.01
LK542463	B31	5.10.2012	23.01.2013	390.34 ± 0.05

Table 3. List of calibration gas cylinders used for the CRDS system in the period from 2010 to 2012 as measured at the Empa laboratory.

Cylinder number	Used as	From	To	Assigned CO_2[ppm] WMO X2007
D621867 (E062)	Cal. gas 1	18.12.2010	now	394.11 ± 0.03
CA06290 (E063)	Cal. gas 2	18.12.2010	now	429.72 ± 0.02
CB555583	Target	18.12.2010	now	426.04 ± 0.04

the World Calibration Center for CH_4, CO_2, CO and O_3 at Empa through comparison with a suite of standards from the WMO/GAW CCL (see Table 3). After measuring the calibration gases, the cylinders are shipped to JFJ.

Lists of the calibration gases used by both systems during the period of intercomparison are included in Tables 2 and 3.

3 Results

3.1 Comparison of the two data sets

The CO_2 measurements of both systems for the period 2010–2012 are shown in Fig. 1. In general, the two data sets show a good agreement. There are two longer gaps in the CRDS record from 11 June to 16 September 2011 and 5 to 18 January 2012. The first gap was caused by technical problems of the Picarro, which was replaced by a new four-channel model. In January 2012 a hard-disk failure was responsible for the roughly 2-week break. The NDIR data set shows four longer gaps from 31 January to 16 February 2011, 21 to 28 December 2011, 14 to 25 April 2012 and 7 to 24 October 2012. The reason for these gaps in the NDIR data set was technical issues with the Maihak analyzer, eventually leading to a replacement of the analyzer with an identical one. Furthermore, a power failure was responsible for the failure of a gas flow controller and, in April and October 2012, the dysfunction of the gas drying system. Throughout the record, small gaps are also present due to system crashes, failures in the regulation of the gas flows, etc. The remaining hourly mean values, where we have overlap of both systems (20 460 common hours), are in good agreement. The data sets show a similar seasonality as well as peaks caused by pollution events (e.g., advection of air masses from the Po basin in Italy or from northeastern Europe).

The average seasonality of the monthly mean values, spanning 36 months, is 10.01 ± 0.33 ppm and 10.05 ± 0.37 ppm for the CRDS and the NDIR data set, respectively. Due to missing data, one monthly value in each data set has to be interpolated. For the complete data sets, the CRDS data show an annual increase of 1.89 ± 0.01 ppm year^{-1}, whereas the NDIR data show a slope of 1.69 ± 0.01 ppm year^{-1}; the uncertainties correspond to the error of the linear increase fitted to the data. Selecting only data points where both systems have overlapping data points, the difference in the slopes decreases slightly as documented by values of the annual increase of CO_2 of 1.91 ppm year^{-1} (CRDS) and 1.72 ppm year^{-1} (NDIR). A correlation plot of the CRDS data set against the NDIR data set reveals a very high correlation with a slope of 0.991 ppm ppm^{-1}, an intercept

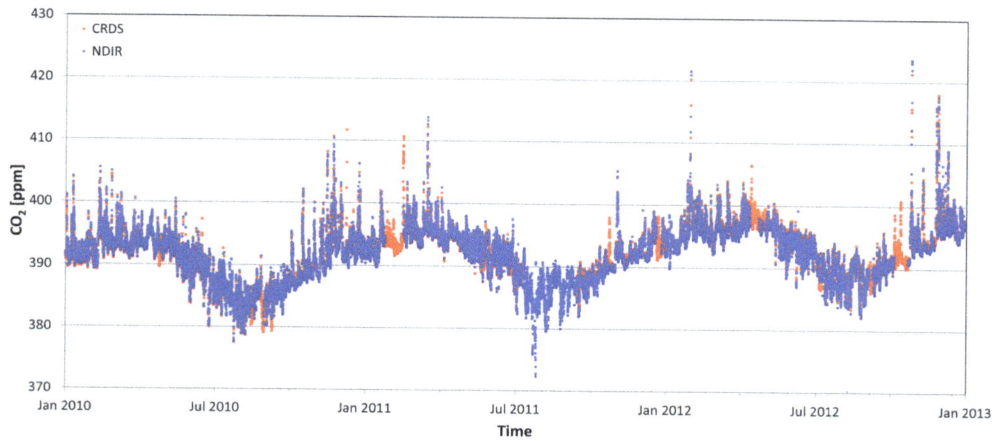

Figure 1. CO_2 measurements at JFJ of the CRDS system (red points) and the NDIR system (blue points) against time. The values represent the hourly averages of the two measurement systems.

3.662 ppm and a R^2 of 0.9909 (Fig. 2, grey points). It can be seen in Fig. 3 (grey points) that the differences of the two data sets are very stable over the 3-year period of the comparison. The average of the difference between the CRDS data set and the NDIR data set is 0.04 ± 0.40 ppm (Fig. 3, grey points); negative and positive differences equal out to almost zero. During periods of rapid CO_2 changes, the standard deviation of the differences is larger than during stable periods and significantly exceeds the precision of both instruments. This point can be emphasized by zooming in to higher-resolution data (115 s averages). For this exercise, the high-resolution CRDS data are aggregated to 115 s averages corresponding to the averaging intervals of the NDIR system. It can be clearly seen that the difference becomes noisier with fast changes (e.g., during the late afternoon in Figs. 4 and 5) and decreases again with more stable conditions (e.g., during the nighttime in Figs. 4 and 5).

The behavior of short-term variations is different for both systems as shown by a change of statistical characteristics when discarding values for which the change rate is above a certain threshold (Table 4). These changes are most probably due to different volumes and gas flow resulting in different residence times and leading to dispersion effects. For example, the NDIR system uses a water trap with a large volume, potentially dampening the CO_2 signal despite the higher flow rate compared to the rather small Nafion dryer of the CRDS system. The cutoff criteria of the change rate is set to 0.75 ppm h^{-1} and values with a higher change rate are discarded. Below 0.75 ppm h^{-1} there is still an improvement of the agreement, but the loss in data points is too high. Additionally, the standard error starts to increase again below a change rate of 0.75 ppm h^{-1} because many data points are omitted by this criteria. By setting the threshold to 0.75 ppm h^{-1}, roughly 20 % of the common data points are omitted. The correlation between the remaining data for both systems gets slightly closer to the ideal 1 : 1 function with a slope of 0.996 ppm ppm^{-1}, an intercept of 1.5696 ppm and a R^2 of 0.9924 (Fig. 3, black points). Also, the differences between the CRDS and the NDIR data set are smaller; considering only the data points of the more stable conditions (Fig. 3, black points), the average of the difference remains almost 0, namely 0.05 ppm $\pm$ 0.32 ppm.

However, there are problematic features worth mentioning. By replacing the working gas B26 with B27 and B28 with B29, small shifts occur that are most likely caused by an inaccurate assignment of the CO_2 values of the working gases (Fig. 3). By excluding the periods of the cylinders B27 and B29, 11 038 common data points remain; the average of the differences, however, stays almost the same, namely −0.03 ppm but with a reduced standard deviation of 0.25 ppm.

In late 2010, the difference between the two data sets increased significantly because of technical issues (see Sect. 3.1). After replacing the analyzer, the difference was again very small and stable (Fig. 3). Furthermore, with the exchange of the Picarro in summer 2011, a small jump in the difference between the two systems occurred (Fig. 3). Since the replacement happened during a single working gas period of the NDIR system (B27), the offset was probably caused by the change of the Picarro analyzer.

3.2 Drift of calibration gases and their corrections

In the case of regular periodic calibrations, the pressure of a calibration gas cylinder decreases relatively linear over its lifetime. Because steel cylinders show pressure-dependent adsorption and desorption effects of gases such as CO_2 or H_2O (Langmuir, 1918), CO_2 continuously desorbs from the cylinder walls and increases the CO_2 mole fraction of the calibration gas during its usage. To avoid a large drift of the CO_2 mole fraction in calibration gases, Keeling et al. (1998) recommends using calibration gas cylinders only to a remaining

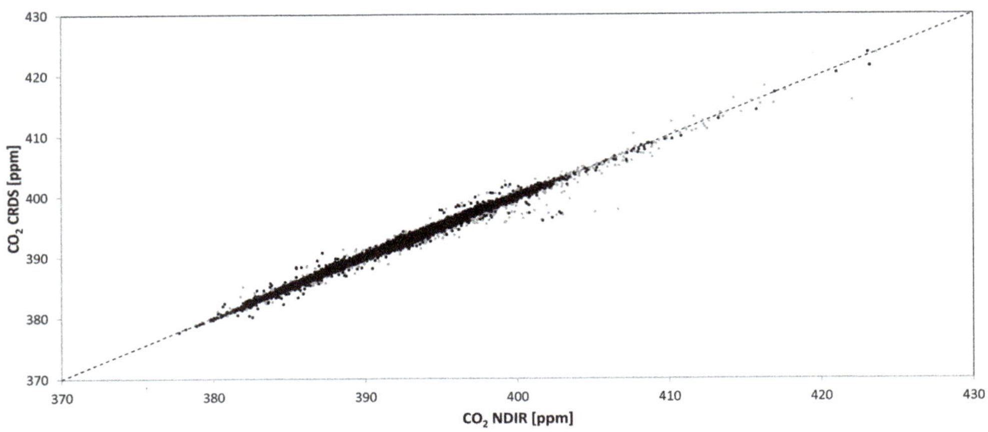

Figure 2. The hourly averages of the CRDS CO_2 measurements versus the hourly averages of the NDIR CO_2 measurements (grey points) and the CRDS CO_2 measurements versus the hourly averages of the NDIR CO_2 measurements without periods of rapid mole fraction changes (black points) over the whole period of comparison. The dashed diagonal represents the ideal 1 : 1 agreement. Considering all data points, the R^2 is 0.9909 with a slope of 0.9908 ppm ppm^{-1} and an intercept of 3.6623 ppm. Excluding periods with rapid mole fraction changes, the agreement of the two data sets is slightly better with a R^2 of 0.9924, a slope of 0.9961 ppm ppm^{-1} and an intercept 1.5696 ppm.

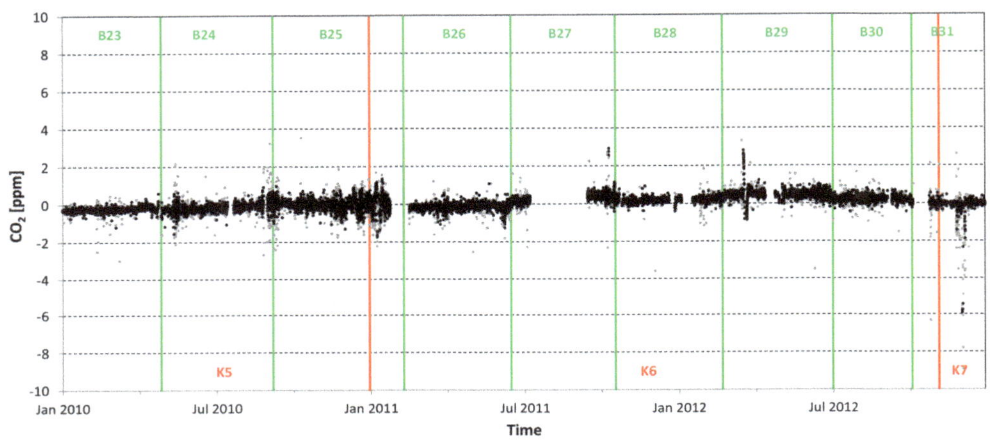

Figure 3. The difference of the CRDS – NDIR CO_2 measurements of all common hourly data points (grey points) and only during stable periods with a CO_2 change of less than 0.75 ppm h^{-1} (black points) against time. The green bars represent changes of the working gases of the NDIR system with the green number indicating the according cylinder and the red bars indicate the changes of the low span with the according number, also in red. The high span was not replaced throughout the entire period.

pressure of 25 to 30 bar. The adsorption of CO_2 to the cylinder wall can be calculated according to Langmuir (Langmuir, 1918) and corrected for by the formula:

$$\mathrm{WG_{corr}}(t) = \mathrm{WG_{meas}}(t) - \left(\frac{a \cdot b \cdot p_0}{b \cdot p_0 + 1} - \frac{a \cdot b \cdot p(t)}{b \cdot p(t) + 1}\right), \quad (1)$$

where $\mathrm{WG_{meas}}$ corresponds to the measured CO_2 value of the working gas [ppm], a and b are constants (related to the Avogadro constant, the number of elementary spaces of the surface, the molecular mass and the temperature), p_0 is the start pressure and $p(t)$ is the pressure at the time of the measurement. This enrichment of CO_2 is more pronounced at lower pressures compared to high pressures, as seen in Eq. (1). For example, at 100 bar cylinder pressure the correction is only about 0.05 ppm, whereas at 30 bar it is already up to 0.3 ppm (with $a = 5\,715\,797$ and $b = 568\,897$, a and b values were derived from measurements of cylinder B23). In steel cylinders, present water vapor would cause an even more pronounced desorption effect. However, all cylinders used here contain dry gas. In Leuenberger et al. (2014), an in-depth discussion of this adsorption/desorption influence of gases on steel and aluminium cylinders is presented, based on dedicated experimental results in our laboratory as well as in a climate chamber.

In April 2010 the working gas tank B23 ran completely empty due to a faulty pressure reader and therefore showed such an enrichment effect. Hence the CO_2 values of the NDIR calibrated with the uncorrected working gas are under-

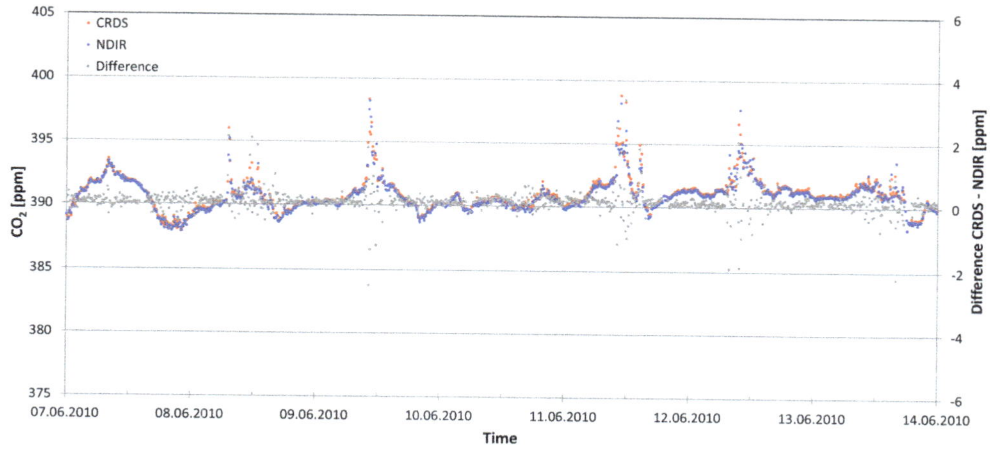

Figure 4. Time series of 115 s averaged CO_2 mole fractions measured with CRDS (red) and NDIR (blue) at JFJ (left-hand *y* axis scale) and the difference between the two systems against time (grey points) (right-hand *y* axis scale) for 1 week in June 2010.

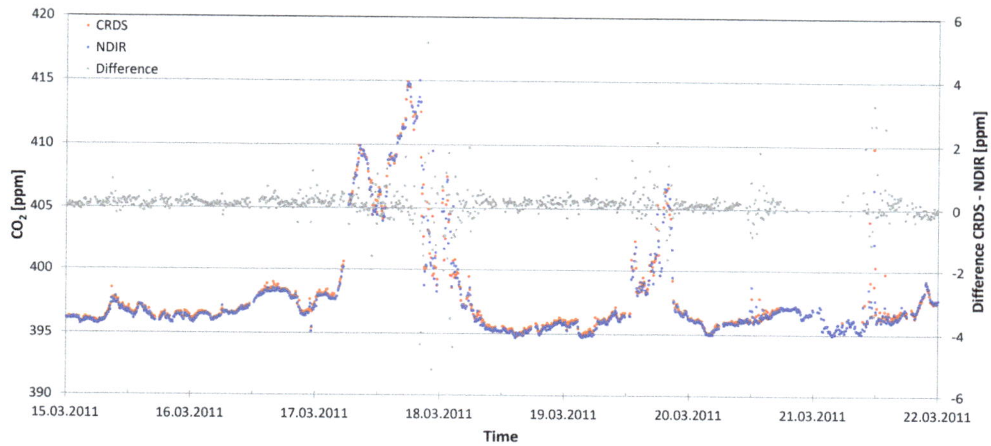

Figure 5. Time series of 115 s averaged CO_2 mole fractions measured with CRDS (red) and NDIR (blue) at JFJ (left-hand *y* axis scale) and the difference between the two systems against time (grey points) (right-hand *y* axis scale) for 1 week in March 2011.

estimated compared to the CRDS measurements. By recalculating measurements of the working gas using only the low and the high span (Fig. 6, black points), it is possible to approximate the evolution of the working gas CO_2 enrichment (Fig. 6, grey points) and correct for it (Fig. 7). The NDIR data are in much better agreement with the CRDS measurements after applying this correction. Comparing the uncorrected values of the B23 period to the corrected period, the R^2 of the NDIR values against the CRDS values increases from 0.89 to 0.99 (Fig. 8). The remaining working gases cylinders are recalculated in the same manner. Most of them show similar behavior as B23, even when they are replaced at a remaining pressure of around 25 bar. The only difference is the intensity of the enrichment, which is weaker than with B23 because they did not run completely empty. Note that for B23 we use time instead of pressure as the dependent variable because the pressure reader did not work properly during this period. Although not preferred, this is possible because the pressure decreased linearly long-term.

3.3 Polynomial correction of the NDIR Maihak measurements

NDIR spectrometers are nonlinear in response and hence require a correction of the data – generally of the polynomial form. Ideally, the analyzer response function is implemented by the producer, mostly for a restricted mole fraction range. This is also the case for the Maihak system. The Maihak measurements at JFJ are done in differential mode; thus the reference cell is always flushed with gas of known CO_2 mole fraction and compared to the cell with the sample. To set the polynomial function to overcome the nonlinearity of the device, the mole fraction of the reference gas has to be entered into the analyzer's system.

Table 4. The statistics of different cutoff criteria of rapid CO_2 mole fraction changes and their influence on the number of points remaining, the average of the hourly differences of the NDIR and the CRDS system, the standard deviation of the differences and the according standard errors. The differences between the cutoff criteria are most probably caused by different volumes and flow rates of the systems and small time lags in the time stamps. The measurement precision of each system is about 10 times better than the standard deviation of the differences.

Cutoff criteria [ppm h^{-1}]	Number of points	Average of differences [ppm]	Standard deviation [ppm]	Standard error [ppm]
All	19779	0.038	0.401	0.0028
< 2	19260	0.044	0.363	0.0026
< 1	17474	0.049	0.336	0.0025
< 0.75	16094	0.053	0.322	0.0025
< 0.5	13602	0.056	0.312	0.0027
< 0.25	8666	0.060	0.295	0.0032

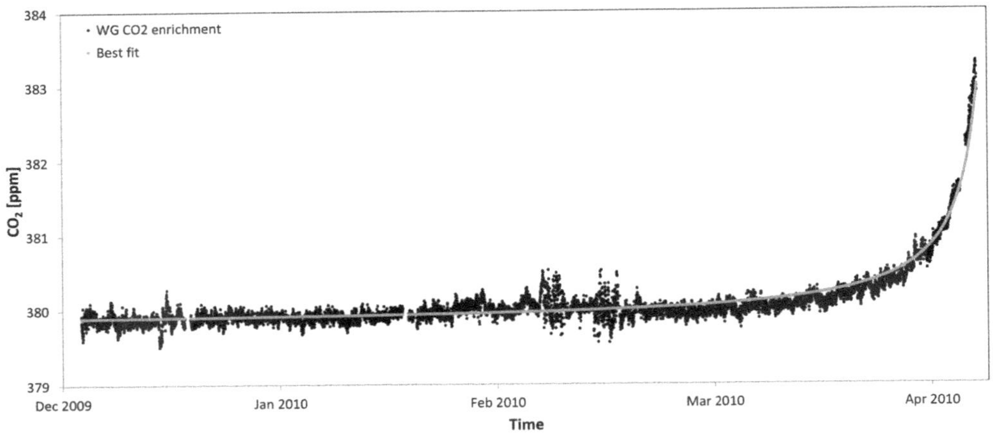

Figure 6. The CO_2 values of the calibration gas cylinder B23 calculated by using only the low and the high span cylinders against time (black points) and the calculated best fit function according to Langmuir (grey points) with $a = 5\,715\,797$ and $b = 568\,897$. The CO_2 values increase with time (corresponding to decreasing pressure). The best fit function was used to correct the measured working gas CO_2 values of the NDIR system.

In fall 2011, 11 used calibration gas cylinders of the KUP system were measured with the CRDS system and remeasured with the NDIR system. The remeasuring of these cylinders reveals that the polynomial function of the Maihak is insufficient. The second order polynomial of the difference between CRDS (Empa) – NDIR (KUP) as a function of CRDS (Empa) yields a R^2 of 0.86. A very similar polynomial dependence is found between the assigned values of NDIR (KUP) and the assigned values according to Table 2 when discarding two cylinders with low remaining pressure (Table 1). We prefer the first polynomial because the cylinders might have already faced desorption effects due to changing pressure between the assignment measurement in Bern (> 2 years before) and the remeasurement at JFJ, whereas the comparison with the CRDS took place within a few days. Since the cavity ring-down technique is linear in the considered mole fraction range, the factors of the polynomial are used to correct the measurement values of the NDIR system, removing this dependence completely (Fig. 9 and Table 1). The correction is only applied to the data measured by the NDIR analyzer actually in use at JFJ. The data of the first NDIR analyzer, which had to be replaced in the beginning of 2011 due to technical problems, remain unaltered since the working points of the two NDIR instruments are different. There is also no possibility to correct the data of the older NDIR analyzer properly, because at the time there were no comparable measurements with a nearly linear measurement system. However, values in the center of this span, where most of the ambient CO_2 mole fractions were measured, should not have been affected strongly; roughly 75 % of all NDIR values are affected less than 0.1 ppm by the polynomial correction. The accuracy of the standard measurements is calculated as the square root of the sum of squared trueness and squared precision (Menditto et al., 2007) and estimated to be 0.08 ppm over the working range from 350 to 450 ppm based on the CRDS – NDIR comparison including both instrumental uncertainties.

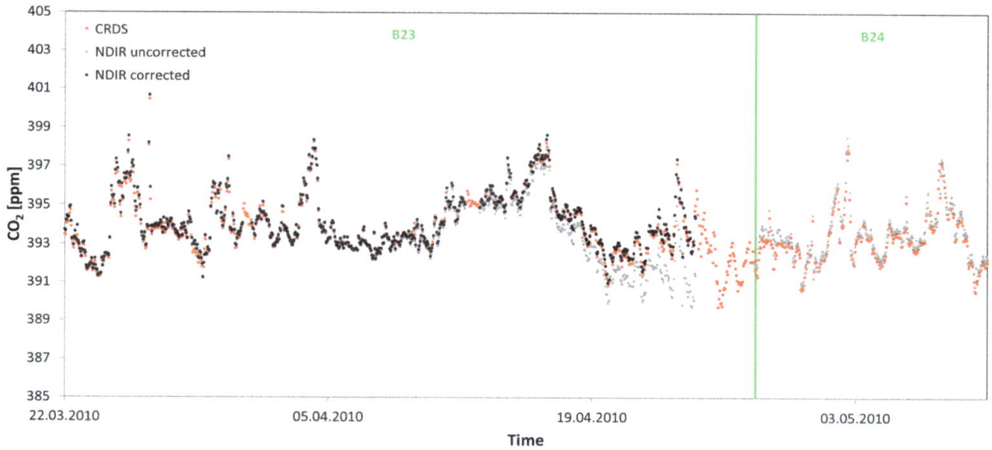

Figure 7. Time series of the CO_2 mole fractions determined with CRDS (red points) and NDIR (uncorrected in grey, after correction in black). The desorption-corrected NDIR CO_2 values show a much better agreement with the CRDS CO_2 values than the uncorrected, which are severely underestimated towards the end of the B23 period. After replacement of the working gas cylinder B23 with B24, the uncorrected NDIR CO_2 values show a much better agreement with the CRDS CO_2 values.

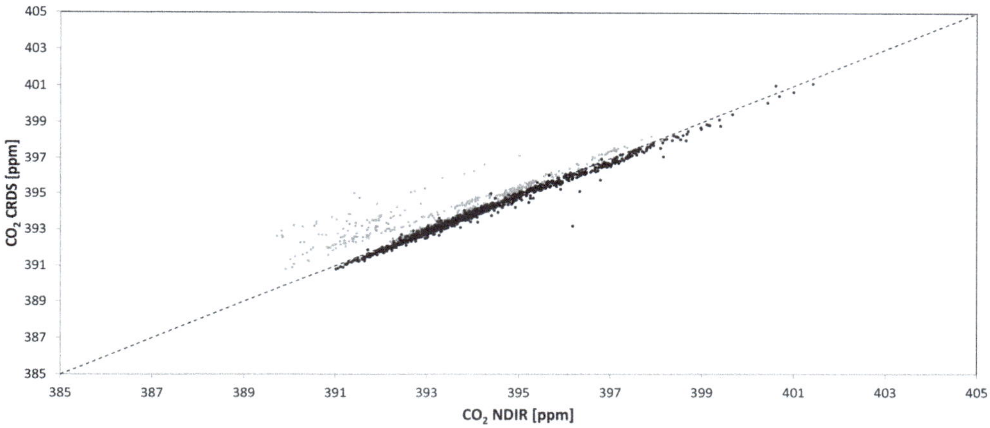

Figure 8. Correlation of ambient CO_2 mole fraction measured by CRDS versus the uncorrected CO_2 mole fraction of the NDIR instrument (grey points) and the desorption-corrected CO_2 NDIR mole fractions (black points), respectively during the period of the working gas cylinder B23. The diagonal represents the ideal 1 : 1 agreement. Due to the CO_2 enrichment in the working gas cylinder B23 the uncorrected KUP values seem to be much lower than the Empa values. By applying the correction, the values of the two systems are in a much better agreement.

4 Discussion

Long periods of overlapping measurements are helpful for assessing the compatibility of different analyzers. In this study we analyze two different data sets which were measured at the same station over 3 years (the simultaneous measurements are still ongoing) to determine the compatibility and quantify differences and biases. The compatibility is challenging since the Empa used the cavity ring-down technique (Picarro Inc.) and the KUP a non-dispersive infrared spectrometer (Maihak) to measure the CO_2 mole fraction of the air. Also, the calibration procedure, given the two techniques used, is different and therefore has to be investigated.

The seasonality based on monthly measured averages (10.01 ± 0.33 ppm and 10.05 ± 0.37 ppm for the CRDS and the NDIR data set) between the two data sets agree nicely and are also in good agreement with the results of flask measurements done at JFJ (10.54 ± 0.18 ppm) (van der Laan-Luijkx et al., 2013). The trend calculation (1.89 and 1.69 ppm year^{-1}) for the complete and common data periods deviates by 0.2 ppm year^{-1}. In the first half of this comparison, the difference of CRDS – NDIR is negative, indicating that the NDIR values are higher than the CRDS values. In the second half the differences are mainly positive, especially during the periods with the working gases B27 and B29. Additionally, changing the CRDS instrument caused a

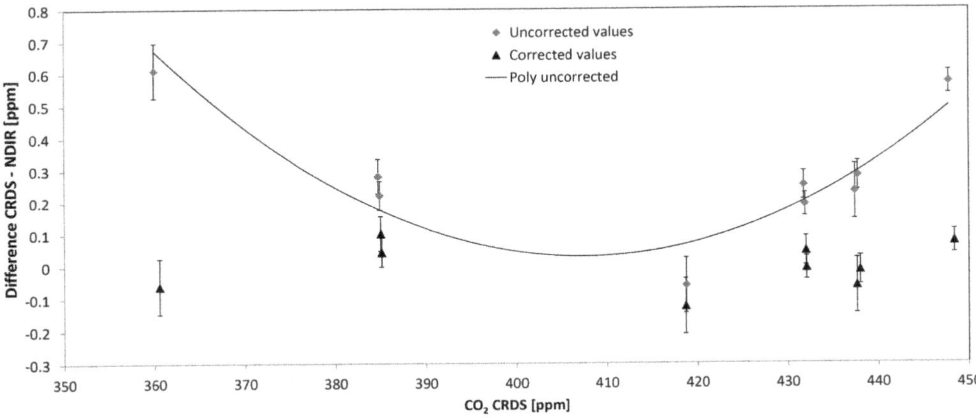

Figure 9. The difference between the Picarro (CRDS) – Maihak (NDIR) CO_2 values as a function of the CRDS values of old KUP calibration cylinders. The data show a polynomial dependency (grey diamonds); by correcting the data points with the according second degree polynomial (black line), the difference between the two systems varies slightly around zero (black triangles). The vertical error bars represent the combined standard deviation of both systems whereas the horizontal error bars represent only the standard deviation of the CRDS system. Because of the scale, the horizontal error bars are not visible.

small offset. This offset, together with the imperfectly assigned working gases, results in a slope of 0.2 ppm $year^{-1}$ for the differences and a slope difference of the same amount between the two CO_2 records. The annual CO_2 increase of the CRDS data set is in good agreement with the average global CO_2 trend of 1.97 ppm $year^{-1}$ for 2003–2012 (Tans, 2014), whereas the NDIR trend is a bit too low. Nevertheless, it is important to mention that for proper trend analysis the period of 3 years is too short. The CO_2 trend of the whole NDIR data set from the end of 2004 until the end of 2013 is 1.99 ppm $year^{-1}$, which is in much better agreement with the global trend of the same period of 2.03 ppm $year^{-1}$ (Tans, 2014).

Despite the fact that it is common knowledge to use well-assigned standard cylinder values for accurate measurements, we find larger offsets for two calibration cylinders, namely B27 and B29 used by the NDIR system. Therefore the calculated CO_2 values of the NDIR data set are a bit too low for these two periods. This effect would have been undetected without the comparison measurements, at least for the B27 cylinder: the offset is only around 0.2 ppm, which is quite small compared to a seasonality of roughly 10 ppm and even more so compared to the daily variations in extreme cases of up to 25 ppm. Experiences of ongoing international comparisons of traveling cylinders highlight similar issues (Zellweger et al., 2011). Therefore, a thorough check of assigned values of calibration and working cylinders is crucial.

Another problem concerning the calibration cylinders is their CO_2 mole fraction. Ideally the low span is lowest in CO_2 mole fraction, the working gas intermediate, preferably close to the expected value of the specimen, and the high span accordingly highest. However, in some periods (e.g., B26, B27, B29) the CO_2 mole fraction of the working gas is higher than, or very close to, the CO_2 mole fraction of the high span (e.g., B28) and therefore not ideal for calibrating the measurements with a very high accuracy. However, because of logistic problems these cylinders have to be used as working gases.

Also, the stability of the CO_2 mole fraction in the calibration cylinders is very important. Adsorption/desorption effects occur when the pressure of steel cylinders drops under 20 to 30 bar (2 to 3 MPa) (Keeling et al., 1998), leading to an enrichment of CO_2 in the cylinder and therefore a CO_2 value different from the assigned. Hence, cylinders should not be used until they are empty. A recalculation of the working gas cylinders with a two-point calibration using the low and the high span show that the effect already starts at roughly 80 to 90 bar (8 to 9 MPa) and has to be corrected for. This finding is supported by Leuenberger et al. (2014). At higher pressures the effects are still very small, but with decreasing pressure the difference between the measured CO_2 mole fraction of the calibration gas and the assigned value increases, thus leading to a decrease in the accuracy of the calculated CO_2 values.

By applying the corrections mentioned above (see Sects. 3.2 and 3.3), discarding two periods of badly assigned working gases (B27 and B29) and excluding periods of fast CO_2 changes, the average difference of the hourly values decreases to -0.03 ppm ± 0.25 ppm. Whereas the average value of almost zero is very good, the standard deviation is still high in comparison to the WMO goal of 0.1 ppm and the precision of each individual system, which is roughly five times better. The rather high standard deviation is probably caused by differing averaging intervals of the hourly values and dispersion of the air parcels in the two systems due to different volumes and flow rates of the air.

Both systems obtain the outside air from the same air inlet at Sphinx observatory at JFJ, but the air is led to the two systems by separate tubing. Due to different flow rates and volumes of the two systems, the origin of an air parcel measured at a certain time is not necessarily exactly the same. For example, the volume of the NDIR drying unit leads to a residence time of 3 minutes, significantly longer than for the CRDS system. Therefore, NDIR system records slightly dampened ambient signals. Furthermore, it is possible that the according air parcels are shifted a little bit, thereby leading to small differences of the CO_2 values of the two systems. This could be the reason for the larger observed differences between the two data sets during fast CO_2 mole fraction changes of the outside air.

Overall, the CO_2 mole fraction measured by the two systems shows the same pattern not only during stable periods but also during extreme events and even in the high-resolution data. Most of the noisier periods can be traced to specific technical problems, whereas systematic shifts are most probably caused by badly assigned calibration gases or drifts of the CO_2 mole fraction in the cylinders due to adsorption/desorption effects. While it is possible to correct for the pressure-dependent drifts of the calibration gases mathematically, it is harder to correct for the shifts caused by badly assigned calibration gases since there are not a lot of independent comparison measurements that can be used for correction.

5 Conclusions

The CO_2 data sets of two different CO_2 measurement systems, a nondispersive infrared analyzer and a cavity ring-down spectrometer, running parallel at the High Altitude Research Station JFJ are compared. The evaluation of the two CO_2 records from the NDIR system of the KUP to the CRDS system of the Empa shows that the two systems are generally in good agreement, not only long-term but also down to the 1 Hz scale despite the fundamentally different measurement principles and gas handling systems. Therefore the two data sets can be used to complement each other. However, the comparison also reveals (i) adsorption/desorption effects in the calibration gas steel cylinders used by NDIR system, (ii) insufficient nonlinearity correction of the NDIR analyzer and (iii) periods of small biases because of badly assigned calibration gas cylinders. Adsorption effects can be corrected for by monolayer adsorption equation (Langmuir, 1918). Nonlinearity of the NDIR analyzer is constrained with a second order polynomial, resulting in better agreement between the two data sets, in particular for values strongly deviating from the average. More research in adsorption/desorption effects of calibration gas cylinders has been done at the KUP and is reported in Leuenberger et al. (2014).

Finally, we like to emphasize that using steel cylinders is not adequate for high-precision trace gas determinations due to gas adsorption/desorption. Therefore, we conclude that all laboratories using steel cylinders for calibration purposes should switch to calibration gases stored in aluminium cylinders to minimize gas composition changes and allow an improved assignment of the cylinder.

It is helpful to continue this comparison exercise for another couple of years to show that once aluminium cylinders are in use for both systems, the compatibility will improve further.

Acknowledgements. This work was financially supported by the Swiss National Science Foundation (SNF-Project 200020_134641) and the Federal Office of Meteorology and Climatology MeteoSwiss in the framework of Swiss GCOS. Empa's CO_2 measurements are part of the Swiss National Air Pollution Monitoring network, a joint responsibility of Empa and the Swiss Federal Office for the Environment. We would like to thank the International Foundation High Altitude Research Stations Jungfraujoch and Gornergrat (HFSJG), especially the custodians Martin Fischer, Felix Seiler and Urs Otz for changing the calibration gases cylinders and other maintenance work on the measurement systems. Additionally, the authors would like to thank Peter Nyfeler for his precious work and help in maintaining and repairing the systems in the laboratory in Bern and also at JFJ.

Edited by: C. Gerbig

References

Arrhenius, S.: On the Influence of Carbonic Acid in the Air upon the Temperature of the Ground, Philosophical Magazine and Journal of Science, Series 5, 41, 237–276, 1896.

Baltensperger, U., Gäggeler, H. W., Jost, D. T., Lugauer, M., Schwikowski, M., Weingartner, E., and Seibert, P.: Aerosol climatology at the high-alpine site Jungfraujoch, Switzerland, J. Geophys. Res., 102, 19707–19715, 1997.

Chevallier, F., Ciais, P., Conway, T. J., Aalto, T., Anderson, B. E., Bousquet, P., Brunke, E. G., Ciattaglia, L. , Esaki, Y., Frohlich, M., Gomez, A., Gomez-Pelaez, A. J., Haszpra, L., Krummel, P. B., Langenfelds, R. L., Leuenberger, M., Machida, T., Maignan, F., Matsueda, H., Morgui, J. A., Mukai, H., Nakazawa, T., Peylin, P., Ramonet, M. Rivier, L., Sawa, Y., Schmidt, M., Steele, L. P. Vay, S.A., Vermeulen, A. T., Wofsy, S., and Worthy, D.: CO_2 surface fluxes at grid point scale estimated from a global 21 year reanalysis of atmospheric measurements, J. Geophys. Res.-Atmos., 115, D21307, doi:10.1029/2010JD013887, 2010.

Henne, S., Brunner, D., Folini, D., Solberg, S., Klausen, J., and Buchmann, B.: Assessment of parameters describing representativeness of air quality in-situ measurement sites, Atmos. Chem. Phys., 10, 3561–3581, doi:10.5194/acp-10-3561-2010, 2010.

Keeling, R. F., Manning, A. C., and McEvoy E. M.: Methods for measuring changes in atmospheric O_2 concentration and their application in southern hemisphere air, J. Geophys. Res., 103, 3381–3397, 1998.

Kohmyr, W. D., Harris, T. B., Waterman, L. S., Chin, J. F. S., and Thoning, K. W.: Atmospheric Carbon Dioxide at Mauna Loa Observatory – 1. NOAA Global Monitoring for Climatic Change Measurements With a Nondispersive Infrared Analyzer, 1974–1985, J. Geophys. Res., 94, 8553–8547, 1989.

Langmuir, I.: The Adsorption of Gases on Plane Surfaces of Glass, Mica and Platinum, Journal of the American Chemical Society, 40, 1361–1403, 1918.

Leuenberger, M. C., Schibig, M. F., and Nyfeler, P.: Gas adsorption and desorption effects on cylinders and their importance for long-term gas records, Atmos. Chem. Phys. Discuss., 14, 19293–19314, doi:10.5194/acpd-14-19293-2014, 2014.

Menditto, A., Patriarca, M., and Magnusson, B.: Understanding the meaning of accuracy, trueness and precision, Accred Qual. Assur., 12, 45–47, 2007.

Peters, W., Krol, M. C.,van der Werf, G. R., Houweling, S., Jones, C. D., Hughes, J., Schaefer, K., Masarie, K. A., Jacobson, A. R., Miller, J. B., Cho, C. H., Ramonet, M., Schmidt, M., Ciattaglia, L., Apadula, F., Helta, D., Meinhardt, F., di Sarra, A. G., Piacentino, S., Sferlazzo, D., Aalto, T., Hatakka, J., Strom, J., Haszpra, L., Meijer, H. A. J., van der Laan, S., Neubert, R. E. M., Jordan, A., Rodo, X., Morgui, J. A., Vermeulen, A. T., Popa, E., Rozanski, K., Zimnoch, M., Manning, A. C., Leuenberger, M., Uglietti, C., Dolman, A. J., Ciais, P., Heimann, M., and Tans, P. P.: Seven years of recent European net terrestrial carbon dioxide exchange constrained by atmospheric observations, Global Change Biol., 16, 1317–1337, 2010.

Peterson, J. T., Kohmyr, W. D., Harris, T. B., and Chin, J. F. S.: NOAA carbon dioxide measurements at Mauna Loa Obersvatory, 1974–1976, Geophys. Res. Lett., 4, 354–356, 1977.

Ramonet, M., Ciais, P., Aalto, T., Aulagnier, C., Chevallier, F., Cipriano, D., Conway, T. J., Haszpra, L., Kazan, V., Meinhardt, F., Paris, J. D., Schmidt, M., Simmonds, P., Xueref-Remy, I., and Necki, J. N.: A recent build-up of atmospheric CO_2 over Europe. Part 1: observed signals and possible explanations, Tellus Ser. B, 62, 1–13, 2010.

Rella, C. W., Chen, H., Andrews, A. E., Filges, A., Gerbig, C., Hatakka, J., Karion, A., Miles, N. L., Richardson, S. J., Steinbacher, M., Sweeney, C., Wastine, B., and Zellweger, C.: High accuracy measurements of dry mole fractions of carbon dioxide and methane in humid air, Atmos. Meas. Tech., 6, 837–860, doi:10.5194/amt-6-837-2013, 2013.

Schimel, D. S., House, J. I., Hibbard, K. A., Bousquet, P., Ciais, P., Peylin, P., Braswell, B. H., Apps, M. J., Baker, D., Bondeau, A., Canadell, J., Churkina, G., Cramer, W., Denning, A. S., Field, C. B., Friedlingstein, P., and Goodale, C.: Recent patterns and mechanisms of carbon exchange by terrestrial ecosystems, Nature, 414, 169–172, 2001.

Tans, P. P.: Annual Mean Global Carbon Dioxide Growth Rates, available at: http://www.esrl.noaa.gov/gmd/ccgg/trends/, last access: 27 May 2014.

Uglietti, C., Leuenberger, M. C., and Valentino, F. L.: Comparison between real time and flask measurements of atmospheric O_2 and CO_2 performed at the High Altitude Research Station Jungfraujoch, Switzerland, Sci. Tot. Environm., 391, 196–202, 2008.

Uglietti, C., Leuenberger, M., and Brunner, D.: European source and sink areas of CO_2 retrieved from Lagrangian transport model interpretation of combined O_2 and CO_2 measurements at the high alpine research station Jungfraujoch, Atmos. Chem. Phys., 11, 8017–8036, doi:10.5194/acp-11-8017-2011, 2011.

van der Laan-Luijkx, I. T., van der Laan, S., Uglietti, C., Schibig, M. F., Neubert, R. E. M., Meijer, H. A. J., Brand, W. A., Jordan, A., Richter, J. M., Rothe, M., and Leuenberger, M. C.: Atmospheric CO_2, $\delta(O_2/N_2)$ and $\delta^{13}CO_2$ measurements at Jungfraujoch, Switzerland: results from a flask sampling intercomparison program, Atmos. Meas. Tech., 6, 1805–1815, doi:10.5194/amt-6-1805-2013, 2013.

WMO: 15th WMO/IAEA Meeting of Experts on Carbon Dioxide, Other Greenhouse Gases and Related Tracers Measurement Techniques, WMO Report No. 194, World Meteorological Organization, Research Department, Atmospheric Research and Environment Branch, Geneva, Switzerland, 2011.

Zellweger, C., Amman, M., Buchmann, B., Hofer, P., Lugauer, M., Rüttimann, R., Streit, N., Weingartner, E., and Baltensperger, U.: Summertime Noy speciation at the Jungfraujoch, 3580 m above sea level, Switzerland, J. Geophys. Res., 105, 6655–6667, 2000.

Zellweger, C., Forrer, J., Hofer, P., Nyeki, S., Schwarzenbach, B., Weingartner, E., Ammann, M., and Baltensperger, U.: Partitioning of reactive nitrogen (NO_y) and dependence on meteorological conditions in the lower free troposphere, Atmos. Chem. Phys., 3, 779–796, doi:10.5194/acp-3-779-2003, 2003.

Zellweger, C., Schwarzenbach, B., Steinbacher, M., and Buchmann, B.: System and performance audit of surface ozone, methane, carbon dioxide and carbon monoxide at the Global GAW station Cape Point South Africa, March 2011, WCC-Empa-Report 11/1, WMO, Geneva, Switzerland, 2011.

3

Retrieval of three-dimensional small-scale structures in upper-tropospheric/lower-stratospheric composition as measured by GLORIA

M. Kaufmann[1], J. Blank[1,*], T. Guggenmoser[1], J. Ungermann[1], A. Engel[6], M. Ern[1], F. Friedl-Vallon[2], D. Gerber[4], J. U. Grooß[1], G. Guenther[1], M. Höpfner[2], A. Kleinert[2], E. Kretschmer[2], Th. Latzko[2], G. Maucher[2], T. Neubert[8], H. Nordmeyer[2], H. Oelhaf[2], F. Olschewski[3], J. Orphal[2], P. Preusse[1], H. Schlager[7], H. Schneider[8], D. Schuettemeyer[5], F. Stroh[1], O. Suminska-Ebersoldt[2], B. Vogel[1], C. M. Volk[3], W. Woiwode[2], and M. Riese[1]

[1]Institute of Energy and Climate Research, Stratosphere (IEK-7), Research Centre Jülich, 52425 Jülich, Germany
[2]Institute for Meteorology and Climate Research, Karlsruhe Institute of Technology, Karlsruhe, Germany
[3]Physics Department, University of Wuppertal, 42097 Wuppertal, Germany
[4]RAL Space, STFC Rutherford Appleton Laboratory, Harwell Oxford, Didcot, OX11 0QX, UK
[5]European Space Agency, Mission Science Division, 2200AG Noordwijk, the Netherlands
[6]Institute for Atmospheric and Environmental Sciences, Goethe University Frankfurt, 60438 Frankfurt am Main, Germany
[7]German Aerospace Center, Institute of Atmospheric Physics, Atmospheric Trace Species, 82234 Oberpfaffenhofen-Wessling, Germany
[8]Central Institute of Engineering, Electronics and Analytics (ZEA-2) – Electronic Systems, Research Centre Jülich, Jülich, Germany
[*]now at: Google Inc., Dublin, Ireland

Correspondence to: M. Kaufmann (m.kaufmann@fz-juelich.de)

Abstract. The three-dimensional quantification of small-scale processes in the upper troposphere and lower stratosphere is one of the challenges of current atmospheric research and requires the development of new measurement strategies. This work presents the first results from the newly developed Gimballed Limb Observer for Radiance Imaging of the Atmosphere (GLORIA) obtained during the ESSenCe (ESa Sounder Campaign) and TACTS/ESMVal (TACTS: Transport and composition in the upper troposphere/lowermost stratosphere, ESMVal: Earth System Model Validation) aircraft campaigns. The focus of this work is on the so-called dynamics-mode data characterized by a medium-spectral and a very-high-spatial resolution. The retrieval strategy for the derivation of two- and three-dimensional constituent fields in the upper troposphere and lower stratosphere is presented. Uncertainties of the main retrieval targets (temperature, O_3, HNO_3, and CFC-12) and their spatial resolution are discussed. During ESSenCe, high-resolution two-dimensional cross-sections have been obtained. Comparisons to collocated remote-sensing and in situ data indicate a good agreement between the data sets. During TACTS/ESMVal, a tomographic flight pattern to sense an intrusion of stratospheric air deep into the troposphere was performed. It was possible to reconstruct this filament at an unprecedented spatial resolution of better than 500 m vertically and 20 × 20 km horizontally.

1 Introduction

The upper troposphere and lower stratosphere has gained increasing attention in recent years due to their importance in the climate system (Solomon et al., 2007; Riese et al., 2012). Quantifying the structure and chemical composition to understand the underlying physical processes are current topics of atmospheric research. The quantification of small-scale structures and dynamical processes is a key topic for the understanding of this atmospheric region (Gettelman et al., 2011) because cross-tropopause transport and mixing around jet streams give an important contribution to stratosphere–troposphere exchange (e.g., Manney et al., 2011; Seo and Bowman, 2002; Olsen et al., 2008) and affect the mixing layers above the tropopause (e.g., Hoor et al., 2002).

Satellite-borne limb observations have enhanced our understanding of the three-dimensional chemical structure and large-scale dynamics of the middle atmosphere significantly in recent years. Our knowledge of small-scale structures is, however, primarily based on in situ observations by means of research aircraft and high-altitude balloons. Airborne limb-emission sensors such as CRISTA-NF (CRyogenic Infrared Spectrometers and Telescopes for the Atmosphere – New Frontiers; Weigel et al., 2012; Ungermann et al., 2013) and MIPAS-STR (Michelson Interferometer for Passive Atmospheric Sounding – STRatospheric Aircraft, Piesch et al., 1996; Woiwode et al., 2012) complemented the satellite and in situ data with their high spatial sampling in two dimensions (vertical and along the flight track).

The quantification of small-scale processes in all three dimensions requires the development of new instruments and new measurement strategies. Ungermann et al. (2010) presented a new approach to obtain trace gas filaments and temperature fluctuations by means of airborne remote sounding utilizing tomographic reconstruction techniques. The two key factors for this measurement strategy are closed or nearly closed flight paths (diameter of a few hundred kilometers) and the ability to make measurements from different viewing angles with respect to the aircraft position.

The Gimballed Limb Observer for Radiance Imaging of the Atmosphere (GLORIA; Riese et al., 2014; Friedl-Vallon et al., 2014) is a newly developed instrument for delivering data for such kinds of tomographic reconstruction. It can measure images of mid-infrared limb emission spectra within a few seconds and has the capability to pan its line of sight from 48 to 118 ° with respect to the aircraft nose.

This work describes results from the first two deployments of the GLORIA instrument, namely during ESSenCe (ESa Sounder Campaign; Kaufmann et al., 2013) and during the TACTS/ESMVal campaign (TACTS: Transport and composition in the upper troposphere/lowermost stratosphere, ESMVal: Earth System Model Validation). The former campaign was conducted to perform a first in-flight testing of the GLORIA instrument and to verify the GLORIA retrieval products. In the latter campaign, several closed flight paths were performed to allow for the application of tomographic reconstruction techniques to real GLORIA measurements for the first time.

The outline of this paper is as follows: After a short description of the GLORIA instrument (Sect. 2), we describe the retrieval strategy for the GLORIA so-called dynamics-mode measurements (Sect. 3) including a short review of the forward and inverse model. Section 4 gives an overview of the ESSenCe campaign and presents the first GLORIA measurements including a discussion of their uncertainties, vertical resolution, and a comparison with collocated measurements and simulations. The first tomographic measurements of a filamentary structure during the TACTS/ESMVal campaign are presented in Sect. 5.

2 GLORIA Instrument

GLORIA is an infrared limb sounder which combines the high horizontal resolution of a nadir sounder (tens of kilometers) with the altitude resolution provided by a limb-sounding instrument. The purpose of GLORIA is to measure temperature and composition in the upper troposphere/lower stratosphere at high spatial resolution (Riese et al., 2014). This is achieved by mounting an imaging Michelson interferometer with a high-resolution two-dimensional infrared detector onto a gimbal, allowing it to point and stabilize the instrument in the azimuthal, elevational, and image rotation directions (Friedl-Vallon et al., 2014). It is designed to be deployed on board different research aircraft, namely the Russian high-altitude research aircraft M55 Geophysica and the German High Altitude and LOng Range (HALO) research aircraft.

GLORIA is able to measure infrared limb or nadir emissions between 780 and 1400 cm^{-1}. The limb images cover a vertical field of view of 0.8 to $-3.3°$ with respect to the horizontal; $-3.3°$ corresponds to a lower tangent altitude of about 4 km. The field of view can be pointed about 10° upward as a quasi-deep space view and into two large area blackbodies for radiometric calibration (Olschewski et al., 2013).

Tomographic measurements are performed in the so-called dynamics mode. In this mode, GLORIA operates at low optical path differences to obtain more spectra at a given time. During ESSenCe, GLORIA recorded spectra at a maximum optical path difference of 1.6 cm in 2.8 s for a pair of forward/backward spectra, yielding a spectral sampling of 0.3125 cm^{-1}. During TACTS/ESMVal, dynamics-mode spectra were recorded at a maximum optical path difference of 0.8 cm in 2 s for a pair of forward/backward spectra, yielding a spectral sampling of 0.625 cm^{-1}. Another measurement mode is the so-called chemistry mode with a spectral sampling of 0.0625 cm^{-1} and 12 s measurement time. GLORIA data recorded in chemistry mode were analyzed by Woiwode et al. (2014). Individual GLORIA images contain 128×48 pixel/spectra. The horizontal resolution of the detector array is used to analyze and filter cloud fractions. After filtering, the spectra are averaged horizontally on a 128×1 grid for further processing. Different viewing angles of the same air mass are obtained by pointing the instrument from 48 to 118° (in 4° steps) with respect to the aircraft nose in a short time frame. The data obtained in this way can be combined and processed in a tomographic way to get a three-dimensional picture of the atmosphere.

3 Data Processing

When operating the GLORIA instrument during field measurement, raw interferograms and housekeeping data are written to a mass storage for later processing. The data rate is about 1 Gbit s^{-1}. In the laboratory, this data is processed in several steps, including different calibration procedures and the retrieval of atmospheric temperature, trace gases, etc. These steps and the corresponding data level products are described below.

3.1 Level 0 and Level 1

The Level 0 and Level 1 processing transforms the measured raw interferograms into radiometrically and spectrally calibrated spectra. The core of the Level 0 processing is the resampling of the interferograms. The raw interferograms are sampled on a time-equidistant grid. Before Fourier transformation, they have to be interpolated onto a space-equidistant grid, taking velocity variations of the interferometer drive into account. The Level 0 processing also includes a quality check of the data, spike detection and correction, nonlinearity correction, correction of phase errors, and the spectral calibration. The Level 1 processing comprises the Fourier transform converting the interferograms to complex spectra and the radiometric calibration. Calibration spectra are generated from blackbody and quasi-deep space measurements using a two-point calibration. Details about the Level 0 and Level 1 processing of GLORIA data are given in Kleinert et al. (2014).

3.2 Level 2

Level 2 processing encompasses the derivation of atmospheric or instrumental parameters from calibrated radiance measurements. This is performed in an iterative way by minimizing the difference between a forward model mapping of the atmospheric state to radiances as they would have been observed by the instrument and real GLORIA measurements. This minimization is performed by the so-called inversion model.

3.2.1 Forward Modeling

The forward model consists of a radiative transfer and an instrument model. The radiative transfer model is based on a combination of the emissivity growth approximation (EGA; Gordley and Russell, 1981) and the Curtis–Godson approximation (CGA; Curtis, 1952; Godson, 1953). Both methods are based on pre-calculated emissivities of homogeneous gas cells for a variety of atmospheric conditions and predefined spectral ranges. The spectral ranges are either individual spectral points or integrated spectral windows combining several spectral points (Riese et al., 1997, 1999). The optical paths have been calculated by means of the line-by-line Reference Forward Model (RFM, Dudhia, 2000) utilizing the HITRAN (High Resolution Transmission)-2012 spectral database (Rothman et al., 2013). The speed up of this method is up to a factor of thousand in comparison to a line-by-line model, and the uncertainties arising from the approximations are generally less than 1 % (Marshall et al., 1994; Francis et al., 2006; Ungermann et al., 2013).

The second part of the forward model is the instrument model. It maps various line-of-sight radiances onto the detector array. To consider field of view effects, several line-of-sight radiances are typically combined to simulate one or more detector pixel.

The retrieval of atmospheric quantities requires the calculation of the Jacobian of the forward model (see next section). This is done by algorithmic differentiation (Lotz et al., 2011; Ungermann et al., 2011), which is highly efficient in our case, because the computational costs for calculating the Jacobian matrix are only a constant multiple of a single execution of the forward model.

3.2.2 Inverse Modeling

The reconstruction of an atmospheric state compatible with a sequence of GLORIA measurements is a nonlinear inversion problem. This is solved by minimizing a cost function σ_y describing the difference between the forward model F: $\mathbb{R}^n \rightarrow \mathbb{R}^m$ of an atmospheric state $\boldsymbol{x} \in \mathbb{R}^n$ (temperature, trace gas abundance, etc.) and the radiance measurement $\boldsymbol{y} \in \mathbb{R}^m$:

$$\sigma_y = (\mathrm{F}(\boldsymbol{x}) - \boldsymbol{y})^T \mathbf{S}_\epsilon^{-1} (\mathrm{F}(\boldsymbol{x}) - \boldsymbol{y}). \quad (1)$$

The difference is weighted by the inverse measurement error/error covariance matrix $\mathbf{S}_\epsilon^{-1}$. In most cases, the solution is not unique and/or under-constrained. Therefore, it has been proven useful to add a regularization term to the cost function:

$$\sigma_x = (\boldsymbol{x} - \boldsymbol{x_a})^T \mathbf{L}^T \mathbf{L} (\boldsymbol{x} - \boldsymbol{x_a}). \quad (2)$$

In our case, **L** the is a Tikhonov-type matrix applying zero- and first-order difference operators (Twomey, 1977). A zero-order Tikhonov regularization constrains the absolute value, while the first-order minimizes the derivative in order to smooth the solution. $\boldsymbol{x}_a$ describes a priori information about the atmospheric state. $\boldsymbol{x}$ and $\boldsymbol{x}_a$ may also contain parameters of the instrument model, such as a radiance offset. The definition of the a priori data and the regularization matrix **L** depends on the atmospheric conditions and the instrument performance and is described later (Sect. 4.3.1).

The overall cost function is then $\sigma = \sigma_y + \sigma_x$, which is minimized using a truncated quasi-Newton minimizer. For three-dimensional tomographic retrieval setups, the size of the state and measurement space is rather large ($n \approx 10^7$, $m \approx 10^6$), and sparse matrix structures for storage and iterative methods for solving linear equation systems are applied. The typical memory and computing requirements for such a setup are demanding and require the utilization of a parallel computer. The retrieval of the tomographic flight discussed in Sect. 5 was performed on six computing nodes with 48 CPU cores and 384 GB main memory.

4 ESSenCe

4.1 Overview

GLORIA was deployed during the ESSenCe campaign for the first time. The main objective of this campaign was the in-flight testing of the GLORIA instrument under stratospheric conditions and the first verification/validation of the GLORIA retrieval products (Kaufmann et al., 2013).

Most favorable conditions for such a test campaign are cloud-free conditions utilizing an aircraft carrier which can fly as high as possible into the stratosphere. For this purpose, the Russian M55 Geophysica was chosen, which has a ceiling altitude of about 21 km and a high probability to find mesoscale structures in the upper troposphere/lower stratosphere. The aircraft is highly flexible in terms of operations, and regarding ground weather and stratospheric conditions. The campaign base was Kiruna in northern Sweden.

The first part of the ESSenCe campaign consisted of ground-based measurements inside and outside of the hangar Arena Arctica to assure the interaction of all subsystems of the GLORIA instrument after transport and final integration. Several certification and calibration measurements were conducted, including electromagnetic tests of GLORIA when mounted on Geophysica.

The second part of the campaign consisted of flights with full instrumentation under stratospheric conditions. Flight patterns were planned such that the functionality and performance of GLORIA could be tested under all major conditions that are expected during later research flights. Particularly, this includes electromagnetic compatibility considerations (under flight conditions), the thermal behavior of the sensor and the gimbal frame, the detector and instrument optics cooling system, the pointing system, and the interferometer control system.

The dynamic conditions of the Arctic upper troposphere and lower stratosphere during the ESSenCe campaign were mostly undisturbed. At the beginning of the campaign in early December 2011, a low pressure system over the Norwegian Sea slowly approached Scandinavia. East of this low, mid-latitudinal air moved polewards, thereby being lifted isentropically. Connected to the low's frontal system, high clouds appeared south of this tongue of air (Fig. 1, blue lines). On the western rim of the low pressure system, air from the lower stratosphere descended deep into the troposphere, coiling up cyclonically. A few days later, the polar vortex strengthened and showed very low core temperatures with values well below 190 K in the lower stratosphere. Due to large-scale disturbances the vortex center moved towards the European sector and air from the vortex edge moved polewards over Siberia, thus forming a broad intrusion of air with low potential vorticity reaching from Siberia across the North Pole to the Norwegian Sea.

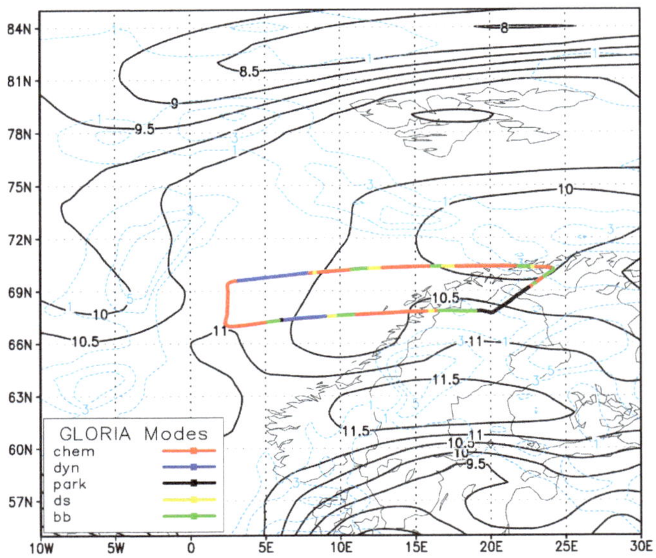

Figure 1. Potential vorticity (in PVU, black solid lines) on a 370 K isentropic surface (at approximately flight level) and cirrus clouds at lower levels (shown by cloud ice water content $[10^5\,\mathrm{kg\,kg^{-1}}]$ on 280 K, blue lines) and the track of the Geophysica aircraft for ESSenCe flight 2 of 16 December 2011. The abbreviations "dyn" and "chem" mark dynamics- and chemistry-mode measurements of GLORIA, respectively; "park" indicates that GLORIA is in parking position; "ds" and "bb" indicate deep space and blackbody calibration measurements, respectively.

4.2 GLORIA operations

Two GLORIA flights were performed during ESSenCe. Flight 1 took place on 11 December 2011, 11:00–15:00 UTC, and flight 2 on 16 December 2011, 14:00–17:50 UTC, respectively. Although the instrument encountered some technical problems, partly related to the extremely cold temperatures in the polar vortex, GLORIA took nominal atmospheric measurements including calibration for more than 1 h during each flight. During flight 1, all measurements were taken in chemistry mode while during flight 2, both chemistry and dynamics-mode measurements were performed.

Due to interferometer slide velocity variations during the first flight, these data were utilized for engineering purposes only. The functionality of the gimbal frame was partly lost in both flights after 1–2 h of flight time. For the remaining flight time it was not possible to further compensate the image rotation. The loss of pointing and image stabilization capabilities impacts data quality significantly such that Level 0 to Level 2 processing was restricted to the first measurement period of the second flight. The flight path of the Geophysica aircraft

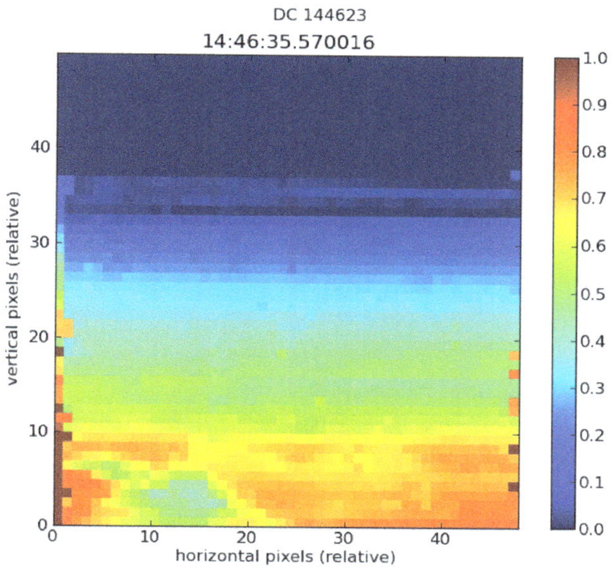

Figure 2. Direct current signal (arbitrary units) for one GLORIA image, before the Fourier transformation is applied. Exposure time is 40 ms and the horizontal extent is about 7.5 km at 10 km tangent altitude.

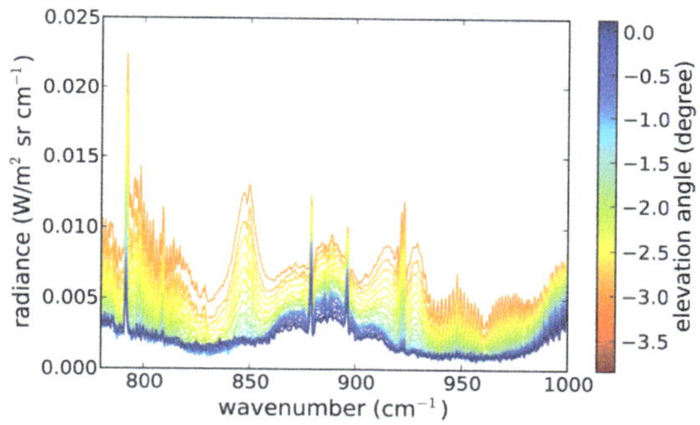

Figure 3. Spectra obtained by averaging over a detector row for a single GLORIA dynamics-mode measurement (Image 72, measured at 15:04:49 UTC at 67° N, 7° E) after pixel weighting and filtering. Corresponding tangent altitudes are 9 km (red) to 15 km (dark blue).

and GLORIA instrument operations during the second flight are illustrated in Fig. 1.

During both ESSenCe flights, one of the internal blackbodies and quasi-deep space spectra were used for two-point radiometric calibration. The latter exhibit remnants of atmospheric radiation, namely signatures of CO_2 below $750\,cm^{-1}$, O_3 around $1050\,cm^{-1}$, and CH_4 and N_2O around $1300\,cm^{-1}$, which affect the radiometric accuracy. These spectral windows are not used in the retrieval.

A typical GLORIA raw data image is illustrated in Fig. 2. It demonstrates the ability of the instrument to resolve spatial structures in the atmosphere as low as 150 m in horizontal and vertical direction. Constantly high signals in the lower two-thirds of the image indicate gray body emissions from clouds or aerosol layers at these altitudes. In contrast to cloud sensing, the observation and quantification of trace gas filaments or temperature fluctuations does not require a horizontal resolution as provided by an individual GLORIA pixel. Although the Fourier transformation is initially applied to all individual detector pixels, only the mean value for each detector row is used in the subsequent Level 2 processing. No averaging in the vertical direction of the image is applied. This means that currently the information used is similar to that of a one-dimensional detector array, which registers a full spectrum in each pixel during an interferometer sweep.

Before detector pixels are averaged in the horizontal domain, broken pixels (deviation of more than 2 times the standard deviation from the line average) or pixels pointing to clouds are removed. Entire detector rows are ignored if more than 75 % of the pixels are marked as broken or cloudy. Weighting of individual pixels is based on the noise analysis of blackbody spectra. Typical GLORIA dynamics-mode spectra for the binned data are illustrated in Fig. 3. The rovibrational emission lines of CO_2 at $792\,cm^{-1}$, CFC-11 at $850\,cm^{-1}$, and HNO_3 at $890\,cm^{-1}$ are clearly visible in the data.

Figure 4 illustrates the horizontal scan pattern of the GLORIA dynamics-mode measurements during ESSenCe. As stated above, the line of sight is panned from 48 to 118° with respect to the airplane nose. GLORIA dynamics-mode data were recorded from 14:58 to 15:10 UTC, yielding 121 individual images at about 2200 spectral points.

4.3 GLORIA Level 2 data analysis

4.3.1 Retrieval Setup

The retrieval setup for GLORIA depends on the instrument performance and atmospheric conditions. It consists of the choice of spectral windows utilized for the radiative transfer forward model, the a priori and regularization parameters utilized in the constrained global fit method, and the error covariance matrix of the measurements.

The selection of appropriate spectral windows for the GLORIA dynamics-mode retrieval is non-trivial because the spectral resolution of this mode does not allow for the separation of individual trace gas emission lines. In addition, the radiometric calibration procedure utilizing "quasi-deep" space spectra (in contrast to "real deep" space spectra as observed from a satellite) causes a non-uniform error pattern in the wavelength domain. Therefore, the selection of spectral windows is performed by an automatic algorithm maximizing

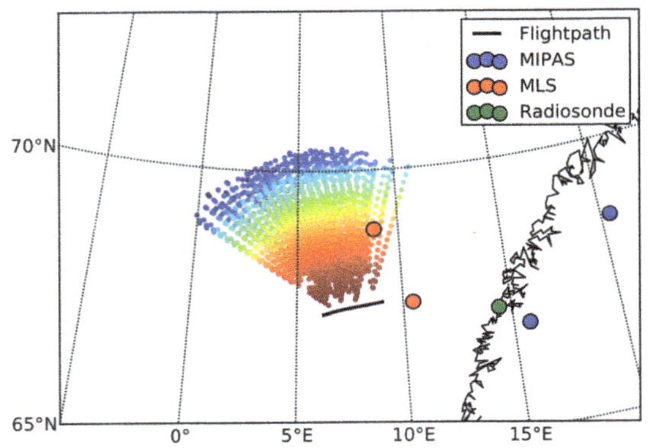

Figure 4. GLORIA tangent points of dynamics-mode measurements during ESSenCe flight 2. Tangent points closest (furthest) from the flight path (black line) are at highest (lowest) altitudes. Corresponding tangent altitudes are 4 km (red) to 15 km (blue). Thick, colored dots mark the geolocation of collocated MIPAS-Envisat and EOS-MLS tangent points and radiosonde positions. MIPAS-Envisat measurements were made on 16 December 2011 at 15:21 and 16:09 UTC, and EOS-MLS measurements at 11:40 and 11:42 UTC, respectively.

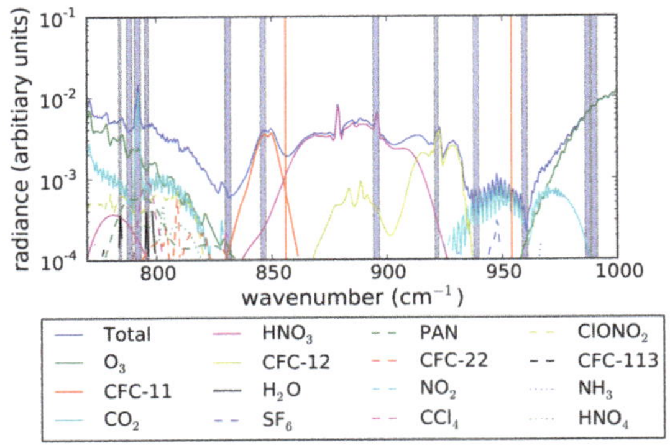

Figure 5. Simulated GLORIA dynamics-mode spectra for 16 km tangent altitude applying the MIPAS reference atmosphere for mid-latitudes. The spectral windows utilized in the retrieval are marked by blue vertical boxes.

the information gain obtained by the measurements. The figure of merit for information gain is determined by the instrumental uncertainties and the a priori knowledge (Kullback and Leibler, 1951). This selection is performed by a newly developed genetic algorithm optimizing the spectral window size and position simultaneously (Blank, 2013). In contrast to previous work (von Clarmann and Echle, 1998; Dudhia et al., 2002), this method does not require pre-selected broader spectral ranges and is able to optimize several spectral windows simultaneously. The latter is most relevant here due to the limited spectral resolution of the GLORIA dynamics-mode measurements.

For ESSenCe, the main retrieval targets are temperature, O_3, HNO_3, and CFC-12. The spectral windows as selected by the genetic algorithm to retrieve these quantities are illustrated in Fig. 5. There are twice as many spectral windows than main retrieval targets for considering cloud/aerosol effects as well as other contaminant gases. A priori data for the constrained global fit are taken from the MIPAS (Michelson Interferometer for Passive Atmospheric Sounding) reference atmospheres (Remedios et al., 2007). Some species (such as CO_2) were updated to 2011 conditions. Pressure and temperature data were taken from the ECMWF ERA (European Centre for Medium-Range Weather Forecasts Re-Analysis)-Interim data set (Dee et al., 2011). The penalty of zero-order regularization (Eq. 2) was suppressed by a factor of 3 in comparison to an optimal estimation method assuming the standard deviations of the MIPAS climatology. For temperature and pressure, an uncertainty of 3 K and 1 % is assumed, respectively. Weighting factors for the first-order regularization in the vertical direction are inversely proportional to the standard deviations of the MIPAS climatology multiplied by a characteristic length scale of 0.25 km for temperature and CFC-12 and 1 km for O_3 and HNO_3. The horizontal weighting factor is 100 km for all species.

The characterization of the GLORIA measurement error covariance matrix is a work in progress. The main source of uncertainty is instrument noise, approximated by a diagonal matrix assuming an uncertainty of 25 nW/(cm^2 cm^{-1} sr) and 2 %, respectively. The transportation and mounting of the GLORIA instrument onto the aircraft, as well as thermal stress during a research flight may cause a misalignment between the GLORIA attitude control system and the interferometer's line of sight. Therefore, it is mandatory to verify the vertical pointing of the instrument and apply some corrections, if necessary. This is done by simulating the CO_2 Q-branch emissions at 792.5 cm^{-1} adopting ECMWF temperature data for the entire flight. Assuming that the real atmosphere during a research flight has a mean temperature similar to ECMWF data, a single elevation angle correction is retrieved. In the case of ESSenCe flight 2 this correction is 0.1955°. Since the dominant uncertainties of the resulting line-of-sight elevation angles are not expected to change during the flight, small- and medium-scale structures in atmospheric temperature data can still be resolved by the measurement. The effect of the remaining pointing uncertainty of 0.023° is considered in the error budget of the retrieval targets (see Sect. 4.3.2).

4.3.2 Results, uncertainties, and comparison to other measurements

Typical limb radiance altitude profiles for a few spectral windows are illustrated in Fig. 6. The intensities are generally increasing towards lower altitudes, as expected. The dynamical range given by maximum and minimum observed radiances in the considered spectral range varies between 50 and 500 % for tangent altitudes between 9 and 17 km, respectively.

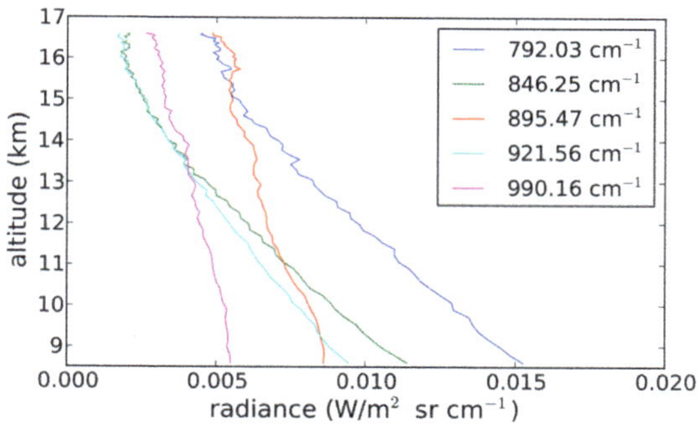

Figure 6. GLORIA integrated radiances for some spectral windows as selected by the genetic algorithm. The data is for Image 72 at 67° N and 7° E on 16 December, 15:05 UTC.

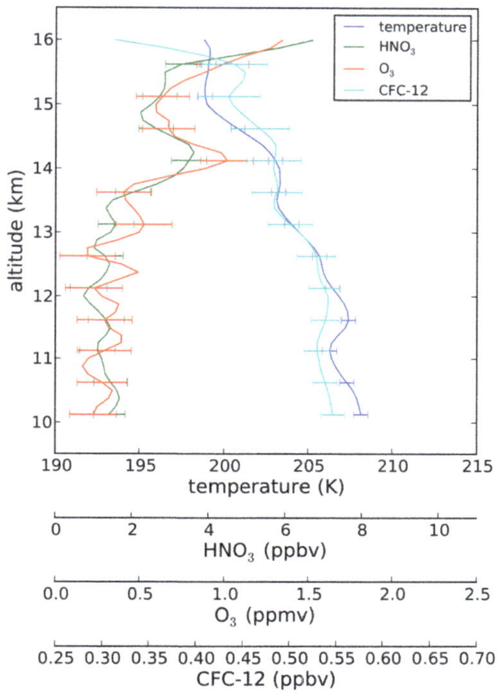

Figure 7. Retrieved temperature, HNO_3, O_3, and CFC-12 for GLORIA Image 72.

Corresponding retrieval results for the main targets of the GLORIA dynamics mode are illustrated in Fig. 7. The most prominent feature in the retrieved quantities is a pronounced enhancement of O_3 and HNO_3 around an altitude of 14 km, which may point to a filamentary structure in the stratosphere. This enhancement extends several hundred kilometers horizontally, as indicated by the two-dimensional cross-sections measured by GLORIA (Fig. 8). The uncertainties of HNO_3 and O_3 (Figs. 9–11) are on the order of 0.25 ppbv and 0.1 ppbv, respectively, and are dominated by detector noise. This means that the trace gas enhancement at 14 km is significant, whereas the small-scale structures below this altitude are within the noise error. The vertical resolution (Fig. 12)

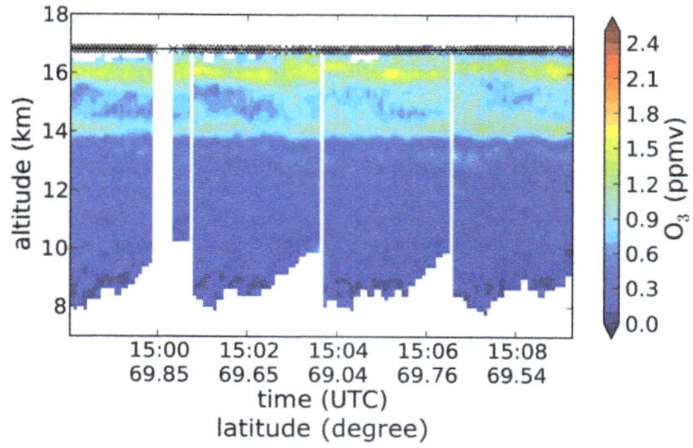

Figure 8. Retrieved ozone volume mixing ratio from GLORIA dynamics-mode data for 16 December 2011. Universal time and geographic latitude (in °) label the *x* axis. The figure shows four horizontal sweeps starting at 48° and ending at 118° with respect to the airplane nose.

is about 500 m for HNO_3 and O_3, and somewhat worse for temperature (750 m) and CFC-12 (600–1400 m). The vertical gradient and (comparatively low) vertical resolution of CFC-12 dynamics-mode data do not allow for the resolution of the filament at 14 km, whereas the high vertical resolution of O_3 and HNO_3 should give a realistic picture of this structure.

To validate these GLORIA measurements the retrieved data are compared to several other instruments. These are the two limb emission sounders mounted on the same airplane as GLORIA, namely the MIPAS-STR (Michelson Interferometer for Passive Atmospheric Sounding – STRatospheric Aircraft) (Woiwode et al., 2012) and MARSCHALS (Millimetre-wave Airborne Receivers for Spectroscopic Characterization in Atmospheric Limb Sounding, Castelli et al., 2013) instruments. Furthermore, two in situ data sets obtained during the ascent and descent of the second ESSenCe flight are used for the comparison, namely measurements of the Whole Atmosphere Sampler (WAS; Laube et al., 2013) and the High Altitude Gas AnalyzeR (HAGAR; Volk et al., 2000; Werner et al., 2010) data sets. In addition, GLORIA measurements are compared to the satellite-borne limb emission measurements of MIPAS (Fischer et al., 2008) on ESA's Environmental Satellite (Envisat) and the Earth Observing System (EOS) Microwave Limb Sounder (EOS-MLS, data version v03.33; Froidevaux et al., 2008) on NASA's EOS AURA satellite. The MIPAS data were derived with the retrieval processor of the Karlsruhe Institute of Technology, data version V5R_O3_221 and V5R_HNO3_221 for ozone and HNO_3 (von Clarmann et al., 2009) and V5R_CFC-11_221 for CFC-11 (Kellmann et al., 2012). Finally, the GLORIA data are compared to simulations of the Chemical Lagrangian Model of the Stratosphere (CLaMS; Konopka et al., 2007; Grooß et al., 2014; Pommrich et al., 2014; Vogel et al., 2014, and references therein). CLaMS runs with full stratospheric chemistry (including heterogeneous chemistry and sedimentation) and

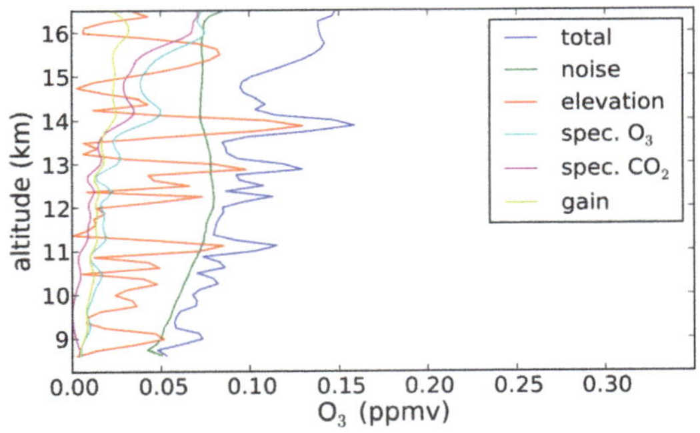

Figure 9. Error budget for O_3 for GLORIA Image 72.

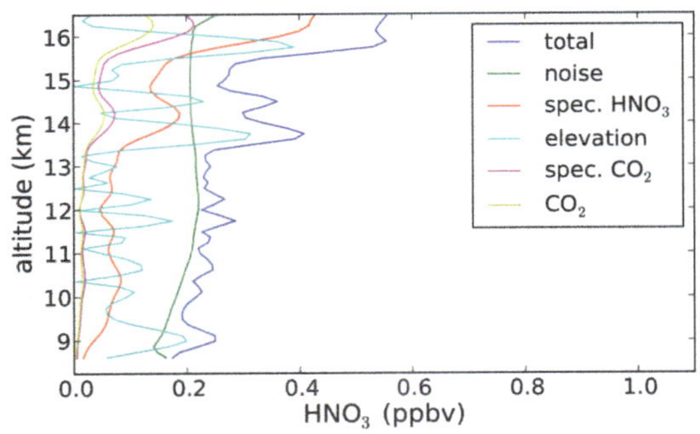

Figure 10. Error budget for HNO_3 for GLORIA Image 72.

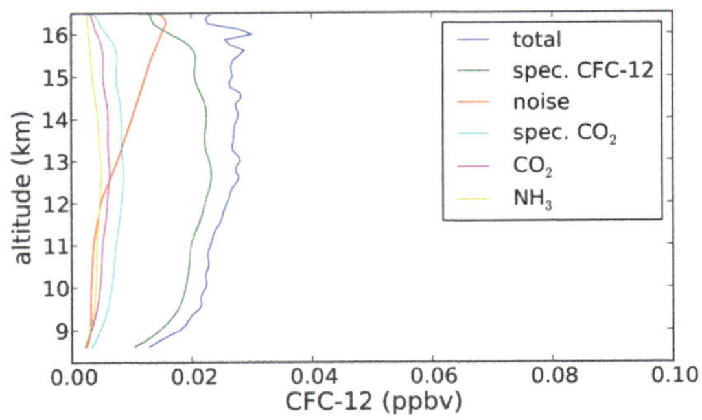

Figure 11. Error budget for CFC-12 for GLORIA Image 72.

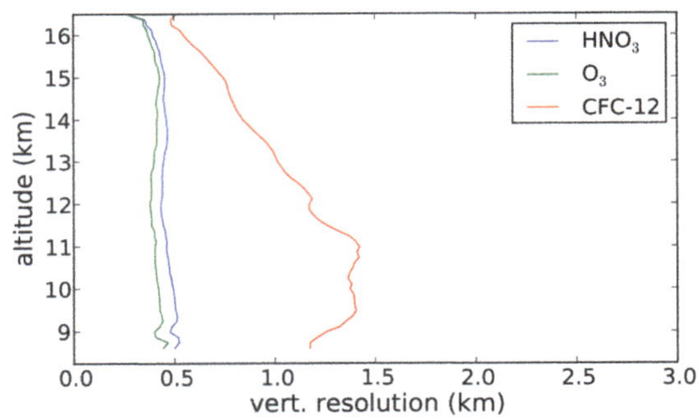

Figure 12. Vertical resolution for HNO_3, O_3, and CFC-12, for GLORIA Image 72.

is driven by ERA-Interim reanalysis data (for the detailed model setup, see Vogel et al., 2014). The simulations cover an altitude range from the surface up to the 900 K potential temperature with a horizontal resolution of approximately 100 km and a maximum vertical resolution of about 400 m around the tropopause. The simulations were initialized on 1 November 2011 (for ESSenCe) and on 1 May 2012 (for the TACTS/ESMVal campaign), respectively, using satellite measurements and tracer–tracer correlations following Grooß et al. (2014).

The mismatch in distance between the comparison data and GLORIA's tangent points is typically a few hundred kilometers (Fig. 4). Since this is larger than the spatial extent of the GLORIA dynamics-mode measurements and since the atmospheric scene is relatively homogeneous in the horizontal direction (Fig. 8), the entire dynamics-mode sequence from 14:58 to 15:10 UTC was averaged for comparison with the other data sets. The error bars of an averaged profile is somewhat smaller than the uncertainties of an individual profile, but larger than the root mean squares error of the single errors of the set of profiles considered for an average.

The satellite instruments' vertical profiles exhibit a much coarser altitude resolution than the GLORIA data. This effect is taken into account by convolving the GLORIA data $\boldsymbol{x}_{\mathrm{gloria}}$ with the averaging kernel of these instruments ($\mathrm{A}_{\mathrm{mipas}}$ and $\mathrm{A}_{\mathrm{mls}}$) considering the respective a priori data ($\boldsymbol{x}^{\mathrm{apr}}_{\mathrm{mipas}}$ and $\boldsymbol{x}^{\mathrm{apr}}_{\mathrm{mls}}$):

$$\boldsymbol{x}^{\mathrm{conv}}_{\mathrm{gloria}} = \boldsymbol{x}^{\mathrm{apr}}_{\mathrm{mipas/mls}} + \mathrm{A}_{\mathrm{mipas/mls}}(\boldsymbol{x}_{\mathrm{gloria}} - \boldsymbol{x}^{\mathrm{apr}}_{\mathrm{mipas/mls}}). \quad (3)$$

Figures 13–15 show comparisons of GLORIA retrieval results with the different data sets described above. The left panels of these figures illustrate the data "as is", and the right panel shows differences between GLORIA and the other data sets, where the effect of the averaging kernels was taken into account for the calculation of the differences. It is obvious that the effect of the averaging kernel is rather large for the comparison with the satellite instruments because the vertical resolution of the satellite data sets is several kilometers in the lower stratosphere.

GLORIA O_3 data show excellent agreement with MARSCHALS data, which were recorded a few minutes (14:52–14:56 UTC) before GLORIA dynamics-mode measurements. MARSCHALS O_3 abundance was retrieved from the Band B data (297–305 GHz). Differences between the two data sets are a few percent and smaller than the combined error bars. Most notable is the observation of the O_3 enhancement at 14 km in both data sets. Technical improvements of the MARSCHALS instrument and a new retrieval setup (Gerber, 2014) allowed the derivation of such a fine structure from MARSCHALS data for the first time. O_3 observations of MIPAS-STR exhibit similar absolute values, but the local enhancement at 14 km is not resolved due to the coarser alti-

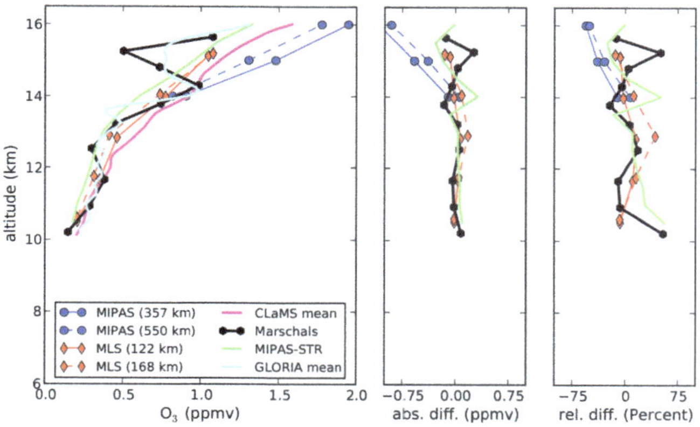

Figure 13. Comparison of O_3 abundance between GLORIA dynamics-mode measurements and various collocated data sets.

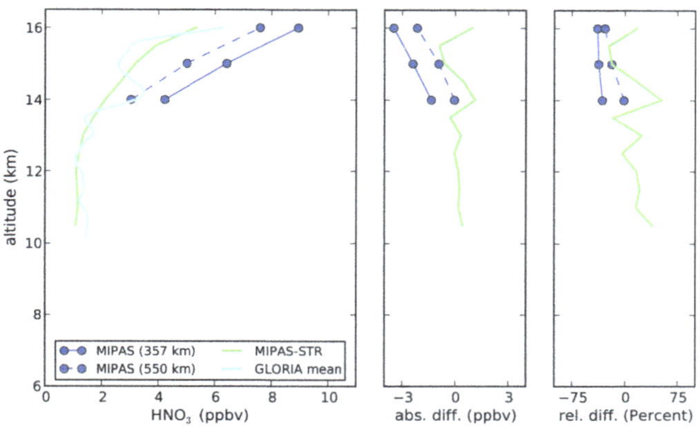

Figure 14. Comparison of HNO_3 abundance between GLORIA dynamics-mode measurements and collocated data sets.

tude resolution of this data set. Differences to the satellite-borne data sets of EOS-MLS are typically less than 15 % and somewhat larger for MIPAS-Envisat, but both data sets cannot resolve the ozone enhancement at 14 km due to their broad averaging kernels. CLaMS model simulations do not show this localized enhancement either, which is consistent with the non-existence of this structure in the ERA-interim data at a resolution of $1° \times 1°$ as utilized for this model run. However, modified potential vorticity data (Müller and Günther, 2003) based upon ECMWF operational analyses at a much higher spatial resolution ($0.1° \times 0.1°$, not shown here) reveals increased values at an altitude of 13.5 km at the location of the GLORIA measurements. This indicates that two different air masses moved close together, and model simulations at higher spatial resolution are needed to resolve such structures.

Mixing ratios of HNO_3 measured from GLORIA and MIPAS-STR show similar values at most altitudes as well – differences are typically less than 15 %, except for the 14 km region, where GLORIA data is enhanced by more than 1 ppbv. MIPAS-Envisat HNO_3 values (spatial separation

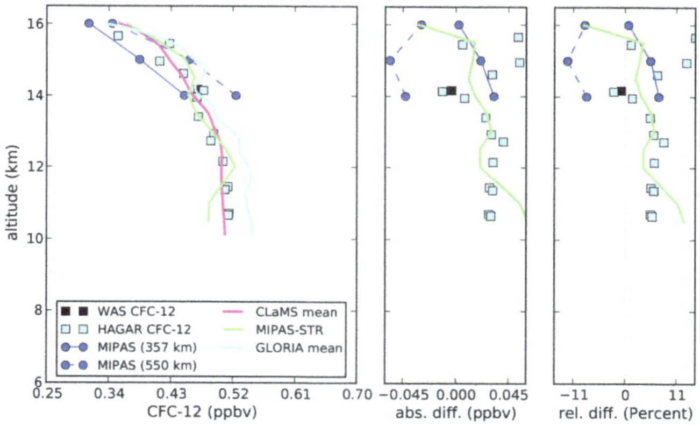

Figure 15. Comparison of CFC-12 abundance between GLORIA dynamics-mode measurements and collocated data sets.

350–550 km) are 20–50 % larger than the other two data sets, which may be due to the differences in the tangent point location and viewing geometry.

For CFC-12, GLORIA data shows a small high bias of about 5 % in the entire altitude region in comparison to collocated MIPAS-STR data. This bias is also observed in comparison to the CFC-12 data of the WAS and HAGAR in situ instruments measured during the ascent and descent of the aircraft. CLaMS results are in good agreement with the in situ measurements. Although this difference between GLORIA and the in situ or model data is within the error margins of the GLORIA instrument, it may point to some instrumental issues to be solved in future missions.

5 The TACTS/ESMVal Campaign

5.1 Overview

The second deployment of GLORIA was during the TACTS/ESMVal campaign in September 2012. This campaign was conducted by means of the new German High Altitude and LOng Range Research Aircraft. It covered a broad geographical region from Spitsbergen at high northern latitudes to Antarctica in the south and from western Africa in the west to the Arabian Sea in the east. The main scientific objective of the TACTS campaign is the measurement of the composition of the lowermost stratosphere during the period from summer to autumn to quantify the transport and mixing in the upper troposphere/lower stratosphere in the vicinity of the subtropical jet. The ESMVal campaign aimed to provide a large set of constituent, aerosol, and cloud data for the validation and improvement of chemistry–climate models. TACTS/ESMVal yielded nearly 90 h of GLORIA measurements at different atmospheric conditions and measurement modes.

The aim of this section is to show some preliminary results regarding the three-dimensional reconstruction of a fil-

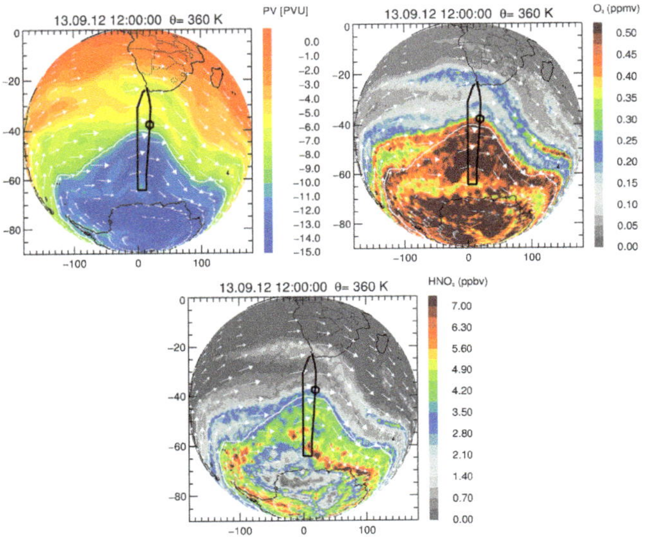

Figure 16. ECMWF potential vorticity (left) at a potential temperature of 360 K (about 12 km altitude) for the TACTS/ESMVal flight on 13 September 2012. The flight track is indicated by a black line. At 45° S, 20° E a tomographic flight pattern (approximate altitude of about 12.5 km) was performed. The lower and right-hand plots show O_3 and HNO_3 volume mixing ratios at 360 K as simulated by the CLaMS model, respectively. The vortex edge is calculated as defined by Nash et al. (1996) and is marked by a white contour line. The horizontal winds are indicated by white arrows.

ament, thereby demonstrating the outstanding capabilities of GLORIA to resolve such small-scale structures in all three dimensions.

5.2 GLORIA measurements

Since implementation during ESSenCe, the GLORIA instrument was subject to several improvements. Problems with the gimbal frame and with some interferometer components have been solved, and instrument configurations have been optimized. The spectral resolution of the dynamics mode during TACTS/ESMVal has been reduced by a factor of two compared to ESSenCe.

During TACTS/ESMVal, the aircraft flew on several closed loop flight tracks to test GLORIA's capabilities to sound the atmosphere tomographically. One of these closed flight path tracks took place on 13 September 2012 south of Africa at the edge of the Antarctic polar vortex (Fig. 16). This region is characterized by strong gradients of potential vorticity, ozone, and HNO_3 mixing ratios. At the edge of the polar vortex, HALO performed a hexagonal flight track such that the GLORIA lines of sight are directed to the center of the hexagon. At this location (45° S, 20° E) the horizontal winds are sheared and tilted as a consequence of a distortion of the jet stream at lower altitudes (Fig. 17). Wind speeds are on the order of 45 m s^{-1}.

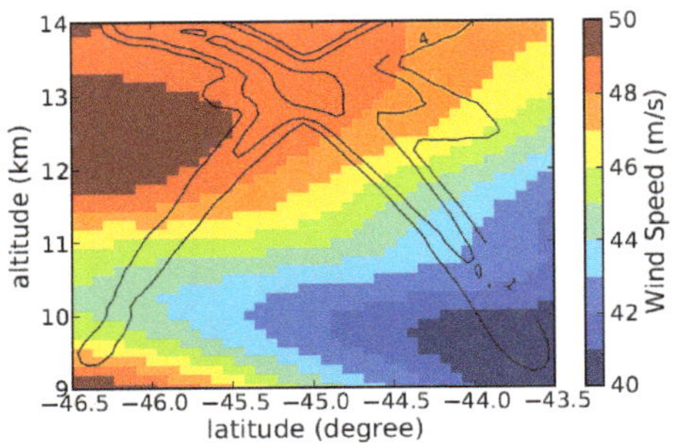

Figure 17. Horizontal wind speeds at 20° E for 13 September 2012, 14:00 UTC. Contour lines indicate the number of tangent points per grid cell. The outer (middle, inner) line marks 2 (4, 16) points/cell.

To retrieve the atmospheric composition within the air volume contained by the hexagon, an appropriate atmospheric grid was defined. The size of the grid cells is based on the vertical and horizontal field of view of the GLORIA instrument, namely 125 m in the vertical and 7.5 km in the horizontal direction, respectively, resulting in about 2×10^6 individual cells. The size of the grid cells increases towards the boundaries of the air volume.

The horizontal and vertical regularization applied to the retrieval targets is similar to that in ESSenCe, as is the selection of spectral windows. The reduced spectral resolution and uncertainties in the instrument calibration led to the elimination of a few spectral windows describing the aerosol background and some "secondary" gases such as PAN or $ClONO_2$, which cannot be retrieved with high confidence for these atmospheric conditions. The most relevant difference to ESSenCe is the omission of the spectral window at 845 cm^{-1} showing strong CFC-11 emissions, which turned out to be contaminated by instrument emissions from the optical window of GLORIA (Kleinert et al., 2014). Cloud indices (Spang et al., 2004) observed during the hexagonal flight were 4–6 between 9 and 14 km, indicating an almost cloud-free atmosphere.

Retrieved temperature and constituent fields exhibit a pronounced intrusion of stratospheric air into the troposphere at about 45° S, 20° E (Figs. 18 and 19). This intrusion is observed between an altitude of 12.5 and 14.5 km and extends about 500 m vertically. The horizontal extent is at least 200 km. Its stratospheric nature is indicated by enhanced HNO_3 and O_3 – at least by a factor of 2 compared to the air parcels just below or above the filament. This enhancement is significantly larger than the uncertainty of these species. Temperature data do not show this filamentary structure and varies by less than 2 K in the vicinity of this structure – in accordance with our expectation that this intrusion is caused by chaotic advection and not by adiabatic movements. It should be noted that CFC-12 does not show a pronounced structure due to its weak vertical gradient at these altitudes.

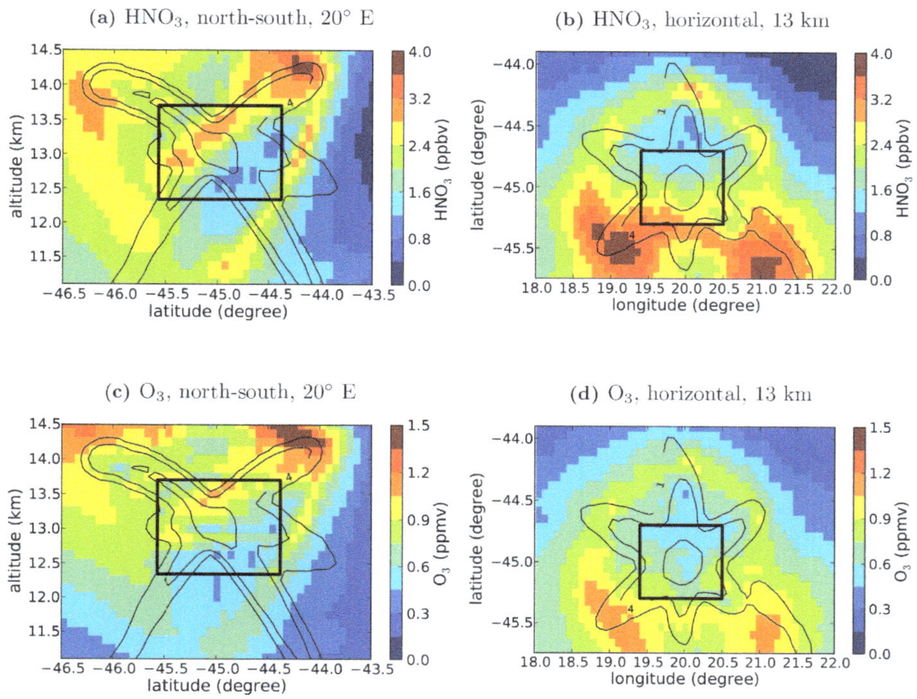

Figure 18. Retrieved HNO_3 and O_3 mixing ratios at 20° E (left) and 13 km altitude (right) for 13:30–14:30 UTC, 13 September 2012. For details see caption of Fig. 17.

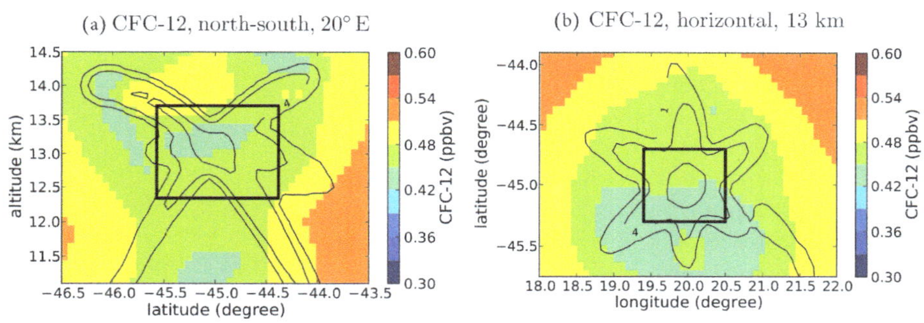

Figure 19. Retrieved CFC-12 mixing ratios at 20° E (left) and 13 km altitude (right) for 13:30–14:30 UTC, 13 September 2012. For details see caption of Fig. 17.

Using three-dimensional tomography, the horizontal resolution is up to 20 km in the core of the measured atmospheric region (Fig. 20). In this region, each atmospheric grid cell is measured from at least four different directions. The vertical resolution of the atmospheric fields is 200–300 m and somewhat larger than for the data obtained from the linear flight tracks during ESSenCe due to the increased instrument performance of GLORIA during the TACTS/ESMVal campaign. Since the calculation of retrieval diagnostics such as averaging kernel and spatial resolution is computationally very demanding (it breaks the sparsity of the computational problem; see, e.g., Ungermann et al., 2010), this information has not been calculated for all atmospheric grids cells considered in the retrieval. Although the horizontal wind speed is rather large in the vicinity of the GLORIA measurements, the alignment of the filamentary structure with respect to the dominant wind direction does not expect to smear out the gradients significantly. However, future data versions will consider the effect of advection in the tomographic reconstruction of small-scale structures.

A preliminary comparison to the CLaMS simulation for this flight (Fig. 16) indicates that the model resolution of 100 km (horizontally) is in turn not sufficient to resolve this intrusion of air as observed by the tomographic sounding.

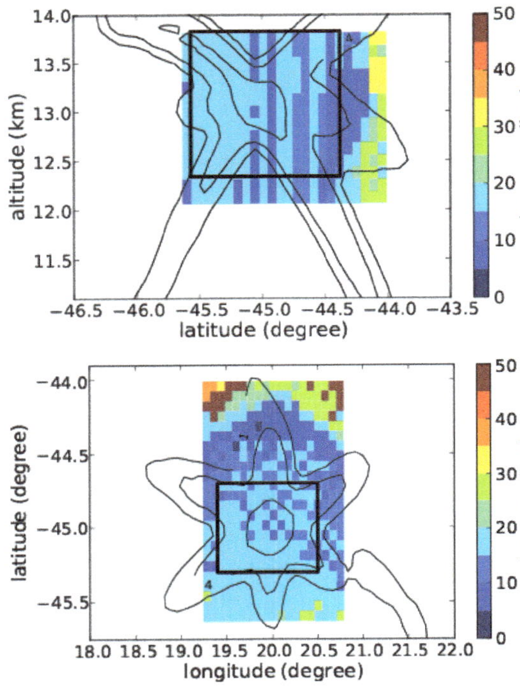

Figure 20. HNO_3 horizontal resolution at 20° E (top) and 13 km altitude (bottom). For details see caption of Fig. 17.

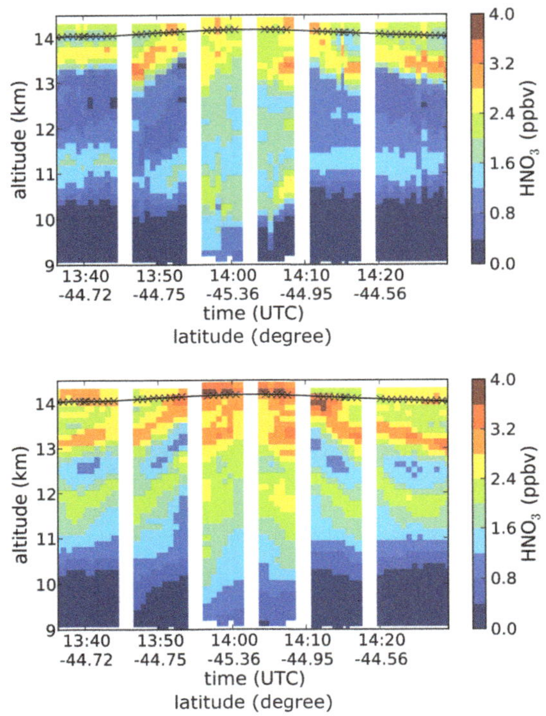

Figure 21. Comparison between a one-dimensional (top) and three-dimensional (bottom) reconstruction of HNO_3 during the GLORIA hexagonal flight. The segments of the hexagon are separated by white stripes. Measurement time (UTC) and latitude (degree) for the beginning of each segment are given on the x axis.

5.3 Advantages of Tomographic Sounding

The major improvement of three-dimensional limb sounding compared to conventional limb sounding not applying tomographic inversion techniques is the reduction of the horizontal (line-of-sight) smearing of these instruments. This is particularly important if the small-scale structure to be observed is smaller than the horizontal averaging kernel or if it is not co-aligned with the instrument's line of sight. Figure 21 illustrates this effect in the case of the filament observed during TACTS/ESMVal. The plot on the right hand side shows the three-dimensional reconstruction of the HNO_3 filament (Fig. 18), sampled at the tangent points of each segment of the hexagon. The plot on the left hand side shows the result of a conventional one-dimensional retrieval approach assuming horizontal homogeneity along the instrument's line of sight. The quality of the match between the retrieved structures and the tilt of the filament is dependent on a good alignment of the line of sight of a given segment and that of the filament. An example of a good match is found at 13:50; whereas a worse case is found at 13:40 UTC. Since the orientation of the filament is generally not known a priori, the tomographic data gives a more realistic picture of the structure.

6 Conclusions

The first two deployments of GLORIA demonstrated the readiness of this instrument to measure upper-tropospheric/lower-stratospheric composition at high spatial resolution. The retrieval of temperature, O_3, HNO_3, and CFC-12 presented no major problems except for a small bias in CFC-12 data. Increased effort in the radiometric calibration procedure will reduce this bias in future data versions. During ESSenCe, GLORIA already demonstrated its ability to measure intrusions of stratospheric air on vertical scales as low as a few hundred meters. However, the outstanding capability of the GLORIA instrument can only reach its full potential when the carrying aircraft flies a closed path around the air volume to be observed as performed several times during the TACTS/ESMVal campaign. In these cases, GLORIA data can provide three-dimensional constituent data at horizontal scales as low as a few tens of kilometers.

GLORIA represents the first airborne realization of the infrared limb imaging technique. The capabilities of this technique were also studied for application from a satellite in low Earth orbit (ESA, 2012). Although the observational mode of the satellite instrument is somewhat different from the airborne version, trace gas fields can be obtained with a similar horizontal resolution as for the tomographic reconstruction of airborne GLORIA dynamics-mode data as presented in this study.

Acknowledgements. We acknowledge financial support by the German Helmholtz society (ATMO, ATMONSYS), by the European Space Agency/Mission Science Division (ESSenCe), and the Deutsche Forschungsgemeinschaft in the priority program 1294 Atmospheric and Earth system research with the High Altitude and Long Range Research Aircraft (HALO). GLORIA retrieval activities were supported under the DFG project RASGLO (HALO-SPP 1294/Ka 2324/1-2) and flight planning by CLaMS model forecasts under the DFG project LASSO (HALO-SPP 1294/GR 3786).

We thank all members of the GLORIA instrument team for their extensive efforts in the development of the GLORIA instrument and the undertaking of the ESSenCe and TACTS/ESMVal campaigns.

We acknowledge the contribution of L. Hoffmann and S. Griessbach (both at Research Center Jülich), G. Geppert (now at Max Planck Institute for Meteorology), Ch. Kaliczinsky (Wuppertal University), and K. Leppkes and J. Lotz (both at RWTH Aachen) for their contribution to the development of the retrieval code.

We would like to thank Th. Roeckmann (Institute for Marine and Atmospheric Research Utrecht, Utrecht University) and J. Laube (School of Environmental Sciences, University of East Anglia) for providing the Whole Atmosphere Sampler data.

We thank J. Wintel (Physics Department, University of Wuppertal) for the measurement of the High Altitude Gas AnalyzeR data.

We thank the satellite group at IMK-ASF for providing the MIPAS data, the MLS team for providing EOS-MLS data, and the CLaMS team for assistance in flight planning.

Finally, we would like to acknowledge the open source community for the development of programming and visualization tools such as scientific python, matplotlib, and swig.

Edited by: J. Notholt

References

Blank, J.: Tomographic retrieval of atmospheric trace gases observed by GLORIA, Ph. D. thesis, Bergische Universität Wuppertal, 2013.

Castelli, E., M. Dinelli, B., Del Bianco, S., Gerber, D., Moyna, B. P., Siddans, R., Kerridge, B. J., and Cortesi, U.: Measurement of the Arctic UTLS composition in presence of clouds using millimetre-wave heterodyne spectroscopy, Atmos. Meas. Tech., 6, 2683–2701, doi:10.5194/amt-6-2683-2013, 2013.

Curtis, A. R.: Discussion of "A statistical model for water vapour absorption" by R. M. Goody, Quart. J. Roy. Meteorol. Soc., 78, 638–640, 1952.

Dee, D. P., Uppala, S. M., Simmons, A. J., Berrisford, P., Poli, P., Kobayashi, S., Andrae, U., Balmaseda, M. A., Balsamo, G., Bauer, P., Bechtold, P., Beljaars, A. C. M., van de Berg, L., Bidlot, J., Bormann, N., Delsol, C., Dragani, R., Fuentes, M., Geer, A. J., Haimberger, L., Healy, S. B., Hersbach, H., Hólm, E. V., Isaksen, L., Kållberg, P., Köhler, M., Matricardi, M., McNally, A. P., Monge-Sanz, B. M., Morcrette, J.-J., Park, B.-K., Peubey, C., de Rosnay, P., Tavolato, C., Thépaut, J.-N. and Vitart, F.: The ERA-Interim reanalysis: configuration and performance of the data assimilation system, Q. J. R. Meteorol. Soc., 137, 553–597, doi:10.1002/qj.828, 2011.

Dudhia, A.: Michelson Interferometer for Passive Atmospheric Sounding (MIPAS) Reference Forward Model (RFM), Software Users Manual, available at: http://eodg.atm.ox.ac.uk/RFM/sum/ (last access: December 2014), 2000.

Dudhia, A., Jay, V. L., and Rodgers, C. D.: Microwindow Selection for High-Spectral-Resolution Sounders, Appl. Opt., 41, 3665–3673, doi:10.1364/AO.41.003665, 2002.

ESA: Report for Mission Selection: PREMIER, vol. SP-1324/3, ESA Communication Production Office, Noordwijk, the Netherlands, 2012.

Fischer, H., Birk, M., Blom, C., Carli, B., Carlotti, M., von Clarmann, T., Delbouille, L., Dudhia, A., Ehhalt, D., Endemann, M., Flaud, J. M., Gessner, R., Kleinert, A., Koopman, R., Langen, J., López-Puertas, M., Mosner, P., Nett, H., Oelhaf, H., Perron, G., Remedios, J., Ridolfi, M., Stiller, G., and Zander, R.: MIPAS: an instrument for atmospheric and climate research, Atmos. Chem. Phys., 8, 2151–2188, doi:10.5194/acp-8-2151-2008, 2008.

Francis, G. L., Edwards, D. P., Lambert, A., Halvorson, C. M., Lee-Taylor, J. M., and Gille, J. C.: Forward modeling and radiative transfer for the NASA EOS-Aura High Resolution Dynamics Limb Sounder (HIRDLS) instrument, J. Geophys. Res.-Atmos., 111, D13301, doi:10.1029/2005JD006270, 2006.

Friedl-Vallon, F., Gulde, T., Hase, F., Kleinert, A., Kulessa, T., Maucher, G., Neubert, T., Olschewski, F., Piesch, C., Preusse, P., Rongen, H., Sartorius, C., Schneider, H., Schönfeld, A., Tan, V., Bayer, N., Blank, J., Dapp, R., Ebersoldt, A., Fischer, H., Graf, F., Guggenmoser, T., Höpfner, M., Kaufmann, M., Kretschmer, E., Latzko, T., Nordmeyer, H., Oelhaf, H., Orphal, J., Riese, M., Schardt, G., Schillings, J., Sha, M. K., Suminska-Ebersoldt, O., and Ungermann, J.: Instrument concept of the imaging Fourier transform spectrometer GLORIA, Atmos. Meas. Tech., 7, 3565–3577, doi:10.5194/amt-7-3565-2014, 2014.

Froidevaux, L., Jiang, Y. B., Lambert, A., Livesey, N. J., Read, W. G., Waters, J. W., Browell, E. V., Hair, J. W., Avery, M. A., McGee, T. J., Twigg, L. W., Sumnicht, G.K., Jucks, K. W., Margitan, J. J., Sen, B., Stachnik, R. A., Toon, G. C., Bernath, P. F., Boone, C. D., Walker, K. A., Filipiak, M. J., Harwood, R. S., Fuller, R. A., Manney, G. L., Schwartz, M. J., Daffer, W. H., Drouin, B. J., Cofield, R. E., Cuddy, D. T., Jarnot, R. F., Knosp, B. W., Perun, V. S., Snyder, W. V., Stek, P. C., Thurstans, R. P., and Wagner, P. A.: Validation of Aura Microwave Limb Sounder stratospheric ozone measurements, J. Geophys. Res.-Atmos., 113, doi:10.1029/2007JD008771, 2008.

Gerber, D.: RAL algorithm for retrievals of vertical trace gas profiles from the ESA UTLS limb sounder MARSCHALS, Atmos. Meas. Tech. Discuss., in preparation, 2014.

Gettelman, A., Hoor, P., Pan, L. L., Randel, W. J., Hegglin, M. I., and Birner, T.: The Extratropical Upper Troposphere And Lower Stratosphere, Rev. Geophys., 49, RG3003, doi:10.1029/2011RG000355, 2011.

Godson, W. L.: The evaluation of infra-red radiative fluxes due to atmospheric water vapour, Quart. J. Roy. Meteorol. Soc., 79, 367–379, 1953.

Gordley, L. L. and Russell, J. M.: Rapid inversion of limb radiance data using an emissivity growth approximation, Appl. Optics, 20, 807–813, doi:10.1364/AO.20.000807, 1981.

Grooß, J.-U., Engel, I., Borrmann, S., Frey, W., Günther, G., Hoyle, C. R., Kivi, R., Luo, B. P., Molleker, S., Peter, T., Pitts, M. C., Schlager, H., Stiller, G., Vömel, H., Walker, K. A., and Müller, R.: Nitric acid trihydrate nucleation and denitrification in the Arctic stratosphere, Atmos. Chem. Phys., 14, 1055–1073, doi:10.5194/acp-14-1055-2014, 2014.

Hoor, P., Fischer, H., Lange, L., Lelieveld, J., and Brunner, D.: Seasonal variations of a mixing layer in the lowermost stratosphere as identified by the CO-O3 correlation from in situ measurements, J. Geophys. Res.-Atmos., 107, 4044, doi:10.1029/2000JD000289, 2002.

Kaufmann, M., Blank, J., Friedl-Vallon, F., Gerber, D., Guggenmoser, T., Höpfner, M., Kleinert, A., Sha, M.K., Oelhaf, H., Riese, M., Suminska-Ebersoldt, O., Woiwode, W., Siddans, R., Kerridge, B., Moyna, B., Rea, S., and Oldfield, M.: Technical Assistance for the Deployment of Airborne Limbsounders during ESSenCe, Tech. rep., European Space Agency, 2013.

Kellmann, S., von Clarmann, T., Stiller, G. P., Eckert, E., Glatthor, N., Höpfner, M., Kiefer, M., Orphal, J., Funke, B., Grabowski, U., Linden, A., Dutton, G. S., and Elkins, J. W.: Global CFC011 (CCl_3F) and CFC-12 (CCl_2F_2) measurements with the Michelson Interferometer for Passive Atmospheric Sounding (MIPAS): retrieval, climatologies and trends, Atmos. Chem. Phys., 12, 11857–11875, doi:10.5194/acp-12-11857-2012, 2012.

Kleinert, A., Friedl-Vallon, F., Guggenmoser, T., Höpfner, M., Neubert, T., Ribalda, R., Sha, M. K., Ungermann, J., Blank, J., Ebersoldt, A., Kretschmer, E., Latzko, T., Oelhaf, H., Olschewski, F., and Preusse, P.: Level 0 to 1 processing of the imaging Fourier transform spectrometer GLORIA: generation of radiometrically and spectrally calibrated spectra, Atmos. Meas. Tech. Discuss., 7, 2827–2878, doi:10.5194/amtd-7-2827-2014, 2014.

Konopka, P., Günther, G., Müller, R., dos Santos, F. H. S., Schiller, C., Ravegnani, F., Ulanovsky, A., Schlager, H., Volk, C. M., Viciani, S., Pan, L. L., McKenna, D.-S., and Riese, M.: Contribution of mixing to upward transport across the tropical tropopause layer (TTL), Atmos. Chem. Phys., 7, 3285–3308, doi:10.5194/acp-7-3285-2007, 2007.

Kullback, S. and Leibler, R. A.: On Information and Sufficiency, The Annals of Mathematical Statistics, 22, 79–86, doi:10.1214/aoms/1177729694, 1951.

Laube, J. C., Keil, A., Bönisch, H., Engel, A., Röckmann, T., Volk, C. M., and Sturges, W. T.: Observation-based assessment of stratospheric fractional release, lifetimes, and ozone depletion potentials of ten important source gases, Atmos. Chem. Phys., 13, 2779–2791, doi:10.5194/acp-13-2779-2013, 2013.

Lotz, J., Naumann, U., and Ungermann, J.: Efficient Discrete Adjoint Computation in a Spectral Simulation Code, Tech. Rep. AIB-2011-05, RWTH Aachen and Forschungszentrum Jülich, 2011.

Manney, G. L., Hegglin, M. I., Daffer, W. H., Santee, M. L., Ray, E. A., Pawson, S., Schwartz, M. J., Boone, C. D., Froidevaux, L., Livesey, N. J., Read, W. G., and Walker, K. A.: Jet characterization in the upper troposphere/lower stratosphere (UTLS): applications to climatology and transport studies, Atmos. Chem. Phys., 11, 6115–6137, doi:10.5194/acp-11-6115-2011, 2011.

Marshall, B. T., Gordley, L. L., and Chu, D. A.: BANDPAK: Algorithms for modeling broadband transmission and radiance, J. Quant. Spectrosc. Ra., 52, 581–599, doi:10.1016/0022-4073(94)90026-4, 1994.

Müller, R. and Günther, G.: A generalized form of Lait's modified potential vorticity, J. Atmos. Sci., 60, 2229–2237, 2003.

Nash, E. R., Newman, P. A., Rosenfield, J. E., and Schoeberl, M. R.: An objective determination of the polar vortex using Ertel's potential vorticity, J. Geophys. Res., 101, 9471–9478, 1996.

Olschewski, F., Ebersoldt, A., Friedl-Vallon, F., Gutschwager, B., Hollandt, J., Kleinert, A., Monte, C., Piesch, C., Preusse, P., Rolf, C., Steffens, P., and Koppmann, R.: The in-flight blackbody calibration system for the GLORIA interferometer on board an airborne research platform, Atmos. Meas. Tech., 6, 3067–3082, doi:10.5194/amt-6-3067-2013, 2013.

Olsen, M. A., Douglass, A. R., Newman, P. A., Gille, J. C., Nardi, B., Yudin, V. A., Kinnison, D. E., and Khosravi, R.: HIRDLS observations and simulation of a lower stratospheric intrusion of tropical air to high latitudes, Geophys. Res. Lett., 35, L21813, doi:10.1029/2008GL035514, 2008.

Piesch, C., Gulde, T., Sartorius, C., Friedl-Vallon, F., Seefeldner, M., Wölfel, M., Blom, C., and Fischer, H.: Design of a MIPAS instrument for high-altitude aircraft, in: Second International Airborne Remote Sensing Conference and Exhibition, 24, 27 p., 1996.

Pommrich, R., Müller, R., Grooß, J.-U., Konopka, P., Ploeger, F., Vogel, B., Tao, M., Hoppe, C. M., Günther, G., Spelten, N., Hoffmann, L., Pumphrey, H.-C., Viciani, S., D'Amato, F., Volk, C. M., Hoor, P., Schlager, H., and Riese, M.: Tropical troposphere to stratosphere transport of carbon monoxide and long-lived trace species in the Chemical Lagrangian Model of the Stratosphere (CLaMS), Geosci. Model Dev., 7, 2895–2916, doi:10.5194/gmd-7-2895-2014, 2014.

Remedios, J. J., Leigh, R. J., Waterfall, A. M., Moore, D. P., Sembhi, H., Parkes, I., Greenhough, J., Chipperfield, M.P., and Hauglustaine, D.: MIPAS reference atmospheres and comparisons to V4.61/V4.62 MIPAS level 2 geophysical data sets, Atmos. Chem. Phys. Discuss., 7, 9973–10017, doi:10.5194/acpd-7-9973-2007, 2007.

Riese, M., Preusse, P., Spang, R., Ern, M., Jarisch, M., Grossmann, U., and Offermann, D.: Measurements of trace gases by the CRyogenic Infrared Spectrometers and Telescopes for the Atmosphere (CRISTA) experiment, Adv. Space Res., 19, 563–566, doi:10.1016/S0273-1177(97)00172-5, 1997.

Riese, M., Tie, X., Brasseur, G., and Offermann, D.: Three-dimensional simulation of stratospheric trace gas distributions measured by CRISTA, J. Geophys. Res.-Atmos., 104, 16419–16435, doi:10.1029/1999JD900178, 1999.

Riese, M., Ploeger, F., Rap, A., Vogel, B., Konopka, P., Dameris, M., and Forster, P.: Impact of uncertainties in atmospheric mixing on simulated UTLS composition and related radiative effects, J. Geophys. Res., 117, D16305, doi:10.1029/2012JD017751, 2012.

Riese, M., Oelhaf, H., Preusse, P., Blank, J., Ern, M., Friedl-Vallon, F., Fischer, H., Guggenmoser, T., Höpfner, M., Hoor, P., Kaufmann, M., Orphal, J., Plöger, F., Spang, R., Suminska-Ebersoldt, O., Ungermann, J., Vogel, B., and Woiwode, W.: Gimballed Limb Observer for Radiance Imaging of the Atmosphere (GLO-

RIA) scientific objectives, Atmos. Meas. Tech., 7, 1915–1928, doi:10.5194/amt-7-1915-2014, 2014.

Rothman, L., Gordon, I., Babikov, Y., Barbe, A., Benner, D. C., Bernath, P., Birk, M., Bizzocchi, L., Boudon, V., Brown, L., Campargue, A., Chance, K., Cohen, E., Coudert, L., Devi, V., Drouin, B., Fayt, A., Flaud, J.-M., Gamache, R., Harrison, J., Hartmann, J.-M., Hill, C., Hodges, J., Jacquemart, D., Jolly, A., Lamouroux, J., Roy, R. L., Li, G., Long, D., Lyulin, O., Mackie, C., Massie, S., Mikhailenko, S., Müller, H., Naumenko, O., Nikitin, A., Orphal, J., Perevalov, V., Perrin, A., Polovtseva, E., Richard, C., Smith, M., Starikova, E., Sung, K., Tashkun, S., Tennyson, J., Toon, G., Tyuterev, V., and Wagner, G.: The HITRAN2012 molecular spectroscopic database, J. Quant. Spectrosc. Ra., 130, 4–50, doi:10.1016/j.jqsrt.2013.07.002, 2013.

Seo, K.-H. and Bowman, K. P.: Lagrangian estimate of global stratosphere-troposphere mass exchange, J. Geophys. Res.: Atmos., 107, ACL 2-1–ACL 2-8, doi:10.1029/2002JD002441, 2002.

Solomon, S., Qin, D., Manning, M., Alley, R., Berntsen, T., Bindoff, N., Chen, Z., Chidthaisong, A., Gregory, J., Hegerl, G., Heimann, M., Hewitson, B., Hoskins, B., Joos, F., Jouzel, J., Kattsov, V., Lohmann, U., Matsuno, T., Molina, M., Nicholls, N., J.Overpeck, Raga, G., Ramaswamy, V., Ren, J., Rusticucci, M., Somerville, R., Stocker, T., Whetton, P., Wood, R. A., and Wratt, D.: Climate Change 2007 – The Physical Science Basis. Contribution of Working Group I to the Fourth Assessment Report of the Intergovernmental Panel on Climate Change, Technical Summary, Cambridge University Press, Cambridge, UK and New York, USA, 19–91, 2007.

Spang, R., Remedios, J. J., and Barkley, M. P.: Colour indices for the detection and differentiation of cloud type in infrared limb emission spectra, Adv. Space Res., 33, 1041–1047, doi:10.1016/S0273-1177(03)00585-4, 2004.

Twomey, S.: Introduction to the Mathematics of Inversion in Remote Sensing and Indirect Measurements, Dover Publications, Mineola, N.Y. 11501, USA, 1977.

Ungermann, J., Kaufmann, M., Hoffmann, L., Preusse, P., Oelhaf, H., Friedl-Vallon, F., and Riese, M.: Towards a 3-D tomographic retrieval for the air-borne limb-imager GLORIA, Atmos. Meas. Tech., 3, 1647–1665, doi:10.5194/amt-3-1647-2010, 2010.

Ungermann, J., Blank, J., Lotz, J., Leppkes, K., Hoffmann, L., Guggenmoser, T., Kaufmann, M., Preusse, P., Naumann, U., and Riese, M.: A 3-D tomographic retrieval approach with advection compensation for the air-borne limb-imager GLORIA, Atmos. Meas. Tech., 4, 2509–2529, doi:10.5194/amt-4-2509-2011, 2011.

Ungermann, J., Pan, L. L., Kalicinsky, C., Olschewski, F., Knieling, P., Blank, J., Weigel, K., Guggenmoser, T., Stroh, F., Hoffmann, L., and Riese, M.: Filamentary structure in chemical tracer distributions near the subtropical jet following a wave breaking event, Atmos. Chem. Phys., 13, 10517–10534, doi:10.5194/acp-13-10517-2013, 2013.

Vogel, B., Günther, G., Müller, R., Grooß, J.-U., Bozem, H., Hoor, P., Krämer, M., Zahn, A., and Riese, M.: Contributions of different boundary layer sources in Asia to the Asian monsoon anticyclone and associated transport pathways to the lowermost Stratosphere over northern Europe in 2012, Atmos. Chem. Phys., in preparation, 2014.

Volk, C. M., Riediger, O., Strunk, M., Schmidt, U., Ravegnani, F., Ulanovsky, A., and Rudakov, V.: In situ Tracer Measurements in the Tropical Tropopause Region During APE-THESEO, Eur. Comm. Air Pollut. Res. Report, 73, 661–664, 2000.

von Clarmann, T. and Echle, G.: Selection of optimized microwindows for atmospheric spectroscopy, Appl. Optics, 37, 7661–7669, 1998.

von Clarmann, T., Höpfner, M., Kellmann, S., Linden, A., Chauhan, S., Funke, B., Grabowski, U., Glatthor, N., Kiefer, M., Schieferdecker, T., Stiller, G. P., and Versick, S.: Retrieval of temperature, H_2O, O_3, HNO_3, CH_4, N_2O, $ClONO_2$ and ClO from MIPAS reduced resolution nominal mode limb emission measurements, Atmos. Meas. Tech., 2, 159–175, doi:10.5194/amt-2-159-2009, 2009.

Weigel, K., Hoffmann, L., Günther, G., Khosrawi, F., Olschewski, F., Preusse, P., Spang, R., Stroh, F., and Riese, M.: A stratospheric intrusion at the subtropical jet over the Mediterranean Sea: air-borne remote sensing observations and model results, Atmos. Chem. Phys., 12, 8423–8438, doi:10.5194/acp-12-8423-2012, 2012.

Werner, A., Volk, C. M., Ivanova, E. V., Wetter, T., Schiller, C., Schlager, H., and Konopka, P.: Quantifying transport into the Arctic lowermost stratosphere, Atmos. Chem. Phys., 10, 11623–11639, doi:10.5194/acp-10-11623-2010, 2010.

Woiwode, W., Oelhaf, H., Gulde, T., Piesch, C., Maucher, G., Ebersoldt, A., Keim, C., Höpfner, M., Khaykin, S., Ravegnani, F., Ulanovsky, A. E., Volk, C. M., Hösen, E., Dörnbrack, A., Ungermann, J., Kalicinsky, C., and Orphal, J.: MIPAS-STR measurements in the Arctic UTLS in winter/spring 2010: instrument characterization, retrieval and validation, Atmos. Meas. Tech., 5, 1205–1228, doi:10.5194/amt-5-1205-2012, 2012.

Woiwode, W., Suminska-Ebersoldt, O., Oelhaf, H., Hösen, E., Höpfner, M., Ebersoldt, A., Friedl-Vallon, F., Grooß, J.-U., Gulde, T., Kaufmann, M., Kleinert, A., Krämer, M., Kretschmer, E., T. Kulessa, G. M., Piesch, C., Riese, M., Rongen, H., Sartorius, C., Schardt, G., Schönfeld, A., Schuettemeyer, D., Sha, M.K., Stroh, F., Volk, C. M., and Orphal, J.: Validation of first chemistry mode retrieval results from new limb-imaging FTS GLORIA with correlative MIPAS-STR observations, Atmos. Meas. Tech. Discuss., submitted, 2014.

4

Dual-channel photoacoustic hygrometer for airborne measurements: background, calibration, laboratory and in-flight intercomparison tests

D. Tátrai[1,2], Z. Bozóki[1,2], H. Smit[3], C. Rolf[4], N. Spelten[4], M. Krämer[4], A. Filges[5], C. Gerbig[5], G. Gulyás[6], and G. Szabó[1,2]

[1]University of Szeged, Department of Optics and Quantum Electronics, Szeged, Hungary
[2]SZTE-MTA Research Group on Photoacoustic Spectroscopy, Szeged, Hungary
[3]Forschungszentrum Jülich, Institute for Energy and Climate Research Troposphere (IEK-8), Jülich, Germany
[4]Forschungszentrum Jülich, Institute for Energy and Climate Research Stratosphere (IEK-7), Jülich, Germany
[5]Max Planck Institute for Biogeochemistry, Department of Biogeochemical Systems, Jena, Germany
[6]Hilase Ltd., Szeged, Hungary

Correspondence to: D. Tátrai (tatraid@titan.physx.u-szeged.hu)

Abstract. This paper describes a tunable diode laser-based dual-channel photoacoustic (PA) humidity measuring system primarily designed for aircraft-based environment research. It is calibrated for total pressure and water vapor (WV) volume mixing ratios (VMRs) possible during airborne applications. WV VMR is calculated by using pressure-dependent calibration curves and a cubic spline interpolation method. Coverage of the entire atmospheric humidity concentration range that might be encountered during airborne measurements is facilitated by applying an automated sensitivity mode switching algorithm. The calibrated PA system was validated through laboratory and airborne intercomparisons, which proved that the repeatability, the estimated accuracy and the response time of the system are 0.5 ppmV or 0.5 % of the actual reading (whichever value is the greater), 5 % of the actual reading within the VMR range of 1–12 000 ppmV and 2 s, respectively. The upper detection limit of the system is theoretically about 85 000 ppmV, limited only by condensation of water vapor on the walls of the 318 K heated PA cells and inlet lines, and was experimentally verified up to 20 000 ppmV. The unique advantage of the presented system is its applicability for simultaneous water vapor and total water volume mixing ratio measurements.

1 Introduction

Volume mixing ratios (VMRs) of both atmospheric water vapor (WV) and total water (TW) (i.e., WV plus cloud droplets and ice crystals) change dynamically in space and time while they influence various meteorological parameters and, as one of the most effective absorbers of infrared radiation, they have a great impact on the global climate (Solomon et al., 2007). Therefore, it is preferable to measure them continuously and simultaneously with high spatial and temporal resolution, and ideally with global coverage. A possible measurement method is to equip research or in-service aircraft with WV and TW VMR measuring instruments. The idea of implementing atmospheric humidity measuring instruments (possibly in combination with other research instruments) onboard in-service aircraft is gaining increasing attention, as they can potentially form a global atmospheric monitoring network. In Europe, the MOZAIC (Measurement of OZone and water vapour by AIrbus in-service airCraft; Marenco et al., 1998) project launched 20 years ago was the first step in this direction. The currently ongoing IAGOS (In-service Aircraft for a Global Observing System; IAGOS, 2014) project equips a growing number of in-service aircraft with instruments measuring atmospheric composition, including WV VMR. Upgrading these instruments to measure TW VMR as well is expected to be a major step in atmospheric research.

Within the CARIBIC (Civil Aircraft for the Regular Investigation of the atmosphere Based on an Instrumented Container; Brenninkmeijer et al., 2007) project, which is now a part of the IAGOS project, a container is in operation with a large number of measuring instruments, and one of them is the instrument developed at the University of Szeged that measures WV and TW VMR simultaneously. That instrument is a predecessor of the system presented here.

Besides conventional requirements like accuracy, selectivity and reliability, an aircraft-operated atmospheric humidity measuring instrument has to fulfill various special requirements as follows:

- Compliance with rigorous safety regulations any aircraft equipment has to meet.
- Fast response time because of the high cruising speed of the aircraft.
- High resistance against mechanical vibrations.
- Fully autonomous and reliable operation.
- Low weight in order to minimize the extra fuel consumption of the aircraft.
- Wide measurement range in order to cover WV- and TW-VMR values that can occur in the atmosphere during typical aircraft flights.

Probably, the last requirement in combination with a fast response time is the most demanding one, as humidity varies from 1–2 ppmV (in dry air at about 12 km in height; CARIBIC, 2014) to 50 000 ppmV (in most humid air at the ground level, typically under tropical conditions). Furthermore, in the low humidity range, the precision of the measurements should be in the sub-ppmV range with high accuracy in order to be able to provide high-quality data.

At the University of Szeged during the last several years, efforts have been made to develop a room temperature diode laser-based photoacoustic (PA) system called WaSul-Hygro into a form optimized for simultaneous WV and TW VMR measurements (Szakáll et al., 2006, 2007; Tátrai et al., 2013). In this paper, the latest version of this system is presented together with its laboratory calibration, a blind laboratory test at Research Center Juelich (FZJ), and blind in-flight comparisons during a measurement campaign where WaSul-Hygro was operated simultaneously with a wavelength-scanned cavity ring-down spectrometer (WS-CRDS; Chen et al., 2010) and a Lyman-α photofragment fluorescence (LPF) based hygrometer (Chen et al., 2010).

2 Experimental

2.1 The photoacoustic system

The PA effect is the conversion of electromagnetic radiation into acoustic waves (Bell, 1880, 1881). Whenever a gas absorbs intensity- or wavelength-modulated light, the absorbed light energy is converted into acoustic waves via non-radiative relaxation processes. These acoustic waves can be considerably amplified by using a properly designed PA cell, through which the analyzed gas sample flows, and by setting the light source modulation frequency to coincide with a selected acoustic resonance frequency of the PA cell. The generated and amplified acoustic wave can be detected by a sensitive microphone. The amplitude of the acoustic signal at the frequency of the modulation of the light source is proportional to the concentration of the light-absorbing molecules (McDonald et al., 1978). This amplitude is usually determined by using the lock-in signal processing technique. In the following, the determined lock-in value is referred to as the PA signal. In the case of weak absorption (when the Lambert–Beer law can be approximated by the linear term of its Taylor series), the PA signal is a linear function of the absorption coefficient, but when the relative absorbed light power in the PA cell (described below) is above approximately 10 %, the sensitivity starts decreasing, which can be observed as nonlinearity during calibration. The PA measurement technique has been discussed in detail in several publications (Castleden et al., 1981; Saarela et al., 2009; Schilt and Thevenaz, 2006; Bozóki et al., 2011).

The layout of WaSul-Hygro, which is placed in a 19 inch 3U height rack and which has a weight of about 15 kg, can be seen in Fig. 1. Its custom controlling electronics is based on a digital signal processor (DSP) and is programmed for signal generation, data acquisition and operational control. Its main parts are a diode laser current driver, a programmable gain dual-channel microphone amplifier, a synchronized sampling-based dual-channel digital lock-in detector, temperature stabilization units, both for the laser and the detection cells, and analog and digital inputs and outputs for sensor signal reading and controlling the supplementary components of the system.

The light source is a room temperature operated, optical fiber coupled, distributed feedback diode laser (NEL NLK1E5E1AA). To stabilize and set the emission wavelength, the laser temperature can be accurately adjusted and stabilized with the help of a Peltier cell and a thermistor, both built into the housing of the laser (Seufert et al., 2004), and by applying a PID (proportional-integral-derivative) control algorithm. The tuning range of the laser extends from 1391 to 1394 nm, where water vapor has several absorption lines, and the measurements were performed on the strongest one, i.e., at 1392.535 nm, where the maximum optical absorption coefficient of 1 ppmV VMR of water vapor is $3.7 \times 10^{-5}\,m^{-1}$ at 1000 hPa pressure (Sharpe et al., 2004). The laser in the system is wavelength modulated: periodically on and off tuned to the peak of the absorption line. Such modulation is optimal for the detection of molecules having narrow absorption lines, while it reduces the level of PA signals generated by interfering broad absorption features of, e.g., large

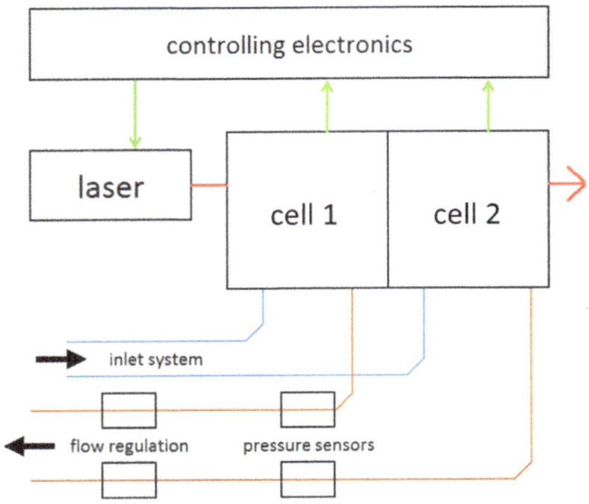

Figure 1. Schematic layout of the photoacoustic system. Green, red and black arrows indicate the electronic communication, laser beam propagation and gas flow directions, respectively. Blue and orange lines represent heated gas input and unheated gas output tubes, respectively.

molecules, atmospheric aerosols or pollutants condensing on the windows of the PA cell.

In order to maximize the sensitivity of the PA system, which is defined as the PA signal change caused by unit change in humidity, the amplitude of the laser modulation has to be optimized for the width of the absorption line (Szakáll et al., 2006), which strongly depends on the total gas pressure (Buldyreva et al., 2011). Taking into account that humidity typically decreases with decreasing air pressure (i.e., with increasing altitude), it is reasonable to optimize the laser modulation at a low pressure (in the current case, at 200 hPa) in order to have maximum sensitivity at low concentrations. Furthermore, the sensitivity of WaSul-Hygro decreases with decreasing pressure due to the acoustic properties of the PA cell. These two effects make the pressure dependence of the sensitivity rather complicated (Sect. 3.1.).

As result of the optimization, the laser is driven by a current waveform: $I(t) = I_0 + A \cdot \sin(2\pi \cdot f \cdot t)$, where $I_0 = 110\,\text{mA}$ and $A = 5\,\text{mA}$. The current tuning coefficient of the laser is $5\,\text{pm}\,\text{mA}^{-1}$; therefore, the offline wavelength is 1392.485 nm, which is at the very side of the absorption line at 200 hPa pressure. For best sensitivity, the modulation frequency has to be the same as the resonance frequency of the applied PA cell (more detail in Sect. 2.2.2). WaSul-Hygro includes two PA cells having the so-called longitudinal differential cell construction (Miklós et al., 2001) milled into a stainless steel block with the dimensions $10\,\text{cm} \times 10\,\text{cm} \times 4\,\text{cm}$ (length by width by height). The cells are fixed together in a way that they have one common window through which the output excitation laser light from the first cell enters directly into the second cell. The cells are thermally isolated from their surroundings and equipped with a resistive heater and a PT100 resistance temperature detector. As sound velocity and, consequently, the acoustic resonance frequency of the PA cell are proportional to the square root of the temperature, in order to avoid uncontrolled variation of the acoustic resonance frequencies, the temperature of the cells is stabilized to $318 \pm 0.1\,\text{K}$ using a pulse-with-modulation algorithm. At this temperature, each cell has an acoustic resonance frequency of 4500 Hz.

The microphone signals are amplified electrically by an INA163 instrumentation amplifier; the gain of the amplifier can be programmatically set to either 60 or 600. The amplified microphone signals from both measuring cells are simultaneously and continuously sampled by the analog-to-digital converters (ADC) of the controlling electronics into sequences having a duration of about 2 s. This temporal resolution of the measurements was selected because 2 s is also the response time of the combination of the PA cells and the inlet system to sudden humidity variation.

As can be seen in Fig. 1, the gas handling of the PA system contains temperature-stabilized (heated to $318 \pm 1\,\text{K}$) inlet lines, pressure sensors (Druck PMP 1400 series, 0–1500 hPa absolute) right after the PA cells, and mass flow controllers (Mykrolis Tylan FC2900, 0–1 sLpm (standard liter per minute)) that set the volume flow rates to 0.4 sLpm. During airborne operation, gas flow through the cells can be maintained either by the ram pressure at the sampling inlet of the aircraft or by a downstream vacuum pump. The former and latter cases are typically realized by a forward and sideward (or backward) facing sampling inlet, which can be used for TW and WV sampling, respectively. Elements of the gas handling are connected by 6 mm outer diameter stainless steel tubes and Swagelok connectors.

The measured PA signals, the pressures within the PA cells, and supplementary measurement information (like temperature values and self-checking results) are transferred to a data-logging PC connected to the electronics via a serial port. Software written in the LabVIEW graphical programming language (National Instruments) running on a PC calculates humidity values in real time from the measured data by using the calibration parameters (Sects. 2.2.4., 2.3., and 3.1.).

2.2 Special features of the photoacoustic system

Compared to its earlier versions, the presented system has several novel features, which were developed in order to improve the reliability of its operation.

2.2.1 Dual-channel operation

The dual-channel construction of the system can be used either to increase the signal-to-noise ratio of a single stream measurement by averaging the VMR readings of the two channels, or to measure humidity simultaneously in two different gas streams. In the latter case, the system can be used for simultaneous VW and TW (after evaporation in the tubes) VMR measurement whenever a proper inlet system is used, like in the CARIBIC project (Brenninkmeijer et al., 2007), or for the validation and comparison of different sampling systems. As most of the setup (controlling electronics, laser) is common to the two cells, small differences between the VMR values in the sampling lines are expected to be more accurately measurable with the presented system rather than by using separate systems for each sampling line, since occasional laser control malfunctions or systematic errors would affect the two channels identically.

2.2.2 Sensitivity mode switching

Proportionality between the PA signal and the measured water vapor concentration holds over at least 5 orders of magnitudes, but the measurement range of the controlling electronics in its actual configuration cannot cover the entire range of interest, because the amplified microphone signal at high concentrations overloads the input of the ADC. To increase the measurement range of the electronics, a sensitivity mode (SM) switching algorithm is implemented in the DSP software, which automatically switches the microphone gain from 600 to 60 (or vice versa from 60 to 600) whenever the PA signal becomes higher (or lower) than a certain limit. Hysteresis between the switching limits is introduced in order to avoid frequent switching generated by small-scale signal fluctuations. The change in the SM makes it possible to extend the upper limit of the measurement range to 30 000 ppmV. Above this concentration, the AD converter saturates again; therefore, the PA signal has to be decreased further. This can be achieved by detuning the laser modulation frequency from the resonance frequency of the cell. This extends the measurement range by at least 1 order of magnitude, but above about, 85 000 ppmV water vapor starts condensing in the PA cell or in the sampling line, although due to occasional colder points can reduce this upper limit. However, such high WV VMR is really unlikely to occur during airborne measurements. Actually, the highest humidity level measured during the tests described in Sects. 2.4 and 2.5 was below 30 000 ppmV, so the off-resonant modulation was not applied.

At SM1 and SM2, the laser is modulated at the resonance frequency of the PA cell to achieve optimal performance, while at SM3, the modulation frequency is increased to 7950 Hz, where the sensitivity of the cell decreases by 1 order of magnitude.

To provide the functions listed above, the SM points are defined as follows: the sensitivity is lowered whenever the PA signal is greater than 66 % of the saturation value of the AD converter, and still can be lowered while increased whenever the PA signal is lower than 10 % of the saturation value of the AD converter, and still can be increased. An example can be found later on in Fig. 3.

2.2.3 Laser wavelength stabilization

The lower the pressure is, the narrower the water vapor absorption lines are, and consequently the wavelength stability of the laser becomes an increasingly critical issue. The wavelength of a properly temperature-stabilized laser can vary spontaneously due to the so-called ageing effects (Woodward et al., 1993); therefore, a wavelength locking method (WLM; Tátrai et al., 2013; Bozóki et al., 2013) with which the laser wavelength can be measured and set with at least 25 fm accuracy with 1 s execution time is applied. This wavelength uncertainty can cause less than 0.1 % error in the measured VMR values at any pressure. As the laser ageing effect becomes significant only on a rather long timescale, it is sufficient to perform the WLM only once during each flight, preferably after the Wasul-Hygro system is switched on and the aircraft is before take-off.

2.2.4 Real-time VMR calculation

The input parameters for the VMR calculation for both cells are the PA signals, the pressures in the cells, the actual SMs, and the calibration curves (Sects. 2.3 and 3.1), which form the calibration surface in the PA signal-pressure–VMR space. To calculate VMR, a two-step approach is used: first, VMRs are calculated from the PA signals for all 10 calibration pressures, and then, from the resulting data set, the actual VMR is determined by interpolation to the actual pressure using a cubic spline algorithm. The execution time of this method is on the order of milliseconds by using a general purpose PC or notebook giving the possibility of real-time VMR calculation.

2.3 Calibration of the PA system

In order to have a simple but reliable tool for the calibration through a wide concentration range and at various pressures, a setup shown in Fig. 2, including a home-made humidity generator (HG) and a reduced pressure stabilizer, was assembled around WaSul-Hygro. Its automatic operation is ensured by including a stepper motor-controlled needle valve, which together with the MFCs inside WaSul-Hygro and with a downstream pump reduces and stabilizes the pressure within the PA cells with ± 1 hPa accuracy. The pressure setting time is 2–4 s. The central part of the HG is a saturator with a spiral of 95 cm in length milled into a 12 cm × 12 cm × 6 cm copper block hermetically sealed from its surrounding. Thanks to an overpressure reduction tube, which is long enough to avoid back-diffusion of water vapor from ambient air and installed to eliminate the non-synchronizability of the MFC-s just before the saturator, the HG is operated close to atmospheric pressure.

At the beginning of the calibration, liquid nitrogen is poured into the HG, which cools down the saturator to around 120–140 K, and then the saturator slowly (during approxi-

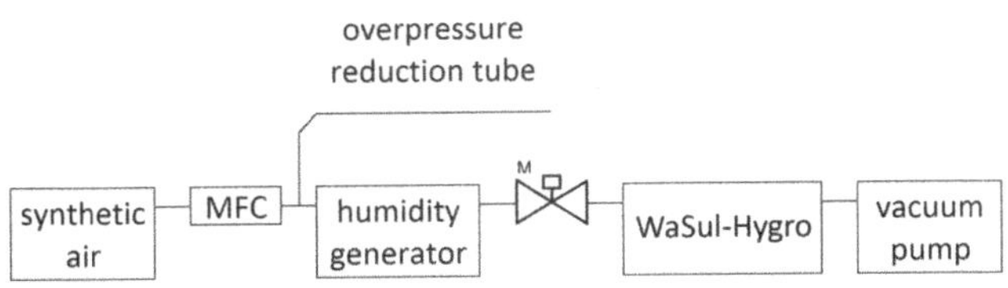

Figure 2. The schematic layout of the calibration system. MFC stands for mass flow controller and M is the stepper motor-controlled needle valve (see text for details).

mately 1.5 days) warms up to room temperature. Inside the saturator there is always ice or liquid water on the walls serving as a source of water vapor during warm-up, so the VMR will always correspond to the saturation pressure of water vapor at the temperature of the saturator. By measuring the temperature of the saturator with a PT100 resistance temperature detector and the gas pressure inside with a pressure sensor (Druck PMP 1400 series 0–1500 hPa absolute), the generated WV VMR can be calculated by using the Goff–Gratch equation (Goff and Gratch, 1946; WMO, 1983). This HG was calibrated against a chilled mirror hygrometer at the Hungarian National Office of Measures in the 1–25 000 ppmV range. During this calibration, an offset of 1.2 K was found between the measured temperature of the saturator and the actual dew/frost point. After correcting this offset, the repeatability of water vapor generation at any temperature was found to be ±0.5 % of the actual VMR or ±0.3 ppmV (the higher value applies).

The slow warm-up of the saturator makes it possible to vary the gas pressure repeatedly between 80, 130, 150, 200, 280, 370, 500, 650, 800 and 970 hPa, and also to repeat the PA measurements with the two different SMs. In this way, during a single warm-up of the HG, 20 different calibrations can be completed for each PA cell: each curve consists of five data points every 5 min.

2.4 Verification of the calibration

The PA system was blind tested at Research Center Jülich (Germany) at the Environmental Simulation Facility (ESF; Smit et al., 2000) chamber. The chamber has a test volume of about 500 L (80 cm × 80 cm × 80 cm), whereby relative humidity can be varied from about 95 down to 2 % over a temperature range between 300 and 200 K and a pressure range between 1000 and 100 hPa. Two reference instruments are connected to the chamber, an LPF hygrometer (Kley and Stone, 1978; Helten et al., 1998), which has an accuracy of ±3 % in the VMR range of 1–1000 ppmV, and a dew-/frost-point hygrometer (General Eastern, Type D1311R), which has a ±0.5 K uncertainty in the saturation temperature above 1000 ppmV.

During the measurements that were performed in the 2–12 000 ppmV VMR range at different pressures, the input lines for the two PA cells were joined by a tee connector. The air from the chamber was sampled through a 1.5 m length 1/8 inch heated stainless steel tube. Its inlet was placed as close as possible to the inlets of the reference instruments.

As the three instruments have different temporal resolutions, the results were interpolated to the time resolution of the PA system using the cubic spline method. From the resulting data sets, the relative deviation values, slopes of the cross plots, and the Pearson correlation values were determined.

2.5 Airborne tests

WaSul-Hygro was tested in a four-flight campaign dedicated to compare various systems measuring WV VMR within the DENCHAR (Development and Evaluation of Novel Compact Hygrometer for Airborne Research) project, which is part of the European Community's Seventh Framework Program EUFAR (European Facility for Airborne Research) funded project. The platform was a Learjet35A (Bombardier) research airplane operated by GFD GmbH (Hohn, Germany). Each flight was performed in the northern European airfield: above northern Germany, northern Poland, Denmark, southern Norway and the North Sea. Data were collected at different flight levels up to 13 km, so WaSul-Hygro was tested under tropospheric and lower stratospheric conditions as well. The PA system was compared to the FISH (Zöger et al., 1999) (Fast In-situ Stratospheric Hygrometer) instrument (measurement frequency of 1 Hz, a noise equivalent mixing ratio of 0.2–0.15 ppmV at 3 ppmV and a lower/upper detection limit of 0.18–0.13 / 1500 ppmV) during 3.6 h and with a commercially available PICARRO G2401-m WS-CRDS system (Picarro Inc., Santa Clara, CA, USA; Crosson, 2008). During the campaign, before and after each flight, FISH was calibrated against a chilled mirror hygrometer (MBW DP30) with the help of a saturation–dilution-based humidity generator. The WS-CRDS system simultaneously measures CO_2, CH_4, CO and H_2O, and it is specifically designed for flight operation; it was operated at 186.6 hPa (140 Torr). Its measurement time is about 2.5 s, and by lowering its sample flow rate down to 0.1 sLpm, it was able to measure at altitude level up to 12.5 km without using a sample pump upstream of the sample cell. The instrument itself is not a factory-calibrated one, but another WS-CRDS instrument was calibrated against a chilled mirror hygrometer (Dewmet, Michell instruments Ltd., UK) at the Max Planck Institute for Biogeochemistry, (Jena, Germany), and the calibration constants were transferred to all subsequently manufactured CRDS instruments by Picarro Inc. (Crosson, 2008). An accuracy better than 10 ppmV for mixing ratios less than 100 ppm and better than 200 ppmV above 10 000 ppmV was estimated from the chilled mirror hygrometer specifications (accuracy 0.2 °C dew point at 20 °C dew point, linearly increasing to 0.4 °C dew point at −60 °C dew point) assuming typical dew-point/pressure combinations. Precision was estimated to be 4 ppm or 1 % (whichever is larger). During the campaign,

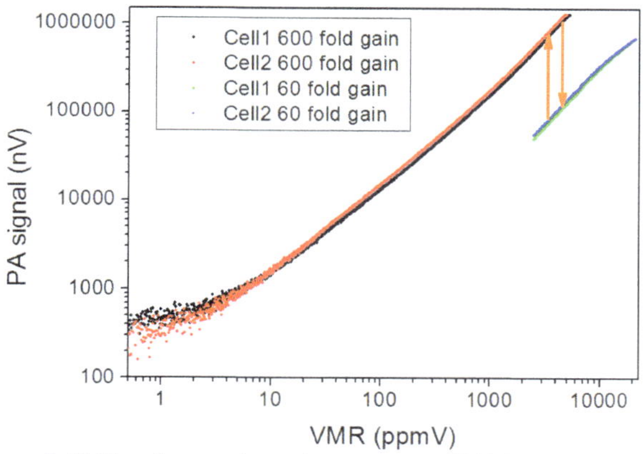

Figure 3. Calibration results at the pressure of 200 hPa. Arrows represent the SM switching points.

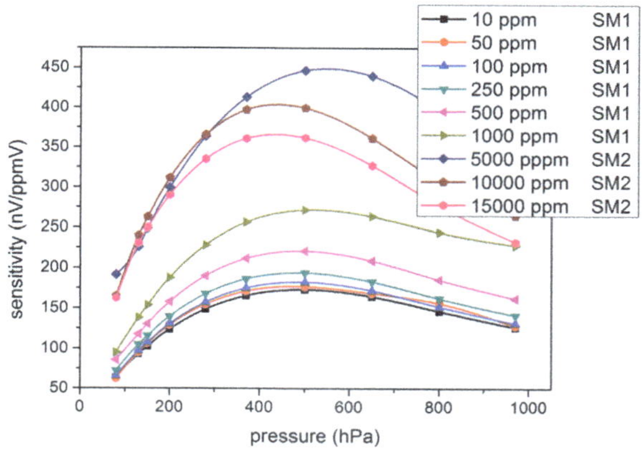

Figure 4. Dependence of the sensitivity of the PA system as the function of pressure at various VMRs. The sensitivities corresponding to 5000, 10 000 and 15 000 ppmV were measured at SM2 and up-scaled with a factor of 10 for better comparability, while the other curves were measured at SM1.

it was compared against the calibration standard used to calibrate FISH, revealing an offset of 14.5 ppmV that was subtracted from the CRDS flight data.

Unfortunately, for logistic reasons, the three instruments took their samples from three different inlets. During the campaign, the PA cells were configured to sample water vapor simultaneously from a single backward-facing aircraft inlet to measure WV VMR. The PA system was connected to the sampler by a heated stainless steel tube having an outer diameter of 1/8 inch and a length of 1.5 m. FISH was connected to a forward-facing inlet through a 0.5 m long, 10 mm inner diameter heated stainless steel tube to measure TW VMR, while the WS-CRDS system was connected to a forward-facing de-iced Rosemount total air temperature housing (model 102BX) through a FEP tube of 1 m in length and 1.58 mm inner diameter to measure WV VMR. For the Wasul-Hygro, the flow was provided by a downstream pump (MD1, Vacuubrand), while the WS-CRDS and FISH systems used the ram pressure provided by the samplers together with a downstream sampling pump.

For various reasons, large portions of the measured data are not suitable for intercomparison purposes, as follows. The central data-logging computer crashed repeatedly, causing data losses of 5 to 10 min periods. Considerable deviations between the readings of the three systems were observed during periods whenever the measured VMR varied rapidly, most probably due to the different response times of the sampling lines. WS-CRDS values below 50 ppmV had to be neglected, as their reliability is under discussion. The performance of FISH was acceptable only during the last two flights, due to a small internal leak that was discovered and fixed only after the second flight. During the campaign, FISH measured cloud-free data in the range 350 to 600 ppmV only for 4 min, and not at all above 600 ppmV. This amount of data above 350 ppmV is not enough for statistic conclusions, so these were neglected from the comparison. During the occasional presence of clouds, the FISH data had to be neglected, as their TW VMR data differed from the WV VMR reading of the other two instruments.

The comparable parts of the measured data were evaluated as done in the case of the ESF chamber comparison.

3 Results and discussion

3.1 Calibration

The calibration curves for the two WaSul-Hygro channels at one of the calibration pressures (200 hPa) together with the programmed SM switching points can be seen in Fig. 3. This curve was measured individually without periodically stepping the pressure for better visualization, but it overlaps the one used to program the calibration surface within the noise level. The measured sensitivities and noise levels (1σ) below 10 ppmV for the two measuring cells at 200 hPa pressure with 600-fold amplification of the microphone signal are 120 nV $ppmV^{-1}$ and 54 nV and 118 nV $ppmV^{-1}$ and 59 nV, respectively. At higher VMRs, similar sensitivities but higher noise levels (typically 0.5 % of the actual VMR reading) were found. Calibration curves at other pressures yield similar sensitivities and noise levels. From these results, it follows that the repeatability of the WV VMR measurements for either cell is 0.5 ppm or 0.5 % of the actual reading (whichever is greater). In Fig. 4, the sensitivity of the PA system as a function of pressure at different VMRs is shown. The dependencies are very similar for the two cells, but at higher VMRs, the sensitivity of the second PA cell is definitely lower due to the light attenuation in the first cell. Previously during the calibration procedure, only one calibration curve at a defined pressure was determined per PA cell, and the pressure-dependent sensitivity was determined at a certain VMR per PA cell (Szakáll et al., 2006). From Fig. 4, it can

Table 1. Results from comparisons of WaSul-Hygro with other instruments during laboratory and flight tests. In the case of the comparison with the WS-CRDS system above 4000 ppmV, there were no comparable numbers of data available, so the upper limits of the 2 and 5 % coincidences are certainly higher.

Reference instrument	Relative difference ranges from reference instruments (ppmV): Within noise level	2 %	5 %	Pearson correlation coefficient	Slope of cross-plot line
ESF chamber	< 150	200–750	15–12 000	0.99948	1.015
WS-CRDS	< 300	1500–4000	100–4000	0.99986	1.047
FISH	< 20	NA	NA	0.9965	0.86

be clearly seen that the use of this simplifying assumption would lead to highly inaccurate VMR determination.

As calibration curves are not completely linear, partially due to the nonlinear behavior causing sensitivity decreases at higher concentrations, they have to be approximated with high (up to 10) order polynomials for both channels, for both SMs, at each pressure. These polynomials form the calibration surface, as mentioned before (Sect. 2.2.4.).

The VMR calculation method applied during laboratory tests showed traceability to the humidity generator within less than 0.5 % ± 0.5 ppmV at any VMR-s and at any pressures for both PA cells.

3.2 Laboratory intercomparison

During the laboratory measurements at the ESF chamber, the VMR values measured by the two channels of the PA system coincided within the noise level or within 0.5 %. These results are practically identical to the observed ones during laboratory calibration. One representative result of the laboratory intercomparison at FZJ is shown in Fig. 5a, where two SM switching points are also indicated. In Fig. 5b, the comparison cross plot between WaSul-Hygro and the reference instrument can be seen. The lower parts of Fig. 5a and b are the measured relative deviations as functions of time or VMR, respectively, with the horizontal line indicating ±5 % limits.

Values measured in the fast cooling and drying periods when the ESF chamber is far from its steady state and values in the case of fast VMR change when the chamber was not in its quasi steady state were omitted from the comparison. An example of these posterior omitted data in the range of 5–7000 ppmV can be seen in the deviation plot (Fig. 5b), where the loopy behavior indicates that the sampling line of the PA system could not follow the fast VMR variation in the chamber as fast as the reference instruments inside. It is important to note that the relative deviation in the presented case is still below 10 %. During the quasi steady-state parts of the measurements, the relative deviation (the difference between the VMR reading of WaSul-Hygro and the reference instrument divided by the WaSul-Hygro reading) is less than 5 %, or is

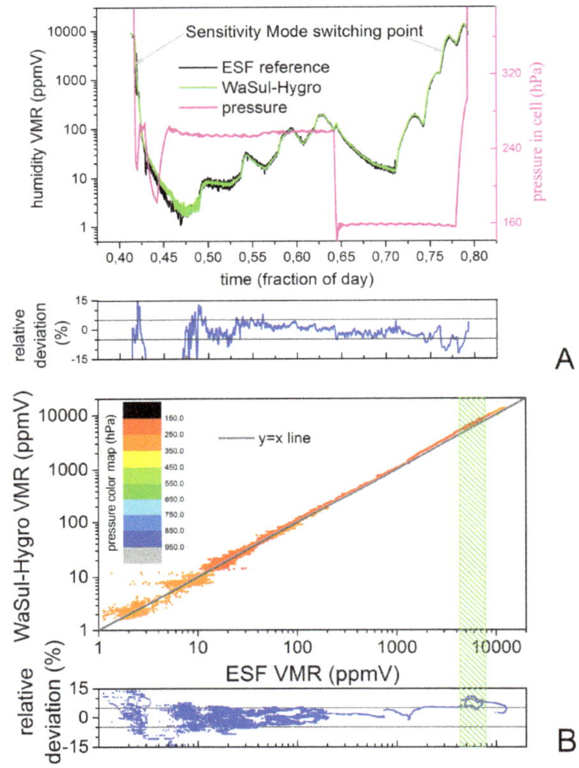

Figure 5. A representative measurement at the ESF chamber (5 May 2011). **(a)** is the time series of the measurements; **(b)** is the corresponding cross plot, where the pressure is coded by color and the green shaded area indicates the omitted loopy part.

noise level in the 15–12 000 ppmV range. Other parameters of comparison can be found in the first line of Table 1.

3.3 Airborne tests

During the flight tests, the noise level of the PA system and the deviation between the two sampling lines were almost the same as during laboratory intercomparison.

The comparisons of WaSul-Hygro with FISH and WS-CRDS instruments can be seen in Figs. 6 and 7, respectively,

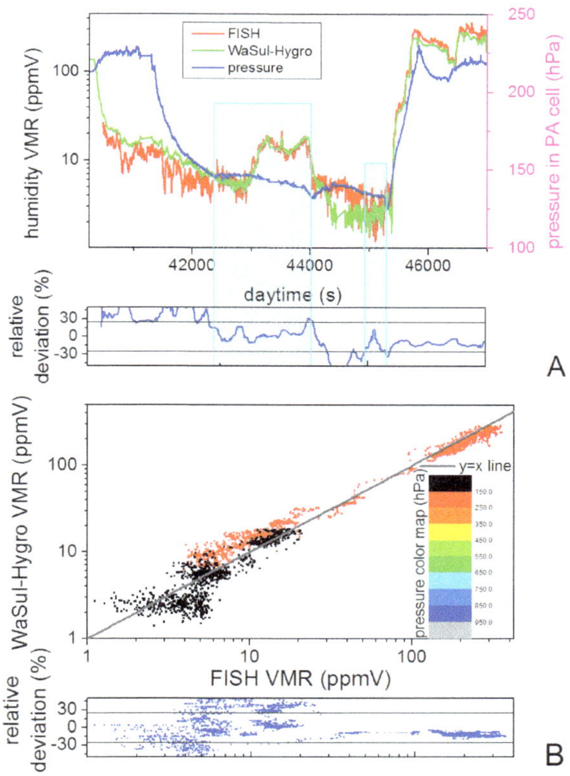

Figure 6. A representative comparison of WaSul-Hygro and FISH (DENCHAR flight, 31 May 2011). **(a)** is the time series of the measurements, where the blue shaded areas indicate when the two systems were measuring the same values within the noise level; **(b)** is the corresponding cross plot, where the pressure is coded by color.

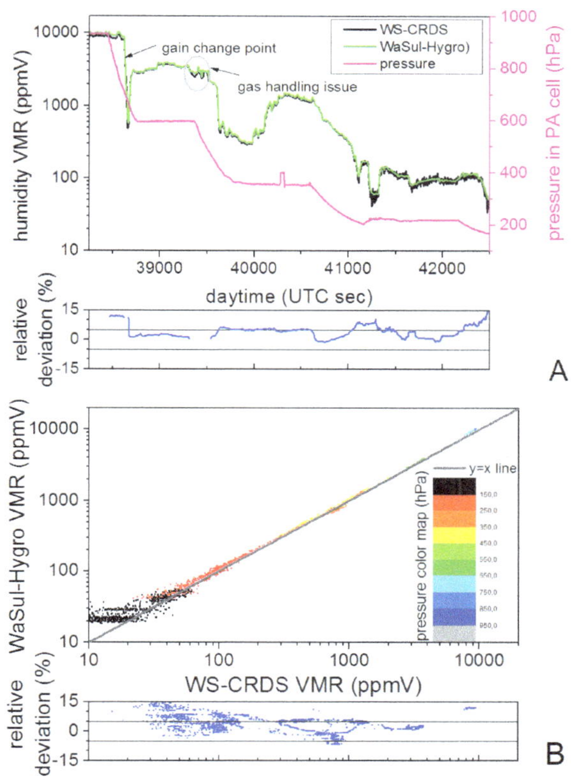

Figure 7. A representative comparison flight measurement of WaSul-Hygro and the WS-CRDS system (DENCHAR flight, 26 May 2011). **(a)** is the time series of the measurements; **(b)** is the corresponding cross plot, where the pressure is coded by color. The relative deviation below ~ 30 ppmV is higher than 15 %, and is not indicated in plot **(b)**.

in a similar way as, previously, the results of the comparison at the ESF chamber, but in the case of the comparison with FISH, ±25 % limits are shown. An SM switching point and a period when the measurement results of the two instruments deviate considerably (most probably due to differences in the sampling lines) are marked in Fig. 7a. The comparison results can be found in detail in Table 1.

The slope of the cross plot between FISH and Wasul-Hygro deviates by about 20 % from the slope of the cross plot between Wasul-Hygro and the WS-CRDS system, which is close to the ideal of 1. The facts that the cross plot in Fig. 6b is close to linear and that the corresponding Pearson correlation coefficient is still high indicate that the deviation of FISH is most probably due not to a random effect or instrument malfunction such as noise. Furthermore, the deviation cannot be related to occasional occurrence of cloud droplets or ice crystals in the gas stream sampled by the TW inlet of FISH, because these events were carefully eliminated from the intercomparison. Therefore, it is rather a phenomenon that indicates that the WV VMR was indeed systematically different for the other sampling inlets, and definitely requires further studies.

3.4 Overall performance and outlook

Besides the performance parameters discussed above in detail, it is important to note that non-validated and therefore not presented tests show that the precision and accuracy values 18 months after the calibration did not vary significantly. These results still have to be verified and validated through more rigorous blind tests.

The actual size and weight of the system is acceptable on research aircraft, but in order to be mounted on an in-service one as a standard, these values still have to be improved. That is why a new controlling electronics is currently being developed, which, besides the significant size reduction, will be capable of real-time VMR calculation without the need for an external PC.

Acknowledgements. The authors would like to honor the memory of Cornelius Schiller (1961–2012), the former leader of the FISH development team and a worldwide recognized leading scientist of atmospheric humidity measurements.

The research leading to these results has received funding from EUFAR contract no. 227159. The project has been partially funded by TÁMOP-4.2.2.A-11/1/KONV-2012-0047 supported by the European Union and co-financed by the European Social Fund. The project has also been partially supported by the Hungarian Research and Technology Innovation Fund (OTKA), project no. NN109679. The research leading to these results has also received funding from the European Community's Seventh Framework Programme (FP7/2007-2013) under grant agreement no. 312311. D. Tátrai was partially supported by TÁMOP 4.2.4.A/2-11-1-2012-0001 supported by the European Union and Hungary and co-financed by the European Social Fund. For the operation of the Picarro WS-CRDS system, funding from IAGOS-ERI (In-service Aircraft for a Global Observing System – European Research Infrastructure, contract no. 212128), part of the EU's Seventh Framework Programme, was used.

Edited by: S. Malinowski

References

Bell, A. G.: On the production and reproduction of sound by light: the photophone, Am. J. Sci., 20, 305–324, 1880.

Bell, A. G.: Upon the production of sound by radiant energy, Philosophical Magazine and Journal of Science, 11, 510–528, 1881.

Bozóki, Z., Pogány, A., and Szabó, G.: Photoacoustic instruments for practical applications: present, potentials, and future challenges, Appl. Spectrosc. Rev., 46, 1–37, doi:10.1080/05704928.2010.520178, 2011.

Bozóki, Z., Tátrai, D., and Szabó, G.: Method and arrangement for wavelength monitoring of wavelength tunable light source and stabilizing based on absorption spectroscopic detecting; Hungarian patent, HU1100719 (A2), 2013.

Brenninkmeijer, C. A. M., Crutzen, P., Boumard, F., Dauer, T., Dix, B., Ebinghaus, R., Filippi, D., Fischer, H., Franke, H., Frieß, U., Heintzenberg, J., Helleis, F., Hermann, M., Kock, H. H., Koeppel, C., Lelieveld, J., Leuenberger, M., Martinsson, B. G., Miemczyk, S., Moret, H. P., Nguyen, H. N., Nyfeler, P., Oram, D., O'Sullivan, D., Penkett, S., Platt, U., Pupek, M., Ramonet, M., Randa, B., Reichelt, M., Rhee, T. S., Rohwer, J., Rosenfeld, K., Scharffe, D., Schlager, H., Schumann, U., Slemr, F., Sprung, D., Stock, P., Thaler, R., Valentino, F., van Velthoven, P., Waibel, A., Wandel, A., Waschitschek, K., Wiedensohler, A., Xueref-Remy, I., Zahn, A., Zech, U., and Ziereis, H.: Civil Aircraft for the regular investigation of the atmosphere based on an instrumented container: The new CARIBIC system, Atmos. Chem. Phys., 7, 4953–4976, doi:10.5194/acp-7-4953-2007, 2007.

Buldyreva, J., Lavrentieva, N., and Starikov, V.: Collisional Line Broadening And Shifting Of Atmospheric Gases; A Practical Guide for Line Shape Modelling by Current Semi-classical Approaches; Imperial College Press: London, UK, 2011.

CARIBIC: CARIBIC website, available at: http://www.caribic-atmospheric.com/, last access: 27 November 2014.

Castleden, S. L., Kirkbright, G. E., and Spillane, D. E. M.: Wavelength modulation in photoacoustic spectroscopy, Anal. Chem., 53, 2228–2231, doi:10.1021/ac00237a019, 1981.

Chen, H., Winderlich, J., Gerbig, C., Hoefer, A., Rella, C. W., Crosson, E. R., Van Pelt, A. D., Steinbach, J., Kolle, O., Beck, V., Daube, B. C., Gottlieb, E. W., Chow, V. Y., Santoni, G. W., and Wofsy, S. C.: High-accuracy continuous airborne measurements of greenhouse gases (CO_2 and CH_4) using the cavity ring-down spectroscopy (CRDS) technique, Atmos. Meas. Tech., 3, 375–386, doi:10.5194/amt-3-375-2010, 2010.

Crosson, E. R.: A cavity ring-down analyzer for measuring atmospheric levels of methane, carbon dioxide, and water vapor, Appl. Phys. B, 92, 403–408, doi:10.1007/s00340-008-3135-y, 2008.

Goff, J. A. and Gratch, S.: Low-pressure properties of water from −160 to 212 F, T. Am. Soc. Heat. Vent. Eng., 52, 95–122, 1946.

Helten, M., Smit, H. G. J., Straeter, W., Kley, D., Nedelec, P., Zöger, M., and Busen, R.: Calibration and performance of automatic compact instrumentation for the measurement of relative humidity from passenger aircraft, J. Geophys. Res., 103, 643–652, doi:10.1175/2007JTECHA975.1, 1998.

IAGOS: IAGOS website, available at: http://www.iagos.org, last access: 27 November 2014.

Kley, D. and Stone, E. J.: Measurement of water vapor in the stratosphere by photodissociation with Ly (a) (1216 A) light, Rev. Sci. Instrum., 49, 691–697, doi:10.1063/1.1135596, 1978.

Marenco, A., Thouret, V., Nédélec, P., Smit, H. G. J., Helten, M., Kley, D., Karcher, F., Simon, P., Law, K., Pyle, J., Poschmann, G., Wrede, R. V., Hume, C., and Cook, T.: Measurement of ozone and water vapor by Airbus in-service aircraft: The MOZAIC airborne program, an overview, J. Geophys. Res.-Atmos., 103, 25631–25642, doi:10.1029/98JD00977, 1998.

McDonald, F. A. and Wetsel Jr., G. C.: Generalized theory of the photoacoustic effect, J. Appl. Phys., 49, 2313–2322, doi:10.1063/1.325116, 1978.

Miklós, A., Hess, P., and Bozóki, Z.: Application of Acoustic Resonators in Photoacoustic Trace Gas Analysis and Metrology, Rev. Sci. Instrum., 72, 1937–1955, doi:10.1063/1.1353198, 2001.

Szakáll, M., Huszár, H., Bozóki, Z., and Szabó, G.: On the pressure dependent sensitivity of a photoacoustic water vapor detector using active laser modulation control, Infrared Phys. Techn., 48, 192–201, doi:10.1016/j.infrared.2006.01.002, 2006.

Szakáll, M., Csikós, J., Bozóki, Z., and Szabó, G.: On the temperature dependent characteristics of a photoacoustic water vapor detector for airborne application, Infrared Phys. Techn., 51, 113–121, doi:10.1016/j.infrared.2007.04.001, 2007.

Tátrai, D., Bozóki, Z., and Szabó, G.: Method for wavelength locking of tunable diode lasers based on photoacoustic spectroscopy, Opt. Eng., 52, 096104, doi:10.1117/1.OE.52.9.096104, 2013.

Saarela, J., Toivonen, J., Manninen, A., Sorvajärvi, T., and Hernberg, R.: Wavelength modulation waveforms in laser photoacoustic spectroscopy, Appl. Optics, 1, 743–748, doi:10.1364/AO.48.000743, 2009.

Schilt, S. and Thevenaz, L.: Wavelength modulation photoacoustic spectroscopy: Theoretical description and experimental results, Infrared Phys. Techn., 48, 154–162, doi:10.1016/j.infrared.2005.09.001, 2006.

Seufert, J., Fischer, M., Koeth, J., Werner, R., Kamp, M., and Forchel, A.: DFB laser diodes in the wavelength range from

760 nm to 2.5 μm, Spectrochim. Acta A, 60, 3243–3247, doi:10.1016/j.saa.2003.11.043, 2004.

Sharpe, S. W., Johnson, T. J., Sams, R. L., Chu, P. M., Rhoderick, G. C., and Johnson, P. A.: Gas-Phase Databases for Quantitative Infrared Spectroscopy, Appl. Spectrosc., 58, 1452–1461, doi:10.1366/0003702042641281, 2004.

Smit, H. G. J., Strater, W., Helten, M., and Kley, D.: Environmental simulation facility to calibrate airborne ozone and humidity sensors, Tech. Rep. Juel. Berichte No. 3796, Forschungszentrum Jülich, available at: http://www.fz-juelich.de/SharedDocs/Downloads/IEK/IEK-8/EN/ESF/ESF.pdf (last access: 27 November 2014), 2000.

Solomon, S., Qin, D., Manning, M., Chen, Z., Marquis, M., Averyt, K. B., Tignor, M., and Miller, H. L.: Climate Change 2007: The Physical Science Basis Contribution of Working Group I to the Fourth Assessment Report of the Intergovernmental Panel on Climate Change; Cambridge University Press, Cambridge, United Kingdom and New York, NY, USA, 2007.

WMO: WMO-Report No. 8, Measurement of atmospheric humidity, Guide to meteorological instruments and methods of observation, 5th Edn., World Meteorological Organization, Geneva, 5.1–5.19, 1983.

Woodward, S. L., Parayanthal, P., and Koren, U.: The effects of aging on the Bragg section of a DBR laser, Photonics Technology Letters, 5, 750–752, doi:10.1109/68.229794, 1993.

Zöger, M., Afchine, A., Eicke, N., Gerhards, M. T., Klein, E., McKenna, D. S., Mörschel, U., Schmidt, U., Tan, V., Tuitjer, F., Woyke, T., and Schiller, C.: Fast in situ stratospheric hygrometers: A new family of balloonborne and airborne Lyman-α photofragment fluorescence hygrometers, J. Geophys. Res., 104, 1807–1816, doi:10.1029/1998JD100025, 1999.

5

Forecast indices from a ground-based microwave radiometer for operational meteorology

D. Cimini[1,2]**, M. Nelson**[3]**, J. Güldner**[4]**, and R. Ware**[3,5]

[1]IMAA-CNR, Potenza, Italy
[2]CETEMPS, University of L'Aquila, L'Aquila, Italy
[3]Radiometrics, Boulder, Colorado, USA
[4]DWD, Meteorological Observatory Lindenberg – Richard Aßmann Observatory, Lindenberg, Germany
[5]UCAR, Boulder, Colorado, USA

Correspondence to: D. Cimini (cimini@imaa.cnr.it)

Abstract. Today, commercial microwave radiometer profilers (MWRPs) are robust and unattended instruments providing real-time, accurate atmospheric observations at ~ 1 min temporal resolution under nearly all weather conditions. Common commercial units operate in the 20–60 GHz frequency range and are able to retrieve profiles of temperature, vapour density, and relative humidity. Temperature and humidity profiles retrieved from MWRP data are used here to feed tools developed for processing radiosonde observations to obtain values of forecast indices (FIs) commonly used in operational meteorology. The FIs considered here include K index, total totals, KO index, Showalter index, T1 gust, fog threat, lifted index, S index (STT), Jefferson index, microburst day potential index (MDPI), Thompson index, TQ index, and CAPE (convective available potential energy). Values of FIs computed from radiosonde and MWRP-retrieved temperature and humidity profiles are compared in order to quantitatively demonstrate the level of agreement and the value of continuous FI updates. This analysis is repeated for two sites at midlatitude, the first one located at low altitude in central Europe (Lindenberg, Germany) and the second one located at high altitude in North America (Whistler, Canada). It is demonstrated that FIs computed from MWRPs well correlate with those computed from radiosondes, with the additional advantage of nearly continuous updates. The accuracy of MWRP-derived FIs is tested against radiosondes, taken as a reference, showing different performances depending upon index and environmental situation. Overall, FIs computed from MWRP retrievals agree well with radiosonde values, with correlation coefficients usually above 0.8 (with few exceptions). We conclude that MWRP retrievals can be used to produce meaningful FIs, with the advantage (with respect to radiosondes) of nearly continuous updates.

1 Introduction

Commercial microwave radiometers profilers (MWRPs) are robust instruments that perform continuous unattended operations and real-time atmospheric observations at ~ 1 min temporal resolution under nearly all weather conditions. MWRPs perform measurement of thermal emission from downwelling brightness temperature (T_b) in the atmosphere. Most common commercial units operate in the 20–60 GHz frequency (0.5 to 1.5 cm wavelength) range, in which atmospheric thermal emission is influenced by atmospheric temperature, humidity, and the presence of hydrometeors. From calibrated MWRP T_b, atmospheric thermodynamic profiles can be retrieved using a variety of inversion methods, as multivariate regression, neural networks, and variational approaches (Solheim et al., 1998; Ware et al., 2003; Löhnert et al., 2004; Cimini et al., 2006, 2010) with an accuracy that is compatible with most of meteorology applications, especially in the lower troposphere (Güldner and Spänkuch, 2001; Hewison, 2007; Cimini et al., 2011; Ware et al., 2013). Due to the increasing distribution, networks of MWRPs are being established for climate and meteorology applications (Cadeddu et al., 2013; Cimini et al., 2014).

Currently operational radiosonde launches are normally performed twice daily by national and/or regional weather services (NWS) at hundreds of sites worldwide. Relatively few sites perform launches 4 times a day. Radiosondes are usually launched at synoptic hours (00:00 and 12:00, or 00:00, 06:00, 12:00, and 18:00 UTC), and the observed thermodynamic profiles are assimilated into numerical weather prediction (NWP) model analysis. In addition, the same radiosonde observations are processed locally at NWS to compute forecast indices (FIs) developed starting from some 60 years ago (Showalter, 1947; Fawbush and Miller, 1954; Galway, 1956; George, 1960) which are still broadly used for operational meteorology and local short-term forecast (Andersson et al., 1989; Haklander and Van Delden, 2003; Holtslag et al., 2010). Although the FIs based on sounding observations are rather simple, they perform reasonably well, especially once optimized for site-specific conditions, and show forecast skills that often outperform the skill of modern numerical weather predication models (Kuhlman, 2006; Holtslag et al., 2010). However, these FIs are available at NWS radiosonde sites at the time of radiosonde flight, i.e. usually 2 to 4 times a day. More recently, FIs at higher temporal resolution ($\sim$ 15 min) became available from geostationary satellite retrievals (König, 2002; König and de Coning, 2009; de Coning et al., 2011). Though they are limited to clear-sky conditions, satellite-based FIs offer the advantage of covering large areas (at continental scale). On the other hand, satellite-based FIs suffer from the lack of quality data in the boundary layer due to the low vertical resolution of satellite observations in the lower troposphere.

The US National Research Council (NRC) recently reported that continuous boundary layer temperature, humidity, and wind observations provide a practical and cost-effective means to improve local high-impact weather forecasting (NRC, 2008, 2010). In fact, the structure and variability of the lower troposphere is currently not well known because vertical profiles of water vapour, temperature, and winds are not systematically observed with sufficient spatial or temporal resolution. This lack of observations results in the planetary boundary layer being the single most important under-sampled part of the atmosphere. Consequently, short-term forecast skill may be poor due to a lack of pertinent data, particularly in the lower troposphere, where severe weather originates.

Conversely, traditional FIs developed for radiosondes can be generated from MWRP retrievals continuously and under nearly all weather conditions. Temperature and humidity profiles are available from MWRP data at $\sim$ 1 min resolution. Note that most of the information content resides within the planetary boundary layer (Cimini et al., 2011; Löhnert and Maier, 2012) and that useful information is provided even under precipitation (Cimini et al., 2011; Xu et al., 2014). The high-temporal-resolution MWRP retrievals have recently been exploited in support of nowcasting low-level windshear at Hong Kong airport (Chan and Lee, 2011), dynamic weather conditions in US (Knupp et al., 2009), and intense convective weather in Hong Kong (Chan and Hon, 2011) and southeastern India (Madhulatha et al., 2013; Venkat Ratnam et al., 2013). A similar approach, though using a ground-based infrared interferometer, is also reported (Feltz and Mecikalski, 2002; Wagner et al., 2008). In addition, MWRPs have been used together with lidar to measure wind and temperature for wind energy applications (Friedrich et al., 2012), revealing the same accuracy as tower measurements, with the advantage of monitoring stability and turbulence.

Thus, the MWRP-retrieved profiles can be used to feed tools for computing FIs developed for radiosondes profiles, providing timely and continuous FI data. These timely FI data are particularly important for thunderstorm forecast in the 0 to 6 h range, as recently shown for two tropical sites (Chan, 2009; Chan and Hon, 2011; Madhulatha et al., 2013; Venkat Ratnam et al., 2013). In this paper we extend the application to two sites at midlatitude and to a few additional FIs, demonstrating the ability of MWRP retrievals to provide quality FIs similar to radiosonde, as well as the added value of continuous FI data.

This paper is organized as follows: Sect. 2 describes the data set under consideration; Sect. 3 summarizes the methodology used for data intercomparison; Sect. 4 reports the results of the data analysis; and Sect 5 summarizes the results and draws final conclusions, suggesting possibilities for additional advancements and application of MWVR observations.

2 Data set

Data collected at two sites are presented in the following. The first site is in Lindenberg (Germany) at the Meteorological Observatory Lindenberg – Richard Aßmann Observatory (MOL-RAO) operated by the German Meteorological Service (Deutscher Wetterdienst, DWD). The second site is Whistler (Canada) at the meteorological station operated by Environment Canada – Meteorological Service of Canada – in support of nowcasting and short-term weather forecasting during the Vancouver 2010 Olympic and Paralympic Winter Games (Mailhot et al., 2010).

2.1 Radiosonde data

The radiosondes used in this analysis are Global Positioning System (GPS) enabled RS92-SGP systems, manufactured by Vaisala, providing vertical profiles of pressure, temperature, relative humidity, dew point temperature, and wind at 2 s resolution. These data are then used to calculate the dew point temperature and the geopotential height according to hydrostatic equilibrium. However, depending upon sensor type, ra-

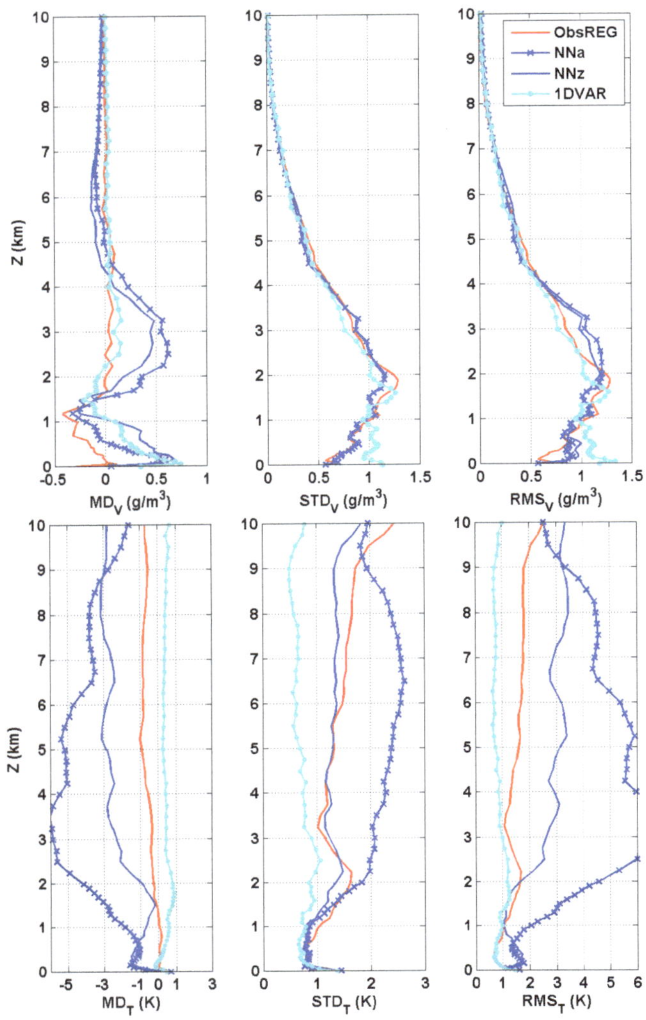

Figure 1. Statistics of water vapour density (top) and temperature (bottom) profiles retrieved with NN, ObsREG, and 1-DVAR, as compared to radiosondes (233 cases). Left: mean difference (radiosonde minus retrievals). Centre: SD difference. Right: rms difference. The legend in the top-right corner indicates the line color/style coding.

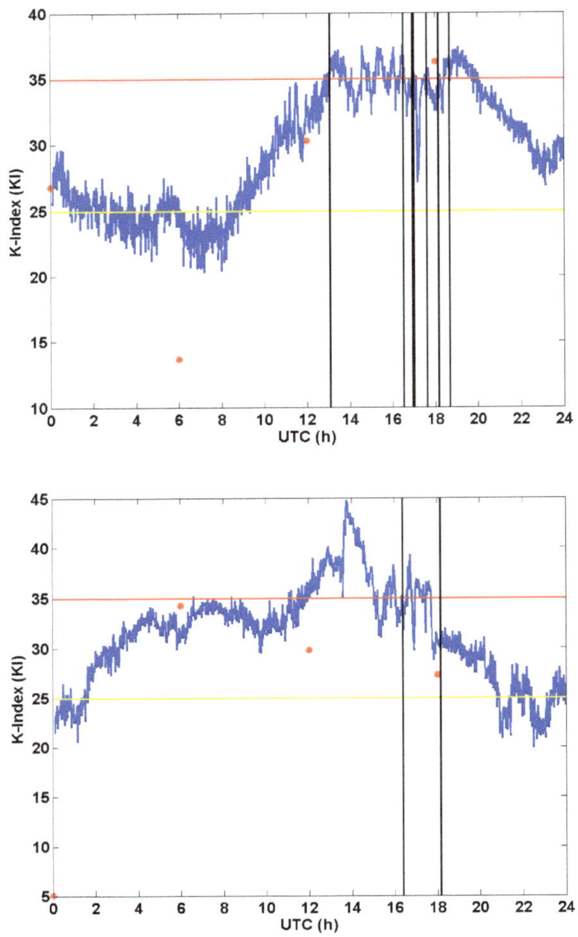

Figure 2. Top: 24 h time series of K index computed from MWRP NNz retrievals (blue line), and radiosonde profiles (red dots) for 12 August 2010. Thunder near Lindenberg started at 13:05 and lasted until 18:40 UTC; vertical black lines indicate the time of thunder detection. Yellow and red lines show the boundaries for weak-to-moderate and moderate-to-strong thunderstorm potential thresholds, respectively. Bottom: case of 15 August 2010. Thunder near Lindenberg was reported between 16:10 and 18:05 UTC.

diosondes experience some errors, including daytime solar radiation dry bias, horizontal drift, time lag, and calibration errors at low temperatures. Nevertheless, radiosondes remain the de facto standard for upper air monitoring and presently provide the most widely available information on the vertical structure of the troposphere and lower stratosphere. For the type used during the experiment, the specifications for total measurement uncertainty in soundings are 1.0 mb for pressure, 0.5 °C for temperature, 5 % for relative humidity, and 0.2 m s^{-1} for wind speed (Vaisala, 2013). Four radiosondes per day were launched at standard synoptic hours (00:00, 06:00, 12:00, and 18:00 UTC) from the two sites considered in this study.

2.2 MWRP data

The MWRP data used in this analysis are provided by two MP-3000A units manufactured by Radiometrics. The MP-3000A units include a scanning multichannel microwave radiometer; a one-channel broadband infrared (IR) radiometer; and surface pressure, temperature, and humidity sensors. The MWRP IR radiometer (one channel covering approximately 9.6–11.5 μm) measures sky IR temperature and gives information on cloud-base temperature. The MWRP meteorology sensors measure temperature (T_s), pressure (P_s), and relative humidity (RH_s) at the instrument level. The multichannel microwave radiometer can observe T_b at up to 35 channels in the 20–60 GHz frequency (0.5 to 1.5 cm wavelength). In this

Table 1. Summary of MWRP characteristics.

Channels	22
Frequencies (GHz)	22.234, 22.5, 23.034, 23.834, 25.0, 26.234, 28.0, 30.0, 51.248, 51.76, 52.28, 52.804, 53.336, 53.848, 54.4, 54.94, 55.5, 56.02, 56.66, 57.288, 57.964, 58.8
Bandwidth (MHz)	300
Elevation	15–90°
Azimuth	Fixed

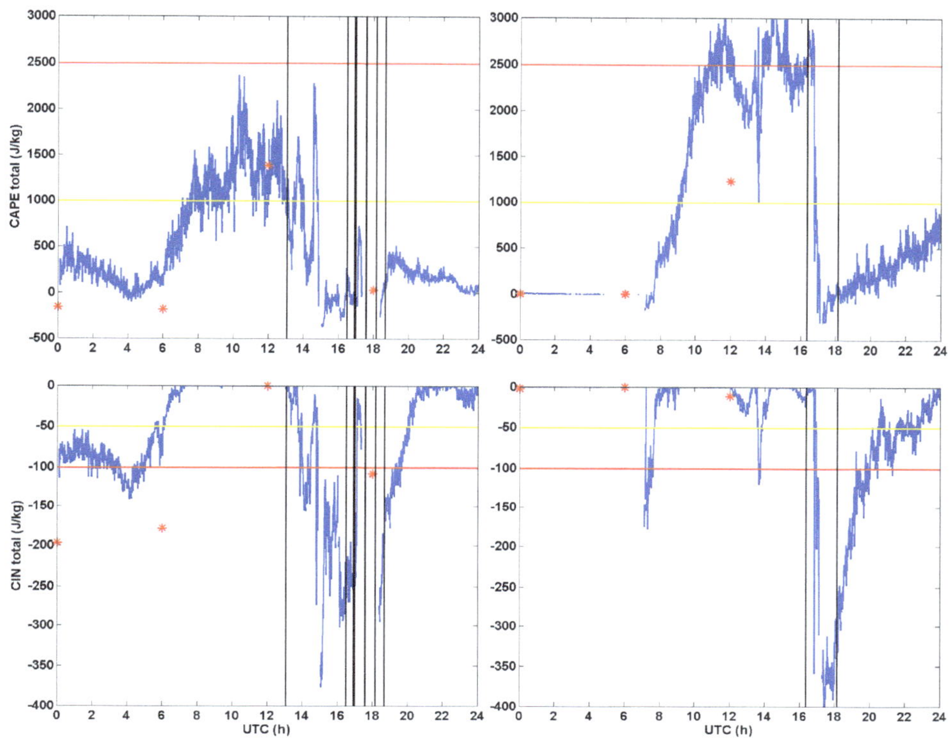

Figure 3. Same as Fig. 2 but for CAPE (top) and CIN (bottom). Yellow and red lines show the thresholds for, respectively, weak and moderate thunderstorm potential (CAPE) and convective inhibition (CIN). Left: 12 August 2010. Right: 15 August 2010.

frequency range, atmospheric thermal emission comes from atmospheric gases (primarily oxygen and water vapour) and hydrometeors (mainly liquid water particles, since ice emission is negligible). The Rayleigh scattering regime applies up to sizes of small raindrops, and in general the scattering contribution is negligible up to light precipitation.

When properly calibrated, a MWRP provides T_b with an absolute accuracy of $\sim$0.3–0.7 K (Cimini et al., 2003; Löhnert and Maier, 2012; Maschwitz et al., 2013). Atmospheric temperature and humidity profiles can be retrieved from MWRP T_b with a variety of inversion methods, including multivariate regression, neural networks, and variational approaches (Solheim et al., 1998; Löhnert et al., 2004; Hewison, 2007; Cimini et al., 2006, 2010). The actual vertical resolution of MWRP retrievals depends on several factors, including inverse method and atmospheric conditions. In general, the vertical resolution degrades approximately linearly with altitude as half of the height (Cimini et al., 2006). Typical root-mean-square (rms) accuracy for tropospheric temperature and absolute humidity profiles during non-precipitating conditions are, respectively, $\sim$0.5–2.0 K and 0.2–1.5 g m^{-3} (Güldner and Spänkuch, 2001; Cimini et al., 2011). Accuracy estimates during precipitation including heavy rain have been reported as equal to 3.1 K and 1.9 g m^{-3} (Xu et al., 2014).

For the data set used in this analysis, the MWRP observed T_b at 22 channels and at 2 elevation angles (zenith and 15° elevation) and 1 fixed azimuth angle. The microwave radiometer is calibrated using noise diode injection to measure the system gain continuously. The noise diode effective temperature is determined by observing an external cryogenic target less frequently (once every 3 to 6 months). The main characteristics of the MWRP microwave radiometer are reported in Table 1.

3 Methodology

3.1 MWRP retrieval techniques

Atmospheric temperature and humidity profiles are retrieved from MWRP observations using any of the above-mentioned inversion methods. The MP-3000A proprietary algorithm deploys a neural networks (NN) method (Solheim et al., 1998), trained with thousands of profiles generated from historical data sets of operational radiosondes. The software offers two versions of the NN retrievals, which may run separately and in parallel, one ingesting zenith observations (NNz) and the other ingesting the available slant observations (NNa). Alternatively, other methods can be applied to retrieve temperature and humidity profiles from MP-3000A data, as multivariate regression or variational approaches. Observation-based multivariate regression (ObsREG), using MWRPs and radiosonde measurements from the past to calculate regression operators, has been successfully demonstrated to remove systematic errors and produce weakly biased retrievals with respect to radiosondes (Güldner and Spänkuch, 2001; Cimini et al., 2006). More recently, the use of a one-dimensional variational (1-DVAR) technique, coupling radiometric observations with outputs from a NWP model, has been demonstrated (Hewison, 2007; Cimini et al., 2006, 2009, 2011). The 1-DVAR method combines observed and forward-modelled T_b with measurement and background error covariance matrices to optimize retrieval accuracy with respect to observation and model uncertainties. The accuracy of temperature and humidity profile retrievals in the boundary layer and lower troposphere it depends primarily on the MWRP observations, while in the upper troposphere depends primarily on the NWP model output. Thus, the 1-DVAR approach avoids the error inherent in methods initialized with local climatology and benefits from recent surface, radiosonde, satellite, radar, and other data assimilated in the local NWP analysis and forecast.

Figure 1 shows the statistical results for water vapour and temperature profiles computed using the above methods as compared with radiosondes profiles. In Fig. 1 are reported the mean difference (MD), standard deviation (SD), and rms difference for simultaneous profiles (from ObsREG, NNa, NNz, and 1-DVAR), using radiosonde as the reference. These results are obtained in the period May–June 2010 for a continental site at midlatitude, the MOL-RAO in Lindenberg, Germany. Figure 1 shows the results from the 233 cases in which all the data sources (radiosondes and MWRPs) are simultaneously available. All observations during all weather conditions are included in the statistical comparison. Note that the radiosonde observations used for validation are not assimilated into NWP at the time of comparison, so they provide a really independent data set. Concerning temperature profiles, Fig. 1 suggests that both ObsREG and 1-DVAR show a small bias (MD within 1 K) relatively to NNa and NNz (MD within 3 and 6 K, respectively, for NNz and NNa). Although the methods show similar SD in the boundary layer, this tends to separate for altitudes higher than 1 km: SD remains within 1 K up to 10 km for 1-DVAR, while it increases with height for ObsREG, NNz, and especially NNa. The resulting rms remains within 1 K up to 10 km for 1-DVAR, within 2 K for ObsREG, and within 3 K for NNz, while it exceeds 6 K between 2.5 and 4 km for NNa. Figure 1 confirms that 1-DVAR provides better agreement with radiosondes than other considered methods for temperature profiling, minimizing both the systematic and random errors by benefiting from recent data assimilated in the NWP model. Concerning humidity profiles, Fig. 1 reports similar performances for the four techniques, though ObsREG and 1-DVAR show smaller bias in the 2–4 km range, and 1-DVAR shows degraded performances in the lowest 1 km layer. Similar results were obtained for other environmental conditions (Cimini et al., 2011; Ware et al., 2013), specifically for the mountain site at Whistler, Canada.

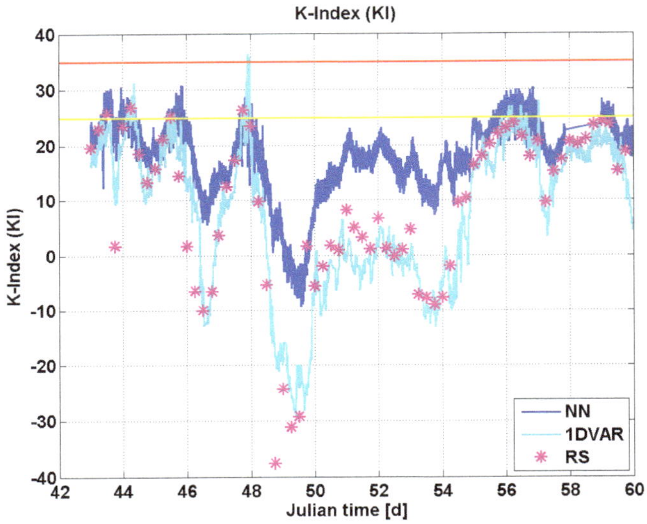

Figure 4. 17-day-long time series of K index in Whistler. K index from radiosondes (RS) and from MWRP NNz and 1-DVAR retrievals is indicated with magenta stars, and a blue and cyan line, respectively. The horizontal yellow and red lines indicate moderate and high thunderstorm potential, respectively.

3.2 Forecast indices

Forecast indices are used at various airports and NWS as tools for operational meteorology, e.g. to provide a quantitative and objective way to assist forecasters in issuing weather hazard alerts (Haklander and Van Delden, 2003). There are a number of FIs that have been developed and are currently used, depending upon the particular climatology of various sites. A quite comprehensive review can be obtained from Miller (1972), Peppier (1998), and Haklander and Van Delden (2003). FIs are usually computed from radiosonde profiles of temperature, humidity, and wind (if available).

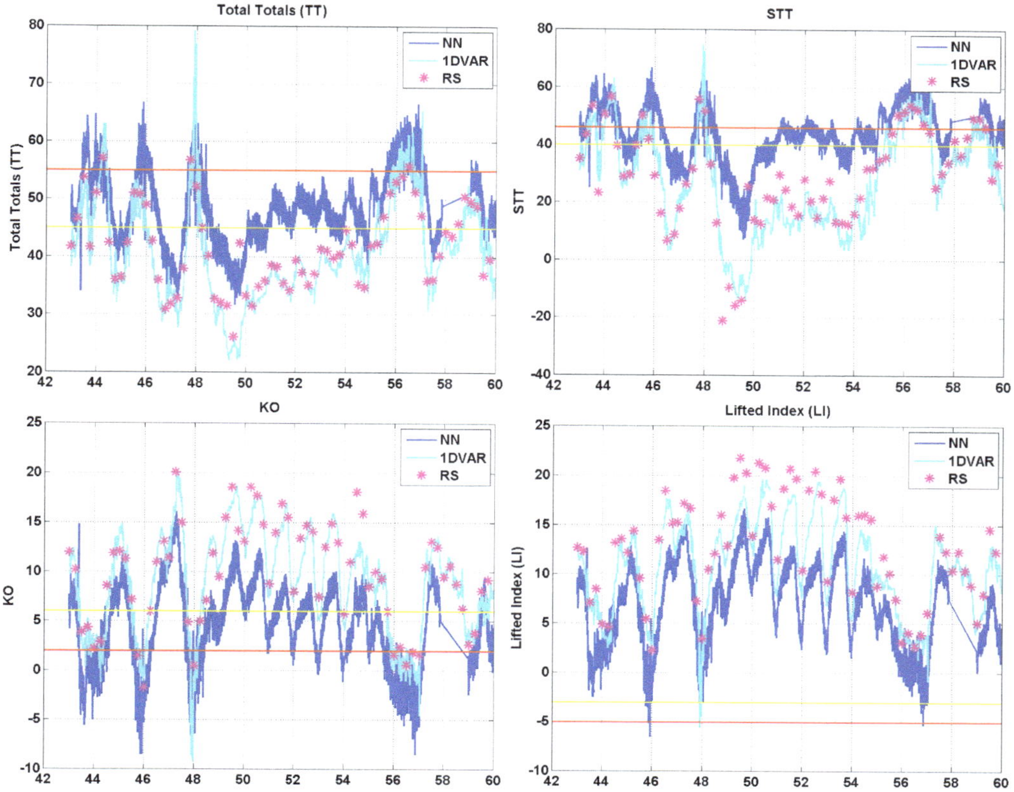

Figure 5. As in Fig. 3 but for (clockwise from top-left panel) total totals, STT, LI, and KO. The horizontal yellow and red lines indicate thresholds for moderate and high potential, respectively.

When only temperature and humidity profiles are available, a reduced set of FIs can be computed. This is the case for PTU (pressure, temperature, relative humidity) radiosondes, which are not equipped for wind estimation, as well as for MWRPs. Therefore, temperature and humidity profiles retrieved by MWRPs may be given in input to forecast tools developed for computing FIs from PTU radiosondes. FIs from temperature and humidity profiles are also estimated from geostationary satellites retrievals (König, 2002; König and de Coning, 2009; de Coning et al., 2011). These satellite FI estimates are produced operationally and cover most of the Earth's surface at some 3–10 km resolution. However, satellite FIs are based on infrared sounding channels, which tend to saturate in the presence of clouds, and thus are only available in clear-sky conditions. Moreover, the reliability of satellite FIs is hampered by the rather coarse vertical resolution of temperature and humidity profile retrievals from geostationary satellites, particularly in the lower troposphere.

More details on the FIs considered in this paper are provided in Appendix A. Note that these simple methods have the advantage of being based on reproduction of actual atmospheric processes, which should be broadly applicable while being relatively easy to understand and implement. The FI values presented in the next sections were computed by processing temperature and humidity profiles with the universal RAwinsonde OBservation (RAOB) program software (www.raob.com). FIs from radiosondes and MWRPs were computed independently. Let us emphasize that FIs from MWRPs are computed from MWRP-retrieved temperature and humidity profiles, i.e. not directly from T_b observations.

Note that it is beyond the scope of this paper to demonstrate the use and the forecast skills of the various FIs. This topic is discussed in the review papers mentioned above (e.g. Haklander and Van Delden, 2003), and it is still the subject of debate within the weather forecast community (e.g. Doswell III and Schultz, 2006). The aim of this paper is to investigate the feasibility of producing good-quality FIs from MWRP observations and to quantify the agreement and correlation with analogous FIs estimated from radiosondes. If the agreement is deemed satisfactory, we shall conclude that MWRPs can deliver additional valuable products that may complement radiosondes, or even supply for the lack thereof, in support of local short-term weather forecasting.

4 Results

This section presents the results of computing FIs from MWRP temperature and humidity profiles and their agree-

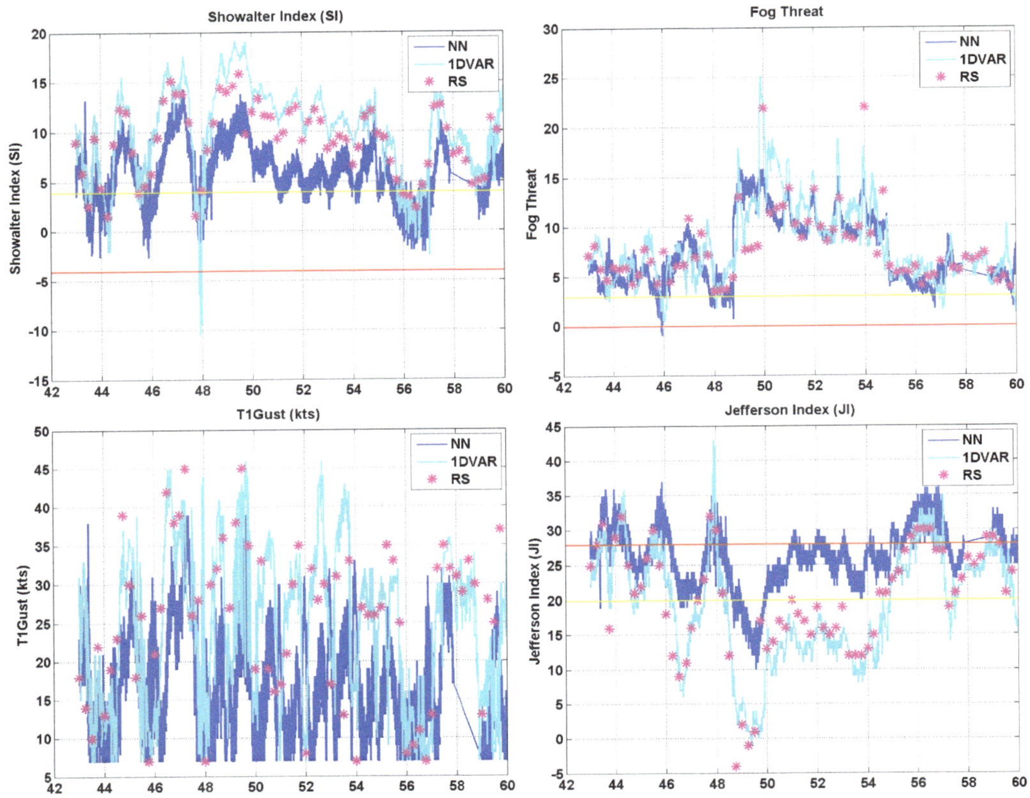

Figure 6. As in Fig. 3 but for (clockwise from top-left panel) Showalter index, fog threat, T1 gust, and Jefferson index. The horizontal yellow and red lines indicate thresholds for moderate and high potential, respectively (where applicable).

ment with the analogous FIs computed from radiosonde profiles. The results are presented for two sites corresponding to different environmental conditions. The MOL-RAO is located at midlatitude (52.17° N, 14.12° E) and low altitude (98 m above sea level (a.s.l.)), in a flat area in central Europe about 50 km to the southeast of Berlin, the capital of Germany. Embedded in this countryside are small and medium-sized lakes. The land use is dominated by forest and agricultural fields. Whistler (Canada) is located at midlatitude (50.09° N, 122.98° W) and relatively high elevation (776 m a.s.l.) in a mountainous environment along the Pacific Ranges of the Coast Mountains in western North America. The quality assessment of the forecast indices in different atmospheric conditions ideally would require data spanning over 1 year and covering all seasons. However, Whistler data include only the 2-week period during the February 2010 Vancouver Winter Olympics. Weather conditions during this time period included multiple days of clear conditions and rain, sleet, and snow days with precipitation rates up to 20 mm h^{-1} (Cimini et al., 2011). Precipitation occurred when southwest winds advected moist maritime air from Vancouver Sound up the Whistler Valley, where it condensed, inflated upper air temperature and humidity, and triggered precipitation (Ware et al., 2013). Conversely, MWRP and radiosonde data in Lindenberg are collected operationally 24/7 throughout the year. However, MWRP 1-DVAR retrievals are only available for the summer 2010 as part of a research project. Lindenberg has a moderate midlatitude climate at the transition between marine and continental influences. During summer, the monthly mean temperature varies between 13 and 18 °C, while the average monthly total precipitation is about 60 mm with 8–9 average days with precipitation per month. Therefore, our analysis is strictly valid for the atmospheric conditions experienced in those two periods.

Two 24 h time series of K index as computed from radiosonde and MWRP profiles in Lindenberg are shown in Fig. 2. These are typical continental summertime cases (12 and 15 August 2010), in which convection starts to develop after the sunrise, generating thunderstorms in the afternoon. As explained in Appendix A, the K index is such that the higher the value, the higher the probability of thunderstorms. In particular, empirical thresholds are often used to separate weak (K index < 25 K) from moderate ($25\text{ K} < K$ index < 35 K) and finally strong thunderstorm potential (K index > 35 K). Although it is beyond the scope of this paper to demonstrate the skills of FIs and the applied thresholds, Fig. 2 shows how increasing values of K index correspond to increasing instability, eventually culminating in a thunder-

Table 2. Statistics of the difference between FIs computed from radiosonde and MWRP retrievals (NNz and 1-DVAR) in Whistler (64 radiosondes). Here are shown average (AVG) FI difference (radiosonde–MWRP), standard deviation (SD), root mean square (rms), and correlation coefficient (COR).

	NNz				1-DVAR			
	AVG	SD	rms	COR	AVG	SD	rms	COR
CAPE ($J\,kg^{-1}$)	−235.67	207.46	326.36	0.57	−49.03	87.34	102.53	0.94
CIN ($J\,kg^{-1}$)	−2.28	21.19	21.33	−0.60	12.50	24.05	27.67	0.95
Fog threat (°C)	0.45	2.37	2.41	0.80	0.45	2.37	2.41	0.80
Jefferson index (°C)	−6.75	5.16	8.54	0.83	1.08	3.33	3.50	0.92
K index (°C)	−10.25	9.23	13.85	0.88	1.05	6.15	6.24	0.91
KO (°C)	5.41	2.28	5.91	0.91	0.60	1.79	1.89	0.94
Lifted index (°C)	5.91	2.36	6.41	0.91	1.76	1.64	2.41	0.96
MDPI ($°C\,kts^{-1}$)	−0.11	0.11	0.16	0.80	0.00	0.11	0.11	0.83
Showalter index (°C)	3.40	1.96	3.95	0.84	−1.10	1.75	2.07	0.92
STT (°C)	−16.13	11.35	19.82	0.81	2.36	7.54	7.91	0.91
T1 gust (kts)	7.71	10.82	13.32	0.24	0.15	7.79	7.79	0.70
Thompson index (°C)	−16.17	10.85	19.58	0.87	−0.73	6.19	6.23	0.94
Total totals (°C)	−7.30	3.94	8.34	0.85	1.16	3.48	3.67	0.91
TQ index (°C)	−5.61	4.41	7.17	0.81	1.07	3.26	3.44	0.90

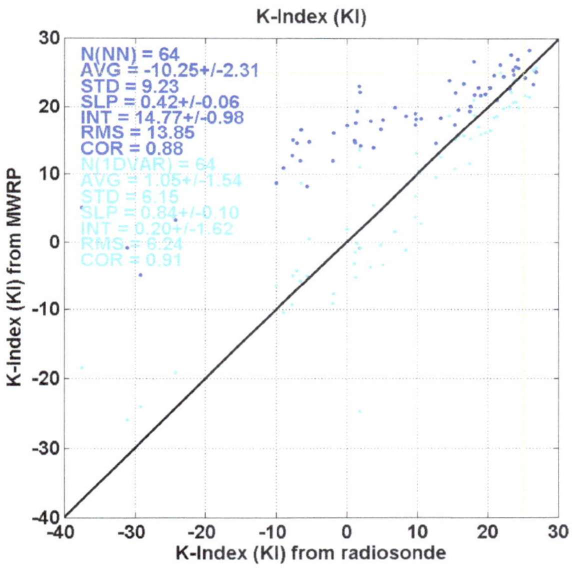

Figure 7. Scatter plot of *K* index values computed from radiosonde and MWRP-retrieved profiles (NNz in blue, 1-DVAR in cyan) and respective statistics: average (AVG), standard deviation (SD), and root-mean-square (rms) differences, correlation coefficient (COR), slope (SLP), and intercept (INT) of a least-square linear fit. AVG, SD, INT, and rms are in kelvin. SLP and COR are unitless. Numbers after the ± sign indicate the 95 % confidence interval.

storm. In fact, we see *K* index values increasing from below the weak-potential threshold in the morning to above the high-potential threshold in the afternoon, and then decaying down in the evening. Multiple thunderstorms were reported near Lindenberg during the afternoon of both days.

These time series suggest that the *K* index computed from MWRP profiles follows quite reasonably the trend given by the radiosondes, launched 4 times a day. Moreover, the value can be appreciated of the nearly continuous (∼ 1 min resolution) *K* index provided by MWRP with respect to the information provided by the radiosondes at a rather coarse time spacing (6 h). For example, considering the trend of *K* index suggested by the radiosondes between 06:00 and 18:00 UTC of 15 August 2010, the thunderstorm potential seems to be decreasing steadily from nearly strong to weak values. By looking at radiosonde *K* index only, a forecaster would have missed the rapid increase in instability after 12:00 UTC indicated by the MWRP, which largely overshoot the strong potential and likely generated the thunderstorm reported right after 16:00 UTC.

Similar considerations apply for other FIs related to convective activity, such as convective available potential energy and convective inhibition (CAPE and CIN, respectively; see Appendix A). Figure 3 reports 24 h time series of CAPE and CIN for the same two cases of Fig. 2. For both cases, the MWRP shows a rapid increase in CAPE between 08:00 and 12:00 UTC, which would have been missed by looking at radiosonde only. On 12 August, CAPE shows a gradual trend of increasing instability until it reaches a peak at 11:00 UTC, i.e. some 2 h before thunderstorm outbreak, and then drops off quickly. This behaviour is quite typical, as reported by Wagner et al. (2008). Correspondingly, low CIN values are reported between 07:00 and 13:00 UTC, while convective inhibition becomes stronger right after the outbreak. On 15 August, the MWRP reports large CAPE values (> 2500 $J\,kg^{-1}$) starting from 11:00 UTC, i.e. more than 5 h before the thunderstorm outbreak. A sharp CAPE drop is detected after

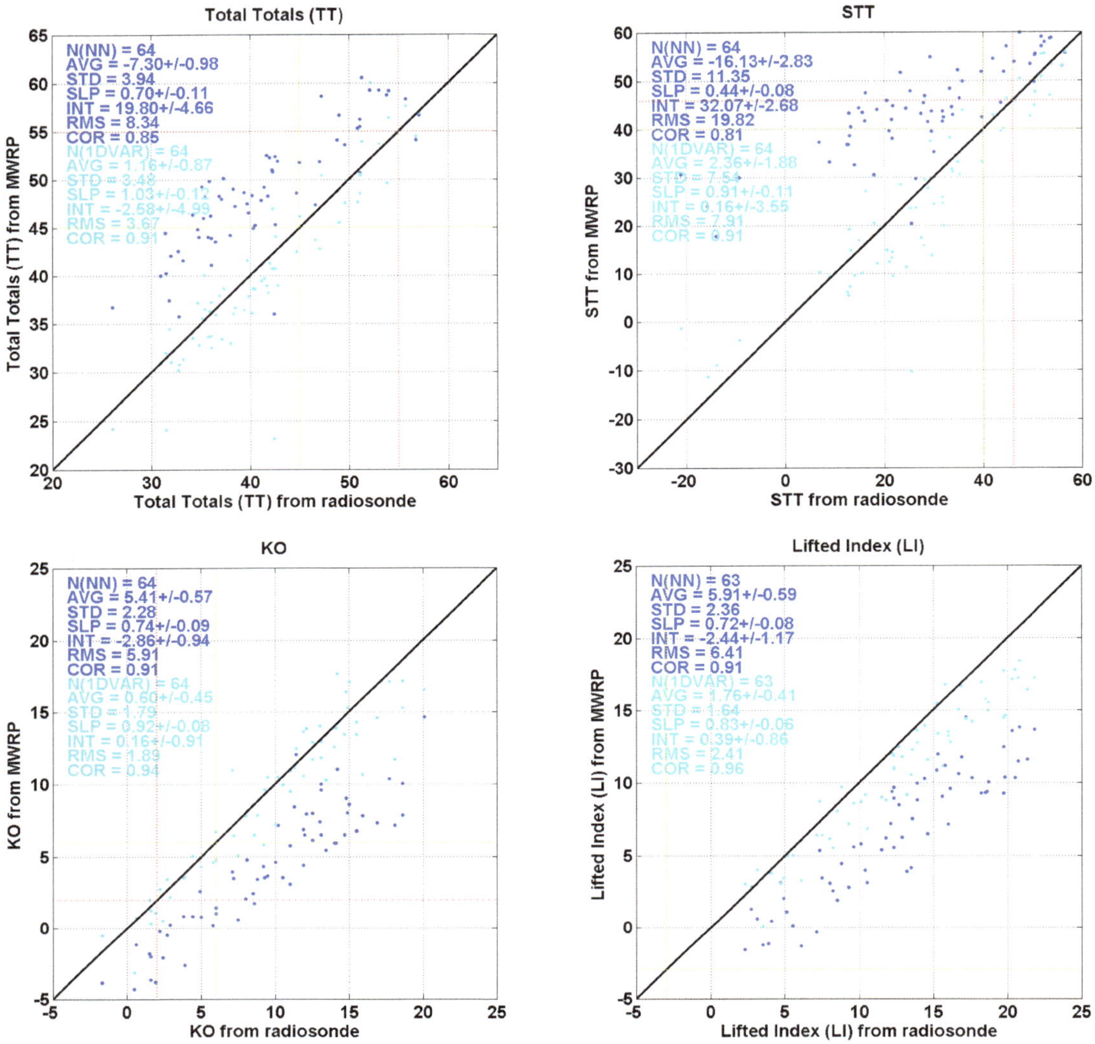

Figure 8. As in Fig. 6 but for (clockwise from top-left panel) total totals, STT, KO, and lifted index. Dotted yellow and red lines indicate thresholds for moderate and high potential, respectively (where applicable).

the outbreak, corresponding to strong convective inhibition (large negative CIN values), indicating the re-establishment of stable conditions for the rest of the day.

For the present analysis, a total of 61 days of MWRP data and 244 radiosonde ascents at 6 h intervals are used for Lindenberg, while a total of 17 days of MWRP data and 68 radiosonde ascents at 6 h intervals are available for Whistler. Figure 4 shows the 17-day time series of K index in Whistler for the period 12–28 February 2010 (Julian day 43–59). In this period the K index computed by radiosondes always reported weak to moderate thunderstorm potential, which is reasonable for Whistler during wintertime. However, the K index computed from MWRP retrievals shows a pick exceeding the high-potential threshold on 16 February (Julian day 47), which was followed by a snowstorm (Cimini et al., 2011; Ware et al., 2013). Other snowstorms were experienced during the periods 12–15 and 24–27 February (Julian day 43–46 and 55–58), in which the K index from both radiosondes and MWRPs occasionally exceeds the moderate-potential threshold. Note that the trend of K index indicated by the radiosonde is followed better by the 1-DVAR than the NNz retrievals. This is especially evident during clear-sky and fair-weather conditions (as for example between Julian day 49 and 55). In fact, these conditions were associated with decreased moisture at 850–700 mb levels and increased static stability between 850 and 500 mb, causing extremely low values of K index. This is a direct consequence of the situation pictured in Fig. 1, which shows lower systematic and random errors for 1-DVAR than for NNz retrievals in the upper air, where the 850–500 mb pressure levels reside. Note also that during clear nocturnal conditions the MWR observations were particularly useful to detect a cold bias in

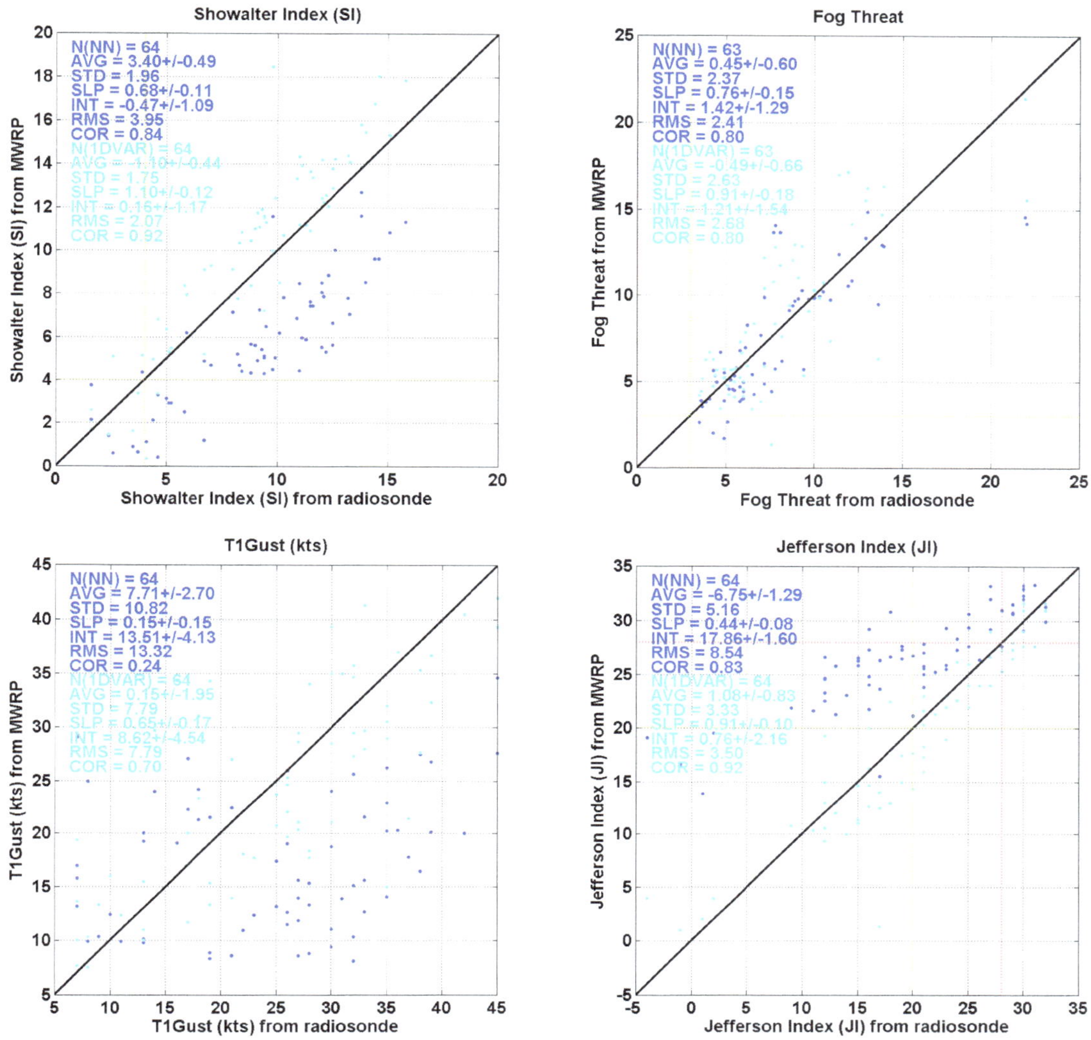

Figure 9. As in Fig. 6 but for (clockwise from top-left panel) Showalter index, fog threat, T1 gust, and Jefferson index. Dotted yellow and red lines indicate thresholds for moderate and high potential, respectively (where applicable).

the analysis boundary layer temperature (Ware et al., 2013), which were used as initial condition for 1-DVAR. This confirmed the large uncertainty associated with gridded analysis in the nocturnal boundary layer (Hart et al., 1998; Hart and Forbes, 1999).

Similarly, Figs. 5 and 6 show the time series for the same period but for other forecast indices, namely the total totals, STT, KO, lifted index, Showalter index, fog threat, T1 gust, and finally Jefferson index. Note that, as detailed in Appendix A, some FIs have an inverted trend; i.e. the lower the value, the higher the potential. For total totals, KO, Showalter index, lifted index, STT, and Jefferson index the same considerations for Fig. 4 apply: the trend indicated by radiosondes is well captured by the MWRP estimates, specially those computed from 1-DVAR retrievals, both for the diurnal and extra-diurnal cycles. More in detail, the lifted index and Showalter index closely resemble the behaviour discussed above for the *K* index, while some other FIs – namely the total totals, KO, STT, and Jefferson index – exceed the high-potential threshold for all the three periods in which snowstorms happened.

For other two of the considered FIs, specifically fog threat and T1 gust, MWRP estimates again follow quite well the trend indicated by radiosondes, but the difference between 1-DVAR and NNz is difficult to appreciate. The fog threat values indicate low potential throughout the period, except for 15 February (Julian day 46), a few hours before a sudden fog event occurred. However, a few other fog/mist events were reported (e.g. Julian days 43, 45, 48, 59), in correspondence to which the fog threat shows values from low to moderate potential. The T1 gust shows maximum wind gust up to 45 knots ($\sim 23\,\mathrm{m\,s^{-1}}$), though reported winds at the radiosonde launching site did not exceed 20 knots ($\sim 10\,\mathrm{m\,s^{-1}}$).

In order to make a quantitative statement on the agreement of the considered FI as computed from radiosonde and MWRP-retrieved profiles, we process the data in order to match the independent data sets and compute statistical scores, such as the average (AVG), SD, and rms differences, the correlation coefficient (COR), and finally the slope (SLP) and intercept (INT) of a least-square linear fit. The statistical scores for K index in Whistler are summarized in Fig. 7. For higher values (K index > 15), it is evident that both NNz and 1-DVAR agree fairly well with the values computed from radiosondes. For lower values the 1-DVAR gives much better agreement than NNz, as anticipated above. Overall, the comparison in Whistler of K index computed from 1-DVAR temperature and humidity retrievals with K index computed from radiosonde profiles shows rms error within 6.1 K (less than 10 % of the total range) with a correlation coefficient better than 0.9. With respect to NNz retrievals, 1-DVAR reduces significantly both the systematic (AVG roughly from 10 to 1 K) and the random (SD by a factor of 1.5, rms by a factor of 2) error components. Note that the results for NNz are still fairly good (0.88 correlation) and tend to agree with 1-DVAR for higher values, where the K index is more interesting for forecasting purposes.

Similarly, Figs. 8 and 9 show the scatter plots and statistical scores for the other forecast indices. For Jefferson index, total totals, KO, Showalter index, lifted index, and STT similar considerations as for K index apply; i.e. the comparison between FIs from radiosonde and MWRP retrievals results in (a) a significant improvement of the statistical scores for 1-DVAR with respect to NNz, and (b) correlation coefficients better than 0.9 and rms within 10 % of the range (considering 1-DVAR). As for the K index, results for NNz are still fairly good (correlation between 0.8 and 0.9 depending on FI) and tend to agree with 1-DVAR in the most significant range. For the fog threat and T1 gust, the correlation coefficients are much lower (respectively 0.8 and 0.7, considering 1-DVAR) and the rms exceeds 10 % of the range (respectively 12 and 19 %, considering 1-DVAR). The improvement brought by 1-DVAR with respect to NNz is significant for T1 gust (a factor of 1.5 for SD and rms, a factor ~ 3 for correlation), while it becomes negligible for the fog threat. The latter result can be explained by looking at the similar results for NNz and 1-DVAR in Fig. 1 for the lower levels, which mainly determine the value of the fog threat index.

The statistics of the difference between FIs computed from radiosonde and MWRP retrievals in Whistler are summarized in Table 2. In addition to the FIs discussed above, Table 2 shows the results for five more FIs, namely the CAPE, the CIN, the microburst day potential index (MDPI), the Thompson index, and finally the TQ index. Note that CAPE and CIN are not always determined by definition, and thus the results are obtained from a smaller data set. For CAPE, the 1-DVAR retrievals show 0.94 correlation coefficient and $102.5\,\mathrm{J\,kg^{-1}}$ rms, with an improvement factor of 1.6 and 4.8, respectively, with respect to NNz. For CIN, the 1-DVAR retrievals show 0.95 correlation coefficient and $12.50\,\mathrm{J\,kg^{-1}}$ rms, while NNz shows negative correlation. For MDPI, results are similar for NNz and 1-DVAR retrievals, showing 0.80–0.83 correlation coefficient, respectively. For Thompson index and TQ index, 1-DVAR retrievals show correlation coefficient exceeding 0.9 and rms within 10 % of the total range, somewhat better than the NNz retrievals. Thus, similarly to other FIs, CAPE, CIN, MDPI, Thompson index, and TQ index show good correlation between radiosondes and MWRP estimates (either NNz or 1-DVAR), with a tangible improvement for 1-DVAR with respect to NNz.

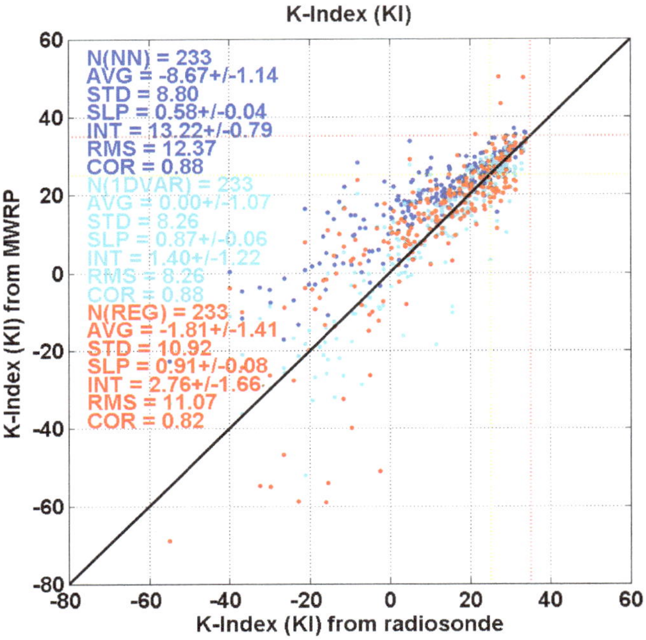

Figure 10. Scatter plot of K index values computed from radiosonde and MWRP-retrieved profiles (NNz in blue, 1-DVAR in cyan, REG in red) and respective statistics: average (AVG), standard deviation (SD), and root-mean-square (rms) differences, correlation coefficient (COR), slope (SLP) and intercept (INT) of a least-square linear fit. AVG, SD, INT, and rms are in kelvin. SLP and COR are unitless. Numbers after the ± sign indicate the 95 % confidence interval.

Similarly, Figs. 10–12 and Table 3 show the results for the analysis at the other site, MOL-RAO in Lindenberg. At MOL-RAO, four types of MWRP retrievals are available, as already seen in Fig. 1. FIs are computed from three of those MWRP retrievals (NNz, 1-DVAR, and ObsREG), and then compared to the FIs computed from radiosondes. Figure 10 shows the results for KI, for which the three MWRP retrievals agree fairly well, with 0.81–0.88 (ObsREG and NNz/1-DVAR, respectively) correlation coefficient with respect to radiosonde values. Figures 11 and 12 show the results for total totals, STT, KO, lifted index, Showalter in-

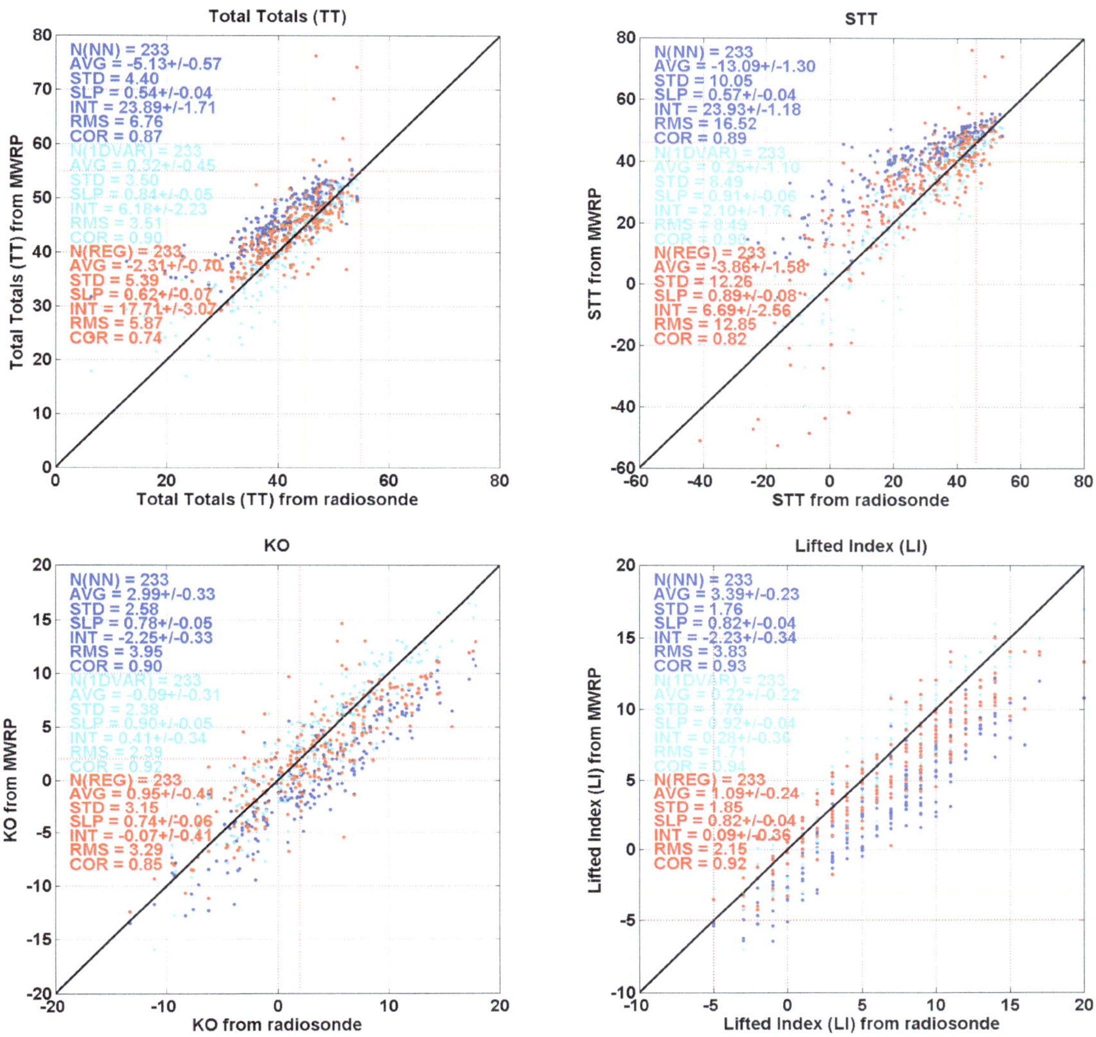

Figure 11. As in Fig. 9 but for (clockwise from top-left panel) total totals, STT, KO, lifted index. Dotted yellow and red lines indicate thresholds for moderate and high potential, respectively (where applicable).

dex, fog threat, T1 gust, and Jefferson index. These results are consistent with the ones in Whistler for the most part. In terms of correlation coefficient, 1-DVAR retrievals usually outperform the other two, though the differences are less significant than in Whistler. However, 1-DVAR shows the lowest correlation of the three retrievals for two FIs (CAPE and fog threat), while it shows lower correlation than NNz for the TQ index. Note that CAPE and CIN show lower correlation than other convective FIs. This may be caused by the different (smaller) data set available for computing the statistics and/or to the MWRP moderate vertical resolution causing uncertainties in the location of the integral boundaries.

Finally, Table 4 presents a summary of the comparison of all FIs as computed from radiosondes and MWRPs for the three retrieval methods and the two data sets considered in this analysis. As a common normalized statistical measure we adopt the normalized rms (NRMS), i.e. rms divided by the range of observed values. NRMS is deemed appropriate to compare all the FIs as it is little sensitive to the occurrence of 0 values (often occurring for CAPE, CIN, *K* index, KO, etcetera). Table 4 shows that NRMS is generally smaller for Lindenberg than for Whistler. In Whistler, NRMS ranges from 8 to 35 % for all FIs but CAPE and CIN, though the latter are affected by a much lower sample size. In Lindenberg, NRMS ranges from 7 to 22 %, depending upon FI (except for CIN with ObsREG). Overall, 1-DVAR shows the best agreement, with NRMS ranging from 7 to 17 %.

5 Summary and conclusions

Forecast indices based on radiosonde soundings have been developed over many decades as local weather prediction

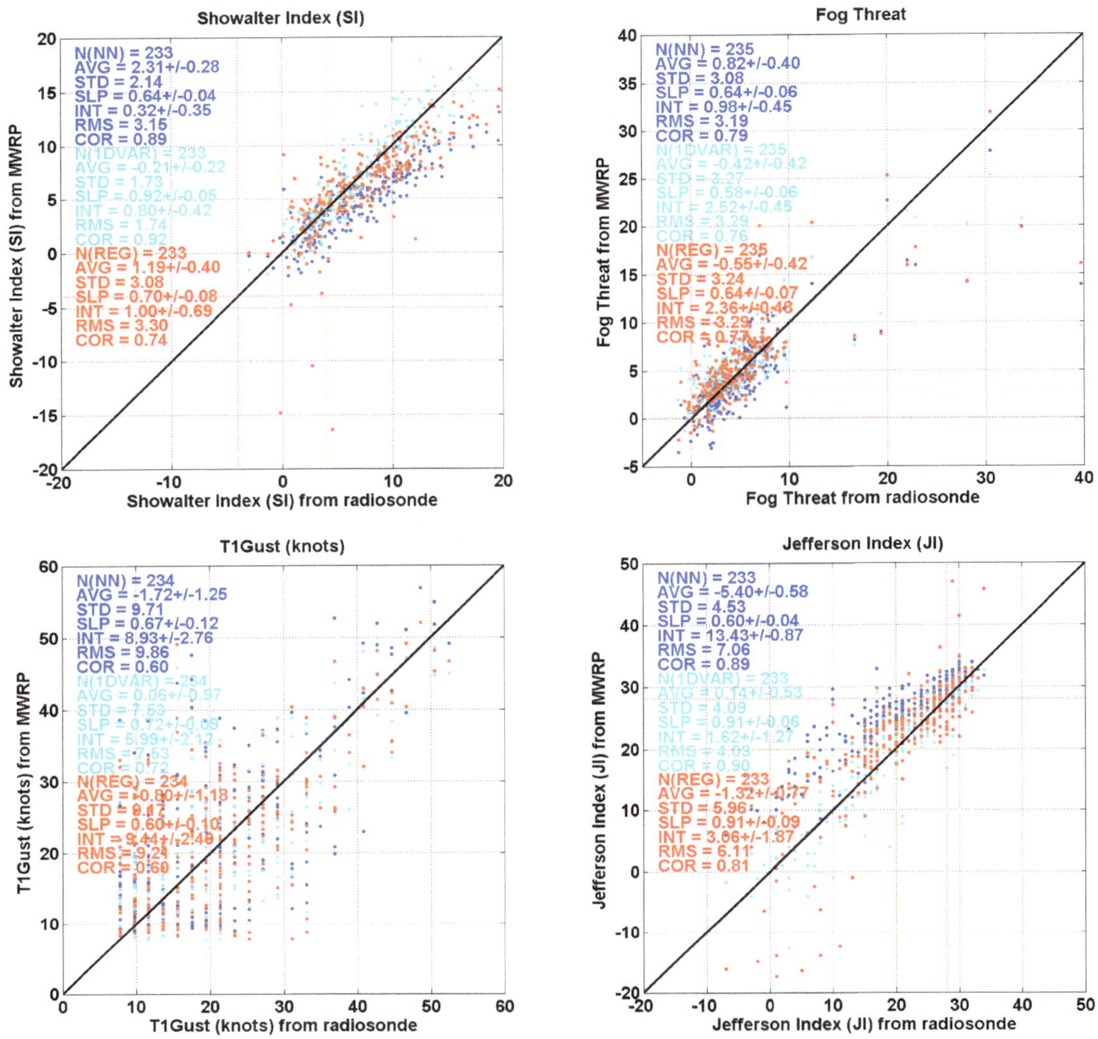

Figure 12. As in Fig. 9 but for (clockwise from top-left panel) Showalter index, fog threat, T1 gust, and Jefferson index. Dotted yellow and red lines indicate thresholds for moderate and high potential, respectively (where applicable).

tools. However, FIs typically lose their value during 6-hour and longer intervals between traditional radiosonde soundings. Continuous tropospheric thermodynamic profiles can be retrieved on a minute timescale from a ground-based MWRP working in the 20–60 GHz range. FIs developed for radiosonde TPU profiles can be derived from the MWRP thermodynamic profiles. The present analysis demonstrates good agreement between FIs derived from MWRP and radiosonde soundings. The analysis is performed for two mid-latitude sites: one residing in a low-elevation, flat rural area in central Europe (Lindenberg, Germany) and the other in a mountainous environment along the Pacific Ranges of the Coast Mountains in western North America (Whistler, Canada).

Though it would be ideal to extend the analysis presented here to a larger data set, possibly covering all different seasons, our analysis revealed that

- there is good agreement between MWRP- and radiosonde-derived FIs at both sites, with correlation coefficients usually exceeding 0.8;

- FIs derived from 1-DVAR retrievals usually outperform neural network and observation-based regression retrievals in terms of correlation, mean, and random difference with respect to FI values derived from radiosondes;

- FI time series derived from MWRP retrievals provide promising new tools for local high-impact weather prediction (Madhulatha et al., 2013) based on uninterrupted surveillance of local tropospheric thermodynamics in all weather conditions (Cimini et al., 2011; Ware et al., 2013; Xu et al., 2014), capturing the entire diurnal cycle and providing fresh and timely data to forecasters.

Table 3. As in Table 2 but for Lindenberg (233 radiosondes). Average (AVG) FI difference (radiosonde–MWRP), standard deviation (SD), and correlation coefficient (COR) are shown.

	NNz			1-DVAR			ObsREG		
	AVG	SD	COR	AVG	SD	COR	AVG	SD	COR
CAPE ($J\,kg^{-1}$)	−217.39	306.68	0.72	−3.38	215.39	0.76	24.58	224.78	0.82
CIN ($J\,kg^{-1}$)	−17.46	46.32	0.74	−8.23	40.80	0.73	19.97	165.17	0.21
Fog threat (°C)	0.82	3.08	0.79	−0.42	3.27	0.76	−0.55	3.24	0.77
Jefferson index (°C)	−5.40	4.53	0.89	0.14	4.09	0.90	−1.32	5.96	0.81
K index (°C)	−8.67	8.80	0.88	0.00	8.26	0.88	−1.81	10.92	0.82
KO (°C)	2.99	2.58	0.90	−0.09	2.38	0.92	0.95	3.15	0.85
Lifted index (°C)	3.39	1.76	0.93	0.22	1.70	0.94	1.09	1.85	0.92
MDPI (°C kts^{-1})	−0.04	0.14	0.83	0.03	0.11	0.90	0.04	0.19	0.68
Showalter index (°C)	2.31	2.14	0.89	−0.21	1.73	0.92	1.19	3.08	0.74
STT (°C)	−13.09	10.05	0.89	0.25	8.49	0.90	−3.86	12.26	0.82
T1 gust (kts)	−1.72	9.71	0.60	0.06	7.53	0.72	−0.80	9.17	0.60
Thompson index (°C)	−12.04	9.35	0.91	−0.20	8.60	0.90	−2.87	11.27	0.85
Total totals (°C)	−5.13	4.40	0.87	0.32	3.50	0.90	−2.31	5.39	0.74
TQ index (°C)	−4.07	3.96	0.87	0.42	3.83	0.85	−0.93	4.35	0.80

Table 4. Summary of normalized rms (NRMS) difference (i.e. rms divided by the range of observed values) between the FIs computed from radiosonde and MWRP retrievals, for all the retrieval methods and data sets given in Tables 2–3. NRMS is expressed in percentage values.

	NNz		1-DVAR		ObsREG
	Lindenberg	Whistler	Lindenberg	Whistler	Lindenberg
CAPE	15.20	135.42	8.70	42.54	9.13
CIN	13.08	64.44	10.96	83.86	43.78
Fog threat	7.77	13.03	8.03	14.47	8.02
Jefferson index	17.22	23.74	9.99	9.73	14.90
K index	13.92	21.55	9.30	9.71	12.46
KO	12.71	27.11	7.67	8.67	10.58
Lifted index	15.30	32.87	6.84	12.36	8.59
MDPI	10.87	17.76	8.75	11.85	14.76
Showalter index	13.89	27.83	7.66	14.59	14.55
STT	17.37	25.51	8.93	10.18	13.51
T1 gust	22.05	35.07	16.85	20.51	20.59
Thompson index	13.51	25.77	7.62	8.20	10.29
Total totals	14.12	26.91	7.33	11.84	12.25
TQ index	12.67	24.74	8.58	11.86	9.90

Therefore, we conclude that MWRP retrievals are able to deliver valuable FIs, with the certain advantage (with respect to radiosondes) of nearly continuous updates. FI time series are promising new tools for high-impact local weather forecasting, complementing and augmenting 6/12 h indices derived from radiosonde soundings.

Appendix A: Definitions of forecast indices

In this appendix we summarize the definitions of the forecast indices considered in our analysis. In the following definitions, T_{xxx} is the temperature at the pressure level xxx (in mb), Td_{xxx} is dew point temperature, θe_{xxx} is the equivalent potential temperature, and θ_{WBxxx} is the wet bulb potential temperature at the same pressure level. The forecast index definitions below were extracted from review papers (Peppier, 1998; Haklander and Van Delden, 2003; de Coning et al., 2011) as well as from American Meteorological Society glossary (AMS, 2013) and RAOB software manual. According to Haklander and Van Delden (2003), the various forecast indices are a combination of three types, the first accounting for pure conditional instability, the second accounting for pure latent instability, and the third accounting for pure potential instability of certain atmospheric layers. In our analysis, we have considered at least one for each type. All temperatures are expressed in °C, unless otherwise indicated.

A1 CAPE

The convective available potential energy (CAPE) defines the maximum energy available to an ascending air parcel, and it is often used to indicate instability and the possibility of thunderstorms. CAPE is also referred to as positive buoyancy. There exist many different definitions of CAPE in the literature (Haklander and Van Delden, 2003). Here, CAPE is calculated by integrating vertically the local buoyancy of a parcel (expressed introducing the virtual temperature of the parcel Tv_p and that of the environment Tv_e) from the level of free convection (LFC) to the equilibrium level (EL):

$$\mathrm{CAPE} = -R_d \int_{\mathrm{LFC}}^{\mathrm{EL}} (\mathrm{Tv_p} - \mathrm{Tv_e})\mathrm{d}(\ln p),$$

where $R_d = 287.05\,\mathrm{J\,kg^{-1}\,K^{-1}}$ is the gas constant for dry air and CAPE is measured in joules per kilogram of air ($\mathrm{J\,kg^{-1}}$). Any value greater than $0\,\mathrm{J\,kg^{-1}}$ indicates instability and the possibility of thunderstorms. CAPE values between 1000 and $2500\,\mathrm{J\,kg^{-1}}$ typically qualify as moderate. Note that CAPE is defined only if atmospheric conditions allow LFC and EL to exist.

A2 CIN

The convective inhibition (CIN) indicates the amount of energy required to overcome the negatively buoyant energy the environment exerts on an air parcel. Conceptually, CIN is the opposite of CAPE (see above) and is also referred to as negative buoyancy. Here, CIN is calculated from the surface (Srf) to the equilibrium level:

$$\mathrm{CIN} = -R_d \int_{\mathrm{Srf}}^{\mathrm{EL}} (\mathrm{Tv_p} - \mathrm{Tv_e})\mathrm{d}(\ln p).$$

CIN is expressed as a negative energy value. Typically, high absolute CIN values correspond to stable conditions and indicate little likelihood of thunderstorm development. In general, CIN values between 0 and $-50\,\mathrm{J\,kg^{-1}}$ indicate weak convective inhibition. CIN values of $-100\,\mathrm{J\,kg^{-1}}$ or less indicate strong convective inhibition.

A3 Fog threat

The fog threat (FT) index indicates the potential for radiation fog and is based on the definition of fog point (FP), which is the temperature at which radiation fog will form. The fog point is determined by following the saturation mixing ratio line from the dew point curve at the lifting condensation level (LCL) to the surface temperature. Then, the fog threat is given by the difference of the wet bulb potential temperature at 850 mb θ_{WB850} and the fog point computed as above:

$$\mathrm{FT} = \theta_{WB850} - \mathrm{FP}.$$

The potential for radiation fog is low for $\mathrm{FT} > 3$, while it becomes high for $\mathrm{FT} < 0$.

A4 Jefferson index

The Jefferson index (JI) was designed originally for maritime and arid areas (Haklander and Van Delden 2003, and references therein). JI is defined as

$$\mathrm{JI} = 1.6 \times \theta_{WB850} - T_{500} - 0.5 \times (T_{700} - \mathrm{Td}_{700}) - 8.$$

Non-frontal thunderstorms can be expected for values above 28. Significant showers with thunderstorm are expected for values above 30.

A5 *K* index

The *K* index is due to George (1960) and is defined by

$$\mathrm{KI} = (T_{850} - T_{500}) + \mathrm{Td}_{850} - (T_{700} - \mathrm{Td}_{700}).$$

The first term is the lapse rate, while the second and third are related to the moisture between 850 and 700 mb, and are strongly influenced by the temperature–dewpoint spread at the 700 mb level. The *K* index increases with decreasing static stability between 850 and 500 hPa, increasing moisture at 850 hPa, and increasing relative humidity at 700 hPa. The higher the *K* index, the higher the probability of thunderstorms. As *K* index increases from a value of 20 or so, the likelihood of showers and thunderstorms is expected to increase. The *K* index was developed for forecasting air mass thunderstorms, particularly useful in predicting non-frontal thunderstorm situations.

A6 KO index

KO index was developed for estimating thunderstorm potential in Europe (Andersson et al., 1989). The KO index describes the potential instability between lower and higher levels of the atmosphere and is thus based on the equivalent potential temperature θe as

$$\mathrm{KO} = ((\theta e_{500} + \theta e_{700}) - (\theta e_{850} + \theta e_{1000}))/2.$$

If the surface level is above 1000 mb, then θe_{srf} is used instead of θe_{1000}. The KO is more sensitive to moisture than other stability indices and is best used in cooler, moist climates. The thunderstorm potential increases as the KO index decreases. Values smaller than 2 generally indicate strong thunderstorm potential.

A7 Lifted index

The lifted index (LI), developed by Galway (1956), is nominally identical to the Showalter index, except for the determination of the level from which the parcel is lifted.

$$\mathrm{LI} = T_{500} - T_{\mathrm{L}}$$

The LI was defined by lifting the parcel adiabatically from the midpoint of the surface layer to 500 mb, where its temperature, considered the updraft temperature within a developing cloud, was compared to that of the environment. The parcel being lifted is defined by the dry adiabat running through the predicted surface afternoon temperature maximum and the mean mixing ratio in the lowest 900 m. The values of this index tend to be somewhat lower than those of the Showalter index.

A8 Microburst day potential index (MDPI)

The microburst day potential index (MDPI) was developed in part on the results from the Microburst and Severe Thunderstorm (MIST) project (Atkins and Wakimoto, 1991). MDPI is based on the vertical profiles of equivalent potential temperature and is defined follows:

$$\begin{aligned}\mathrm{MDPI} =&(\mathrm{Max}\theta e(\mathrm{Sfc} - 850\ \mathrm{mb}) - \\ &\mathrm{Min}\theta e(660 - < 500\ \mathrm{mb})/30\ \mathrm{kts}.\end{aligned}$$

MDPI was designed to determine likely and unlikely environments for downbursts. For MDPI values greater than or equal to 1, microbursts are likely.

A9 Showalter index

The Showalter index (SI) was designed originally for thunderstorm forecasting in the southwestern US (Showalter, 1947). It estimates the potential instability of the 850–500 mb layer by measuring the buoyancy at 500 mb of an air parcel lifted to that level. Thus, it is defined as

$$\mathrm{SI} = T_{500} - T_{\mathrm{L}},$$

where T_{L} is the temperature (° C) of a parcel lifted from 850 to 500 mb, dry-adiabatically to saturation and moist-adiabatically above that. As the index decreases to 0 and below, the likelihood of showers and thunderstorms is considered to increase. SI values $\leq +3$ are indicative of possible thunderstorm activity, while values ≤ -3 are associated with severe convective activity. The SI has been one of the most frequently applied stability indices.

A10 *S* index (STT)

The *S* index (STT) was introduced by the German Military Geophysical Office as an improvement on the total totals index (Haklander and Van Delden, 2003, and references therein) including a variable parameter A based on vertical totals (VT). The *S* index is defined as

$$\begin{aligned}\mathrm{STT} =&\mathrm{TT} - (T_{700} + \mathrm{Td}_{700}) - A = T_{850} + \mathrm{Td}_{850} \\ &- 2 \times T_{500} - (T_{700} - \mathrm{Td}_{700}) - A,\end{aligned}$$

where TT indicates total totals index (see below) and A depends on VT such that it penalizes cases with low values of VT (i.e. low vertical temperature gradients):

$$A = 0\ (\text{if VT} > 25);\ 2\ (\text{if } 22 <= \text{VT} <= 25);\ 6(\text{if VT} < 22).$$

Note that the *S* index takes the same variables into account as the *K* index, but in other proportions, and thus it can also be written as

$$\mathrm{STT} = \mathrm{KI} - T_{500} - A.$$

A11 T1 gust

The T1 gust (also called dry stability index) method was developed with the intent of giving forecasters a way to estimate maximum thunderstorm wind gusts (Fawbush and Miller, 1954). The T1 method is still used to predict the maximum wind gust in air mass thunderstorms, relying on evidence that most convective wind gusts at the surface appear to result from downdraft air originating in the lower portion of a thunderstorm. Thus, the T1 method is computed as the difference between the temperature of a parcel of moist surface air raised moist-adiabatically to 600 mb (T_{MA600}) and the observed dry bulb temperature at the 600 mb level.

$$\mathrm{T1} = T_{\mathrm{MA600}} - T_{600} \tag{A1}$$

In the case of a temperature inversion within 150–200 hPa above the surface, the moist adiabat is followed from the warmest point in the inversion to 600 mb (Miller, 1972). Finally, the speed of maximum wind gust (in knots) is given by the empirical formula $\mathrm{T1gust} = 6.737 + 3.503 \times \mathrm{T1} - 0.0463 \times \mathrm{T1}^2$.

A12 Thompson index

The Thompson index (TI) is primarily used to determine thunderstorm potential in the Rocky Mountains. It is defined as

$$\mathrm{TI} = \mathrm{KI} - \mathrm{LI}.$$

The Thompson index should be an improvement of KI, since KI neglects latent instability below 850 hPa. Thunderstorm potential is considered weak for values below 30, while it becomes strong for values greater than 35.

A13 Total totals

The total totals (TT) index is attributable to Miller (1972) and is defined by the combination of the vertical totals ($VT = T_{850} T_{500}$) and the cross totals ($CT = Td_{850} - T_{500}$), resulting in

$$TT = T_{850} + Td_{850} - 2 \times T_{500}.$$

The TT index is commonly used as a severe weather indicator. The higher the number, the more unstable the atmosphere. Values lower than 45° C generally indicate weak thunderstorm potential, while values larger than 55° C indicate high possibility of severe thunderstorms, though value interpretation varies with season and location.

A14 TQ index

The TQ index (TQ) is used to assess the potential for low-topped convection. It is defined as

$$TQ = (T_{850} + Td_{850}) - 1.7 \times (T_{700}).$$

Values larger than 12 indicate unstable lower troposphere, where thunderstorm and rainfall are possible outside of stratiform clouds; while for values larger than 17, thunderstorm and rainfall are possible in the presence of stratiform clouds.

Acknowledgements. Part of this work was stimulated through the EU COST Actions ES0702 EG-CLIMET and ES1303 TOPROF. Environment Canada and the German Weather Service (Deutscher Wetterdienst) conducted microwave profiler and radiosonde data collection in Whistler and Lindenberg, respectively, and provided grant support for data analysis.

Edited by: U. Wandinger

References

American Meteorological Society Glossary of Meteorology (AMS): available at: http://glossary.ametsoc.org/wiki/Main_Page, last access: 6 November 2013.

Andersson, T., Andersson, M., Jacobsson, C., and Nilsson, S.: Thermodynamic indices for forecasting thunderstorms in southern Sweden, Meteorol. Mag., 116, 141–146, 1989.

Atkins, N. T. and Wakimoto, R. M.: Wet microburst activity over the southeastern United States: Implications for Forecasting, Weather Forecast., 6, 470–482, 1991.

Cadeddu, M. P., Liljegren, J. C., and Turner, D. D.: The Atmospheric radiation measurement (ARM) program network of microwave radiometers: instrumentation, data, and retrievals, Atmos. Meas. Tech., 6, 2359–2372, doi:10.5194/amt-6-2359-2013, 2013.

Chan, P. W.: Performance and application of a multiwavelength, ground-based microwave radiometer in intense convective weather, Meteorol. Z., 18, 253–265, 2009.

Chan, P. W. and Hon, K. K.: Application of ground-based, multichannel microwave radiometer in the nowcasting of intense convective weather through instability indices of the atmosphere, Meteorol. Z., 20, 431–440, 2011.

Chan, P. W. and Lee, Y. F.: Application of a ground-based, multichannel microwave radiometer to the alerting of low-level windshear at an airport, Meteorol. Z., 20, 423–429, 2011.

Cimini, D., Westwater, E. R., Han, Y., and Keihm, S. J.: Accuracy of Ground-based Microwave Radiometer and Balloon-Borne Measurements During WVIOP2000 Field Experiment, IEEE Trans. Geosci. Remote Sens., 41, 2605–2615, 2003.

Cimini, D., Hewison, T. J., Martin, L., Güldner, J., Gaffard, C., and Marzano, F.: Temperature and humidity profile retrievals from groundbased microwave radiometers during TUC, Meteorol. Z., 15, 45–56, 2006.

Cimini, D., Westwater, E. R., and Gasiewski, A. J.: Temperature and humidity profiling in the Arctic using millimeter-wave radiometry and 1-DVAR, IEEE Trans. Geosci. Remote Sens., 48, 1381–1388, doi:10.1109/TGRS.2009.2030500, 2010.

Cimini, D., Campos, E., Ware, R., Albers, S., Giuliani, G., Oreamuno, J., Joe, P., Koch, S., Cober, S., and Westwater, E.: Thermodynamic Atmospheric Profiling during the 2010 Winter Olympics Using Ground-based Microwave Radiometry, IEEE Trans. Geosci. Remote Sens., 49, 4959–4969, doi:10.1109/TGRS.2011.2154337, 2011.

Cimini, D., Caumont, O., Löhnert, U., Alados-Arboledas, L., Bleisch, R., Huet, T., Ferrario, M. E., Madonna, F., Haefele, A., Nasir, F., Pace, G., and Posada, R.: A data assimilation experiment of temperature and humidity profiles from an international network of ground-based microwave radiometers, Proc. Microrad 2014, Pasadena, USA, 24–27 March 2014.

de Coning, E., Koenig, M., and Olivier, J.: The combined instability index: a new very-short range convection forecasting technique for southern Africa, Meteorol. Appl., 18, 421–439, doi:10.1002/met.234, 2011.

Doswell III, C. A. and Schultz, D. M.: On the Use of Indices and Parameters in Forecasting Severe Storms, Electronic Journal of Severe Storms Meteorology, 1, 1–22, available at: http://www.ejssm.org/ojs/index.php/ejssm/article/view/11/10, 2006.

Fawbush, E. J. and Miller, R. C.: A Basis for Forecasting Peak Wind Gusts in Non-Frontal Thunderstorms, Bull. Am. Meteorol. Soc., 35, 14–19, 1954.

Feltz, W. F. and Mecikalski, J. R.: Monitoring High-Temporal-Resolution Convective Stability Indices Using the Ground-Based Atmospheric Emitted Radiance Interferometer (AERI) during the 3 May 1999 Oklahoma–Kansas Tornado Outbreak, Weather Forecast., 17, 445–455, 2002.

Friedrich, K., Lundquist, J. K., Aitken, M., Kalina, E. A., and Marshall, R. F.: Stability and turbulence in the atmospheric boundary layer: A comparison of remote sensing and tower observations, Geophys. Res. Lett., 39, L03801, doi:10.1029/2011GL050413, 2012.

Galway, J. G.: The lifted index as a predictor of latent instability, Bull. Am. Meteorol. Soc., 37, 528–529, 1956.

George, J. J.: Weather Forecasting for Aeronautics. Academic Press, New York, 673 pp., 1960.

Güldner, J. and Spänkuch, D.: Remote sensing of the thermodynamic state of the atmospheric boundary layer by ground-based microwave radiometry, J. Atmos. Oceanic Technol., 18, 925–933, 2001.

Haklander, A. J. and Van Delden, A.: Thunderstorm predictors and their forecast skill for the Netherlands, Atmos. Res., 67–68, 273–299, doi:10.1016/S0169-8095(03)00056-5, 2003.

Hart, R. E. and Forbes, G. S.: The Use of Hourly Model-Generated Soundings to Forecast Mesoscale Phenomena, Part II: Initial Assessment in Forecasting Nonconvective Strong Wind Gusts, Weather Forecast., 14, 461–469, 1999.

Hart, R. E., Forbes, G. S., and Grumm, R. H.: The Use of Hourly Model-Generated Soundings to Forecast Mesoscale Phenomena. Part I: Initial assessment in forecasting warm-season phenomena, Weather Forecast., 13, 1165–1185, 1998.

Hewison, T.: 1D-VAR retrievals of temperature and humidity profiles from a ground-based microwave radiometer, IEEE Trans. Geosci. Remote Sens., 45, 2163–2168, 2007.

Holtslag, M. C., Steeneveld, G. J., and Holtslag, A. A. M.: Fog forecasting: "old fashioned" semi-empirical methods from radio sounding observations versus "modern" numerical models, Proc. 5th Int. Conf. Fog and Dew (FOGDEW2010), available at: http://meetingorganizer.copernicus.org/FOGDEW2010/FOGDEW2010-69.pdf (last access: 8 November 2013), Münster, Germany, 25–30 July 2010.

Knupp, K., Ware, R., Cimini, D., Vandenberghe, F., Vivekanandan, J., Westwater, E. R., and Coleman, T.: Ground-based passive microwave profiling during dynamic weather conditions, J. Atmos. Ocean. Technol., 26, 1057–1073, doi:10.1175/2008JTECHA1150.1, 2009.

König, M.: Atmospheric Instability Parameters Derived from MSG SEVIRI Observations, EUMETSAT Technical Memorandum No. 9, January 2002.

Koenig, M. and de Coning, E.: The MSG Global Instability Indices product and its use as a nowcasting tool, Weather Forecast., 24, 272–285, 2009.

Kuhlman, C. J.: Evaluation of Convective Wind Forecasting Methods During High Wind Events, Master's Thesis, Naval Postgraduate School, Monterey, California, March 2006.

Löhnert, U. and Maier, O.: Operational profiling of temperature using ground-based microwave radiometry at Payerne: prospects and challenges, Atmos. Meas. Tech., 5, 1121–1134, doi:10.5194/amt-5-1121-2012, 2012.

Löhnert, U., Crewell, S., and Simmer, C.: An integrated approach towards retrieving physically consistent profiles of temperature, humidity, and cloud liquid water, J. Appl. Meteorol., 43, 1295–1307, 2004.

Madhulatha, A., Rajeevan, M., Venkat Ratnam, M., Bhate, J., and Naidu, C. V.: Nowcasting severe convective activity over southeast India using ground-based microwave radiometer observations, J. Geophys. Res. Atmos., 118, 1–13, doi:10.1029/2012JD018174, 2013.

Mailhot, J., Belair, S., Charron, M., Doyle, C., Joe, P., Abrahamowicz, M., Bernier, N. B., Denis, B., Erfani, A., Frenette, R., Giguere, A., Isaac, G. A., McLennan, N., McTaggart-Cowan, R., Milbrandt, J., and Tong, L.: Environment Canada's experimental numerical weather prediction systems for the Vancouver 2010 Winter Olympic and Paralympic games, Bull. Amer. Meteorol. Soc., 11, 1073–1085, 2010.

Maschwitz, G., Löhnert, U., Crewell, S., Rose, T., and Turner, D. D.: Investigation of ground-based microwave radiometer calibration techniques at 530 hPa, Atmos. Meas. Tech., 6, 2641–2658, doi:10.5194/amt-6-2641-2013, 2013.

Miller, R. C.: Notes on analysis and severe storm forecasting procedures of the Air Force Global Weather Central. Technical Report 200(R), AWS, USAF, Scott AFB, IL, 1972.

National Research Council Committee on Developing Mesoscale Meteorological Observational Capabilities to Meet Multiple Needs (NRC): Observing Weather and Climate from the Ground Up: A Nationwide Network of Networks, ISBN: 978-0-309-12986-2, 250 pp., 2008.

National Research Council Committee on Progress and Priorities of U.S. Weather (NRC): Research and Research-to-Operations Activities, When Weather Matters: Science and Service to Meet Critical Societal Needs, ISBN: 978-0-309-15249-5, 198 pp., 2010.

Peppier, R. A.: A Review of Static Stability Indices and Related Thermodynamic Parameters, Illinois State Water Survey Division, SWS Miscellaneous Publication 104, available at: http://www.isws.illinois.edu/pubdoc/MP/ISWSMP-104.pdf (last access: 7 November 2013), October, 1988.

Showalter, A. K.: A stability index for forecasting thunderstorms, Bull. Am. Meteorol. Soc., 34, 250–252, 1947.

Solheim, F., Godwin, J., Westwater, E., Han, Y., Keihm, S., Marsh, K., and Ware, R.: Radiometric profiling of temperature, water vapor, and cloud liquid water using various inversion methods, Rad. Sci., 33, 393–404, 1998.

Vaisala: Radiosonde RS92-SGP, Ref. B210358EN-F Vaisala 2013, available at: http://www.vaisala.com/VaisalaDocuments/BrochuresandDatasheets/RS92SGP-Datasheet-B210358EN-F-LOW.pdf, last access: 6 November 2013.

Venkat Ratnam, M., Durga Santhi, Y., Rajeevan, M., and Vijaya Bhaskara Rao, S.: Diurnal variability of stability indices observed using radiosonde observations over a tropical station: Comparison with microwave radiometer measurements, Atmos. Res., 124, 21–33, doi:10.1016/j.atmosres.2012.12.007, 2013.

Wagner, T. J., Feltz, W. F., and Ackerman, S. A.: The temporal evolution of convective indices in storm-producing environments, Weather Forecast., 23, 786–794, 2008.

Ware, R., Solheim, F., Carpenter, R., Güldner, J., Liljegren, J., Nehrkorn, T., and Vandenberghe, F.: A multi-channel radiometric profiler of temperature, humidity and cloud liquid, Rad. Sci., 38, 1–13, 2003.

Ware, R., Cimini, D., Campos, E., Giuliani, G., Albers, S., Nelson, M., Koch, S. E., Joe, P., and Cober, S.: Thermodynamic and Liquid Profiling during the 2010 Winter Olympics, Atmos. Res., 132–133, 278–290, doi:10.1016/j.atmosres.2013.05.019, 2013.

Xu, G., Ware, R., Zhang, W., Feng, G., Liao, K., and Liu, Y.: Effect of off-zenith observation on reducing the impact of precipitation on ground-based microwave radiometer measurement accuracy in Wuhan, Atmos. Res., 140–141, 85–94, 2014.

6

A two-channel, tunable diode laser-based hygrometer for measurement of water vapor and cirrus cloud ice water content in the upper troposphere and lower stratosphere

T. D. Thornberry[1,2], A. W. Rollins[1,2], R. S. Gao[1], L. A. Watts[1,2], S. J. Ciciora[1], R. J. McLaughlin[1,2], and D. W. Fahey[1,2]

[1]NOAA ESRL Chemical Sciences Division, Boulder, Colorado, USA

[2]Cooperative Institute for Research in Environmental Sciences, University of Colorado, Boulder, Colorado, USA

Correspondence to: T. D. Thornberry (troy.thornberry@noaa.gov)

Abstract. The recently developed NOAA Water instrument is a two-channel, closed-path, tunable diode laser absorption spectrometer designed for the measurement of upper troposphere/lower stratosphere water vapor and enhanced total water (vapor + inertially enhanced condensed phase) from the NASA Global Hawk unmanned aircraft system (UAS) or other high-altitude research aircraft. The instrument utilizes wavelength-modulated spectroscopy with second harmonic detection near 2694 nm to achieve high precision with a 79 cm double-pass optical path. The detection cells are operated under constant temperature, pressure, and flow conditions to maintain a constant sensitivity to H_2O independent of the ambient sampling environment. An onboard calibration system is used to perform periodic in situ calibrations to verify the stability of the instrument sensitivity during flight. For the water vapor channel, ambient air is sampled perpendicular to the flow past the aircraft in order to reject cloud particles, while the total water channel uses a heated, forward-facing inlet to sample both water vapor and cloud particles. The total water inlet operates subisokinetically, thereby inertially enhancing cloud particle number in the sample flow and affording increased cloud water content sensitivity. The NOAA Water instrument was flown for the first time during the second deployment of the Airborne Tropical TRopopause EXperiment (ATTREX) in February–March 2013 on the NASA Global Hawk UAS. The instrument demonstrated a typical in-flight precision (1 s, 1σ) of better than 0.17 parts per million (ppm, 10^{-6} mol mol^{-1}), with an overall H_2O vapor measurement uncertainty of 5 % ± 0.23 ppm. The inertial enhancement for cirrus cloud particle sampling under ATTREX flight conditions ranged from 33 to 48 for ice particles larger than 8 μm in diameter, depending primarily on aircraft altitude. The resulting ice water content detection limit (2σ) was 0.023–0.013 ppm, corresponding to approximately 2 μg m^{-3}, with an estimated overall uncertainty of 20 %.

1 Introduction

Water in the upper troposphere and lower stratosphere (UT/LS) plays an important role in Earth's climate system through aspects of radiative transfer, cirrus cloud formation, and stratospheric ozone chemistry (Kirk-Davidoff et al., 1999; Forster and Shine, 2002; Solomon et al., 2010). The dominant transport of H_2O into the stratosphere occurs in the tropics and is controlled to first order by the cold temperatures of the tropical tropopause (Liu et al., 2010; Schoeberl and Dessler, 2011; Randel and Jensen, 2013). However, uncertainty remains in both the microphysics of dehydration in the tropical tropopause layer (TTL) and the potential contributions of other processes (e.g., overshooting convection, transport from the extratropics) to the water budget of the lower stratosphere (Sherwood and Dessler, 2001; Fueglistaler et al., 2009; Randel and Jensen, 2013). Cirrus clouds occur with high frequency and large spatial extent in the TTL, and those occurring near the thermal tropopause facilitate the final dehydration of stratosphere-bound air parcels. The net role of TTL cirrus clouds in the climate system and how cloud effects may change as the climate changes remains uncertain (Dessler, 2010; Zelinka and Hartmann, 2011; Randel and Jensen, 2013). Further accurate measurements of water vapor in the tropical UT/LS are needed to improve understanding of the dynamical and microphysical processes that control the dehydration of air transported into the stratosphere. In contrast to water vapor, fewer in situ measurements of cirrus ice water content (IWC) exist for the UT/LS. IWC is an important metric for estimating both the radiative impact of cirrus clouds and their role in the dehydration of tropospheric air in the TTL.

A number of different in situ measurement techniques have been used to measure water vapor in the UT/LS. These include chilled-mirror hygrometry (Brewer et al., 1948; Mastenbrook and Oltmans, 1983; Vömel et al., 2007), Lyman-α photofragment fluorescence (Kley and Stone, 1978; Weinstock et al., 1994; Zöger et al., 1999), both open- and closed-path tunable diode laser (TDL) absorption spectroscopy (May, 1998; Diskin et al., 2002; Buchholz et al., 2013; Sargent et al., 2013), and chemical ionization mass spectrometry (Thornberry et al., 2013; Kaufmann et al., 2014).

Persistent disagreements among collocated measurements of H_2O in the UT/LS have produced uncertainties in our understanding of the cirrus cloud microphysics that control dehydration of tropospheric air in the tropical tropopause region and determine the amount of H_2O that reaches the lower stratosphere (Kley et al., 2000; Jensen et al., 2005; Peter et al., 2006; Weinstock et al., 2009). These disagreements among collocated in situ measurements have prompted community efforts to determine the sources of error in the H_2O measurements, including the chamber-based AquaVIT-1 intercomparison in 2007 (Fahey et al., 2014) and a recent in situ intercomparison during the Mid-latitude Airborne Cloud Properties EXperiment (MACPEX) in 2011 (Rollins et al., 2014). In the AquaVIT-1 intercomparison, a number of the primary UT/LS in situ H_2O instruments reported values that generally agreed within 20 % at mixing ratios down to 1 part per million (ppm, 10^{-6} mol mol^{-1}) under static conditions similar to those in the TTL. This was a significant improvement over the 30–100 % differences observed at low mixing ratios in a number of previous UT/LS field measurements (Kley et al., 2000; Jensen et al., 2005; Weinstock et al., 2009), suggesting that additional factors affect instrument performance when operating on aircraft in field campaigns. During the MACPEX campaign, the observed differences among instruments at low mixing ratios were typically within 20 %, similar to the AquaVIT-1 results. A careful analysis of the data found that all of the instruments were potentially affected by variable background artifacts that became significant at mixing ratios below 10 ppm (Rollins et al., 2014). Variations at this level in long-term data sets are sufficient to preclude the early identification of trends in UT/LS H_2O that are predicted to occur with changes in Earth's climate (Kley et al., 2000; Fueglistaler et al., 2013).

Measurement of TTL cirrus IWC is even more challenging because it combines the issues of water vapor measurement at low mixing ratios with the need to characterize the sampling efficiency of cloud particles. A number of different instruments have been developed and deployed for cirrus IWC determination, using both TDL (Davis et al., 2007a; Dorsi et al., 2014) and Lyman-α (Brown and Francis, 1995; Weinstock et al., 2006a; Schiller et al., 2008) detection. Typically these instruments measure the sum of water vapor and evaporated/sublimated condensed-phase water, and rely on a second instrument to measure only water vapor in order to derive a value for the IWC. The counterflow virtual impactor (CVI) instrument described by Twohy et al. (1997) eliminates this need by inertially separating the cloud particles into a dry synthetic air stream, allowing the particles to sublimate, and measuring the resulting water vapor. The CVI produces a lower-limit size cut-off for sampled particles and therefore generally will not afford detection of the IWC of sufficiently small ice crystals. Total water sampling inlets are operated either isokinetically (Weinstock et al., 2006a, b) to simplify the characterization of the particle sampling, or subisokinetically (Davis et al., 2007a; Schiller et al., 2008) to enhance the particle concentration sampled into the instrument for improved IWC sensitivity. Davis et al. (2007b) report on a comparison of three in situ IWC measurements made during the NASA Midlatitude Cirrus Experiment in 2004. The measurements demonstrated reasonable agreement at high IWC loadings, but the differences were significant at values below 5 mg m^{-3}, a value more than an order of magnitude larger than typically found in TTL cirrus (Davis et al., 2007b; Schiller et al., 2008).

To take advantage of the TTL sampling capabilities of the NASA Global Hawk UAS platform, we have developed a new, compact, two-channel instrument for the simultaneous measurement of water vapor and total water in the UT/LS, and integrated it onboard the Global Hawk during the recent NASA Airborne Tropical TRopopause EXperiment (ATTREX) mission. Here we describe details of the instrument design and validation, and present instrument performance from recent measurements in the UT/LS obtained during ATTREX.

2 The NOAA Water instrument

2.1 General description

The NOAA Water instrument is a two-channel, closed-path, TDL absorption spectrometer for the measurement of UT/LS water vapor (WV) and enhanced total water (eTW, vapor + inertially enhanced condensed phase) from the NASA Global Hawk UAS or other high-altitude research aircraft. The instrument utilizes wavelength modulation spectroscopy with second harmonic (2f) detection at a wavelength near 2694 nm to achieve high sensitivity and precision at low mixing ratios. The instrument is calibrated on the ground by comparison to a reference chilled-mirror hygrometer (model 373LX, MBW Calibration Ltd., Wettingen, Switzerland) and the stability of

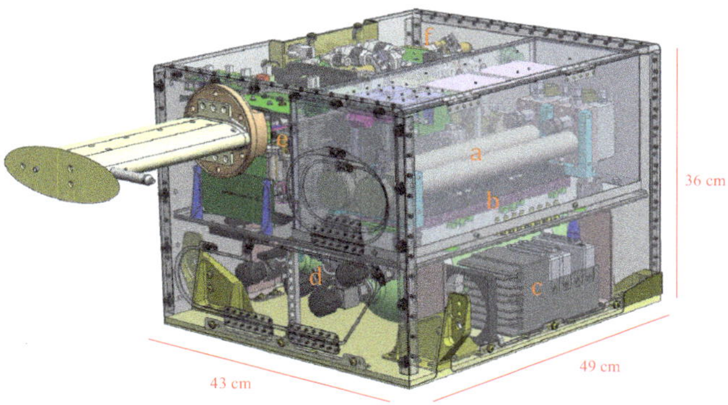

Figure 1. Schematic of the NOAA Water instrument showing the inlet pylon, optics sub-enclosure, and some of the principal instrument components: **(a)** absorption cells within the optics sub-enclosure, **(b)** optical sub-enclosure heaters, **(c)** CompactRIO data system, **(d)** gas bottles, **(e)** sample mass flow control valves, and **(f)** cell pressure control valves. The enclosure structure and pylon are fabricated from aircraft-grade aluminum.

Table 1. NOAA Water instrument specifications.

Size	49 cm × 43 cm × 36 cm	
Mass	40 kg	
Power	70 W DC, 400 W AC	
Data rate	1 Hz	
	WV	TW
Accuracy	5 % ±0.15 ppm	6 % ±0.20 ppm
Precision (1 s, 1σ)	0.17 ppm	0.25 ppm
Uncertainty (1 s)	5 % ±0.23 ppm	6 % ±0.32 ppm
IWC uncertainty (1 s)		15–30 %

the calibration during flight is assessed periodically by addition of water vapor to one of the channels. The absorption cells are operated under conditions of constant pressure, temperature, and flow to maintain constant sensitivity independent of pressure and temperature changes in the payload compartment or in the ambient sampling environment. This approach eliminates the need for highly accurate knowledge of the pressure- and temperature-dependent spectral line parameters and consequently complex data reduction in exchange for a reduced absolute absorption signal that results from operating the cells at pressures below the lowest ambient pressure encountered in flight.

A summary of the instrument physical and performance specifications appears in Table 1. The main instrument enclosure (Fig. 1) measures 49 cm × 43 cm × 36 cm and houses the optical absorption cells, calibration system, custom electronics, and data acquisition and control computer. The inlet pylon is attached to an interface flange at the fuselage surface and extends outside the fuselage perpendicular to the free-stream flow. Two small scroll pumps, their associated controllers, and a 24 VDC power supply are located external to the main instrument enclosure. The total instrument mass, including pumps and inlet pylon, is 40 kg. During flight, typical power consumption is 70 W of DC (28 V) and 400 W of AC (120 V, 400 Hz). The AC power draw varies depending on the power required by the heaters used to maintain instrument component temperatures.

2.2 Sample flow system

A schematic of the instrument flow system is shown in Fig. 2. Sample flow through the instrument is produced by a pair of small scroll pumps (model V12H20N2.5, Air Squared Inc., Broomfield, CO, USA). The sample mass flow through each channel is controlled using a custom butterfly valve (BV) based on the design described by Gao et al. (1999) and similar to those used previously with the NOAA CIMS H_2O instrument (Thornberry et al., 2013). The BV is located in the inlet line upstream of the absorption cell and is servo-controlled by a mass flow meter (MFM, model D6F-03A3, Omron Corp., Kyoto, Japan) located in the exhaust line from the absorption cell. The mass flow through each cell was maintained at 600 standard (273.15 K, 1013.25 hPa) $cm^3 min^{-1}$ (sccm). The flow value was chosen to be as large as possible, within the capacity of the vacuum pumps, in order to reduce the potential for hysteresis arising from the adsorption/desorption of water to inlet tubing and cell surfaces. The pressure in each absorption cell is controlled using a second BV located in the exhaust line from the cell, immediately downstream of the sample flow MFM, which is servo-controlled by a pressure sensor (model 19C015PA4K, Honeywell Sensing and Control, Minneapolis, MN, USA) located at the laser/detector end of the absorption cell. The absorption cell pressures were set to 45 and 50 hPa in the WV and TW channels, respectively, in order for the pressures to be sufficiently lower than the inlet sampling pressures at the highest flight altitudes for Global Hawk operation to enable cell pressures and flows to be held constant at all aircraft flight levels. The control pressure difference between the two channels is due to a combination of the additional ram pressure at the opening of the forward-facing TW inlet and the reduced sampling pressure at the side-facing opening of the WV inlet.

2.3 Absorption cells

The absorption cells for the NOAA Water instrument were designed and manufactured by Port City Instruments (Wilmington, NC, USA). A schematic of an optical absorption cell is shown in Fig. 3. Each cell consists of a 35 cm long, 1.9 cm ID (3.2 cm OD) Ni-plated aluminum cylinder with a Ni-plated aluminum block housing the inlet port and a gold mirror on one end and another block housing the outlet port, laser, detector, and pressure port on the other. This single-reflection arrangement produces a 78.6 cm round-trip optical

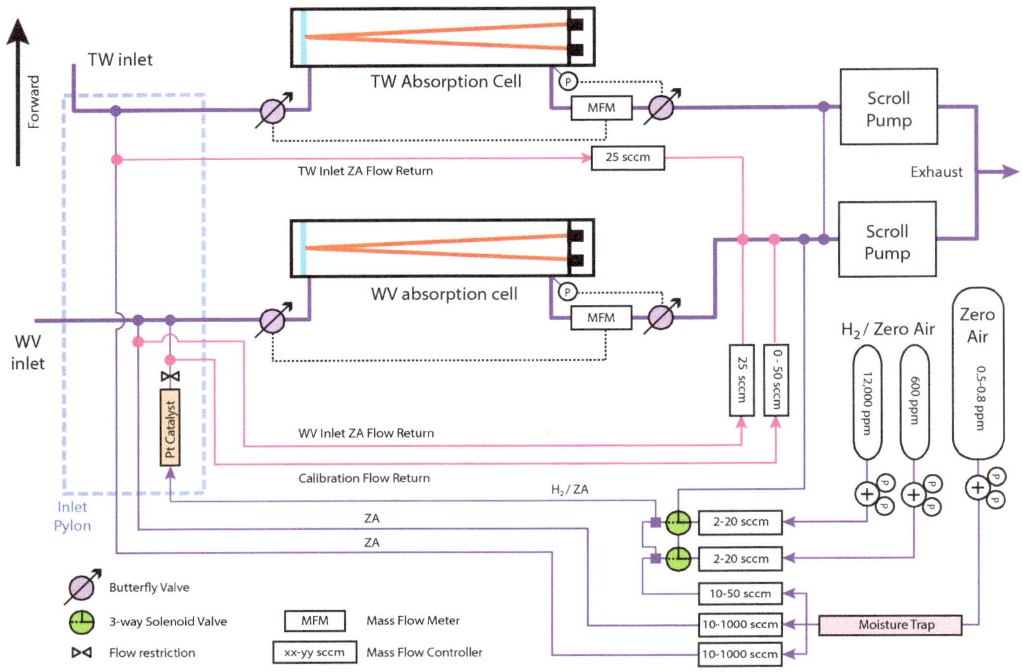

Figure 2. Schematic of the NOAA Water instrument showing the components of the dual-channel flow system. The custom butterfly valves located upstream and downstream of each cell are used to control the mass flow (upstream) and pressure (downstream) in the optical cells to constant values independent of variations in sampling pressure as the aircraft changes altitude. The numbers in the mass flow controller symbols indicate the flow value or range used during flight operation.

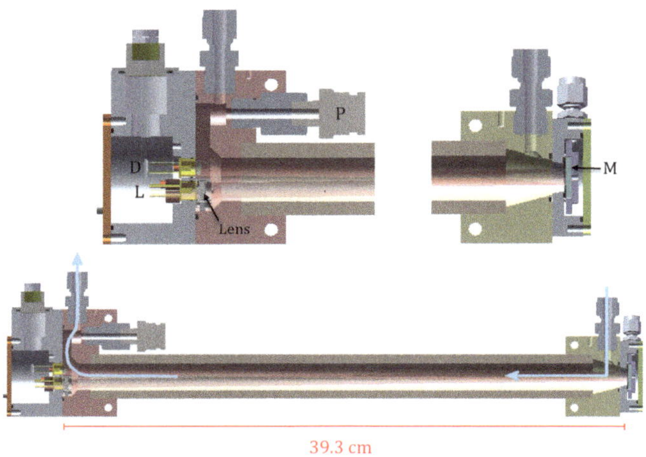

Figure 3. Schematic of a NOAA Water instrument optical absorption cell. Arrows indicate the direction of flow through the cell. The laser (L), detector (D), collimating lens assembly (Lens), and pressure transducer (P) are visible in the detail on the left (outlet), and the mirror (M) assembly is indicated in the detail on the right (inlet).

absorption path length. The internal cell volume is approximately 120 cm^3, yielding a sample residence time of less than 0.52 s at a flow rate of 600 sccm, cell pressure of 50 hPa, and cell temperature of 40 °C. The inlet, outlet, and pressure ports were designed for O-ring face-seal fittings (VCO, Swagelok Inc., Solon, OH, USA) to minimize potential for leaks into the cells.

In order to thermally stabilize the optical cavity and minimize the migration of interference fringes through the spectrum, the absorption cells are housed in a temperature-controlled sub-enclosure within the overall instrument enclosure. The temperature of the sub-enclosure is stabilized by controlling the air temperature within the sub-enclosure by means of forced circulation of air over temperature-controlled heat sinks and by separately controlling the temperatures of the six individual aluminum panels comprising the sub-enclosure. Three thermistors, one at each end and one at the mid-point of the cell, are used to monitor the thermal stability of the absorption cells.

The laser and detector are mounted directly into the block at the outlet end of the absorption cell to minimize the potential for any volume in the optical path that is not actively swept by the sample flow through the cell. A lens mounted in an open structure in front of the laser is used to collimate the beam. The gold mirror is sealed against the inlet end of the absorption cell using an O-ring face seal. Four small screws in the mirror assembly that compress the mirror against the O-ring allow for adjustment of the mirror angle for alignment of the laser beam.

2.4 Lasers and detectors

The lasers used in the instrument are distributed feedback (DFB) lasers (nanoplus GmbH, Gerbrunn, Germany) mounted on thermoelectric coolers (TEC) and sealed in TO5 packages with antireflection-coated, wedged windows. The laser packages are evacuated and backfilled with N_2 at 1 bar pressure and contain less than 30 ppm residual H_2O. The sealed optical path between the diode facet and TO5 package window is 0.5 mm, resulting in a maximum optical depth due to residual H_2O in the laser packages equivalent to 0.375 ppm in the sample cell at 50 hPa. Lasing threshold current and tuning characteristics are inherently variable with DFB lasers; however the useful tunable range accessible in a single current scan for the lasers used here was typically 2–3 cm^{-1}.

The InAs detectors (J12TE1-37S-R01M, Teledyne Judson Technologies, Montgomeryville, PA, USA) used are mounted on TECs in sealed packages. These packages were evacuated, baked, and backfilled with N_2 during fabrication to 50 hPa with a maximum of 5 ppm residual H_2O. The optical path between the photodiode and window is approximately 2 mm, resulting in a signal from trapped water equivalent to less than 0.014 ppm water in the detection cell.

For measurement of H_2O, a pair of absorption lines near 2694 nm was identified from the HITRAN 2008 database (Rothman et al., 2009). To access the selected H_2O absorption features, the laser TEC temperatures are set to values that yield output near the target wavelength for each individual laser and the laser drive currents are ramped to scan the lasers across a spectral range of approximately 1 cm^{-1}. The primary absorption line used to detect low mixing ratios of H_2O is from the asymmetric stretch mode $((0,0,1) \leftarrow (0,0,0))$ at 3712.20 cm^{-1} (2693.82 nm) with a line strength of 1.49×10^{-19} cm^2 $molecule^{-1}$ cm^{-1} . This line is approximately an order of magnitude stronger than strong lines in the H_2O overtone/combination band near 1400 nm that have typically been used for UT/LS H_2O measurement. A weaker line at 3711.88 cm^{-1} (2694.06 nm) due to the symmetric stretching mode $((1,0,0) \leftarrow (0,0,0))$ with a line strength of 4.41×10^{-21} cm^2 $molecule^{-1}$ cm^{-1} is used at higher H_2O mixing ratios to provide a linear response up to and exceeding 2000 ppm. Two prominent CO_2 absorption lines at 3711.62 cm^1 (2694.24 nm) and 3711.75 cm^{-1} (2694.15 nm) also fall within the scan window and could potentially be used for additional instrument diagnostics.

The 1 cm^{-1} spectral sweep is performed at 10 Hz and the spectra are coaveraged to produce 1 s data. Improved precision is achieved by use of 2f detection, which is employed by adding a 30 kHz sine wave modulation to the diode laser current ramp, producing a 0.08 cm^{-1} modulation depth of the scan, and demodulating the signal from the photodiode detector at 60 kHz. DC spectra are concurrently measured by passing the detector signal through a low-pass filter to remove the modulation. Figure 4 presents the spectral region scanned showing calculated transmission through the optical cell using line parameters from the HITRAN 2008 database along with the measured DC and 2f spectra of ambient air containing 100 ppm H_2O. The strong and weak line 2f signals are measured as the peak-to-peak amplitude in the spectrum in a defined window around each line. The asymmetry observed in the peaks in the 2f spectrum arises from contributions to the second harmonic signal from the first and third harmonics due to co-modulation of laser intensity with the applied wavelength modulation (Kluczynski and Axner, 1999). To account for variations in laser power that would affect the instrument sensitivity, the 2f signals are normalized to the laser power at each spectral feature (N2f: 2f peak-to-peak amplitude/laser power at line center) before calculating water mixing ratios. Laser power at the center of each line is determined from a polynomial fit to the corresponding DC spectrum baseline.

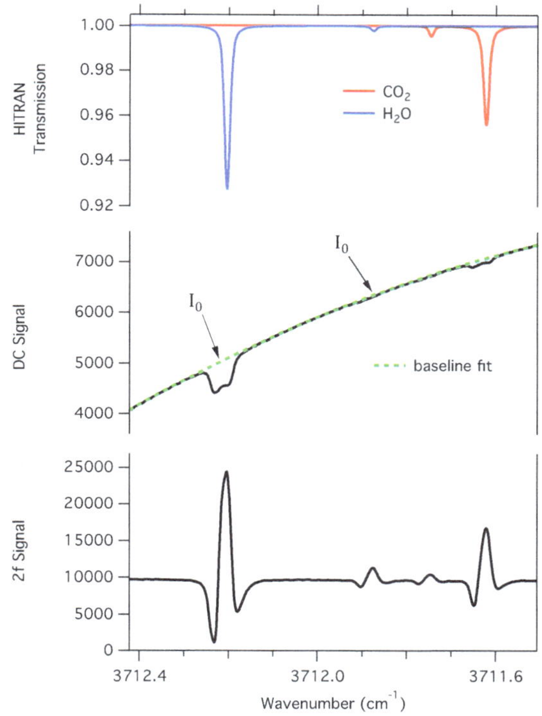

Figure 4. Top: HITRAN calculation of transmission through the detection cell with 100 ppm H_2O (blue) and 370 ppm CO_2 (red). Middle: 1 s coaveraged DC spectrum in the WV channel from the final descent on 27 February 2013 at an ambient H_2O mixing ratio of approximately 100 ppm. The green dashed line is the polynomial fit to the baseline for determination of the I_0 values at the strong and weak H_2O absorption lines. Bottom: corresponding 1 s coaveraged 2f spectrum. Note that the scan is plotted as it is recorded in time and the wave number scale decreases from left to right.

2.5 Inlets

The inlets for the two channels of the NOAA Water instrument are housed in an airfoil-shaped pylon assembly that ex-

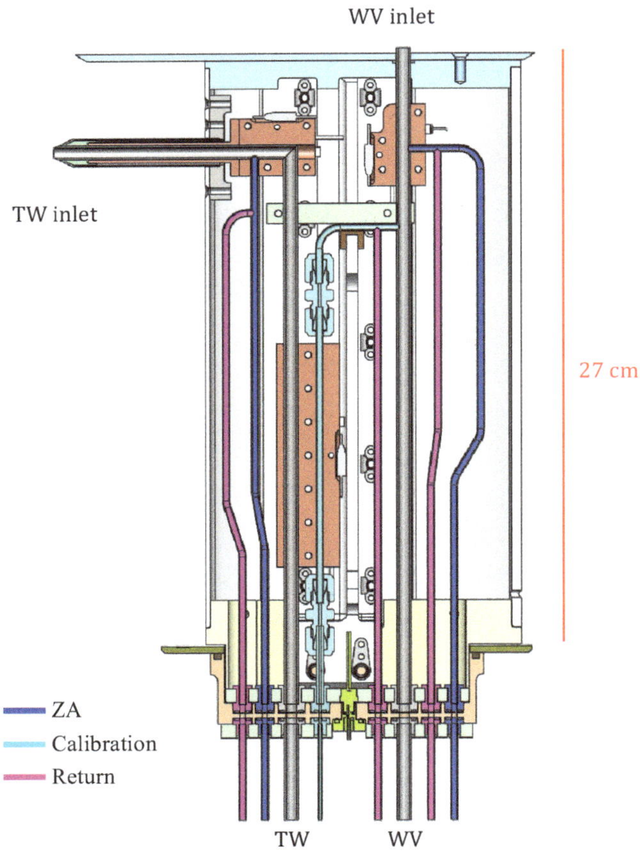

Figure 5. Schematic of the NOAA Water instrument inlet pylon showing the WV and TW sample inlet tubes, the inlet line and catalyst heater blocks (orange), the zero air (ZA) delivery lines (blue) to each inlet, the Pt catalyst and calibration delivery line (cyan) to the WV inlet, and the ZA and calibration return lines (magenta).

tends 27 cm perpendicular to the skin of the aircraft in order to sample in the free-stream flow around the aircraft. The pylon is 14 cm wide (fore–aft) and 3.5 cm thick. The major axis of the airfoil is oriented parallel to the aircraft longitudinal axis (i.e., no offset angle). A cross section of the inlet pylon is presented in Fig. 5 showing the arrangement of the inlet components and calibration tubing. A photograph of the instrument installed onboard the NASA Global Hawk during ATTREX appears in Fig. 6. Gas flows between the inlet pylon and the instrument pass through custom O-ring face-seal fittings at the interface flange. This design provides a good, reproducible seal while allowing the inlet pylon to be easily installed and removed when necessary for accessing the instrument.

2.5.1 Water vapor inlet

The sampling point for the WV channel is located at the end of the inlet pylon, oriented perpendicular to the flow past the aircraft (Fig. 5). This sampling arrangement, combined with

Figure 6. Photo of the NASA Global Hawk showing the location of the NOAA Water instrument inlet during the ATTREX mission (photo: T. Tschida, NASA; inset photo: T. Thornberry).

the thin plate at the end of the pylon, allows for sampling of ambient water vapor while rejecting cloud particles. The design is similar to that described in Popp et al. (2004) for sampling gas-phase nitric acid while excluding nitric acid trihydrate particles, and that described by Perring et al. (2013) for sampling submicron aerosol while rejecting cloud particles. A similar inlet was used with the NOAA CIMS H_2O instrument (Thornberry et al., 2013) for water vapor measurements during the MACPEX campaign, where the measurements showed no detectable contamination from cloud water during extensive cirrus sampling. The WV sample line is constructed of 0.46 cm ID electropolished stainless steel tubing (WinTech 10, Winter Technologies, Pacific, MO, USA), which was found in our previous laboratory experiments to produce little hysteresis from rapid changes in water vapor at mixing ratios between 1 and 50 ppm (Thornberry et al., 2013). A section of the WV inlet line closest to the tip is heated to 50 °C to prevent condensation in the inlet.

2.5.2 Total water inlet

The sampling point for the TW channel is located at the end of a forward-facing 0.46 cm ID tube that affords sampling of both water vapor and condensed-phase water. The inlet tube is oriented parallel to the aircraft longitudinal axis at a distance of 22.5 cm from the fuselage surface and extends 6.7 cm forward from the leading edge of the pylon and tapering (20°) to a knife edge at the tip. Inside the pylon, 10.5 cm from the tip, the inlet tube makes a sharp (mitered) 90° turn to facilitate impaction of large ice particles on the heated inlet wall. Nine centimeters of the forward facing sample line is enclosed in a 0.2 cm thick copper sheath extending from a copper block enclosing the bend that is maintained at 180 °C. After the bend, the sample line passes through the catalyst heater block, also controlled to 180 °C, before connecting to the heated (30 °C) interface flange at the fuselage surface and entering the instrument.

The controlled inlet flow of 600 sccm causes the sampling to be subisokinetic, i.e., the sample air velocity inside the inlet (U) is lower than the free-stream velocity (U_0) at sampling altitudes and typical high-altitude research aircraft speeds (140–200 m s^{-1}). This subisokinetic sampling leads to an inertially enhanced aspiration efficiency for particles into the inlet (Belyaev and Levin, 1974; Krämer and Afchine, 2004; Eddy et al., 2006; Davis et al., 2007a). As a result, cirrus particle number densities in the sample flow are enhanced by a factor, EF, defined as the ratio of particle number concentration entering the inlet tube to that in the free stream. The EF is a function of the inlet geometry, aircraft attitude, the ratio of the free-stream flow to the inlet flow (U_0/U), and the Stokes numbers (Stk) of the particles. Since the mass flow through the TW inlet is constant, the volumetric flow rate and therefore the sample flow velocity, U, is a function of the ambient pressure and temperature and varies significantly with altitude. The parameterization developed by Eddy et al. (2006, Eq. 15) based on computational fluid dynamic simulations of various inlet geometries was used to calculate the particle-size-dependent EF for the TW inlet. Figure 7 shows calculated EF curves as a function of particle diameter for typical TTL operating conditions of the Global Hawk UAS. A density of 0.7 g cm^{-3} was used for the calculations based on the reported effective density of small (< 50 μm) cirrus particles by Cotton et al. (2013). EF curves calculated for a particle density of 0.9 g cm^{-3} (close to the bulk density of ice) are displayed as well to show the sensitivity of the calculated EF to the assumed density. The subplot in Fig. 7 shows that the potential error associated with using a given value in this range for the density is < 3 % for particles greater than 5 μm, but becomes large (35–45 %) for particles near 1 μm. As expected, the value of the EF is near unity for small particles (< 0.2 μm) and increases with particle size. For particles larger than approximately 7 μm in diameter, typical of cirrus particles, the EF for the TW inlet is within 2 % of the maximum value of U_0/U, which is in the range of 30 to 50 for Global Hawk TTL flight conditions. This subisokinetic sampling enhancement results in a significantly increased sensitivity to cirrus IWC.

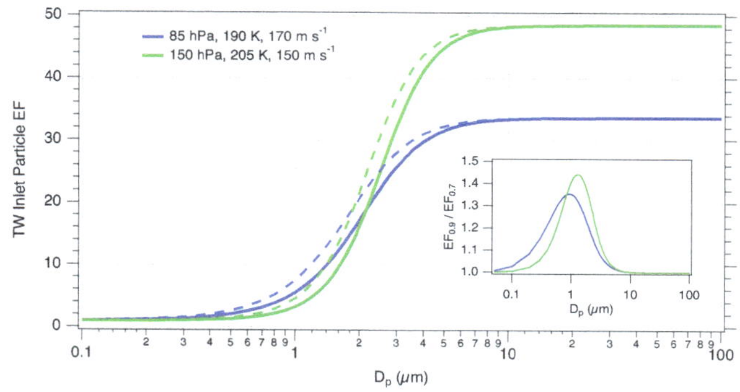

Figure 7. TW inlet particle aspiration enhancement factor (EF) as a function of particle diameter (D_p) calculated using the formulation of Eq. (15) in Eddy et al. (2006) for the TW inlet geometry and assumed ice particle densities of 0.7 g cm^{-3} (solid) and 0.9 g cm^{-3} (dashed). The curves represent conditions near the tropical tropopause (approximately 17.6 km, blue) and near the bottom (approximately 14.4 km, green) of the typical NASA Global Hawk operational altitude range during the ATTREX mission. The inset plot shows the ratio $EF_{0.9}/EF_{0.7}$ as a function of particle diameter to illustrate the effect of assumed particle density on the inlet EF.

After cloud particles are sampled into the inlet, the condensed-phase water must be vaporized in order for it to be measured by the instrument. The high-temperature section of the TW inlet consists of 9 cm of the forward-facing tube plus an additional 19 cm length following the 90° bend. Calculations of the sublimation of ice particles (assumed to be solid spheres with density 0.7 g cm^{-3}) in the inlet at a pressure of 85 hPa indicate that particles less than 7 μm diameter will vaporize before reaching the bend, and particles less than 15 μm diameter would completely vaporize within the 28 cm high-temperature region, even without contacting the heated walls. Particles larger than 15 μm initial diameter have sufficient residual inertia that they will impact the tubing at the 90° bend, potentially sticking and greatly increasing heat transfer or shattering into small fragments (Korolev et al., 2013a), accelerating subsequent vaporization. Calculations at 150 hPa indicate that particles up to 21 μm would sublimate in the sample flow within the high-temperature region, while particles with an initial diameter > 20 μm will impact the tubing wall at the bend. With repetition of the calculations for an ice crystal density of 0.9 g cm^{-3}, the diameters for both complete sublimation and impaction decrease slightly (~ 15 %), but it remains the case that the diameter of particles large enough to not fully vaporize in the heated sample flow have sufficient remaining inertia at the 90° bend to insure impaction on the hot tubing wall. Adequate models do not exist to provide a full description of ice particle shattering/sticking and collisional heat transfer from the walls in order to accurately calculate the details of vaporization of large ice crystals and their fragments. However, based on our calculations of sublimation within the high-temperature zone of the inlet, the rapid vaporization that will occur from contact with the hot tubing surfaces, and the additional 40 cm of warm (20–40 °C) inlet tubing through which the sample flow passes prior to reaching the absorption cell, we believe it is reasonable to assume that all condensed-phase water entering the TW inlet is vaporized prior to measurement.

The direct measurement of the water mixing ratio in the TW channel is defined as enhanced total water (eTW), which is composed of ambient water vapor (H_2O_{vap}) and evaporated, inertially enhanced IWC (eIWC) as

$$eTW = H_2O_{vap} + EF \cdot IWC = H_2O_{vap} + eIWC. \quad (1)$$

The actual IWC of a sampled cirrus cloud is then determined as

$$IWC = (eTW - H_2O_{vap})/EF, \quad (2)$$

where H_2O_{vap} is measured in the WV channel, and the overall EF is determined by convolving the calculated EF curve for the immediate sampling conditions with an assumed cirrus ice particle size distribution or that provided by a co-measuring cloud probe instrument. Due to the EF size dependence for particles $< 7\,\mu$m, uncertainty in the assumed or measured particle size distribution for cirrus with a significant mass fraction in small particles will lead to uncertainty in the determination of IWC.

2.6 Calibration system

The NOAA Water instrument uses an onboard calibration system to evaluate the stability of the sensitivity and background of the WV channel during flight. The calibration system is similar to the one used previously in the NOAA CIMS H_2O instrument (Thornberry et al., 2013) and is based on the catalytic oxidation of H_2 standards to produce known H_2O mixing ratios (Rollins et al., 2011). The current system (see Fig. 2) consists of one dry synthetic air (zero air, ZA) cylinder (M22A, Luxfer Gas Cylinders, Riverside, CA, USA) and two cylinders (M09B, Luxfer) containing different mixing ratios of H_2 in air. The ZA is passed through a 1 cm ID by 25 cm long stainless steel tube containing a molecular sieve desiccant (BMSR-1, Agilent Technologies Inc., Santa Clara, CA, USA), which reduces the residual water vapor mixing ratio in the ZA. This moisture trap is located downstream of the ZA pressure regulator (Series 4300-N, Premier Industries, Blaine, MN, USA) to capture water from as many potential leaks as possible before the ZA is directed to the inlets. The catalyst assembly, located in the inlet pylon, consists of a 14 cm long, 0.21 cm ID, electropolished stainless steel tube containing two 2.5 cm × 5 cm pieces of tightly rolled Pt mesh (part no. 298093-1.7G, Sigma-Aldrich Co., St. Louis, MO, USA). The catalyst is maintained at a temperature of 180 °C and a pressure above 2000 hPa, which was demonstrated to yield complete conversion of H_2 to H_2O for a similar geometry and flow rates to those used in the NOAA Water instrument (Rollins et al., 2011).

Mass flow controllers (MC-Series, Alicat Scientific Inc., Tucson, AZ, USA), which were tested and found to be insensitive to changes in ambient enclosure pressure, are used to control the ZA flows to the inlets and the ZA and H_2/air flows through the Pt catalyst.

To perform a calibration, the ZA flow to the WV inlet is set to 1000 sccm, overflowing the inlet with ZA. The H_2/air flows through the catalyst are adjusted in a series of steps between 2 and 20 sccm and the output of the catalyst is added to the ZA in the sample flow to produce a series of known H_2O mixing ratios, which are used periodically to determine the sensitivity of the WV channel. Due to space and weight constraints and the desire to use a minimum number of fittings (potential leaks), only the WV channel is calibrated in flight. The stability of the sensitivity of the TW channel during flight is assessed by comparing its signal with that of the WV channel when sampling in cloud-free air. The calibrations performed during flight are compared with identical calibration sequences run on the ground to verify the consistency of the calibration system during flight.

To mitigate the potential effects of parasitic water from infinitesimal leaks into the calibration system, 10 sccm flows of ZA are used to continuously flush the ZA and calibration system tubing out to the inlet addition points and back through return flow controllers (see Figs. 2 and 5). The return flow controllers are set to additionally pull small (10–15 sccm) flows of sample air from the inlets to prevent ZA from reaching the sample flow during ambient sampling.

2.7 Data system

Instrument control, data acquisition, and communication with the aircraft are accomplished using a custom LabView program running on a CompactRIO system (National Instruments Inc., Austin, TX, USA). Custom Arduino-based electronics boards are used for temperature control of instrument components and operation of the flow- and pressure-control butterfly valves.

Instrument temperatures, pressures, and flows (housekeeping) along with laser parameters, N2f signals, and calculated H_2O_{vap} and eTW mixing ratios are recorded at 1 Hz ($\sim$4 MB h^{-1}) for post-flight data processing and instrument performance evaluation. A subset of this data can be transmitted to the ground for real-time reporting of mixing ratios and monitoring of instrument performance during flight if communication is available. The 1024-point DC and 2f spectra for each channel are acquired at 10 Hz, coaveraged, and saved at 1 Hz (29.5 MB h^{-1}). The instrument can operate entirely autonomously or with user interaction, depending on the communications capability of the aircraft platform.

3 Instrument performance

The first field deployment of the NOAA Water instrument occurred in the second phase of the ATTREX mission in January–March 2013, during which the instrument was integrated on the NASA Global Hawk UAS. During this mission, the Global Hawk conducted six research flights flight with durations of 24 ($\pm$0.3) hours, allowing for transit south from NASA Dryden Flight Research Center in California to the central and eastern tropical Pacific with substantial time at low latitudes to conduct extensive sampling in the TTL. Measured water vapor mixing ratios in the TTL reached values as low as 1.5 ppm and cirrus clouds with a wide range of IWC were encountered.

Operational procedures for the NOAA Water instrument during ATTREX involved calibrating the instrument on the ground between flights, as well as extensively in the laboratory before and after the mission. These calibrations were conducted with sample flows containing H_2O mixing ratios in the range between 0.5 and 2500 ppm as quantified by the MBW frost point hygrometer. The calibration flows

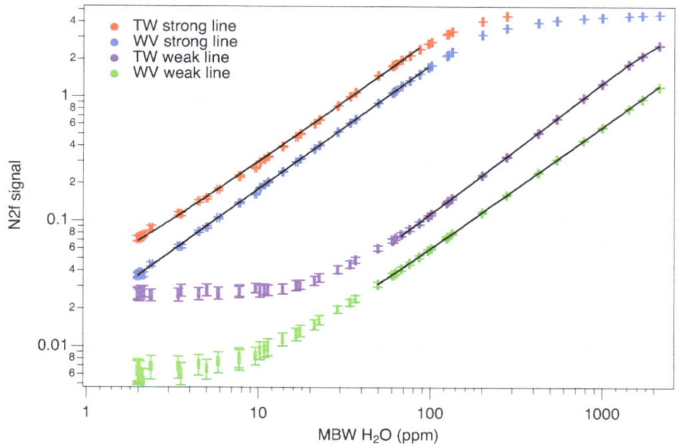

Figure 8. Laboratory calibration curves showing the laser power normalized second harmonic (N2f) signals from the strong and weak absorption lines (2693.82 and 2694.06 nm, respectively) in the WV and TW channels as a function of the H_2O mixing ratio in the sample flow measured by the MBW reference hygrometer. Each data point represents an average of 100–300 s of data at a constant mixing ratio. The error bars are the standard deviation of the N2f signal in each interval. The black lines are the weighted orthogonal linear regression fits to the data, which are used to determine instrument sensitivity for each absorption line. The differences in N2f values between channels at the same MBW value are due to intrinsic performance and configuration differences between the channels.

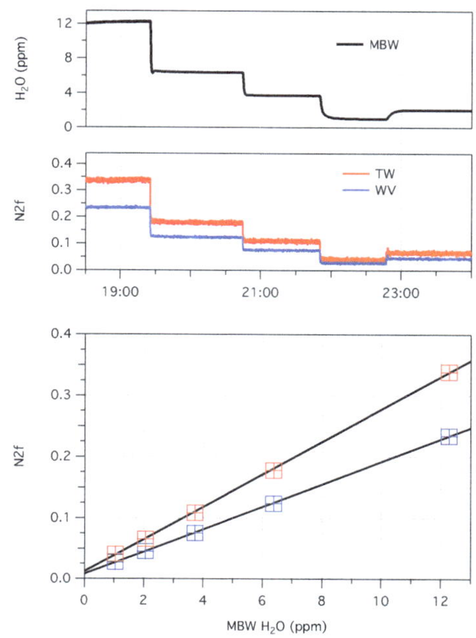

Figure 9. Laboratory calibration of the WV and TW channel strong absorption line signals at low H_2O mixing ratios. Top: time series of water vapor values in the sample flow as measured by the MBW reference hygrometer. Middle: corresponding time series of the strong absorption line laser power normalized second harmonic (N2f) signals in each channel. Bottom: linearity of the N2f signals at low mixing ratios. Repeated long-duration calibrations at low mixing ratios were used to quantify and verify the stability of the background water signal from residual water in the laser and detector packages.

were generated by an external H_2/Pt catalytic H_2O source (Rollins et al., 2011) and distributed to the instrument and the MBW through a common manifold. Figure 8 presents results from one laboratory calibration over the range 2–2140 ppm showing the calibration ranges for both weak and strong lines of both channels. The nonlinearity in the N2f signal that arises from high optical depth is apparent at mixing ratios above 100 ppm in the strong absorption lines, while electronic noise and optical fringing produce nonlinearity in the WV and TW weak absorption line signals below 40 and 60 ppm, respectively. Repeated calibrations performed with many hours of continuous operation at mixing ratios less than 20 ppm were used to assess the linearity of the strong absorption line N2f signals at these low mixing ratios and to quantify the background water signal in each channel (Fig. 9). Fringes in the TW channel were somewhat larger than those in the WV channel and led to larger variations over time in the TW N2f signal at low mixing ratios. The background equivalent water value in the WV channel due to residual trapped water in the laser and detector packages was determined to be 0.3 ± 0.1 ppm, while that in the TW channel was 0.3 ± 0.2 ppm.

During the ATTREX flights, H_2O vapor and eTW mixing ratios were calculated in real time on a 1 Hz basis from the N2f signals in the WV and TW channels and transmitted to the ground when communication was available. Because data reduction is performed by empirically calibrating the N2f signal against a H_2O reference standard (MBW), the measurement accuracy does not depend on highly accurate knowledge of the laser modulation amplitude. Further, uncertainties due to changes in spectral line shape and line strength with changes in gas temperature and pressure do not contribute to measurement uncertainty as long as these parameters do not change between ground-based calibrations and in-flight measurements. Mixing ratios were calculated from the N2f signals in the low optical depth regime for each line, where the nonlinearity in the instrument response is less than 0.5 %. Only the TW channel measured values high enough to require use of the weak absorption line, which was used for mixing ratios above 80 ppm.

During flights, the detection cell pressures were maintained to within ± 0.7 % of the pressure at which the channel was calibrated in the laboratory, and the absorption cell temperatures were maintained within a ± 0.03 °C range after the optical cell sub-enclosure reached its temperature set point. Therefore variations in the instrument sensitivity due to changes in these parameters are negligible. To directly verify the stability of the instrument response during flight operation, the ambient flow in the WV channel was periodically replaced with calibration flows produced using the Pt/H_2/ZA

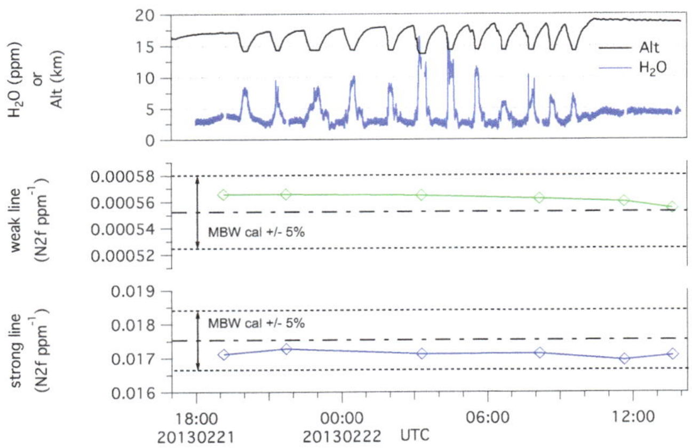

Figure 10. WV channel sensitivity measured during in-flight calibrations from a flight on 21–22 February 2013. Top panel: water vapor mixing ratios and aircraft altitude during a series of profiles through the TTL. Middle panel: calibration slopes (sensitivity) for the weaker absorption line (3711.9 cm^{-1}). Bottom panel: calibration slopes for the strong H_2O absorption line (3712.2 cm^{-1}). Each data point represents a calibration procedure in which dry synthetic air displaced the ambient flow for 8.7 min and a series of 8 H_2O mixing ratios between 2 and 400 ppm were added in time intervals of 40–60 s. In the middle and bottom panels, the dash-dot horizontal lines indicate the sensitivity measured in the laboratory prior to the flight. The dotted lines indicate the ± 5 % range surrounding the pre-flight laboratory value. The solid lines simply connect the data points as an aid to the reader.

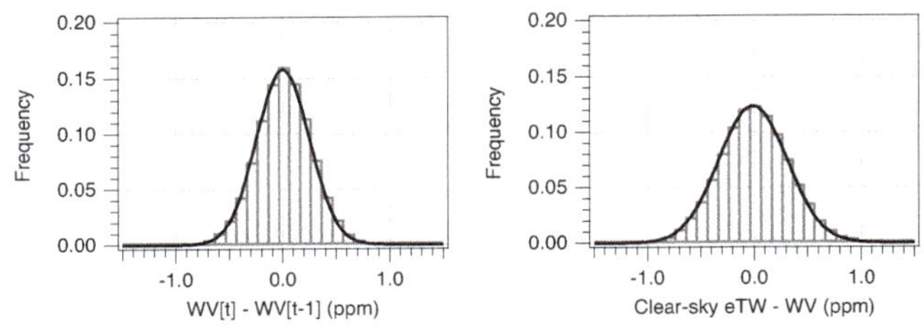

Figure 11. Left: histogram of 1 s changes in the WV channel (strong line) during the 21–22 February 2013 flight. The width of the Gaussian fit to the distribution is 0.34, indicating the signal noise is normally distributed with $\sigma = 0.17$ ppm. Right: histogram of the difference between 1 s measurements in the TW and WV channels during cloud-free sampling periods of the 21–22 February 2013 flight. The difference is normally distributed around 0 with a width of 0.45 ppm. This scatter yields a detection threshold (2σ) for inertially enhanced ice water content, eIWC, of 0.641 ppm.

system. Figure 10 shows results from the in-flight calibrations during the flight on 21–22 February 2013. The measured instrument response (sensitivity) for both H_2O spectral lines in individual calibrations during the flight agreed with those measured in the laboratory to better than 4 %, which is within the uncertainty expected of the onboard calibration system. These results are typical of the in-flight calibrations on all of the ATTREX flights.

In-flight measurement precision was evaluated by calculating the distribution of 1 s changes in the WV channel measurements ($\Delta WV = WV[t] - WV[t-1]$). Figure 11 shows a histogram of ΔWV from the 21–22 February 2013 flight with a Gaussian fit indicating that the noise is normally distributed with a width of 0.34 ppm, corresponding to a 1σ precision of 0.17 ppm. Noise in the TW channel was typically slightly higher than in the WV channel (0.25 for the 21–22 February flight). The observed in-flight precision is independent of signal amplitude and compares well, given a contribution to the 1 s changes from real atmospheric variability, with the typical value of 0.1 ppm observed while sampling stable mixing ratios in the laboratory.

Also shown in Fig. 11 is the histogram of the difference between 1 s TW and WV channel signals during cloud-free sampling. This difference was normally distributed and centered near zero (-0.012 ppm). The detection limit for eIWC is defined as twice the standard deviation (2σ) in the clear-sky difference of the signal. For the 21–22 February flight this results in an eIWC detection limit of 0.641 ppm. Assuming the cirrus particles are greater than 8 μm effective diameter, the calculated EFs of 33–48 yield an IWC mixing ratio detection limit from 0.023 to 0.013 ppm, depending on altitude. Converting these numbers to mass, the IWC detection limit was approximately 2 $\mu g\,m^{-3}$ across the sampling altitude range. Figure 12 shows a sample time series of WV and eTW mixing ratios during a vertical profile through the TTL from the 21–22 February flight. Cirrus encounters are clearly visible on both the ascent and descent. This data sample illustrates the typical instrumental precision in the measurements, as well as the good agreement between the WV and TW channels.

3.1 Water vapor measurement uncertainty

The principal sources of uncertainty in the determination of ambient H_2O mixing ratios in the NOAA Water instrument are (1) the accuracy of the MBW reference frost point hygrometer, (2) uniformity of the H_2O mixing ratio in the external calibration flows to the MBW and the NOAA Water instrument (analytical split), (3) stability of the absorption cell pressures, (4) temperature of the sample gas in the absorption cells, (5) linearity of the calibration fit, and (6) variation of optical fringes in the wavelength scan.

The NIST-traceable accuracy of the MBW frost point determination is 0.1 °C, which at 850 hPa (typical ambient pressure in Boulder, CO) corresponds to an uncertainty of 1.6 % at 1 ppm decreasing to approximately 1 % at 1000 ppm. Potential issues with analytical split of the external calibration between the MBW and the instrument were addressed by examining the repeatability of the laboratory calibrations with a number of different plumbing configurations and found to be negligible. The uncertainty arising from changes to the sensitivity (H_2O absorption) with changes in cell pressure are similar in magnitude to the variations in the cell pressure for

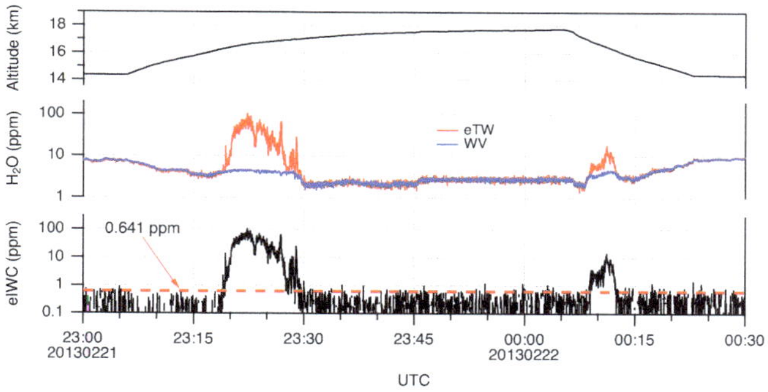

Figure 12. Time series of measured H_2O vapor (WV) and enhanced total water (eTW) including two cloud encounters during a vertical profile maneuver through the TTL during the flight of 21–22 February 2013. The top section shows the aircraft altitude. The middle section shows the measured H_2O_{vap} (blue) and eTW (red) mixing ratios. The bottom section shows the calculated enhanced ice water content, eIWC (=eTW − H_2O_{vap}). The dashed red line indicates the detection threshold for eIWC of 0.641 ppm determined from the precision (2σ) in the difference between the two channels during cloud-free sampling.

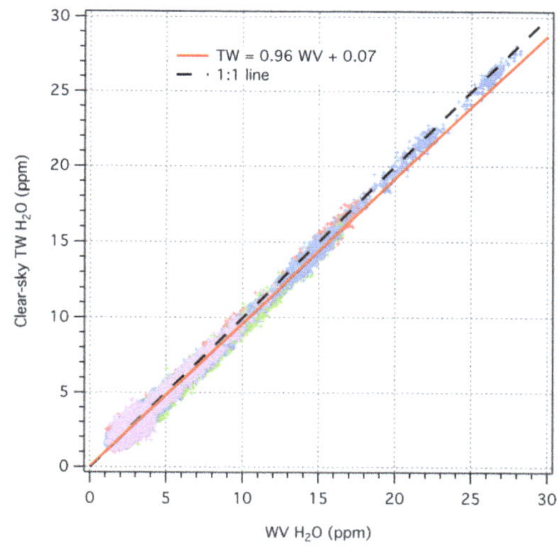

Figure 13. Correlation of H_2O measured in the TW channel with that measured in the WV channel for cloud-free sampling conditions during ATTREX 2013 flights. The slope of the correlation calculated using linear orthogonal distance regression was 0.96. Individual flight correlation slopes ranged from 0.95 to 0.98 and the r^2 values of the correlations ranged from 0.92 to 0.99.

small changes, and were typically less than 0.7 %. The temperature of the sample gas in the absorption cells was not directly measured, but the active heating of the inlets and interface flange, the small range of temperature in the main instrument enclosure during flight (10–20 °C), and thermal equilibration with the 20 cm of inlet tubing within the temperature-controlled sub-enclosure result in small (< 1 °C) calculated variations in the temperature of the sample gas entering the absorption cells. Based on HITRAN calculations, a 1 °C change in gas temperature would result in less than a 0.4 % change in the absorption line strength. The insensitivity of the measurement to changes in external temperature is supported by the lack of any observed difference in the signals in the laboratory at constant H_2O mixing ratio with and without the inlet and interface flange heaters operating. The uncertainty due to changes to the N2f signal caused by variations in the optical fringes in the spectrum that arise from small changes in the optical path with fluctuations in cell temperature was assessed by analyzing extended (up to 12 h) sampling of constant mixing ratios. The observed variations in the signal due to shifting optical fringes were equivalent to less than 0.1 ppm H_2O for temperature fluctuations similar to those observed in flight (±0.03 °C). During the 2013 ATTREX deployment, including pre- and post-mission calibrations and calibrations conducted between flights, the goodness of the individual linear fits to the calibration data was better than 4 %, and the average difference between individual calibrations of the WV channel was described by the linear function $0.1 + 0.005 \cdot H_2O$. Combining the uncertainty arising from these factors yields an accuracy of 5 % ± 0.15 ppm for the WV channel. Including the typical precision value of 0.17 ppm, the overall uncertainty in the H_2O mixing ratio in the WV cell is 5 % ± 0.23 ppm for a 1 s measurement. For the eTW H_2O measurement in the TW channel, the overall uncertainty was slightly higher at 6 % ± 0.35 ppm due to contributions from larger optical fringes in the spectrum and slightly higher electronic noise (precision). The overall good agreement between TW and WV channels during periods of cloud-free sampling in all of the ATTREX 2013 flights is shown in the correlation plot in Fig. 13. An orthogonal linear regression fit to all of the data yielded a slope of 0.96 and the correlation had an r^2 value of 0.97. Regressions for individual flights ranged from 0.95 to 0.98, indicating there was some variation in relative sensitivity over time. This variation appears to arise primarily from optical fringes in the TW channel that affected the signal at lower mixing ratios. No consistent offset between the two channels in flight was observed.

3.2 Ice water content measurement uncertainty

A number of additional sources of uncertainty affect the IWC measurement. These include (1) the particle aspiration efficiency of the inlet, (2) size distribution of the ice particles, (3) completeness of ice particle vaporization within the inlet, (4) inlet geometric sampling area, and (5) disturbance of the ambient cirrus ice particle distribution at the sampling point by the presence of the aircraft. Conservative calculations of ice particle vaporization indicate that an assumption of complete vaporization of all ice particles within the TW inlet prior to measurement is reasonable and incomplete vaporization therefore contributes negligibly to the overall IWC measurement uncertainty. Calculating the TW inlet EF requires knowledge of a number of atmospheric (pressure, temperature) and aircraft (true air speed, pitch) and instrument (inlet flow) parameters that all contribute some uncertainty. For ATTREX, the atmospheric and aircraft parameters were measured by the Meteorological Measurement System (NASA Ames Research Center, T. V. Bui, PI), which reports uncertainties of ± 0.3 hPa, ± 0.3 K and $\pm 1.0\,\mathrm{m\,s^{-1}}$ for temperature, pressure, and true air speed, respectively. These uncertainties are sufficiently small to contribute only about 1 % uncertainty to the calculated value of the free-stream flow, U_0. The uncertainty in inlet flow measured by the TW sample mass flow meter is 3 %. Eddy et al. (2006) report that the EF parameterization presented (Eq. 15) agreed with their detailed computational fluid dynamics simulations to within 20 % for most Stk values, so the uncertainty in EF contributes significantly to the overall uncertainty. The uncertainty in IWC determination arising from the size dependence of the EF depends on the fraction of ice mass in crystals with diameters less than 7 μm, where the EF begins to decrease significantly from the asymptotic value of U/U_0. If all ice mass were in particles of 5 μm diameter, assuming the U_0/U value for the EF would result in an underestimate of IWC by less than 10 %, while if the particles were all 3 μm, the underestimate would rise to 33 %. Applying the calculated EF size dependence to a co-measured ice particle size distribution, if one is available, can reduce the magnitude of this uncertainty. The typical angle of attack of the Global Hawk while flying in the TTL was $< 4°$, so the uncertainty associated with the reduction in geometric inlet sampling area due to the flow angle is negligible. The effects of cloud ice particle shattering on probe and aircraft surfaces have been a significant issue in measurements of cirrus size distribution and habit (e.g., Korolev et al., 2013b). Shattering on the knife edge of the TW inlet does not present a major uncertainty for the IWC determination since it results in < 3 % uncertainty in the inlet geometric area, and shattered particles entering the inlet are vaporized prior to measurement. Combining these uncertainty factors results in an overall uncertainty for IWC determination of 15–30 %, depending on the mass-weighted cloud particle size distribution.

Estimating the additional IWC uncertainty associated with potential sampling artifacts caused by the presence of the aircraft is difficult. During the ATTREX mission, the NOAA Water instrument was mounted in a forward fuselage bay on the right side of the aircraft, approximately 3 m behind the nose (see Fig. 6). Potential ice crystal bounce and deflection effects on sampling have not been investigated for the Global Hawk using computational fluid dynamic simulations, although in the future they could potentially yield some insight into these issues. In the comparison of total water measurements during the MidCiX campaign reported by Davis et al. (2007b), the reasonable agreement between three instruments sampling from different locations on the NASA WB-57F aircraft suggest that the effects for IWC measurement of the type reported here may be small.

4 Summary

The NOAA Water instrument was developed for accurate in situ measurements of water vapor and cirrus IWC in the TTL and elsewhere in the UT/LS, which are important for our understanding of the dynamical and microphysical processes that regulate the H_2O distributions in this climatically important region. The instrument is based on TDL absorption spectroscopy of H_2O in two single-reflection absorption cells with an absorption path length of 78.6 cm. Wavelength modulation spectroscopy with a strong H_2O absorption line near 2694 nm is used to achieve high sensitivity and good measurement precision at low-ppm H_2O mixing ratios. A second, weaker absorption line is used to provide dynamic range for measurement of mixing ratios up to 2500 ppm. Data reduction is simplified by operating the absorption cells at constant pressure, temperature, and flow, independent of changes in the ambient sampling environment. The instrument is calibrated against a reference hygrometer on the ground and utilizes an onboard calibration system to verify the stability of the measurement sensitivity during flight. The in-flight calibrations are compared with identical calibration sequences run on the ground to verify the consistency of the calibration system during flight.

The dual-cell design of the instrument allows for simultaneous measurement of water vapor and enhanced total water for the determination of cirrus IWC. Air for the measurement of H_2O vapor is sampled through an inlet oriented perpendicular to the free-stream flow past the aircraft in order to reject ice crystals, while a forward-facing, subisokinetic inlet produces inertially enhanced sampling of cirrus particles to yield high sensitivity to IWC.

The instrument was successfully deployed in multiple flights in the tropical UT/LS onboard the NASA Global Hawk UAS during the second deployment of the ATTREX mission in 2013. Based on flight performance, the overall uncertainty in the measurement of water vapor (1 s, 1σ) is 5 % ± 0.23 ppm, while uncertainty in the determination of cirrus IWC is estimated to be approximately 20 % (15–30 %). The enhancement factor for ice particles larger than 7 μm was calculated to be in the range of 33–48 depending on aircraft altitude, which results in an IWC detection limit of approximately 2 $\mu\mathrm{g\,m^{-3}}$. Measurements of H_2O mixing ratio and cirrus IWC in the Pacific TTL during the ATTREX mission will be used to investigate the dynamic processes and microphysics related to the dehydration of air as it is transported through this region.

Acknowledgements. The authors thank Randy May of Port City Instruments for contributing his experience and expertise to the development of the NOAA Water instrument, and the ground and flight crews of the NASA Global Hawk UAS for their support during the ATTREX mission. This work was supported with funding from the NASA Upper Atmosphere Research Program, the NASA Radiation Sciences Program, the NASA Airborne Tropical

TRopopause EXperiment, and the NOAA Climate Program Office.

Edited by: T. F. Hanisco

References

Belyaev, S. P. and Levin, L. M.: Techniques for collection of representative aerosol samples, J. Aerosol Sci., 5, 325–338, doi:10.1016/0021-8502(74)90130-X, 1974.

Brewer, A. W., Cwilong, B., and Dobson, G. M. B.: Measurement of absolute humidity in very dry air, Proc. Phys. Soc. London, 60, 52–70, doi:10.1088/0959-5309/60/1/307, 1948.

Brown, P. R. A. and Francis, P. N.: Improved Measurement of the Ice Water Content in Cirrus Using a Total-Water Probe, J. Atmos. Ocean. Technol., 12, 410–414, 1995.

Buchholz, B., Kühnreich, B., Smit, H. G. J., and Ebert, V.: Validation of an extractive, airborne, compact TDL spectrometer for atmospheric humidity sensing by blind intercomparison, App. Phys. B, 110, 249–262, doi:10.1007/s00340-012-5143-1, 2013.

Cotton, R. J., Field, P. R., Ulanowski, Z., Kaye, P. H., Hirst, E., Greenaway, R. S., Crawford, I., Crosier, J., and Dorsey, J.: The effective density of small ice particles obtained from in situ aircraft observations of mid-latitude cirrus, Q. J. R. Meteorol. Soc., 139, 1923–1934, doi:10.1002/qj.2058, 2013.

Davis, S. M., Hallar, A. G., Avallone, L. M., and Engblom, W.: Measurement of Total Water with a Tunable Diode Laser Hygrometer: Inlet Analysis, Calibration Procedure, and Ice Water Content Determination, J. Atmos. Ocean. Technol., 24, 463–475, doi:10.1175/JTECH1975.1, 2007a.

Davis, S. M., Avallone, L. M., Weinstock, E. M., Twohy, C. H., Smith, J. B., and Kok, G. L.: Comparisons of in situ measurements of cirrus cloud ice water content, J. Geophys. Res., 112, D10212, doi:10.1029/2006JD008214, 2007b.

Dessler, A. E.: A determination of the cloud feedback from climate variations over the past decade, Science, 330, 1523–1527, doi:10.1126/science.1192546, 2010.

Diskin, G. S., Podolske, J. R., Sachse, G. W., and Slate, T. A.: Open-path airborne tunable diode laser hygrometer, Proc. SPIE 4817, Diode Lasers and Applications in Atmospheric Sensing, 196 (September 24, 2002), doi:10.1117/12.453736, 2002.

Dorsi, S. W., Kalnajs, L. E., Toohey, D. W., and Avallone, L. M.: A fiber-coupled laser hygrometer for airborne total water measurement, Atmos. Meas. Tech., 7, 215–223, doi:10.5194/amt-7-215-2014, 2014.

Eddy, P., Natarajan, A., and Dhaniyala, S.: Subisokinetic sampling characteristics of high speed aircraft inlets: A new CFD-based correlation considering inlet geometries, J. Aerosol Sci., 37, 1853–1870, doi:10.1016/j.jaerosci.2006.08.005, 2006.

Fahey, D. W., Gao, R.-S., Möhler, O., Saathoff, H., Schiller, C., Ebert, V., Krämer, M., Peter, T., Amarouche, N., Avallone, L. M., Bauer, R., Bozóki, Z., Christensen, L. E., Davis, S. M., Durry, G., Dyroff, C., Herman, R. L., Hunsmann, S., Khaykin, S. M., Mackrodt, P., Meyer, J., Smith, J. B., Spelten, N., Troy, R. F., Vömel, H., Wagner, S., and Wienhold, F. G.: The AquaVIT-1 intercomparison of atmospheric water vapor measurement techniques, Atmos. Meas. Tech., 7, 3177–3213, doi:10.5194/amt-7-3177-2014, 2014.

Forster, P. M. de F. and Shine, K. P.: Assessing the climate impact of trends in stratospheric water vapor, Geophys. Res. Lett., 29, 10-1–10-4, doi:10.1029/2001GL013909, 2002.

Fueglistaler, S., Dessler, A. E., Dunkerton, T. J., Folkins, I., Fu, Q., and Mote, P. W.: Tropical Tropopause Layer, Rev. Geophys., 47, RG1004, doi:10.1029/2008RG000267, 2009.

Fueglistaler, S., Liu, Y. S., Flannaghan, T. J., Haynes, P. H., Dee, D. P., Read, W. J., Remsberg, E. E., Thomason, L. W., Hurst, D. F., Lanzante, J. R., and Bernath, P. F.: The relation between atmospheric humidity and temperature trends for stratospheric water, J. Geophys. Res., 118, 1052–1074, doi:10.1002/jgrd.50157, 2013.

Jensen, E. J., Smith, J. B., Pfister, L., Pittman, J. V., Weinstock, E. M., Sayres, D. S., Herman, R. L., Troy, R. F., Rosenlof, K., Thompson, T. L., Fridlind, A. M., Hudson, P. K., Cziczo, D. J., Heymsfield, A. J., Schmitt, C., and Wilson, J. C.: Ice supersaturations exceeding 100 % at the cold tropical tropopause: implications for cirrus formation and dehydration, Atmos. Chem. Phys., 5, 851–862, doi:10.5194/acp-5-851-2005, 2005.

Kaufman, S., Voigt, C., Jeßberger, P., Jurkat, T., Schlager, H., Schwarzenboeck, A., Klingbiel, M., and Thornberry, T.: In situ measurements of ice saturation in young contrails, Geophys. Res Lett., 41, 702–709, doi:10.1002/2013GL058276, 2014.

Kirk-Davidoff, D. B., Hintsa, E. J., Anderson, J. G., and Keith, D. W.: The effect of climate change on ozone depletion through changes in stratospheric water vapour, Nature, 402, 399–401, doi:10.1038/46521, 1999.

Kley, D. and Stone, E. J.: Measurement of water vapor in the stratosphere by photodissociation with Ly α (1216 Å) light, Rev. Sci. Inst., 49, 691–697, doi:10.1063/1.1135596, 1978.

Kley, D., Russell III, J. M., and Phillips, C. (Eds.): SPARC Assessment of Upper Tropospheric and Stratospheric Water Vapour, World Clim. Res. Prog., Geneva, Switzerland, 2000.

Kluczynski, P. and Axner, O.: Theoretical description based on Fourier analysis of wavelength-modulation spectrometry in terms of analytical and background signals, App. Optics, 38, 5803–5815, 1999.

Korolev, A., Emery, E., and Creelman, K.: Modification and tests of particle probe tips to mitigate effects of ice shattering, J. Atmos. Oceanic Technol., 30, 690–708, 2013a.

Korolev, A., Emery, E. F., Strapp, J. W., Cober, S. G., and Isaac, G. A.: Quantification of the effects of shattering on airborne ice particle measurements, J. Atmos. Oceanic Technol., 30, 2527–2553, 2013b.

Krämer, M. and Afchine, A.: Sampling characteristics of inlets operated at low U/U-0 ratios: new insights from computational fluid dynamics (CFX) modeling, J. Aerosol Sci., 35, 683–694, doi:10.1016/j.jaerosci.2003.11.011, 2004.

Liu, Y. S., Fueglistaler, S., and Haynes, P. H.: Advection-condensation paradigm for stratospheric water vapor, J. Geophys. Res., 115, D24307, doi:10.1029/2010JD014352, 2010.

Mastenbrook, H. J. and Oltmans, S. J: Stratospheric Water Vapor Variability for Washington, DC/Boulder, CO: 1964-82, J. Atmos. Sci., 40, 2157–2165, doi:10.1175/1520-0469(1983)040<2157:SWVVFW>2.0.CO;2, 1983.

May, R. D.: Open-path, near-infrared tunable diode laser spectrometer for atmospheric measurements of H_2O, J. Geophys. Res., 103, 19161–19172, doi:10.1029/98JD01678, 1998.

Perring, A. E., Schwarz, J. P., Gao, R. S., Heymsfield, A. J., Schmitt, C. G., Schnaiter, M., and Fahey, D. W.: Evaluation of a Perpendicular Inlet for Airborne Sampling of Interstitial Submicron Black-Carbon Aerosol, Aerosol Sci. Technol., 47, 1066–1072, doi:10.1080/02786826.2013.821196, 2013.

Peter, T., Marcolli, C., Spichtinger, P., Corti, T., Baker, M. B., and Koop, T.: When dry air is too humid, Science, 314, 1399–1402, doi:10.1126/science.1135199, 2006.

Popp, P. J., Gao, R. S., Marcy, T. P., Fahey, D. W., Hudson, P. K., Thompson, T. L., Kärcher, B., Ridley, B. A., Weinheimer, A. J., Knapp, D. J., Montzka, D. D., Baumgardner, D., Garrett, T. J., Weinstock, E. M., Smith, J. B., Sayres, D. S., Pittman, J. V., Dhaniyala, S., Bui, T. P., and Mahoney, M. J.: Nitric acid uptake on subtropical cirrus cloud particles, J. Geophys. Res., 109, D06302, doi:10.1029/2003JD004255, 2004.

Randel, W. J. and Jensen, E. J.: Physical processes in the tropical tropopause layer and their roles in a changing climate, Nat. Geosci., 6, 169–176, doi:10.1038/NGEO1733, 2013.

Rollins, A. W., Thornberry, T. D., Gao, R.-S., Hall, B. D., and Fahey, D. W.: Catalytic oxidation of H_2 on platinum: a robust method for generating low mixing ratio H_2O standards, Atmos. Meas. Tech., 4, 2059–2064, doi:10.5194/amt-4-2059-2011, 2011.

Rollins, A. W., Thornberry, T. D., Gao, R. S., Smith, J. B., Sayres, D. S., Sargent, M. R., Schiller, C., Krämer, M., Spelten, N., Hurst, D. F., Jordan, A. F., Hall, E. G., Vömel, H., Diskin, G. S., Podolske, J. R., Christensen, L. E., Rosenlof, K. H., Jensen, E. J., and Fahey, D. W.: Evaluation of UT/LS hygrometer accuracy by intercomparison during the NASA MACPEX mission, J. Geophys. Res., 119, 1915–1935, doi:10.1002/2013JD020817, 2014.

Rothman, L. S., Gordon, I. E., Barbe, A., Benner, D. C., Bernath, P. F., Birk, M., Boudon, V., Brown, L. R., Campargue, A., Champion, J.-P., Chance, K., Coudert, L. H., Dana, V., Devi, V. M., Fally, S., Flaud, J.-M., Gamache, R. R., Goldman, A., Jacquemart, D., Kleiner, I., Lacome, N., Lafferty, W.J., Mandin, J.-Y., Massie, S. T., Mikhailenko, S. N., Miller, C. E., Moazzen-Ahmadi, N., Naumenko, O. V., Nikitin, A. V., Orphal, J., Perevalov, V. I., Perrin, A., Predoi-Cross, A., Rinsland, C. P., Rotger, M., Šimečkova, M., Smith, M. A. H., Sung, K., Tashkun, S. A., Tennyson, J., Toth, R. A., Vandaele, A.C., and Vander Auwera, J.: The HITRAN 2008 molecular spectroscopic database, J. Quant. Spec. Rad. Transf., 110, 533–572, doi:10.1016/j.jqsrt.2009.02.013, 2009.

Sargent, M. R., Sayres, D. S., Smith, J. B., Witinski, M., Allen, N. T., Demusz, J. N., Rivero, M., Tuozzolo, C., and Anderson, J. G.: A new direct absorption tunable diode laser spectrometer for high precision measurement of water vapor in the upper troposphere and lower stratosphere, Rev. Sci. Inst., 84, 074102, doi:10.1063/1.4815828, 2013.

Schiller, C., Krämer, M., Afchine, A., Spelten, N., and Sitnikov, N.: Ice water content of Arctic, midlatitude, and tropical cirrus, J. Geophys. Res., 113, D24208, doi:10.1029/2008JD010342, 2008.

Schoeberl, M. R. and Dessler, A. E.: Dehydration of the stratosphere, Atmos. Chem. Phys., 11, 8433–8446, doi:10.5194/acp-11-8433-2011, 2011.

Sherwood, S. C. and Dessler, A. E.: A Model for Transport across the Tropical Tropopause, J. Atmos. Sci., 58, 765–779, doi:10.1175/1520-0469(2001)058<0765:AMFTAT>2.0.CO;2, 2001.

Solomon, S., Rosenlof, K. H., Portmann, R. W., Daniel, J. S., Davis, S. M., Sanford, T. J., and Plattner, G.-K.: Contributions of Stratospheric Water Vapor to Decadal Changes in the Rate of Global Warming, Science, 327, 1219–1223, 2010.

Thornberry, T. D., Rollins, A. W., Gao, R. S., Watts, L. A., Ciciora, S. J., McLaughlin, R. J., Voigt, C., Hall, B., and Fahey, D. W.: Measurement of low-ppm mixing ratios of water vapor in the upper troposphere and lower stratosphere using chemical ionization mass spectrometry, Atmos. Meas. Tech., 6, 1461–1475, doi:10.5194/amt-6-1461-2013, 2013.

Twohy, C. H., Schanot, A. J., and Cooper, W. A.: Measurement of condensed water content in liquid and ice clouds using an airborne counterflow virtual impactor, J. Atmos. Ocean. Technol., 14, 197–202, doi:10.1175/1520-0426(1997)014<0197:MOCWCI>2.0.CO;2, 1997.

Vömel, H., David, D. E., and Smith, K.: Accuracy of tropospheric and stratospheric water vapor measurements by the cryogenic frost point hygrometer: Instrumental details and observations, J. Geophys. Res., 112, D08305, doi:10.1029/2006JD007224, 2007.

Weinstock, E. M., Hintsa, E. J., Dessler, A. E., Oliver, J. F., Hazen, N. L., Demusz, J. N., Allen, N. T., Lapson, L. B., and Anderson, J. G.: New fast response photofragment fluorescence hygrometer for use on the NASA ER-2 and the Perseus remotely piloted aircraft, Rev. Sci. Inst., 65, 3544–3554, doi:10.1063/1.1144536, 1994.

Weinstock, E. M., Smith, J. B., Sayres, D., Spackman, J. R., Pittman, J. V., Allen, N., Demusz, J., Greenberg, M., Rivero, M., Solomon, L., and Anderson, J. G.: Measurements of the Total Water Content of Cirrus Clouds. Part I: Instrument Details and Calibration, J. Atmos. Ocean. Technol., 23, 1397–1409, 2006a.

Weinstock, E. M., Smith, J. B., Sayres, D., Pittman, J. V., Allen, N., and Anderson, J. G.: Measurements of the Total Water Content of Cirrus Clouds. Part II: Instrument Performance and Validation, J. Atmos. Ocean. Technol., 23, 1410–1421, 2006b.

Weinstock, E. M., Smith, J. B., Sayres, D. S., Pittman, J. V., Spackman, J. R., Hintsa, E. J., Hanisco, T. F., Moyer, E. J., St. Clair, J. M., Sargent, M. R., and Anderson, J. G.: Validation of the Harvard Lyman-? in situ water vapor instrument: Implications for the mechanisms that control stratospheric water vapor, J. Geophys. Res., 114, D23301, doi:10.1029/2009JD012427, 2009.

Zelinka, M. D. and Hartmann, D. L.: The observed sensitivity of high clouds to mean surface temperature anomalies in the tropics, J. Geophys. Res., 116, D23103, doi:10.1029/2011JD16459, 2011.

Zöger, M., Afchine, A., Eicke, N., Gerhards, M. T., Klein, E., McKenna, D. S., Mörschel, U., Schmidt, U., Tuitjer, F., Woyke, T., and Schiller, C.: Fast in situ hygrometers: A new family of balloon-borne and airborne Lyman alpha photofragment fluorescence hygrometers, J. Geophys. Res.-Atmos., 104, 1807–1816, doi:10.1029/1998JD100025, 1999.

7

Observations of volcanic SO_2 from MLS on Aura

H. C. Pumphrey[1], W. G. Read[2], N. J. Livesey[2], and K. Yang[3]

[1]School of GeoSciences, The University of Edinburgh, Edinburgh, UK
[2]Jet Propulsion Laboratory, California Institute of Technology, Pasadena, CA, USA
[3]Department of Atmospheric and Oceanic Science, University of Maryland, College Park, MD, USA

Correspondence to: H. C. Pumphrey (hugh.pumphrey@ed.ac.uk)

Abstract. Sulfur dioxide (SO_2) is an important atmospheric constituent, particularly in the aftermath of volcanic eruptions. These events can inject large amounts of SO_2 into the lower stratosphere, where it is oxidised to form sulfate aerosols; these in turn have a significant effect on the climate. The MLS instrument on the Aura satellite has observed the SO_2 mixing ratio in the upper troposphere and lower stratosphere from August 2004 to the present, during which time a number of volcanic eruptions have significantly affected those regions of the atmosphere. We describe the MLS SO_2 data and how various volcanic events appear in the data. As the MLS SO_2 data are currently not validated we take some initial steps towards their validation. First we establish the level of internal consistency between the three spectral regions in which MLS is sensitive to SO_2. We compare SO_2 column values calculated from MLS data to total column values reported by the OMI instrument. The agreement is good (within about 1 DU) in cases where the SO_2 is clearly at altitudes above 147 hPa.

1 Introduction

Sulfur dioxide (SO_2) is an important minor constituent of the atmosphere. Natural tropospheric sources include volcanoes, while anthropogenic sources include combustion of fossil fuels and smelting of sulfur-containing metal ores. Tropospheric emission of SO_2 has a variety of detrimental effects on air quality and ecosystems; in particular it can be a major contributor to acid rain (Likens and Bormann, 1974). The high solubility of SO_2 in water which leads to acid rain means that very little of the SO_2 emitted at the surface reaches the stratosphere. Sulfur dioxide in the stratosphere may be produced in situ by chemical reactions involving other, less soluble, sulfur-containing molecules – mostly carbonyl sulfide (OCS) (Brühl et al., 2012). Sulfur dioxide may also reach the upper troposphere or stratosphere if it is emitted directly from the Earth's surface by a process which is sufficiently energetic to loft it to that altitude. A volcanic eruption is the only such process which has been observed to lead to enhanced SO_2 in the lower stratosphere.

Once in the stratosphere, SO_2 becomes an important component of the climate system (Robock, 2000). It is oxidised on a timescale of about a month, becoming aerosol particles which have a lifetime in the stratosphere of over a year. These particles alter the albedo of the Earth, reflecting a fraction of sunlight back into space and thereby reducing the Earth's temperature. Based on six of the largest eruptions of between 1875 and 1991, Robock and Mao (1995) show that the cooling can be on the order of 0.1 to 0.2 °C.

A variety of techniques exist for the remote sensing of atmospheric SO_2 from satellites. Nadir sounding provides good horizontal resolution but little or no vertical resolution. Both thermal emission in the infrared (IR) (Clarisse et al., 2012) and backscattered sunlight in the ultraviolet (UV) (Yang et al., 2007) can be used. Note that SO_2 layer heights may also be retrieved from hyper-spectral UV observations when the columns are sufficiently large (Yang et al., 2010). Limb-sounding instruments provide vertically resolved profiles but with limited horizontal resolution. Thermally emitted radiation in both the infrared (Höpfner et al., 2013) and microwave (Read et al., 1993) regions can be used for limb sounding.

The Microwave Limb Sounder, or MLS (Waters et al., 2006), is one of the four instruments on NASA's Aura satellite (Schoeberl et al., 2006). MLS measures the concentrations of a suite of 14 chemical species in the upper troposphere and middle atmosphere. Sulfur dioxide (SO_2) is one of the species measured. The sensitivity of the measurement is not sufficient to detect the background levels of SO_2, but the enhanced levels present following a sufficiently large volcanic eruption are detected; it is these observations which we report in this paper. We describe the instrument and the measurement process in more detail in Sect. 2, give an overview of the volcanic events observed in Sect. 3, and examine three of the larger events in more detail in Sect. 4. In Sect. 5 we estimate the total mass of SO_2 injected into the stratosphere by the larger volcanic eruptions of the last decade. In Sect. 6 we make an initial attempt to validate the data.

2 Data

2.1 The MLS instrument

Aura (Schoeberl et al., 2006) was launched in July 2004, and the MLS instrument (Waters et al., 2006) has operated with little interruption from August 2004 to date. The satellite orbits at an altitude of 705 km, performing approximately 14.5 orbits per day. The MLS instrument consists of a 1.6 m parabolic dish antenna feeding heterodyne radiometers operating at 118, 190, 240 and 640 GHz. A separate small antenna feeds another radiometer operating at 2.5 THz. The output of the radiometers is analysed by banks of filters. The antenna looks forward from the Aura platform, in the plane of the orbit, and is scanned across the Earth's limb 240 times per orbit. As the orbit is inclined at 98° to the Equator, the instrument observes a latitude range from 82° S to 82° N every day. The observations are of thermal emission from the atmosphere and can therefore be made day and night. The orbit is sun-synchronous, so the observations are always made at the same two local times for a given latitude. The radiances reported by the filter banks are used as input to a software package (Livesey et al., 2006) which estimates profiles of temperature, and of the mixing ratios of the target chemical species. Most MLS estimated profiles, including SO_2, are reported on pressure levels spaced at six levels per pressure decade, a spacing of approximately 2.7 km in altitude. The estimated profiles are spaced 1.5° (167 km) apart along the orbit track. The software produces an estimate of the precision of every quantity estimated. All mixing ratios in this paper are produced by version 2 (V2) of this software package. The precisions and accuracies of these data are summarised in a data quality document (Livesey et al., 2007).

2.2 The MLS SO_2 measurement

The SO_2 molecule, like the H_2O and O_3 molecules, is a nonlinear triatomic molecule. A combination of this shape, a moderately large dipole moment (similar to that of water) and a large moment of inertia leads to a spectrum with a large number of strong lines and an even larger number of weak ones. There are reasonably strong SO_2 lines present in the passband of all MLS radiometers. The software makes separate attempts to estimate the SO_2 mixing ratio from the 190, 240 and 640 GHz radiometers, but only the 240 GHz product is of sufficient quality for general use.

Three versions of the MLS data have been released to the public: V1.5, V2 and V3. Although V3 is an improvement on V2 in most respects, CO and SO_2 data in the upper troposphere are distinguished less well from clouds in V3 than in V2. For this reason we use the V2 data in this paper. The pressure levels on which SO_2 is reported and where significant amounts have been observed are 316, 215, 147, 100, 68 and 46 hPa. At altitudes below the 316 hPa level no retrieval is attempted. Retrieval is attempted between 316 hPa and 1 hPa; the sensitivity and vertical resolution of the retrieval become rapidly poorer at altitudes above 10 hPa. Between 215 and 10 hPa the vertical resolution is close to the spacing between the pressure levels, about 3 km, and the estimated precision is approximately 3 ppbv. Although retrieval is attempted at 315 hPa the data at this level are not recommended for general use. We present them in this paper but are cautious about drawing any conclusions from them. Further details on the data quality are given in Livesey et al. (2007).

Figure 1 shows data from a single pressure level, for all profiles on two different days. On 7 August 2008 the SO_2 mixing ratio, as observed by MLS, was very small compared to the measurement noise. The additional scatter seen in the tropics is caused by high clouds interfering with the retrieval. The standard deviation of the data is generally 2–4 times larger than the estimated precision, suggesting that there are contributions to the random error which are not accounted for in the retrieval software. On 11 August 2008 there are a collection of points with large and positive values; as we show in more detail below these points represent SO_2 emitted from the eruption of Kasatochi on 8 August 2008. Those points which are more than 7.7 standard deviations above the mean are marked in red. The mean and standard deviation used are for the unmarked points only; an iterative procedure was used to separate the marked points. The choice of 7.7 standard deviations as a cut-off is arbitrary and was chosen by trial and error in order to have only a small number of false positives while still identifying the major volcanic events. Figure 1 suggests that there are a number of points which are affected by volcanic SO_2 but which are below this threshold. The 7.7 standard deviation threshold is typically in the 50–100 ppbv range; this is about 100 times larger than the background SO_2 mixing ratio as measured by other instruments (see e.g. Höpfner et al., 2013).

Figure 1 also suggests that the daily zonal mean MLS SO_2 mixing ratio is not quite zero. Inspection of a time series (not shown) shows that it varies somewhat throughout the year. At most altitudes the annual cycle is largest at the poles, where it has an amplitude of about 2 ppbv. The data therefore have a bias which is dependent upon both time and latitude. We do not consider this feature of the data to be due to SO_2 in the

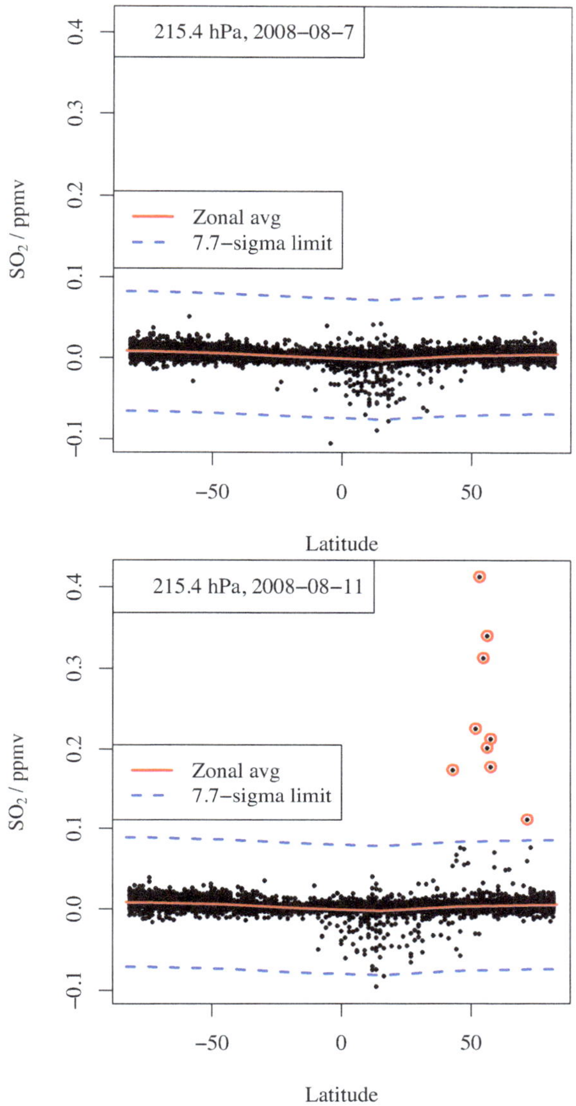

Figure 1. Scatter plot of MLS SO_2 mixing ratio against latitude for 7 and 11 August 2008 at 215 hPa. All profiles (and hence the entire 82° S to 82° N latitude range) are shown. An eruption of the volcano Kasatochi in the Aleutian Islands occurred on 8 August. Points that are more than 7.7 standard deviations above the zonal mean, and which therefore lie within the SO_2 plume caused by the eruption, are marked with red circles.

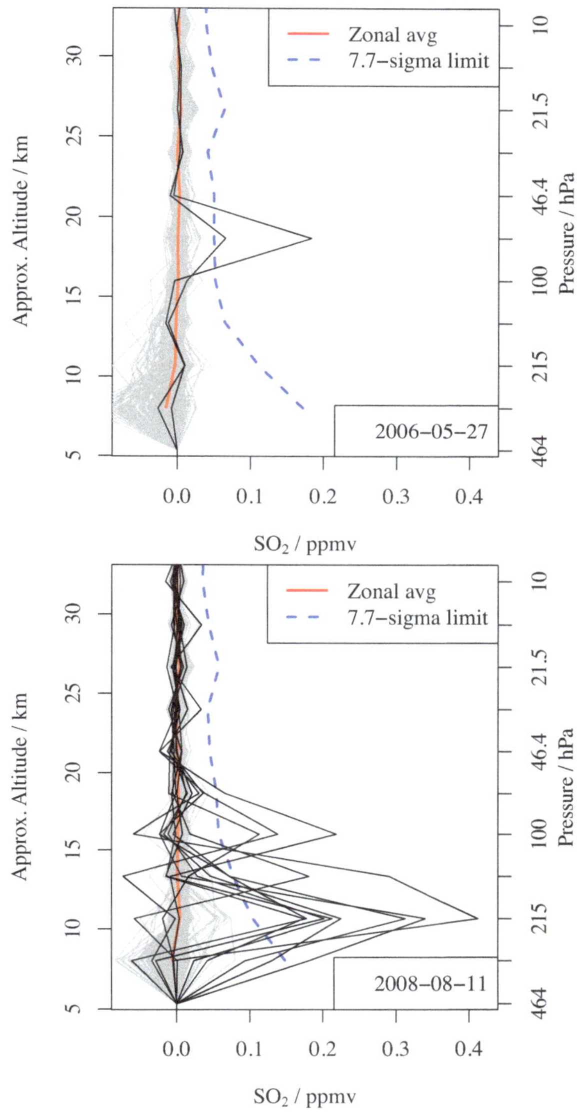

Figure 2. Profiles of SO_2 mixing ratio for a day shortly after the eruption of Soufrière hills, Montserrat, in 2006, and a day shortly after the eruption of Kasatochi in 2008. The Montserrat profiles are from 0 to 20° N; the Kasatochi profiles are from 40 to 60° N. Profiles with unusual amounts of SO_2 as defined in the text are shown in black, the other profiles in grey. The mean of the grey profiles is shown in red; the dashed blue line is 7.7 standard deviations above this mean.

atmosphere for several reasons. Firstly, recent observations of the background SO_2 amounts by Höpfner et al. (2013) show that the true background levels are much smaller than this. Secondly, the seasonal cycles in MLS SO_2 show many features in common with the seasonal cycles in MLS measurements of ozone and nitric acid; these species have strong spectral lines in the passband of the 240 GHz radiometer. We conclude that the seasonal cycles seen in the zonal mean SO_2 data are essentially leakage of information from the O_3 and HNO_3 measurements. While the MLS data appear promising for the study of enhanced levels of SO_2 caused by volcanoes, it is not currently possible to average them down in an attempt to study seasonal variability of the non-volcanic background or other features with amplitudes below about 2–3 ppbv.

In Fig. 2 we show some profiles for 2 days for latitude bands affected by volcanic eruptions. Points with mixing ratios greater than 7.7 standard deviations above the mean are identified as for Fig. 1, and profiles containing such a point

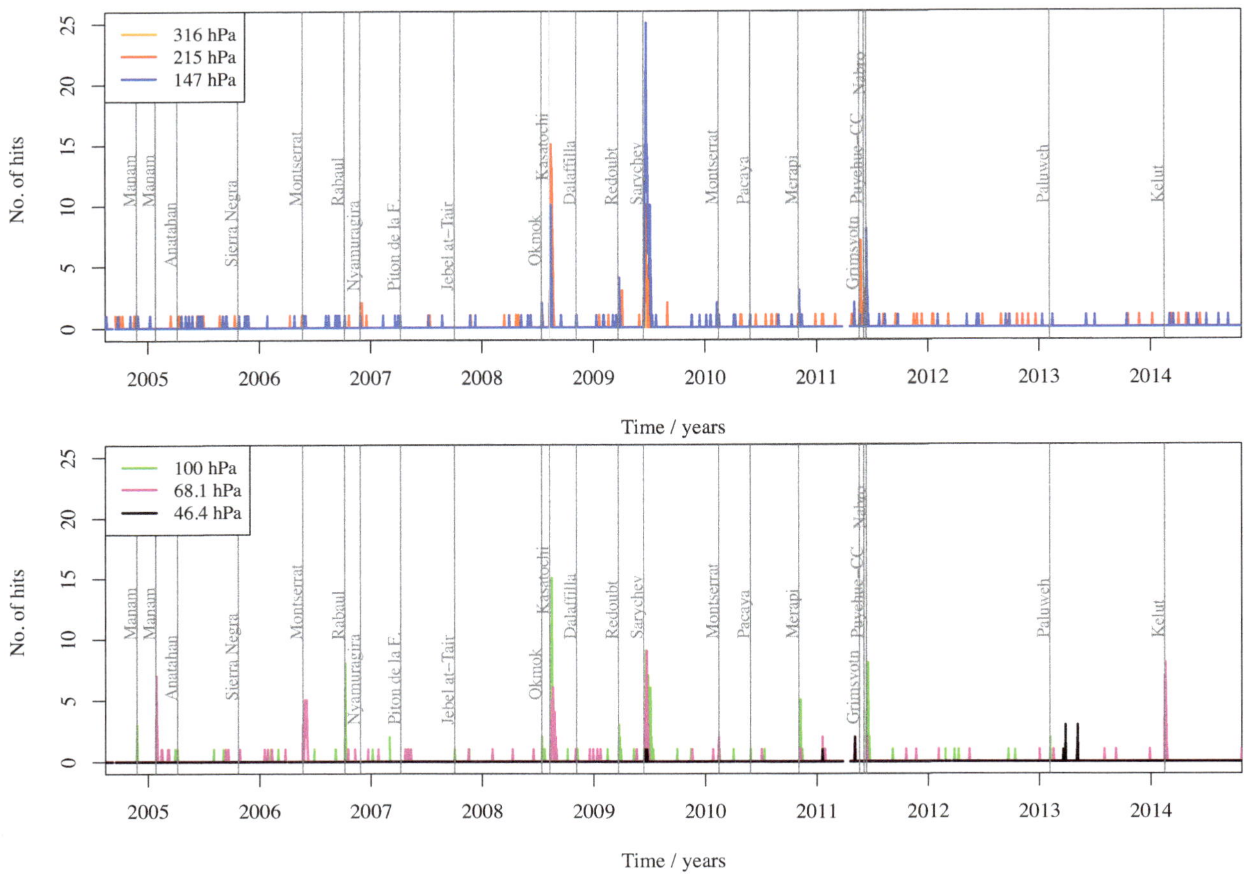

Figure 3. Number of observations of SO_2 per day for the entire mission. For the purposes of this figure an "observation" is a point in the MLS profile which is more than 7.7 standard deviations above the usual zonal mean value for that pressure and latitude.

are shown in black. The two events shown differ from each other in the vertical distribution of SO_2: in one event the SO_2 is all at the 68 hPa level, and in the other it is distributed between 68 and 316 hPa with the largest amounts at 215 hPa.

3 Event detection

We apply the detection procedure used for Figs. 1 and 2 to the entire mission; the result is shown in Fig. 3. The times of volcanic eruptions that are detected by MLS are marked on the figure. The larger events are very obvious. Some of the smaller ones are not obvious; detection was confirmed by a combination of independent reports of an eruption and co-located observations of enhanced SO_2 by the OMI instrument (also on Aura). The events marked on Fig. 3 are summarised in Table 1. Much of the information about names and locations of volcanoes, and times of eruptions, was provided by the Smithsonian Global Volcanism Program (http://www.volcano.si.edu). Figure 4 shows the typical mixing ratios observed in the volcanic plumes; where there is no plume the maximum value observed that day is shown. Figures 3 and 4 show a few spikes in 2011 and 2013 which on closer inspection appear to be faults or glitches in the data and do not appear to be associated with any volcanic activity.

4 Major events in more detail

In this section we examine in more detail several of the volcanic eruptions shown in Fig. 3 and Table 1.

4.1 Sarychev, June 2009

The Sarychev Peak volcano is located in the Russian Kuril island chain, between Japan and the Kamchatka Peninsula. A major eruption of the volcano occurred between 11 and 16 June 2009; the eruption is described by Rybin et al. (2011) and a detailed chronology of the explosion is given by Matoza et al. (2011). The most energetic phases of the eruption were between 12 June and 16 June. We show the time series of MLS observations of elevated SO_2 in Fig. 5 and a map of the locations of these observations in Fig. 6.

Figure 5 is produced in the same manner as Fig. 3 Note that the greatest number of observations occurs at the 147 hPa

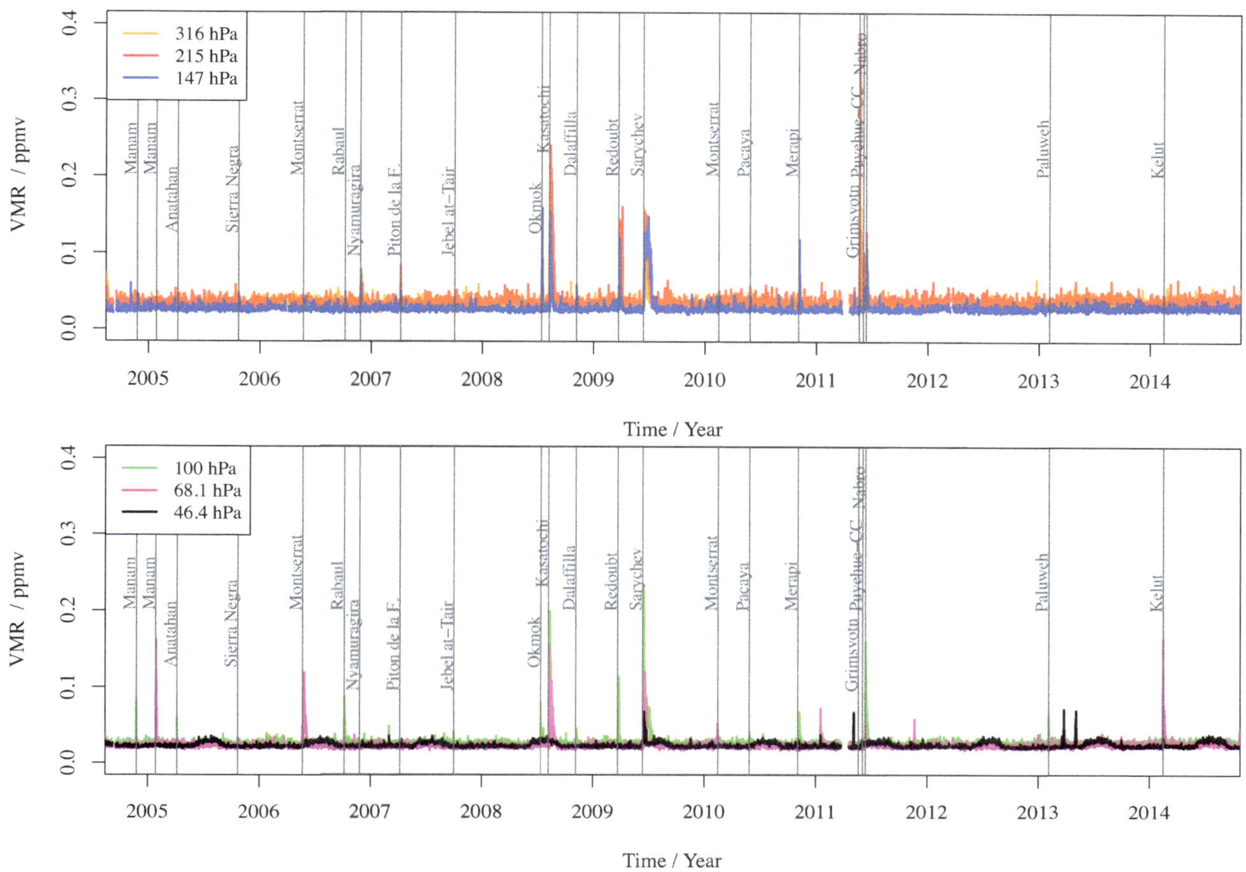

Figure 4. Average SO_2 mixing ratio for "observations" of SO_2 as defined in Fig. 3. Where there are insufficient observations, the three highest mixing ratios observed that day are used instead. This choice means that there is little artificially induced difference in the plotted data between eruption and non-eruption periods. The data plotted when there is no volcanic SO_2 observed are not typical values or daily average values either for the MLS data or for the real background SO_2.

level, but that the first observations (days 164 and 165; 13 and 14 June) are at 215 hPa. Figure 6 suggests that most of the SO_2 travelled eastwards away from the volcano, curving northwards over Canada and Alaska and dividing into two parts: one of which travelled westwards to disperse over eastern Russia, the other of which travelled eastwards across northern Canada and Greenland. A portion of the SO_2 travelled westwards away from the volcano; closer inspection suggests that this is the SO_2 that was emitted towards the end of the eruption.

4.2 Kasatochi, August 2008

The Kasatochi volcano in the Aleutian island chain erupted rather unexpectedly on 7–8 August 2008. The eruption is described by Waythomas et al. (2010) and the evolution of the SO_2 plume as observed by OMI is described by Krotkov et al. (2010).

We show the time series of MLS observations of elevated SO_2 in Fig. 7 and a map of the locations of these observations in Fig. 8. The SO_2 plume appears to be at a slightly lower altitude than that from Sarychev; there are no detections at 46 hPa and the largest number occurs at 215 hPa rather than 146 hPa. All of the observations are to the east of the volcano; the upper level winds were presumably westerly at all altitudes at which MLS observed any volcanic SO_2. The plume experiences considerable wind shear once over North America; the SO_2 at 215 hPa travelling rapidly eastwards across the North Atlantic while that at 100 and 68 hPa remains over North America. The observations persist for a longer time at these three levels than at the intervening 147 hPa level.

4.3 Montserrat, May 2006

The Soufrière Hills volcano on the island of Montserrat in the West Indies underwent a long phase of eruptive activity between 1995 and 2010. The eruption was characterised by the repeated growth and collapse of lava domes (see Wadge et al., 2014, Loughlin et al., 2010, and references therein). Most of this activity had little impact on regions of the at-

Table 1. Table summarising MLS observations of SO_2 during various volcanic eruptions. The eruptions are sorted roughly in order of impact. For the purposes of this table an "observation" is a point in the MLS profile which is more than 7.7 standard deviations above the usual zonal mean value for that pressure and latitude.

Volcano	Location long/lat	Date of first observation	Days observed	Total no. of observations	Highest volume mixing ratio (in ppmv)
Sarychev, Kuril Islands	153.2° W, 48.1° N	14 Jun 2009	31	455	0.55 at 100 hPa
Kasatochi, Aleutian Islands	175.5° W, 52.2° N	8 Aug 2008	23	268	0.46 at 215 hPa
Soufrière Hills, Montserrat	62.2° W, 16.7° N	23 May 2006	17	39	0.18 at 68 hPa
Nabro, Eritrea	41.7° E, 13.4° N	14 Jun 2011	9	73	0.29 at 100 hPa
Kelut, Java	112.3° E,7.93° S	14 Feb 2014	7	48	0.4 at 68 hPa
Grímsvötn, Iceland	17.3° W, 64.4° N	22 May 2011	8	30	0.5 at 215 hPa
Redoubt, Alaska	152.7° W, 60.5° N	23 Mar 2009	13	25	0.2 at 147 hPa
Okmok, Aleutian Islands	168.1° W, 53.4° N	13 Jul 2008	10	16	0.34 at 147 hPa
Manam, Papua New Guinea	145.0° E, 4.1° S	28 Jan 2005	4	22	0.28 at 68 hPa
Rabaul (Tavurvur), New Britain	152.2° E, 4.3° S	8 Oct 2006	4	18	0.16 at 100 hPa
Nyamuragira, Dem. Rep. Congo	29.2° E, 1.4° S	28 Nov 2006	7	7	0.15 at 147 hPa
Manam, Papua New Guinea	145.0° E, 4.1° S	24 Nov 2004	6	7	0.19 at 100 hPa
Puyehue-Cordón Caulle, Chile	72.1° W, 40.6° S	5 Jun 2011	3	6	0.15 at 215 hPa
Dalaffilla (a.k.a. Gabuli), Ethiopia	40.55° E, 13.8° N	5 Nov 2008	4	4	0.1 at 147 hPa
Soufrière Hills, Montserrat	62.2° W, 16.7° N	13 Feb 2010	2	3	0.087 at 68 hPa
Merapi, Indonesia	110.4° E, 7.5° S	6 Nov 2010	4	7	0.18 at 147 hPa
Pacaya, Guatemala	90.6° W, 14.4° N	27 May 2010	2	4	0.086 at 215 hPa
Piton de la Fournaise, Réunion	55.7° E, 21.2° S	7 Apr 2007	1	2	0.13 at 215 hPa
Paluweh, Indonesia	121.7° E, 8.3° S	4 Feb 2013	1	2	0.11 at 100 hPa
Anatahan, Mariana Islands	145.7E° N, 16.4° N	7 Apr 2005	1	1	0.12 at 100 hPa
Sierra Negra, Galápagos Islands	91.2° W, 0.8° S	25 Oct 2005	1	1	0.048 at 147 hPa
Jabal al-Tair, Yemen	41.8° E, 15.6° N	2 Oct 2007	1	1	0.06 at 100 hPa

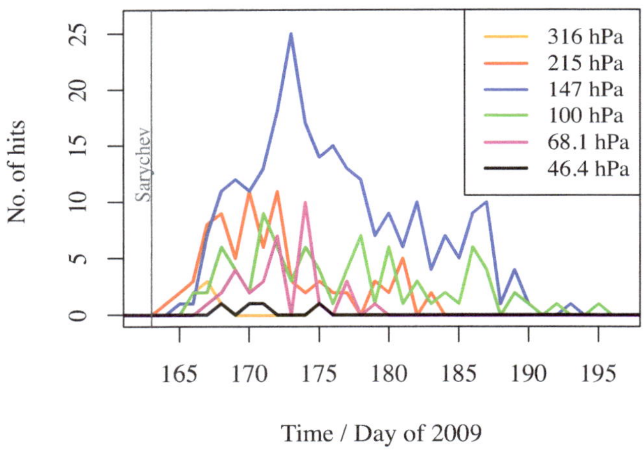

Figure 5. Number of observations of volcanic SO_2 in the days following the eruption of Sarychev. An observation is defined in the same way as in Fig. 3 and Table 1. The eruption is marked as occurring on day 163 (12 June 2009).

mosphere observable by MLS; the main exception was the dome collapse on 20 May 2006. The collapse is described by Loughlin et al. (2010) and the SO_2 release by Carn and Prata (2010) and by Prata et al. (2007).

We show the time series of MLS observations of elevated SO_2 in Fig. 9 and a map of the locations of these observations in Fig. 10. Note that the SO_2 is confined to a narrow layer, affecting only the 68 hPa level in the MLS data. This is in agreement with the modelling results in Prata et al. (2007), which show the SO_2 forming a layer centred at a height of 20 km and full width at half maximum of 2 km. It is some days after the eruption before MLS observes any volcanic SO_2. Inspection of data from the OMI instrument, which has a higher horizontal resolution, shows that the plume was of a small size and fell between the MLS orbits between 20 May and 22 May.

5 Total SO_2 burden

We calculated the total mass of SO_2 in a suitable latitude region for each major eruption. To do this, we first calculated daily zonal means of mixing ratio. These were then integrated vertically as in Sect. 6.2.1 to give zonal mean column amounts. Finally, the column amounts are weighted by area and summed for a suitable range of latitudes. Both the latitude range and the highest pressure used in the vertical integration are chosen separately for each eruption in order to encompass all of the profiles and levels showing enhanced SO_2 from that eruption.

The results are in most cases dominated by the spurious seasonal cycle described in Sect. 2.2. As this varies smoothly in time we can remove it by fitting annual and semi-annual sinusoidal cycles to the data for the time unaffected by the eruption; a typical example is shown in Fig. 11.

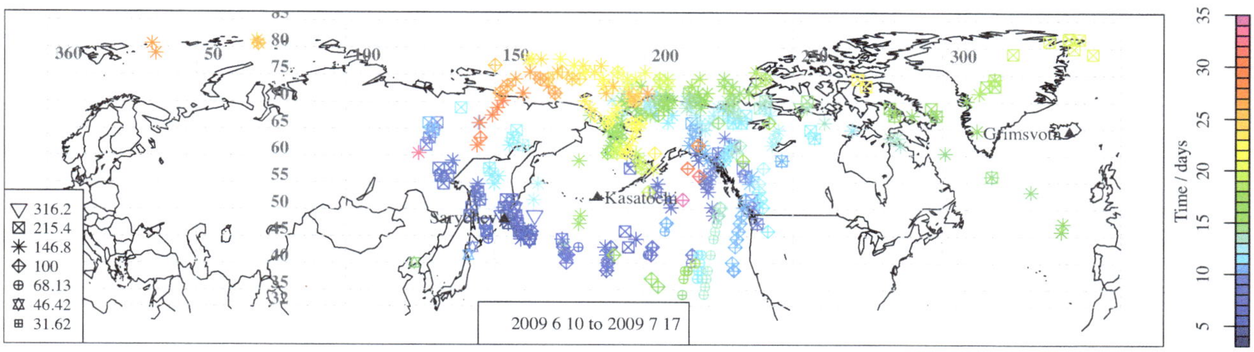

Figure 6. Locations where unusual levels of SO_2 were recorded in the days following the eruption of Sarychev which began on 11 June 2009. Colours represent time in days; day 1 is 10 June 2009. A different symbol is used for each of the MLS pressure levels.

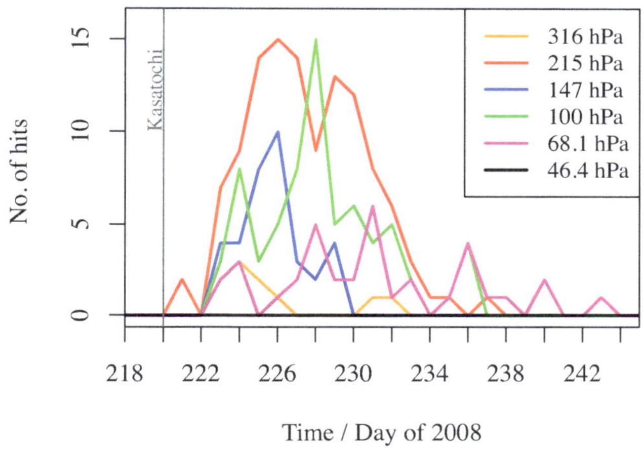

Figure 7. As Fig. 5 but for the eruption of Kasatochi in 2008 which began on 7 August 2008 (day 220).

The excess mass, M_t, above the spurious background can be adequately fitted as a function of time, t, by a decaying exponential:

$$M_t = M_0 \exp\left(\frac{-(t-t_0)}{\tau}\right)$$

or, equivalently

$$\log_e M_t = \log_e M_0 - \frac{(t-t_0)}{\tau},$$

where M_0 is the total mass injected by the volcano at time t_0, and τ is the e-folding time for the conversion of SO_2 to sulfate aerosol. An example of such a fit is shown in Fig. 11, and a summary of the largest events observed by MLS is shown in Fig. 12. Table 2 shows the two fitted parameters M_0 and τ for each of these events.

The agreement between injected mass estimates from MLS and those found in the literature is generally good. Clear discrepancies are the much larger values reported by Clarisse et al. (2012) for Grímsvötn and Nabro. For Grímsvötn this is unsurprising as most of the SO_2 observed by MLS is at 316 and 215 hPa; it seems likely that a fraction of the plume was at rather low altitudes, where MLS would be unable to observe it. For Nabro, most of the SO_2 observed by MLS is at 100 and 147 hPa, where we would expect the MLS observations to be reasonably good. However, Nabro is at a rather low latitude, where the ability of MLS to observe at lower altitudes is more likely to be adversely affected by high levels of water vapour in the upper troposphere. A more recent paper (Clarisse et al., 2014) gives a mass of 650 Gg above 10 km for the initial phase of the eruption; this is in somewhat better agreement with the MLS estimate. It should be noted that all of the eruptions during the Aura mission to date have been small compared to the eruption of Mount Pinatubo in 1991; Read et al. (1993) give a total mass of $M_t = 17\,000$ Gg for that eruption. The e-folding time for Pinatubo was approximately 33 days, in reasonable agreement with the values in Table 2.

6 Initial validation of the MLS data

The data we have described have not been validated. They appear reasonable at a first glance in that the values are smaller than the measurement noise except shortly after the eruption of volcanoes which have been observed erupting from the ground and from other satellites. Further validation from in situ measurements is not straightforward as for most of the time, and at most places there is not sufficient SO_2 for MLS to detect. Volcanic eruptions happen with little warning so it is not usually possible to plan aircraft or balloon campaigns to coincide with them. The most tractable approach is to cross-validate the various satellite SO_2 measurements against each other. While this process can not be considered watertight, we can feel some increased confidence if several satellite measurements using different spectral ranges and observation geometries are in agreement with each other.

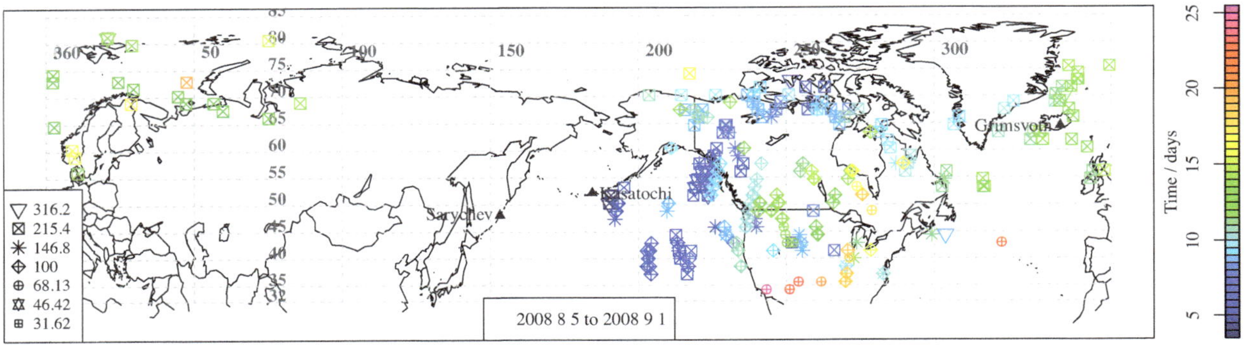

Figure 8. As Fig. 6 but for the eruption of Kasatochi which began on 7 August 2008. Colours represent time in days; day 1 is 5 August 2008.

Table 2. Total injected SO_2 mass M_0 and decay time τ for a selection of the volcanic events observed by MLS. Estimates of M_0 from other sources are shown for comparison. Krotkov et al. (2010) also give an estimate of $\tau = 9$ days for Kasatochi. For Sarychev we show two estimates for two choices of highest pressure used (p_{max}). The 215 hPa result is more use for comparison with nadir-sounder data but is less satisfactory as it proved difficult to remove the seasonal background for this case.

Name	p_{max}/ hPa	τ/days	M_0/Gg	M_0/Gg (from other sources)
Sarychev	147	27 ± 2	571 ± 42	
Sarychev	215	17 ± 3	1160 ± 180	1200 ± 200 (Haywood et al., 2010), 900 (Clarisse et al., 2012)
Kasatochi	215	27 ± 1	1350 ± 38	1373 (D'Amours et al., 2010), 2200 (Krotkov et al., 2010), 500–2500 (various references cited by Krotkov et al., 2010)
Nabro	147	20 ± 2	543 ± 45	1500 (Clarisse et al., 2012), 650 (above 10 km) (Clarisse et al., 2014)
Grímsvötn	215	17 ± 2	108 ± 11	350–400 (Clarisse et al., 2012)
Kelut	100	34 ± 7	144 ± 12	200 (http://so2.gsfc.nasa.gov/pix/special/2014/kelut/Kelut_summary_Feb14_2014.html)
Rabaul	100	34 ± 5	190 ± 14	230 (Carn et al., 2008)
Montserrat	68.1	22 ± 4	139 ± 24	123–233 (Carn and Prata, 2010)
Manam (2005)	68.1	20 ± 4	99 ± 13	

In this section we compare the standard MLS product against SO_2 estimates from the 190 and 640 GHz radiometers on the same instrument. We then compare integrated column values calculated from the MLS data to column values from the OMI instrument.

6.1 Internal consistency

As we noted in Sect. 2.2, MLS makes three partly independent estimates of SO_2: from the 190, 240 and 640 GHz radiometers. Some errors are common to all these estimates: these include errors from the temperature/pointing retrieval. Other errors, such as those from spectroscopic data, should be more-or-less uncorrelated between the three estimates. The field of view of the three radiometers is very similar, to the extent that we can consider the three measurements to be made at exactly the same time and place.

We compare the 190 and 640 GHz products against the standard 240 GHz product in Fig. 13. In order to avoid plotting an excessive number of points which are effectively zero, we consider only periods of time after a major eruption; a different colour is used for each such period. We also show only points from profiles which contain a point at some level

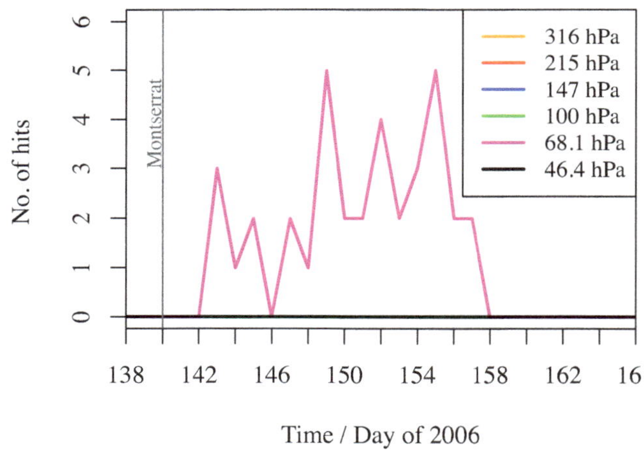

Figure 9. As Fig. 5 but for the eruption of the Soufrière hills volcano, Montserrat, on 20 May 2006.

which is more than 5.5 standard deviations above the mean for that level; this criterion is less restrictive than that used in earlier sections. For each scatter plot in Fig. 13 we fit a straight line to the data in two ways: first we assume that the

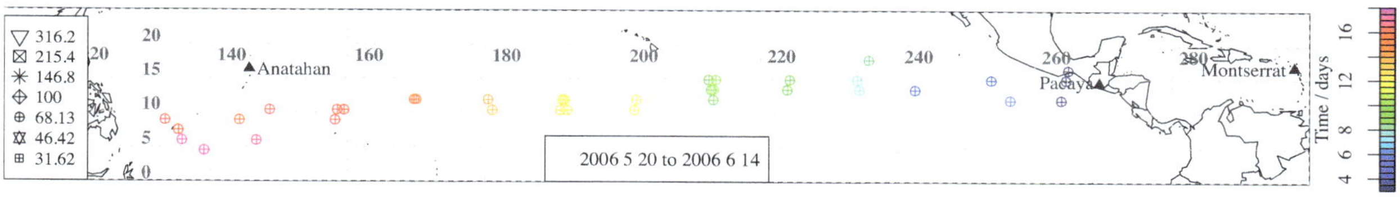

Figure 10. As Fig. 6 but for the eruption of the Soufrière hills volcano, Montserrat, on 20 May 2006. Colours represent time in days; day 1 is 20 May 2006.

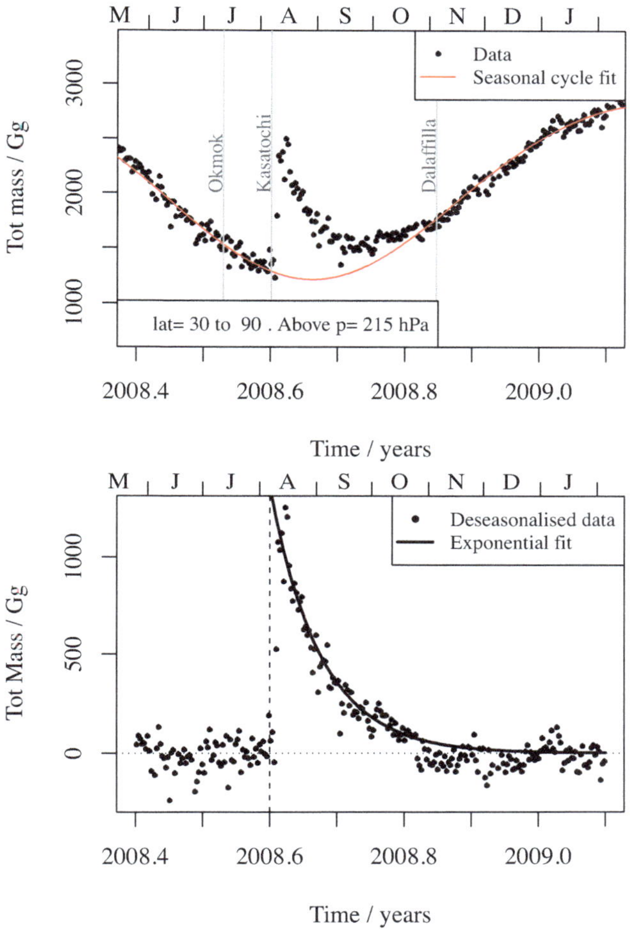

Figure 11. Isolating the volcanic SO_2 from the background value (with its spurious seasonal cycle). The example shown is for the eruption of Kasatochi.

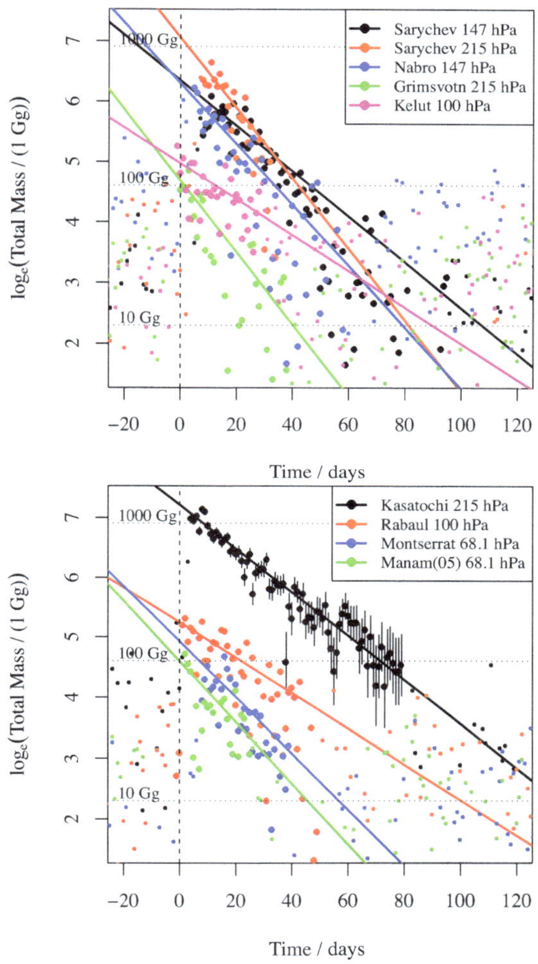

Figure 12. Mass M_t of SO_2 above background as a function of time after an eruption. The quantity plotted is $\log_e M_t$ so that the exponential decay becomes a straight line. The parameters M_0 and τ were determined by fitting the straight lines shown on this figure: the points were weighted with the inverse square of errors which are constant in M_t and hence not so in $\log(M_t)$. These errors are shown for Kasatochi. Only the large dots are used in the fit.

standard product is the independent variable and that all the errors are in the 640 or 190 GHz product (dashed line $y \sim x$ in the figure). Next, we assume that all the errors are in the standard product (dot-dash line $x \sim y$). We note that in general the 640 GHz SO_2 appears to underestimate the standard 240 GHz SO_2; the slopes of the fit lines tend to be considerably less than 1. In contrast, the 190 GHz SO_2 tends to overestimate the standard product. At 46 hPa there is very little correlation between the different SO_2 products, but we note that the four points with mixing ratios above 0.06 ppmv lie close to the 1 : 1 line. At 68 and 100 hPa the correlation is much stronger and a larger fraction of the points are responsible for it.

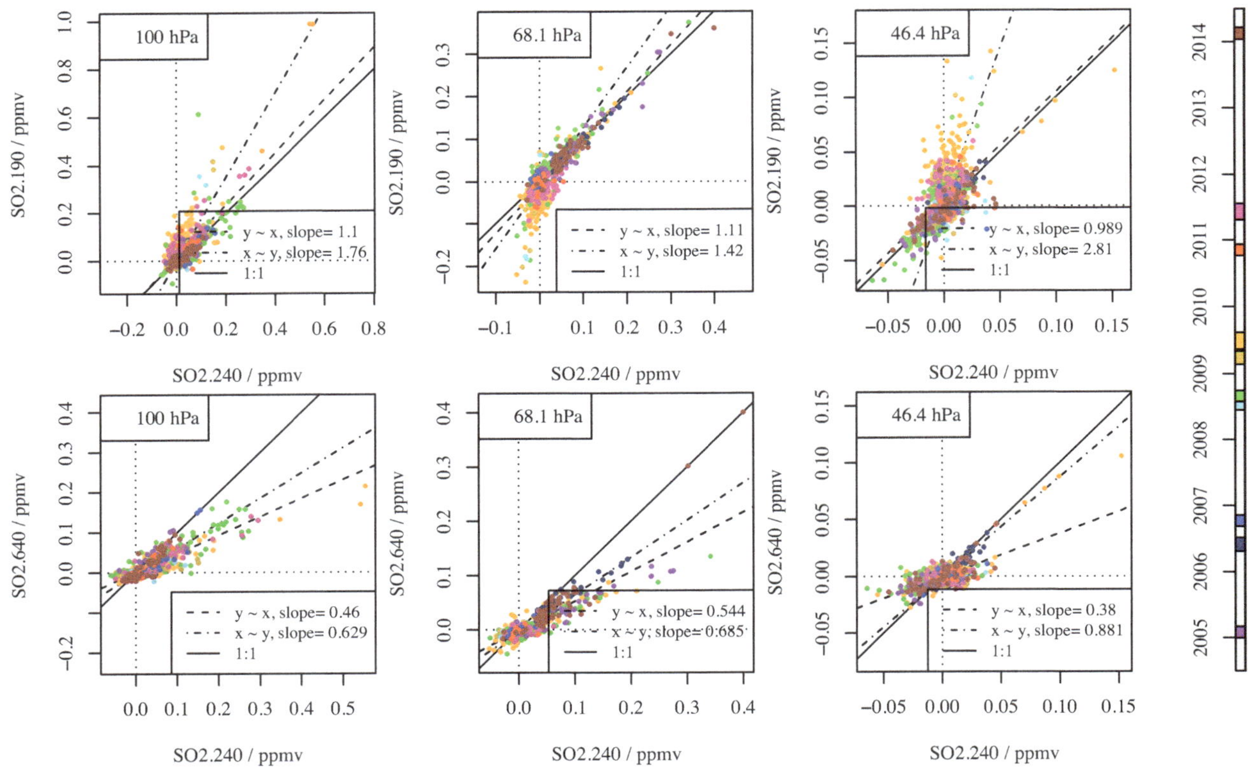

Figure 13. Scatter plots of the 240 GHz SO_2 against the estimates of the same molecule from the 190 (top row) and 640 GHz (bottom row) radiometers. We show only the three pressure levels at which all three radiometers provide usable data. Data are shown only from periods with enhanced volcanic SO_2; a different colour is used for each such period. See Table 1 for details of which volcano was the cause of each period of enhanced SO_2.

6.2 Comparison with OMI

The OMI instrument is described by Levelt et al. (2006). OMI is a nadir UV/visible imaging spectrometer. In its usual operating mode it observes a 2600 km wide swath with 60 image pixels across the width of the swath. The pixels are 24 km across at nadir, becoming wider towards the edges of the swath. In the along-track direction the pixels are 13 km across. The algorithm used to derive total column SO_2 from OMI measurements is described by Yang et al. (2007). It is suitable for most conditions but underestimates the total column where that column is very large. The error can be as large as 70 % for a column of 400 DU dropping to 20 % for a column of 100 DU. The formula used requires that the vertical distribution of SO_2 is specified. The data files contain four separate estimates of column SO_2, each with a different assumption about the vertical profile. The estimates are called PBL, TRL, TRM and STL, corresponding to centre-of-mass altitudes of 0.9, 2.5, 7.5 and 17 km respectively. The README file supplied with the OMI data states that the STL data are to be used when studying explosive volcanic eruptions where the SO_2 is placed in the upper troposphere or lower stratosphere and that differences in actual centre of mass from 17 km produce only small errors. As the useful SO_2 measurements from MLS are at 10 km or above, we compare them only to the STL product from OMI.

The OMI instrument has been affected by a somewhat mysterious problem known as the "row anomaly". This anomaly first appears in the data on 25 June 2007 and affects an increasing number of pixels over the subsequent years.

6.2.1 Method

As the MLS data are vertical profiles of mixing ratio and the OMI data are total column amounts, it is necessary to integrate the MLS profiles with respect to a vertical co-ordinate to form a column amount in order to compare the two data sets. The MLS column value will always be a partial column as the instrument can not observe SO_2 at altitudes below the 315 hPa level. The formula used is derived in Appendix A of Livesey and Snyder (2004), assuming that, between the standard MLS pressure levels, the mixing ratio varies linearly in $\log(p)$. The column amount due only to the SO_2 value at the jth pressure level is

$$N_j = \frac{A}{Mg} f_j \left[\frac{\Delta^+ p}{\Delta^+(\log(p))} - \frac{\Delta^- p}{\Delta^-(\log(p))} \right],$$

where p is pressure, A is Avogadro's number, g is the acceleration due to gravity, M is the average mole mass of air and

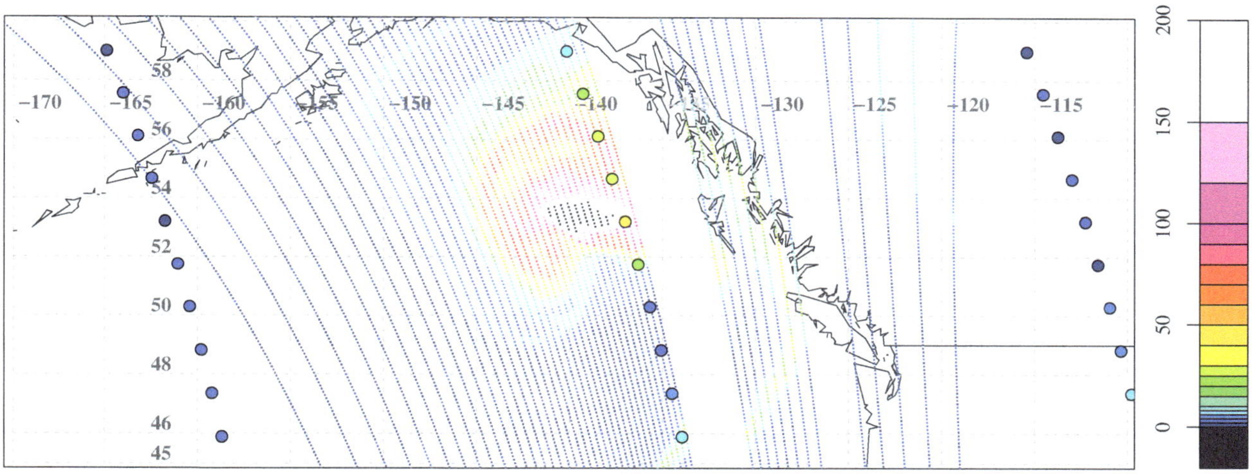

Figure 14. MLS (circles) and OMI (dots) column SO_2 in Dobson units for part of an orbit on 11 August 2008. MLS points are about 160 km apart along the orbit track; two other orbits are shown. OMI pixels are 13 km apart along the orbit track and 24 km apart at nadir across the track. The large gap in the OMI swath to the east of the MLS points is caused by the row anomaly.

f_j is the retrieved volume mixing ratio at the jth pressure level. We use the notation that Δ^+x means the change in x between the jth and the $j+1$th level; Δ^-x is the change in x between the jth and the $j-1$th level. The column value for MLS is obtained by summing the N_j values over a suitable range of j. We use SI units for g and p so that we obtain a value of N_j in molecules m^{-2}; this is then divided by a factor of 2.687×10^{20} m^{-2} in order to convert it to Dobson units. If the jth pressure level is the highest pressure level used, the resulting value is essentially an estimate of the column above a pressure halfway in altitude between level j and level $j-1$. The error in the resulting column is a combination of the errors in the mixing ratios at all pressure levels used. As we note in Sect. 2.2, the random errors in the data appear to be larger than the precision estimates supplied with the data. We therefore estimate the error in the MLS partial columns by calculating the standard deviation of the partial columns in a region or day not affected by volcanic SO_2.

The OMI data are provided as column values in Dobson units. As OMI is a nadir sounder and is on the same platform as MLS, its measurement at a given location is made at 425 ± 10 s after the MLS measurement. This 7 min delay is small enough that we do not attempt to correct for it. As the MLS horizontal field of view is narrower than an OMI pixel, we only consider those OMI pixels which are less than 18 km from the line joining successive MLS profile positions. For each MLS profile we form an average of those OMI pixels which meet this criterion and which are closer to that MLS profile than to any other. For most MLS profiles this means that the single MLS column is compared to a mean value of between 12 and 26 OMI pixels. We also calculate the standard deviation of this set of pixels. The slight oblateness of the Earth means that the coincident pixels are not all in the same pixel row of the OMI swath; the coincident row varies from row 31 (of 60) near the poles to row 39 near the Equator. During the polar summer there may be usable OMI data on the descending half of the orbit; the coincidences for such data may be in a pixel row below the 31st. Figure 14 shows both MLS and OMI column values for a region in the northeast Pacific. Both instruments show a region of large column values; this is due to the eruption of Kasatochi which occurred a few days earlier.

6.2.2 Results

Comparisons as described above can not be performed after January 2009 as the OMI row anomaly affects pixel rows 28–38 from that time onward. We therefore perform such comparisons for periods immediately following the eruptions of Kasatochi on 8 August 2008, Montserrat, on 20 May 2006, Rabaul on 8 October 2006 and Manam on 28 January 2005. The results are shown in Figs. 15 to 18.

The agreement between MLS and OMI is reasonable for Rabaul, Manam and Montserrat, where the SO_2 profile peaks at 68 or 100 hPa. In all three cases, the differences are on the order of 1 DU. This is about 4 times larger than the 0.25 to 0.35 DU variability of the MLS data in the absence of volcanic SO_2, suggesting that a large part of the differences observed is caused by the spatial variability of the volcanic SO_2. For Kasatochi, where there are measurable amounts of SO_2 down to (and presumably below) 315 hPa, the MLS column tends to underestimate the OMI column by many tens of DU, or 100–300 %. These differences are very much larger than the 0.8 DU scatter in MLS columns calculated in the same manner in the absence of volcanic SO_2 and are almost certainly due to the presence of SO_2 in the mid-troposphere where OMI is sensitive to it and MLS is not.

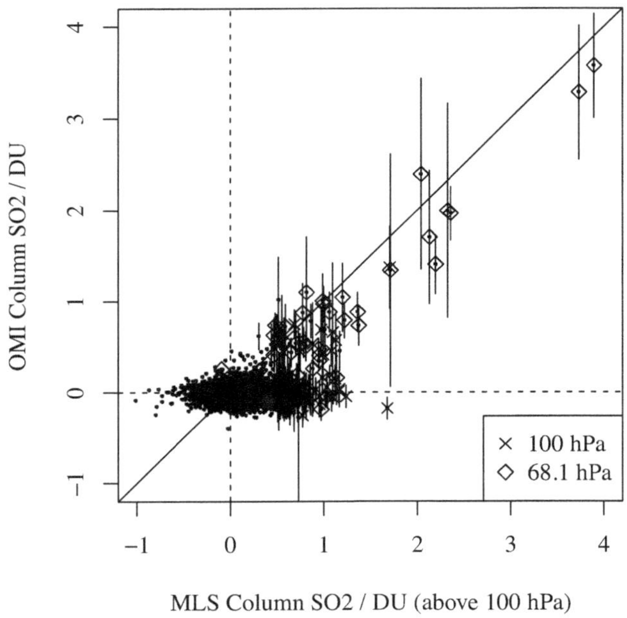

Figure 15. Scatter plot of OMI vs. MLS column SO_2 for the period 20 May 2006 to 9 June 2006 and for the latitude range 9 to 16° N. SO_2 above background levels at this time is attributed to an eruption of the Soufrière Hills volcano in Montserrat. All points are shown by a small dot. Where the MLS SO_2 profile has a clear peak above the background noise, the points are shown with an extra symbol corresponding to the pressure level at which the peak occurs. The error bars represent the standard deviation of the OMI pixels which are averaged to give the values plotted. Note that "above 100 hPa" means that we use the MLS data from this pressure level and levels at higher altitudes. The error in the MLS columns is about 0.25 DU. The solid line shows the ideal 1 : 1 relationship.

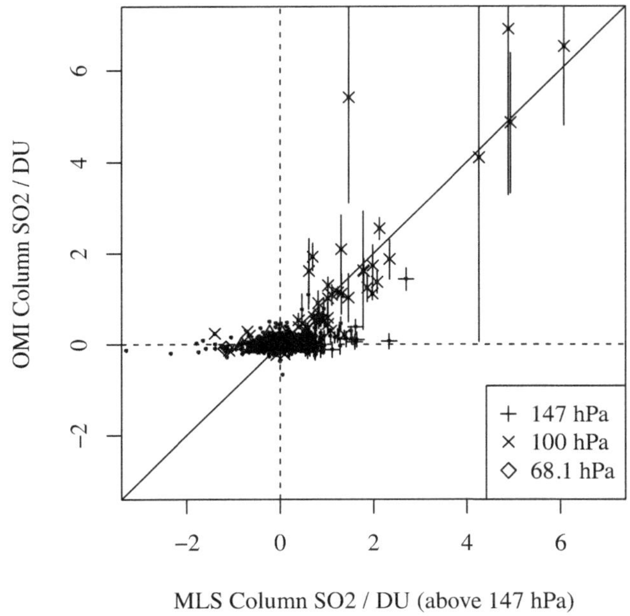

Figure 16. As Fig. 15 but for the period 7–15 October 2006 and for the latitude range 6° S to 8° N. SO_2 above background levels at this time is attributed to an eruption of Rabaul in New Britain, Papua New Guinea. The error in the MLS columns is about 0.35 DU.

6.3 Further validation possibilities

6.3.1 ACE-FTS

The ACE-FTS instrument (Bernath et al., 2005) measures a wide range of chemical species in the upper troposphere and middle atmosphere using the solar occultation technique at visible and near-infrared wavelengths. The technique provides great sensitivity at the cost of limited geographical coverage; data are only available at 15 sunrises and 15 sunsets per day. The latitude at which these events occur changes slowly throughout the year. SO_2 is not a standard ACE-FTS data product, but an experimental SO_2 data set has been produced (Doeringer et al., 2012). We have attempted to compare this data set to the MLS data with no success; the limited coverage of ACE-FTS ensures that on no occasion do the ACE-FTS measurement latitudes coincide with a region of SO_2 which is concentrated enough to be observable by MLS.

6.3.2 MIPAS

MIPAS (Fischer et al., 2008) was an infrared limb-sounding instrument on Envisat. The MIPAS spectrometer was a Fourier transform type, producing spectra with a high spectral resolution. Only a few species were retrieved from these spectra on an operational basis, but experimental retrievals have been produced for a number of other molecules of which SO_2 is one.

Monthly zonal means of SO_2 have been produced by Höpfner et al. (2013). Comparisons of these data to MLS have not proved to be useful as the non-volcanic background in the MLS data is, as we have noted earlier, dominated by systematic errors. The small contribution to the monthly zonal mean from volcanoes affects too small a number of months and latitude bins to provide any kind of useful statistics.

A single-profile retrieval of SO_2 from MIPAS is in development (M. Höpfner, personal communication, 2014). A forthcoming paper (Höpfner et al., 2015) describing that data set will include a comparison with MLS.

6.3.3 IASI

Several recent papers (Clarisse et al., 2012, 2014; Carboni et al., 2012) have demonstrated the potential of the IASI instrument to provide useful SO_2 measurements. In particular, Clarisse et al. (2014) compare the centre of mass of MLS pro-

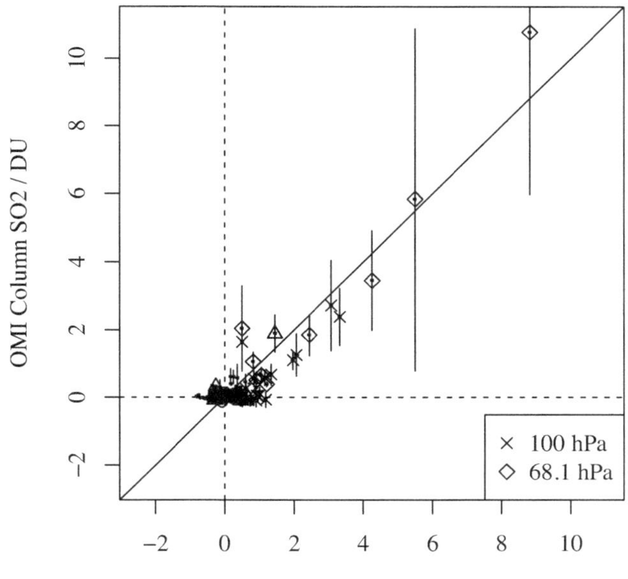

Figure 17. As Fig. 15 but for the period 28–31 January 2005 and for the latitude range 12° S to 7° N. SO_2 above background levels at this time is attributed to an eruption of Manam in Papua New Guinea. The error in the MLS columns is about 0.25 DU.

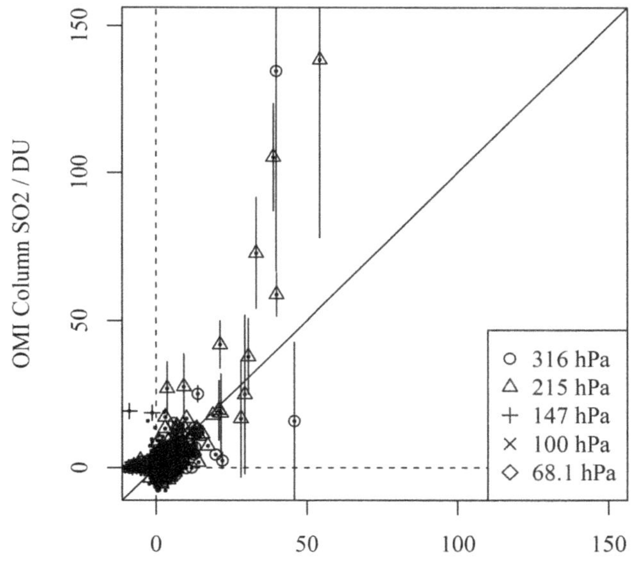

Figure 18. As Fig. 15 but for the period 9–25 August 2008 and for the latitude range 29 to 83° N. SO_2 above background levels at this time is attributed to an eruption of Kasatochi in the Aleutian Islands. The error in the MLS columns is about 0.85 DU.

files with plume heights estimated from IASI and find very good agreement.

7 Discussion

Although the progress we have made in validating the MLS SO_2 data is limited, the results so far are encouraging in that the measurements agree well with OMI, and total masses of SO_2 estimated from the MLS data are in line with values previously published by a variety of authors.

As a tool for studying the dispersion of volcanic SO_2, MLS has both advantages and disadvantages over other currently operating satellite instruments. Its main advantage is that it provides vertical profiles (see Fig. 2) and not just column amounts. The altitude of the SO_2 is therefore measured directly rather than requiring to be inferred from the known meteorology. Its main disadvantage is that the horizontal coverage is very sparse compared to that of most nadir sounders, as shown in Fig. 14. Designs for a future MLS-like instrument (Livesey et al., 2008) provide for viewing directions at a number of angles to the orbit track, allowing for global coverage at a resolution of 50 km; such an instrument would combine the advantages of MLS with those of nadir-sounding instruments and would be particularly valuable for the study of volcanic SO_2.

8 Conclusions

The MLS instrument on Aura has observed enhanced SO_2 mixing ratios following a number of volcanic eruptions of various sizes. Total injected masses of SO_2 calculated from the MLS data agree with previously published values in most cases.

The total column SO_2 calculated from the MLS profiles agrees well with the total column reported by the OMI instrument under the right circumstances. The agreement is good for events where most of the SO_2 is clearly in the stratosphere (Montserrat and Rabaul in 2006, Manam in 2005). Agreement is less good for events where some of the SO_2 is at lower altitudes. This may be because there are significant amounts at altitudes below 215 hPa, where the MLS sensitivity to SO_2 is reduced or zero.

The MLS V2 data show a seasonal cycle with an amplitude of about 2 ppbv which is thought to be spurious. The seasonal effect is smaller than the random error in an individual profile but becomes obvious if sufficient profiles are averaged. This seasonal cycle needs to be removed with some care if calculating the total mass of SO_2 due to a volcanic eruption. Its presence means that the MLS V2 data can not currently be used to study any seasonal cycle which might exist in the non-volcanic SO_2 background.

Acknowledgements. The authors thank Michael Höpfner and Chris Boone for providing preliminary SO_2 retrievals from MIPAS and ACE-FTS for comparison. The authors thank the RCUK open access publication fund for paying publication charges. Work on MLS in the UK has been funded by NERC. MLS data used in this research were produced by the Jet Propulsion Laboratory, California Institute of Technology, under contract with the National Aeronautics and Space Administration.

Edited by: G. Stiller

References

Bernath, P. F., McElroy, C. T., Abrams, M. C., Boone, C. D., Butler, M., Camy-Peyret, C., Carleer, M., Clerbaux, C., Coheur, P.-F., Colin, R., DeCola, P., DeMaziere, M., Drummond, J. R., Dufour, D., Evans, W. F. J., Fast, H., Fussen, D., Gilbert, K., Jennings, D. E., Llewellyn, E. J., Lowe, R. P., Mahieu, E., McConnell, J. C., McHugh, M., McLeod, S. D., Michaud, R., Midwinter, C., Nassar, R., Nichitiu, F., Nowlan, C., Rinsland, C. P., Rochon, Y. J., Rowlands, N., Semeniuk, K., Simon, P., Skelton, R., Sloan, J. J., Soucy, M.-A., Strong, K., Tremblay, P., Turnbull, D., Walker, K. A., Walkty, I., Wardle, D. A., Wehrle, V., Zander, R., and Zou, J.: Atmospheric Chemistry Experiment (ACE): Mission overview, Geophys Res. Lett, 32, L15S01, doi:10.1029/2005GL022386, 2005.

Brühl, C., Lelieveld, J., Crutzen, P. J., and Tost, H.: The role of carbonyl sulphide as a source of stratospheric sulphate aerosol and its impact on climate, Atmos. Chem. Phys., 12, 1239–1253, doi:10.5194/acp-12-1239-2012, 2012.

Carboni, E., Grainger, R., Walker, J., Dudhia, A., and Siddans, R.: A new scheme for sulphur dioxide retrieval from IASI measurements: application to the Eyjafjallajökull eruption of April and May 2010, Atmos. Chem. Phys., 12, 11417–11434, doi:10.5194/acp-12-11417-2012, 2012.

Carn, S. A. and Prata, F. J.: Satellite based constraints on explosive SO_2 release from Soufrière Hills Volcano, Montserrat, Geophys. Res. Lett, 37, L00E22, doi:10.1029/2010GL044971, 2010.

Carn, S. A., Krueger, A. J., Krotkov, N. A., Yang, K., and Evans, K.: Tracking volcanic sulfur dioxide clouds for aviation hazard mitigation, Nat. Hazards, 51, 325–343, doi:10.1007/s11069-008-9228-4, 2008.

Clarisse, L., Hurtmans, D., Clerbaux, C., Hadji-Lazaro, J., Ngadi, Y., and Coheur, P.-F.: Retrieval of sulphur dioxide from the infrared atmospheric sounding interferometer (IASI), Atmos. Meas. Tech., 5, 581–594, doi:10.5194/amt-5-581-2012, 2012.

Clarisse, L., Coheur, P.-F., Theys, N., Hurtmans, D., and Clerbaux, C.: The 2011 Nabro eruption, a SO_2 plume height analysis using IASI measurements, Atmos. Chem. Phys., 14, 3095–3111, doi:10.5194/acp-14-3095-2014, 2014.

D'Amours, R., Malo, A., Servranckx, R., Bensimon, D., Trudel, S., and Gauthier-Bilodeau, J.-P.: Application of the atmospheric Lagrangian particle dispersion model MLDP0 to the 2008 eruptions of Okmok and Kasatochi volcanoes, J. Geophys. Res., 115, D00L11, doi:10.1029/2009JD013602, 2010.

Doeringer, D., Eldering, A., Boone, C. D., González Abad, G., and Bernath, P. F.: Observation of sulfate aerosols and SO2 from the Sarychev volcanic eruption using data from the Atmospheric Chemistry Experiment (ACE), J. Geophys. Res., 117, D03203, doi:10.1029/2011JD016556, 2012.

Fischer, H., Birk, M., Blom, C., Carli, B., Carlotti, M., von Clarmann, T., Delbouille, L., Dudhia, A., Ehhalt, D., Endemann, M., Flaud, J. M., Gessner, R., Kleinert, A., Koopman, R., Langen, J., López-Puertas, M., Mosner, P., Nett, H., Oelhaf, H., Perron, G., Remedios, J., Ridolfi, M., Stiller, G., and Zander, R.: MIPAS: an instrument for atmospheric and climate research, Atmos. Chem. Phys., 8, 2151–2188, doi:10.5194/acp-8-2151-2008, 2008.

Haywood, J. M., Jones, A., Clarisse, L., Bourassa, A., Barnes, J., Telford, P., Bellouin, N., Boucher, O., Agnew, P., Clerbaux, C., Coheur, P., Degenstein, D., and Braesicke, P.: Observations of the eruption of the Sarychev volcano and simulations using the HadGEM2 climate model, J. Geophys. Res., 115, D21212, doi:10.1029/2010JD014447, 2010.

Höpfner, M., Glatthor, N., Grabowski, U., Kellmann, S., Kiefer, M., Linden, A., Orphal, J., Stiller, G., von Clarmann, T., Funke, B., and Boone, C. D.: Sulfur dioxide (SO_2) as observed by MIPAS/Envisat: temporal development and spatial distribution at 15–45 km altitude, Atmos. Chem. Phys., 13, 10405–10423, doi:10.5194/acp-13-10405-2013, 2013.

Höpfner, M., Boone, C. D., Funke, B., Glatthor, N., Grabowski, U., Gunther, A., Kellmann, S., Kiefer, M., Linden, A., Pumphrey, H. C., Read, W. G., Roiger, A., Stiller, G., Schlager, H., von Clarmann, T., and Wissmuller, K.: Sulfur dioxide (SO_2) from MIPAS in the upper troposphere and lower stratosphere, Atmos. Chem. Phys. Discuss., in preparation, 2015.

Krotkov, N. A., Schoeberl, M. R., Morris, G. A., Carn, S., and Yang, K.: Dispersion and lifetime of the SO_2 cloud from the August 2008 Kasatochi eruption, J. Geophys. Res., 115, D00L20, doi:10.1029/2010JD013984, 2010.

Levelt, P. F., van den Oord, G. H. J., Dobber, M. R., Mälkki, A., Visser, H., de Vries, J., Stammes, P., Lundell, J., and Saari, H.: The Ozone Monitoring Instrument, IEEE T. Geosci. Remote, 44, 1093–1101, doi:10.1109/TGRS.2006.872333, 2006.

Likens, G. E. and Bormann, F. H.: Acid Rain: A Serious Regional Environmental Problem, Science, 184, 1176–1179, doi:10.1126/science.184.4142.1176, 1974.

Livesey, N., Santee, M., Stek, P., Waters, J., Levelt, P., Veefkind, P., Kumer, J., and Roche, A.: A future "Global Atmospheric Composition Mission" (GACM) concept, in: 2008 IEEE Aerospace Conference, IEEE, doi:10.1109/AERO.2008.4526243, 2008.

Livesey, N. J. and Snyder, W. V.: EOS MLS Retrieval Processes Algorithm Theoretical Basis, Tech. Rep. JPL D-16159, JPL, available at: http://mls.jpl.nasa.gov/data/eos_algorithm_atbd.pdf (last access: 5 January 2015), version 2.0, 2004.

Livesey, N. J., Snyder, W. V., Read, W. G., and Wagner, P. A.: Retrieval algorithms for the EOS Microwave Limb Sounder (MLS) instrument, IEEE T. Geosci. Remote, 44, 1144–1155, 2006.

Livesey, N. J., Read, W. G., Lambert, A., Cofield, R. E., Cuddy, D. T., Froidevaux, L., Fuller, R. A., Jarnot, R. F., Jiang, J. H., Jiang, Y. B., Knosp, B. W., Kovalenko, L. J., Pickett, H. M., Pumphrey, H. C., Santee, M. L., Schwartz, M. J., Stek, P. C., Wagner, P. A., Waters, J. W., and Wu, D. L.: Earth Observing System (EOS) Aura Microwave Limb Sounder (MLS) Version 2.2 and 2.3 Level 2 data quality and description document., Tech. Rep. JPL D-33509, NASA Jet Propulsion Laboratory California Institute of Technology, Pasadena, California, 91109-8099, available at: http://mls.jpl.nasa.gov (last access: 5 January 2015), 2007.

Loughlin, S. C., Luckett, R., Ryan, G., Christopher, T., Hards, V., Angelis, S. D., Jones, L., and Strutt, M.: An overview of lava dome evolution, dome collapse and cyclicity at Soufrière Hills Volcano, Montserrat, 2005–2007, Geophys. Res. Lett., 37, L00E16, doi:10.1029/2010GL042547, 2010.

Matoza, R. S., Pichon, A. L., Vergoz, J., Herry, P., Lalande, J.-M., il Lee, H., Che, I.-Y., and Rybin, A.: Infrasonic observations of the June 2009 Sarychev Peak eruption, Kuril Islands: Implications for infrasonic monitoring of remote explosive volcanism, J. Volcanol. Geoth. Res., 200, 35–48, 2011.

Prata, A. J., Carn, S. A., Stohl, A., and Kerkmann, J.: Long range transport and fate of a stratospheric volcanic cloud from Soufrière Hills volcano, Montserrat, Atmos. Chem. Phys., 7, 5093–5103, doi:10.5194/acp-7-5093-2007, 2007.

Read, W. G., Froidevaux, L., and Waters, J. W.: Microwave Limb Sounder measurement of stratospheric SO_2 from the Mt. Pinatubo volcano, Geophys. Res. Lett., 20, 1299–1302, doi:10.1029/93GL00831, 1993.

Robock, A.: Volcanic eruptions and climate, Rev. Geophys., 38, 191–219, doi:10.1029/1998RG000054, 2000.

Robock, A. and Mao, J.: The Volcanic Signal in Surface Temperature Observations, J. Climate, 8, 1086–1103, 1995.

Rybin, A., Chibisova, M., Webley, P., Steensen, T., Izbekov, P., Neal, C., and Realmuto, V.: Satellite and ground observations of the June 2009 eruption of Sarychev Peak volcano, Matua Island, Central Kuriles, B. Volcanol., 73, 1377–1392, doi:10.1007/s00445-011-0481-0, 2011.

Schoeberl, M. R., Douglass, A. R., Hilsenrath, E., Bhartia, P. K., Barnett, J., Beer, R., Waters, J., Gunson, M., Froidevaux, L., Gille, J., Levelt, P. F., and DeCola, P.: Overview of the EOS Aura Mission, IEEE T. Geosci. Remote, 44, 1066–1074, 2006.

Wadge, G., Robertson, R., and Voight, B.: Eruption of Soufriere Hills Volcano, Montserrat from 2000 to 2010, in: Eruption of Soufriere Hills Volcano, Montserrat from 2000 to 2010, edited by: Wadge, G., Robertson, R. E. A., and Voight, B., Vol. 39 of Geological Society Memoirs, 1–501, doi:10.1144/M39.0, 2014.

Waters, J. W., Froidevaux, L., Harwood, R., Jarnot, R., Pickett, H., Read, W., Siegel, P., Cofield, R., Filipiak, M., Flower, D., Holden, J., Lau, G., Livesey, N., Manney, G., Pumphrey, H., Santee, M., Wu, D., Cuddy, D., Lay, R., Loo, M., Perun, V., Schwartz, M., Stek, P., Thurstans, R., Boyles, M., Chandra, S., Chavez, M., Chen, G.-S., Chudasama, B., Dodge, R., Fuller, R., Girard, M., Jiang, J., Jiang, Y., Knosp, B., LaBelle, R., Lam, J., Lee, K., Miller, D., Oswald, J., Patel, N., Pukala, D., Quintero, O., Scaff, D., Snyder, W., Tope, M., Wagner, P., and Walch, M.: The Earth Observing System Microwave Limb Sounder (EOS MLS) on the Aura satellite, IEEE T. Geoscience Remote, 44, 1106–1121, 2006.

Waythomas, C. F., Scott, W. E., Prejean, S. G., Schneider, D. J., Izbekov, P., and Nye, C. J.: The 7–8 August 2008 eruption of Kasatochi Volcano, central Aleutian Islands, Alaska, J. Geophys. Res., 115, B00B06, doi:10.1029/2010JB007437, 2010.

Yang, K., Krotkov, N. A., Krueger, A. J., Carn, S. A., Bhartia, P. K., and Levelt, P. F.: Retrieval of large volcanic SO_2 columns from the Aura Ozone Monitoring Instrument: Comparison and limitations, J. Geophys. Res., 112, D24S43, doi:10.1029/2007JD008825, 2007.

Yang, K., Liu, X., Bhartia, P. K., Krotkov, N. A., Carn, S. A., Hughes, E. J., Krueger, A. J., Spurr, R. J. D., and Trahan, S. G.: Direct retrieval of sulfur dioxide amount and altitude from spaceborne hyperspectral UV measurements: Theory and application, J. Geophys. Res., 115, D00L09, doi:10.1029/2010JD013982, 2010.

8

Using digital image processing to characterize the Campbell–Stokes sunshine recorder and to derive high-temporal resolution direct solar irradiance

A. Sanchez-Romero[1], J. A. González[1], J. Calbó[1], and A. Sanchez-Lorenzo[1,2]

[1]Department of Physics, University of Girona, Girona, Spain
[2]Pyrenean Institute of Ecology, Spanish National Research Council (CSIC), Zaragoza, Spain

Correspondence to: A. Sanchez-Romero (alejandro.sanchez@udg.edu)

Abstract. The Campbell–Stokes sunshine recorder (CSSR) has been one of the most commonly used instruments for measuring sunshine duration (SD) through the burn length of a given CSSR card. Many authors have used SD to obtain information about cloudiness and solar radiation (by using Ångström–Prescott type formulas), but the burn width has not been used systematically. In principle, the burn width increases for increasing direct beam irradiance. The aim of this research is to show the relationship between burn width and direct solar irradiance (DSI) and to prove whether this relationship depends on the type of CSSR and burning card. A method of analysis based on image processing of digital scanned images of burned cards is used. With this method, the temporal evolution of the burn width with 1 min resolution can be obtained. From this, SD is easily calculated and compared with the traditional (i.e., visual) determination. The method tends to slightly overestimate SD, but the thresholds that are used in the image processing could be adjusted to obtain an improved estimation. Regarding the burn width, experimental results show that there is a high correlation between two different models of CSSRs, as well as a strong relationship between burn widths and DSI at a high-temporal resolution. Thus, for example, hourly DSI may be estimated from the burn width with higher accuracy than based on burn length (for one of the CSSR, relative root mean squared error is 24 and 30 %, respectively; mean bias error is -0.6 and $-30.0\,\mathrm{W\,m^{-2}}$, respectively). The method offers a practical way to exploit long-term sets of CSSR cards to create long time series of DSI. Since DSI is affected by atmospheric aerosol content, CSSR records may also become a proxy measurement for turbidity and atmospheric aerosol loading.

1 Introduction

Sunshine duration (SD) is a useful indicator of the amount of solar radiation arriving on the earth and a key variable for various sectors, including tourism, public health, agriculture, and energy. According to the World Meteorological Organization (WMO, 2008), the SD for a given period is defined as the total time length of those sub-periods for which the direct solar irradiance exceeds $120\,\mathrm{W\,m^{-2}}$. For climatological purposes, the units used are "hours per day" as well as percentage quantities such as "relative daily sunshine duration", where SD is divided by the maximum possible SD (i.e., as if sky was clear and a bright sun was present during the entire day, from sunrise to sunset).

One of the most used instruments to measure SD is the Campbell–Stokes sunshine recorder (CSSR). It was invented in the late 19th century to provide a measurement of the duration of bright sunlight by making a burn mark on a piece of specially treated cardboard. The measurement of the length of the burn for a given card gives daily SD. For details on the history of the CSSR, we refer to Stanhill (2003) and Sanchez-Lorenzo et al. (2013). In brief, the main parts of a current CSSR are a sphere made of transparent glass and a rounded metal frame placed behind the sphere. The glass sphere is designed to focus the Sun's rays onto a piece of recording cardboard. The metal frame part has three overlapping sets of grooves to hold the recording cards for the winter, summer, and spring/autumn periods. The recording card has to be replaced daily after sunset. Different designs of cards exist with hourly and half-hourly divisions marked across these cards, enabling determination of the times of sunshine, and an estimated resolution of 0.1 h. For further details on the instrument and instructions for obtaining uniform results, as well as other traditional instruments for measuring SD, see Middleton (1969) and WMO (2008).

Over the last few years, various automated instruments and other methods for obtaining SD have been developed, which are summarized in WMO (2008). One of these is the pyrheliometric method, which is based on direct irradiance measurements (e.g., Hinssen, 2006; Hinssen and Knap, 2007; Vuerich et al., 2012). Another way of determining SD is by means of automatic instruments specifically designed to this end, which have become commercially available (Wood et al., 2003; Kerr and Tabony, 2004; Matuszko, 2014). These instruments detect direct solar radiation and count the time interval during which the irradiance exceeds a certain threshold. Progressively, many weather stations have changed traditional manual instruments (such as CSSR and Jordan photographic recorders) to these automatic systems. Moreover, different methods exist nowadays to estimate SD from geostationary satellite data, which potentially provide improved spatial coverage and representativeness (Olseth and Skartveit, 2001; Good, 2010; Kothe et al., 2013).

Despite the fact that new models of sunshine recorders do not require daily attention by an observer and their data reduction (i.e., the process of recording and storing the SD data) is faster and more accurate, there is a consensus with regard to preserving CSSR type instruments at long-established (in some cases more than 120 years ago), well-maintained, and freely exposed meteorological stations (Stanhill, 2003; Wood and Harrison, 2011; Sanchez-Lorenzo et al., 2013).

A change of the instrument used to measure SD can affect the homogeneity of the data series. This may lead to significant errors when evaluating trends and may hinder the possibility of determining long-term secular trends (Powell, 1983; Steurer and Karl, 1991; Brázdil et al., 1994; Stanhill and Cohen, 2008). Among the studies that analyze long series of SD, some works choose to restrict the period in order to avoid instrumental changes, therefore not encompassing the entire period of observation (Angell et al., 1984). Other studies assess the homogeneity of the SD series, such as research conducted in the United States (Cerveny and Balling, 1990; Stanhill and Cohen, 2005), United Kingdom (Kerr and Tabony, 2004), Iberian Peninsula (Guijarro, 2007; Sanchez-Lorenzo et al., 2007), Japan (Katsuyama, 1987; Stanhill and Cohen, 2008), China (Xia, 2010), and Switzerland (Sanchez-Lorenzo and Wild, 2012). The results of these studies point towards the need for further research, including homogenization of the long-term SD series and the necessity of simultaneous measurements by both traditional and automatic instrumentation (Aguilar et al., 2003).

Differences between types of SD measurements might be attributed to their particular characteristics and limitations. As the longer SD series are generally measured using CSSR, the errors connected with this instrument are well-described (Painter, 1981; Brázdil et al., 1994). The two major problems with CSSR when comparing their measurements with other methods or instruments lie in the variability of the level of direct irradiance which produces a burn and the overburning of the card in conditions of intermittent high irradiance (Stanhill, 2003). These difficulties can be added to the obvious element of subjectivity in measuring the burn length on the CSSR cards (Brázdil et al., 1994). The problem of overburning is very difficult to evaluate as one small burst of high direct irradiance causes a burn apparently lasting far longer than the few seconds of its actual duration; standard methods have therefore been proposed to take into account this fact when evaluating the burn lengths (WMO, 2008). Despite these rules, during events of very broken cloudiness a measurement of SD by means of CSSR can be significantly overestimated (Painter, 1981; Kerr and Tabony, 2004).

Regarding the first problem, defining the direct irradiance value that produces burning ("burning threshold") is a well-known issue. One proof is the variability of values that the WMO has given: in 1971, WMO suggested that the threshold can vary between 70 and 280 $W\,m^{-2}$; in 1976, WMO recommended a threshold value of 200 $W\,m^{-2}$; finally, in 1981, a value of 120 $W\,m^{-2}$ was recommended (WMO, 2008). Gueymard (1993) tried to give some scientific justification to this value. Nevertheless, Bider (1958), Jaenicke and Kasten (1978), and Roldán et al. (2005) showed a large variety of burning thresholds for different CSSR. Similarly, Helmes and Jaenicke (1984) described the effects of using different types of recording cards. In addition, the measurement of threshold values indicated that there were notable losses of record, which must be attributed to dew or other water deposits on the glass sphere. Painter (1981) obtained monthly averaged threshold values ranging from 16 to 142 $W\,m^{-2}$ despite some thresholds up to 400 $W\,m^{-2}$ that were obtained in particular conditions. From the above discussion it is clear that long time series of SD include errors of several kinds and that their removal is a complicated issue. It is also important to stress that the problem of thresholds is not exclusive to the CSSR; it has been studied for other instruments such as the Foster sunshine recorder (Baker and Haines, 1969; Benson et al., 1984; Michalsky, 1992).

SD series also provide additional embedded information on other magnitudes. For example, many authors relate SD for a period of time with direct and global (the sum of the solar direct and diffuse contributions) irradiance with the use of the so-called Ångström–Prescott type formulas (e.g., Sears et al., 1981; Benson et al., 1984; Martínez-Lozano et al., 1984; Stanhill, 1998; Power, 2001; Suehrcke, 2000; Bakirci, 2009), which were first proposed by Ångström (1924) and further modified by Prescott (1940). In addition, R. Jaenicke and L. Helmes developed a series of pioneering studies presenting a method to determine atmospheric turbidity from SD records and cloud cover data (Jaenicke and Kasten, 1978; Helmes and Jaenicke, 1984, 1985, 1986) that has recently aroused interest (for a review, see Sanchez-Romero et al., 2014). SD data are also used in other fields such as agriculture (Monteith, 1977; Stanhill and Cohen, 2001) or hydrological modeling (Döll et al., 2003).

This paper describes an analysis method to derive direct solar irradiance (DSI) from the CSSR burned cards by using digital image processing. The idea we assume here was first proposed by Wright (1935): the size of the burn at any point is related to the strength of the DSI focused on the card at that time. Galindo Estrada and Fournier D'Albe (1960) applied a similar approach: they compared daily values of the mass loss of burned cards (which is related to the perforation of the card and thus to mean burn width) with pyrheliometer readings. The hypothesis of using the burn width of CSSRs to obtain DSI data has recently been revisited, and several studies have shown the potential of the burn width of CSSR cards as a proxy for DSI (e.g., Wood and Harrison, 2011; Horseman et al., 2013); however, here a more complete research is presented, comparing burn width and DSI for a relatively long period of time (2 years) and investigating results at high-temporal resolution. As long-term records of SD cards are available at some historical meteorological stations, e.g., Blue Hill Observatory in United States (Magee et al., 2014; Mike Iacono, personal communication, 2014), the reconstruction of DSI for these sites can reach more than 100 years from the present day. Section 2 provides a description of data and instruments used in this research. In Sect. 3 we describe the semiautomatic method proposed for determining the width of the burned traces. In Sect. 4 we show the application of the new treatment of the CSSR cards, first to determine the daily SD and second to estimate the DSI at hourly resolution. Finally, conclusions of this study and suggestions for further research are presented in Sect. 5.

2 Instruments and data

Exposed cards from CSSRs and meteorological and radiative measurements used in the present study come from a weather and radiometric station located on the roof of a building of the University of Girona (41.962° N, 2.829° E, 115 m a.s.l.). Girona is a city located in the northeast of the Iberian Peninsula, about 30 km from the Mediterranean Sea. It has a Mediterranean climate, with moderate winters and hot summers and maximum precipitation during autumn and spring. The mean height of the horizon for the CSSRs is 4.2°, with a minimum of 0.2° at azimuth angle 86° W and a maximum of 9.2° at azimuth angle 21° E (Fig. 1). Note that the Sun rises over the horizon at angular heights greater than 5° (except during summer period), while in the evening the

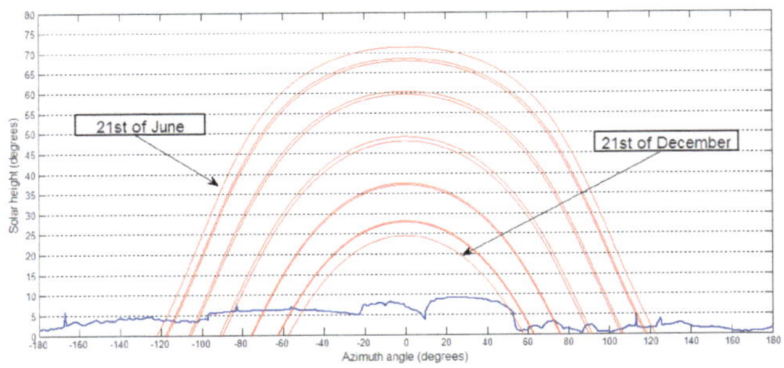

Figure 1. Solar path for the 21st of each month (red lines) and horizon height for each azimuth angle (blue line).

Sun sets at much lower angular heights during most months. This means that the SD records and DSI measurements are slightly affected in the morning, but we assume that this issue does not affect the relationship between both variables.

Two different models of CSSR have been used (Fig. 2, top panel) with three different cards for each CSSR (Fig. 2, bottom panel), which are used for different periods along the year. Hours and half-hours are printed on every card, as also shown in Fig. 2 (bottom panel). One of the CSSR and cards is from Thies Clima (hereafter referred to as CSSR1); the other CSSR is from Negretti & Zambra, manufactured in the 1980s with cards of the model Mod.98 (hereafter referred to as CSSR2), formerly used by the Spanish Meteorological Agency (AEMET). DSI is measured by a pyrheliometer (CH-1 model) from Kipp & Zonen mounted on a Sun tracker. The DSI is measured every second, and values are stored as 1 min averages. The station is also equipped with meteorological sensors to measure temperature, humidity, wind speed, precipitation, etc., as well as a ceilometer, a multifilter rotating shadowband radiometer, and other radiometric sensors. In addition, a whole sky camera is continuously taking pictures of the sky dome.

We have processed 239 cards covering the period from January 2012 to January 2014 for each type of CSSR. We have 94 winter cards (from 16 October to 28/29 February), 56 equinoctial cards (from 1 March to 15 April and from 1 September to 15 October), and 89 summer cards (from 16 April to 31 August). The number of cards is limited because we did not add cards every day, and we removed the cards that were too damaged due to the rain or had no traces of burns.

3 Semiautomatic method for reading the burned traces

We have developed a semiautomatic method to retrieve information from the burned cards. We have considered "burned" not only the perforated part but also the black-grey area (scorched area) present on the edges of the burn. The method is very effective in detecting the burned areas of the CSSR cards and can be summarized in four steps. First, we scan

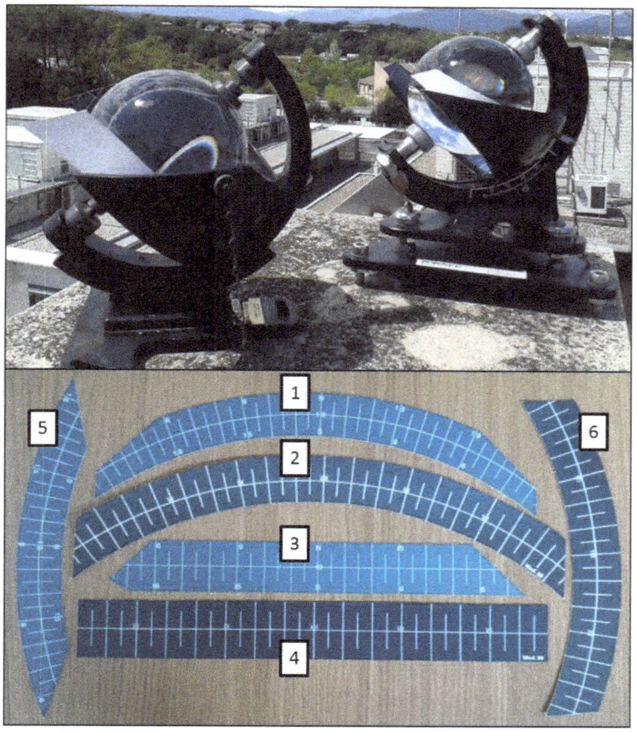

Figure 2. (Top) Details of the two different CSSRs mounted in Girona (NE Spain), and (bottom) an example of the 3 types of cards used during summer, winter, and equinoctial seasons, respectively (the longer the daylight hours, the longer the card): (1) summer card of Thies Clima; (2) summer card of Mod.98; (3) equinoctial card of Thies Clima (note that although the card edges are not symmetric, the markings are); (4) equinoctial card of Mod.98; (5) winter card of Thies Clima; and (6) Winter card of Mod.98.

each card on a suitable background. Second, we apply a digital process that increases the contrast of the burned area. Third, the middle of the day, 12 true solar time (TST), on the card is determined. Finally, the burn width is measured, with the help of a computer program, along cross sections spaced every minute. A schematic illustration of the different steps can also be seen in Fig. 3. With this, we obtain daily evolution of the burn width on a resolution of 1 min; the length of the burn (i.e., SD) can be easily determined. These four steps are detailed in the following sections. Our method is similar to the method of Horseman et al. (2013), but there are also significant differences that will be highlighted below.

3.1 Image capture

Similar to Horseman et al. (2013), scanning the burned card is the first step of our method and the more time-consuming manual part. For this purpose we use a commercial scanner (model HP Scanjet 5590). As sensors in image scanners alter with time and use, a method to calibrate the scanner must be applied regularly (Horseman et al., 2013). We use a standard image format, 24 bit RGB BMP (bitmap), and the same dimension (2340 × 1700 pixels) for all scans. Unlike Horseman et al. (2013), we use a green background (Fig. 3a) for scanning the cards in order to obtain a contrast with the card, which is blue (face) and white (hour markers and other information). The positioning of the card on the scanner does not need to be precise, which simplifies and speeds up the card scanning process. Before the scanning, an evaluation is done to remove those cards that do not present any burn or to control for those that present any anomalies (stains on the card face, deformations due to the rain, etc.).

3.2 Image treatment

On used cards, the edges of the burned or scorched traces are, for the most part, black-brown, occasionally with some grey ash. As stated by Fan and Zhang (2013), there are no abrupt changes in grey intensity between the weak scorch and the card face, but their signatures on the RGB color coordinates are noticeably different. This fact will be the basis for identifying the burned parts of the card by using the Image Processing Toolbox from MatLab on the RGB image. First, we build an image called Im1 (Fig. 3b.1), in which the pixels corresponding to the white markers in the card (hour and half-hour markers) have the value "1" and the rest of the pixels have "0". To distinguish the white pixels, a threshold of $R > 200$ is applied on the red component (which ranges from 0 to 255). Then we build a second image, called Im2 (Fig. 3b.2), distinguishing the blue area of the card. The selection is made applying the condition $B - R < 20$ on the red and blue components. In Im2, the pixels in blue areas are labeled as "0" and all the other pixels (green, black, grey, and white) are labeled as "1". Next, by subtracting Im1 from Im2 and setting all pixels to 0 where the difference is negative, an image with pixels labeled "1" for the burn and the background (black, grey, and green) and labeled "0" for the unburned part of the card (blue and white) is obtained. There is still some "noise" in this image (that is, pixels of one kind surrounded by eight pixels of the other kind); this noise is removed from the image to obtain the final processed image (Fig. 3c; 1 pixels in white, 0 pixels in black). Note that this process is entirely automatic. The choice of the thresholds ($R > 200$ for Im1 and $B - R < 20$ for Im2) has been empirically determined after performing various checks and will certainly affect the identification of the burn but, as long as they are used consistently throughout the two card archives, their exact absolute values are not very important.

3.3 Image positioning

Unlike Horseman et al. (2013), the positioning of the card within the image also requires some manual intervention. This is convenient for checking the scan and the image treatment and necessary to convert the image of the burned pixels into length (millimeters of width) and into time (minutes along the day). Although this step slows down the process,

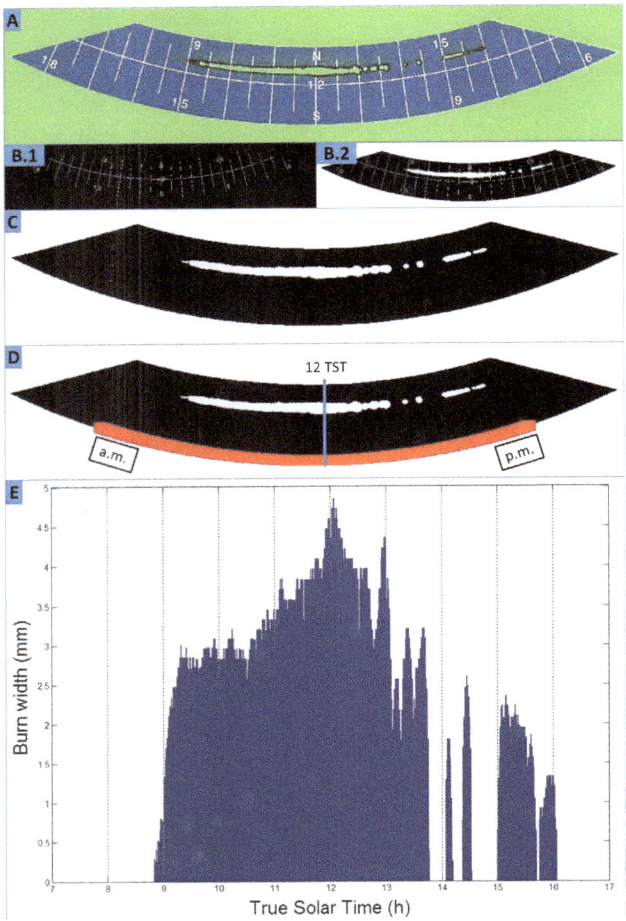

Figure 3. Steps of the semiautomatic method for retrieving information from the CSSR burned cards and for a certain day: **(a)** captured image, **(b)** treated image, **(c)** positioned image, and **(d)** measured burn width over time.

it allows using our method for any type of card. We need to manually identify two or three points (depending on the shape of the card) to find the middle of the day (12 TST). More specifically, in curved cards (summer and winter cards) we mark three points (two on the ends of the outer arc and the third close to the center of the outer arc) to find the center and the radius of the outer circumferences (the same process could be applied for the inner arc). Then, as these cards are symmetric about the midday marker, it is easy to find all points in the image belonging to the outer or inner arc and the point corresponding to noon (Fig. 3d). In the equinoctial cards, finding the midday point is even easier than in the previous ones: knowing two points (at both ends of the outer or inner contour) provides sufficient information to locate the midday point.

3.4 Measurement of burning width

The Horseman et al. (2013) method rectifies the image of each card (that is, images are transformed to a representation with a straight burned trace) before extracting the burn width, thus simplifying geometrical calculations. Contrarily, we consider here continuous radial sections that cover the whole card and measure the burn width (i.e., the length from the first to the last "burned" pixels) along each section. Because the size and shape of each card type is known and standardized, we know the radii of the outer and inner boundaries and their distance; thus there is no need to cross the whole image, rather just the area where the card is placed. For the equinoctial cards, parallel sections are performed instead of radial sections.

We continue by defining the relation between angular (or linear) displacement and time: as we know the angle (or distance) between the two edges that we defined in the previous step, and we also know the interval of time that corresponds to these two points (this time is fixed in each type of card), the relation between angle (or distance) and time is immediate. For example, for winter and summer cards, a 1 min displacement corresponds to an angular displacement of 0.064°; in the case of equinoctial cards, 1 min displacement corresponds to a linear displacement of 0.294 mm (0.317 mm) for CSSR1 (CSSR2). If we consider a resolution of one pixel, the resolution of the burn width is 0.126 mm in all cards (for the scanner used).

As we know the point corresponding with 12 TST, consecutive angular displacements (corresponding to a 1 min temporal resolution) towards the left are applied, the radial sections between the inner radius to the outer radius are inspected, and the distance between the first and last "1" pixel in the radial section is computed. Thus, we obtain the temporal evolution of the burn width for the morning. We proceed in a similar way (but towards the right) to obtain the afternoon evolution of the burn width. With this, the daily evolution of the burn width in each card is finally obtained (Fig. 3e). Note that all processes in this last step are fully automatic.

4 Applications of the measurements of the burn width

The semiautomatic method proposed to process the burned cards can help in the assessment of different types of errors that can affect the SD measurements (Sect. 4.1). The method can also be useful in order to provide a record of DSI, since the burn width can be used as a proxy for pyrheliometer measurements (Sect. 4.2).

4.1 Assessment of the measurement of SD

As summarized by Brázdil et al. (1994), the sources of errors in the SD series related to the use of CSSRs are (i) the ageing of the glass ball, which with increasing operation time

Table 1. The first two columns are the mean value of SD by manual and semiautomatic methods, respectively. The third column is the value of SD from pyrheliometric method by using a DSI threshold (in parenthesis) that would approach the corresponding mean SD using the semiautomatic method. Recall that the "true" mean SD (by using 120 W m^{-2} as the DSI threshold) is 7.16 h.

	Manual	Semiautomatic	DSI threshold
SD1	7.31 h	7.53 h	7.49 h (55 W m^{-2})
SD2	7.10 h	7.22 h	7.21 h (110 W m^{-2})

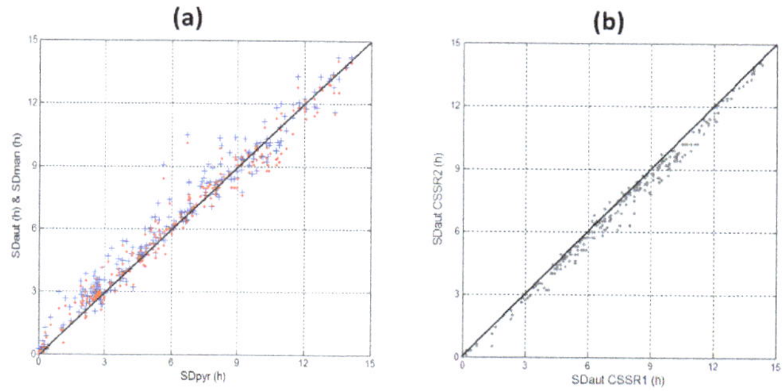

Figure 4. Scatter plots of the **(a)** daily SD obtained for CSSR1 from both SDaut (blue crosses) and SDman (red points) methods against SDpyr, considering the threshold of 120 W m^{-2} in DSI; and **(b)** the daily SD obtained by the two CSSR using the automatic method. In each graphic we also represent the 1 : 1 line (black line).

becomes less transparent, (ii) the replacement of the recorder by a new device manufactured by different companies, and (iii) the variability of the recording card (e.g., different quality, color, or material). In order to homogenize the worldwide measurements of SD, a specific design of the CSSR was recommended as the reference in the 1960s (WMO, 1962). This, however, did not eliminate all the problems, especially when long-term series with records before and after the 1960s are used to study SD trends. Thus, we will here quantify these errors/differences in the observations by using our digital method applied to two different CSSRs that use different burning cards.

One of the major problems when comparing SD measurements from different CSSR devices and card types is the variability of the level of DSI that produces a burn (e.g., Stanhill, 2003; Sanchez-Romero et al., 2014). Here we define the SDpyr method as considering a threshold of 120 W m^{-2} in DSI in order to calculate SD; that is, counting the minutes when DSI is greater than 120 W m^{-2}. By doing so, the mean value found for our database is 7.16 h, which can be taken as the reference (correct value) for other estimations. Table 1 shows the mean value obtained when using different methods of estimating SD applied both to CSSR1 and CSSR2. To obtain SD with the semiautomatic method described in Sect. 3 (SDaut), all minutes showing burn are counted. As shown in Fig. 4a for CSSR1, retrieved SD using the SDaut and the SDpyr methods agree very well (i.e., correlation coefficients are higher than 0.98 for both instruments), although a minor overestimation is apparent. The semiautomatic method gives a mean SD deviation with respect to SDpyr of 0.37 h (4.9 %) and 0.06 h (0.8 %) for CSSR1 and CSSR2, respectively. Although both instruments give a slight overestimation of SD, it is more pronounced for CSSR1, which shows a somewhat higher sensitivity than CSSR2. The instrument sensitivities can also be quantified by searching for the DSI threshold values that would give the corresponding mean SDaut: 55 and 110 W m^{-2} for CSSR1 and CSRR2, respectively; these threshold values are lower than the 120 Wm^{-2} suggested by the WMO.

Therefore, there is some overestimation in the SD given by SDaut, which was expected because some instructions and recommendations (WMO, 2008) are not applied in the method. For example, when SD is retrieved by reading the cards manually (SDman), the time length is reduced at each end by an amount equal to half the radius of curvature of the end of the burn in the case of a clear burn with round ends; on the other hand, an amount of 0.1h is subtracted from the total length of the burned segment in the case of a clear burn that is temporarily reduced in width by at least one-third. Contrarily, SDaut accounts strictly for all minutes where a burn (or a scorch) is detected, without any further correction, so it tends to (slightly) overestimate SD. Among the possible improvements in the SDaut method, the introduction of the advice proposed by WMO (WMO, 2008) in the algorithm would reduce the differences between SDaut and SDpyr.

Table 1 also shows the mean SD that is obtained by the SDman method (we processed all cards in the usual visual way). As expected they are slightly lower than values found with the SDaut method, but the general agreement between the SDaut and SDman methods is very good for both instruments: the mean deviation of SDaut with respect SDman is 0.22 h (2.9 %) for CSSR1 and only 0.12 h (1.7 %) for CSSR2. Figure 4a shows (for CSSR1) the excellent agreement between daily SD obtained from both SDaut and SDman methods with SDpyr. It is important to notice that, for low values of SD, SDman gives also higher values than SDpyr (as it does SDaut); this must be related to the overburning of the card in conditions of broken cloudiness (Stanhill, 2003; Sanchez-Romero et al., 2014).

From the above results, we confirm that 120 W m^{-2} is a suitable DSI threshold. Nevertheless, it is worth noting that there is a high variability in the exact threshold that gives the best agreement of SD for each particular card along the year (not shown). This large seasonal variability was pointed out by Painter (1981), Michalsky (1992), and Roldán et al. (2005). The position of the card on the CSSR, the humidity conditions (Bider, 1958; Painter, 1981), and the poor horizon at the location of the instruments (especially in the morning between October and April) are factors that could explain such high variability.

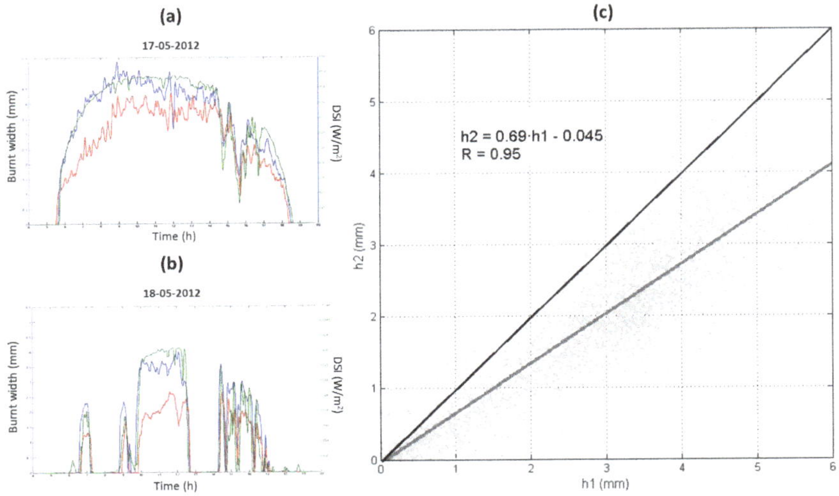

Figure 5. (a, b) Daily evolution of burn width (blue line for CSSR1 and red line for CSSR2) and DSI (green line) for 2 different days. **(c)** Scatter plot of the burn width in CSSR2 (h_2) versus that of CSSR1 (h_1) on hourly resolution. We also represent the 1 : 1 line (black line) and the linear regression fit (green line).

The differences between instruments are evident in Fig. 4b and in Table 1. CSSR1 gives slightly greater values than CSSR2 for both SDaut and SDman methods. It could be due to the ageing of the glass sphere, as the latter is an instrument from the 1980s and the former was brand new at the beginning of our research, or be related to the different quality and colors of the recording cards used for each device (Brázdil et al., 1994). In this sense, note that the semiautomatic method uses the same threshold values in the image segmentation when defining the burned regions for both types of CSSR and cards; an improved method could tune a particular color threshold for each CSSR or card in order for SDaut to match almost exactly SDpyr (or SDman if pyrheliometric measurements are not available).

All mean differences found are similar to the uncertainty usually assigned to the measurement of SD (i.e., 0.1 h), and even in the case of the automatic method, the difference between CSSR is much lower than the maximum errors of around 7 % suggested by Brázdil et al. (1994), referring to different instruments or to the ageing of the glass sphere. In fact, other radiometric variables have similar or even higher uncertainties: global solar radiation has instrumental errors of around 5 and 2 % for the monthly and annual means, respectively (Gilgen et al., 1998; for more information about instrumental errors of radiation data, see for example Vignola et al., 2012). With the improvements proposed above, not only the linear-relationship between SDman and SDaut would be even better, but also the differences between two different CSSR would be reduced to almost nil.

4.2 Burn width and its relationship with DSI

Figure 5a and b shows the daily evolution of the burn width for the two CSSR and for 2 different days (one mostly sunny and the other with scattered and broken cloudiness). In this figure and hereafter, we refer to the burn width for CSSR1 and CSSR2 as h_1 and h_2, respectively. It is evident that the evolution of the burn width shows an excellent agreement for both recorders (and also with DSI), with a correlation coefficient higher than 0.90 when all the 1 min records are taken. Nevertheless, for some days there might be a shift of up to a few minutes between CSSR and DSI data. This is likely the result of a slight misalignment of the cards from their correct position in the CSSR device. Thus, if analysis with less than a few minutes resolution is to be performed, we would need to consider this misalignment; in the present study, hourly averages of burn width and of DSI will be used onwards. As an example of the strong relationship, Fig. 5c shows the scatter plot, the linear fit, and the correlation coefficient of h_2 against h_1 on hourly basis. Note that the slope of the linear regression is lower than 1, i.e., burn width on CSSR1 is notably higher than that on CSSR2, as was expected given the higher sensitivity of CSSR1 found above. The same analyses have been performed separately for each set of seasonal cards (not shown), revealing that the correlation coefficients are almost 1 for all card types and the slopes of the linear regressions are also very similar for all seasons. In other words, the relationship between h_1 and h_2 is almost constant during the year.

The mean difference between h_1 and h_2 is 1.03 mm at 1 min resolution, which is a relative difference of 32 % with respect to h_1 (much higher than the difference in daily SD). This value agrees with the slope of the fit of hourly averages (0.69), which points to a relative difference of 31 %. Wood and Harrison (2011) already stated that the burns in one site could be thinner than those in another site although DSI conditions were the same (e.g., due to differences between in-

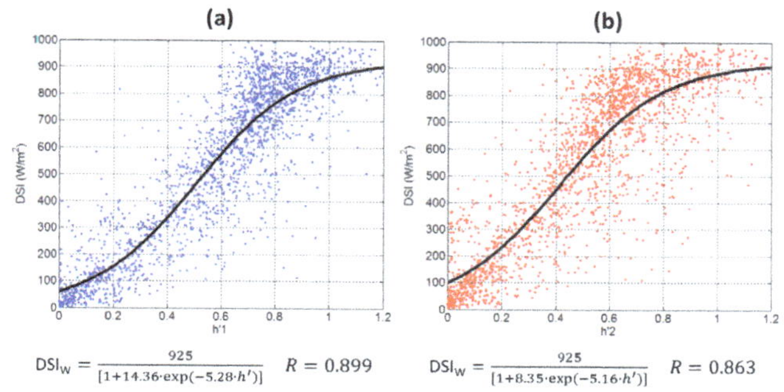

Figure 6. (a) Scatter plots in hourly resolution of the DSI against normalized width for CSSR1. The fitted logistic function and the correlation coefficient are also shown (black line). **(b)** The same but for CSSR2.

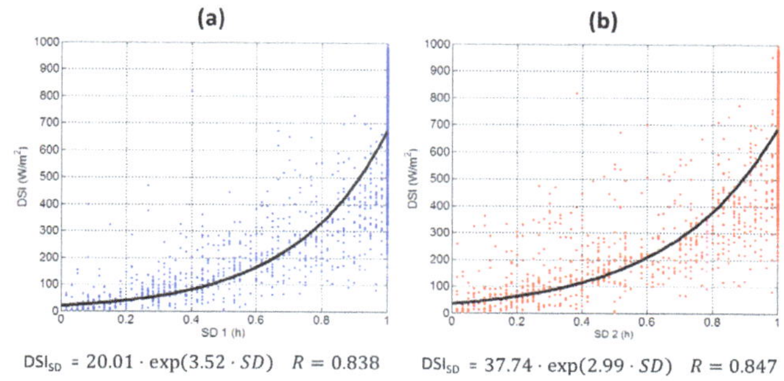

Figure 7. (a) Scatter plots in hourly resolution of the DSI against SD for CSSR1. The exponential regression (black line) and the correlation coefficient are also displayed. **(b)** The same but for CSSR2.

struments). Similarly, in our study we find that h_1 and h_2 are different (although highly correlated with each other). Then, when estimating the DSI from burn width data, it will be important to know which instrument has been used. This is similar to what happens with Ångström–Prescott equations that relate global solar radiation with SD, where parameters depend on local calibration and time interval (Martínez-Lozano et al., 1984). Note, however, that the latter parameters must be adjusted for different climates (typical atmospheric turbidity), while the former depend on the specific instrument.

In Fig. 5a and b we can also see the daily evolution of DSI. The evolution of DSI, h_1, and h_2 shows a strong relationship, also displayed in Fig. 6, that suggests a fit between burn width and DSI. Wright (1935) and Lally (2008) proposed an exponential fit for the estimation of DSI from burn width ($\mathrm{DSI_W}$). These studies did not consider many days and that is probably why they did not find that increasing burn width tends toward a certain maximum value of DSI as a horizontal asymptote. The effect of DSI on burn width decreases, and other factors have a role in the measure of burn width (e.g., the atmospheric or the card conditions). This is seen in Fig. 6: from a certain value of burn width, the value of DSI does not vary too much and tends toward a maximum value. Thus, in order to do a fit with consideration of an exponential growing and a horizontal asymptote, a logistic function is proposed:

$$\mathrm{DSI_W} = \frac{L}{1 + Ke^{-Gh'}}. \qquad (1)$$

In this equation, h' is the normalized burn width (i.e., the value of burn width divided by its 95-percentile value, for each CSSR). The function depends on some parameters having physical meanings: L is fixed to the 95-percentile value of DSI, K is related to the threshold DSI (i.e., the $\mathrm{DSI_W}$ value in the y intercept), and G is related to the growth ratio. We consider K and G as the free-parameters for the fit. Figure 6 shows the scatter plot between burn width and DSI, the logistic fit ($\mathrm{DSI_W}$), and the correlation coefficient for the two CSSRs, on an hourly basis. We can see that the growth ratio (G value) is similar in both CSSRs and that the K value is higher in CSSR1 than CSSR2; this is consistent with the fact that CSSR1 has a lower DSI threshold than CSSR2 because CSSR1 is more sensitive than CSSR2.

We compare our method based on measuring the burn width with the estimations of DSI using hourly SD, i.e., counting the minutes with burn within each hour ($\mathrm{DSI_{SD}}$). Stanhill (1998) presented a similar approach and suggested both linear and quadratic regressions, but for our data we have found it more appropriate to fit exponential functions, which are shown in Fig. 7 for both CSSRs. It is important to note that values of DSI for SD equal to 1 h (that is, for sun shining along the whole hour) range from 200 to $1000\,\mathrm{W\,m^{-2}}$, thus making it difficult to obtain a good estimation of DSI for completely clear-sky hours. In addition, the cases of SD equal to 1 h are the most probable (more than 62 % of all the cases for the two CSSRs).

As seen in Fig. 5a and b, DSI affects burn width at very short time scales (1 min); the advantages of using the burn width instead of SD to estimate DSI even at hourly resolution are shown in Table 2 and in Fig. 8. In Table 2 we can see different statistical indexes comparing the estimated DSI with the measured DSI for each method along with the parameters describing the relationship between both variables. It is clear that the method based on burn width gives better results, especially if we look at the columns of the mean bias error, intercept value, and relative root mean squared error. These results show that the burn width gives more information about the variation of DSI high-resolution temporal data (1 h), whereas counting the time length of the burn for short periods only gives information as to whether the Sun is visible (i.e., whether DSI is higher than a certain value). It is remarkable that the correlation between DSI and burn width at hourly resolution is as good as other relationships found by other authors at lower temporal resolution (daily, monthly, seasonally, or annually) using only SD and linear and quadratic correlations (Benson et al., 1984; Louche et al., 1991; Nfaoui and Buret, 1993; Maduekwe and Chendo, 1995; Stanhill, 1998; Power, 2001).

Table 2. Statistical parameters of estimations of hourly DSI when compared with DSI measurements by different methods and instruments (from mean burn width, h_1, h_2 and hourly sunshine duration, SD1, SD2; Stanhill, 1998). MBE is the mean bias error; RMSE is the root mean square error; RRMSE is the relative root mean square error; R^2 is the coefficient of determination; A is the slope of the regression line; and B is the y intercept of the regression line.

METHOD	MBE ($W m^{-2}$)	RMSE ($W m^{-2}$)	RRMSE (%)	R^2	A	B ($W m^{-2}$)
DSI_W CSSR1	−0.6	129	23.9	0.81	1.00	−3.20
DSI_W CSRR2	6.6	147	27.1	0.75	1.01	−1.23
DSI_{SD} CSSR1	−30.0	165	30.5	0.70	0.98	−16.7
DSI_{SD} CSSR2	−29.0	159	29.0	0.71	0.98	−20.7
Stanhill (linear)	–	–	48.4	0.74	–	–
Stanhill (quadratic)	–	–	44.7	0.78	–	–

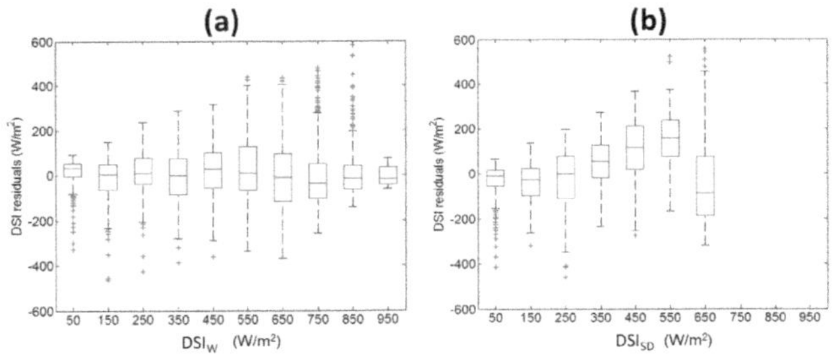

Figure 8. Box plots of the differences between estimated and true DSI for **(a)** estimation based on burn width and **(b)** estimation based on SD. It is applied for CSSR1.

In Fig. 8, box plots for the residuals between the estimated and measured DSI, in bins of $100\,W\,m^{-2}$ width, are represented for the two methods: (a) DSI_W and (b) DSI_{SD} (only results for CSSR1 are presented, as they are similar for CSSR2). Again, the burn width method gives better mean values and less dispersion than the SD method. Note that the burn width method can give values of DSI up to $925\,W\,m^{-2}$ (the parameter L in Eq. 1) covering nearly the whole DSI range, while the SD method cannot give values above $673\,W\,m^{-2}$ (for CSSR1, according to the exponential fit found for our data). This is a consequence of a previous statement: for hourly SD close to 1 h it is not possible to distinguish values of DSI ranging from 200 to $1000\,W\,m^{-2}$.

5 Conclusions and future research

In this study we use a semiautomatic method to obtain the temporal evolution of the burn width in Campbell–Stokes sunshine recorder (CSSR) cards. This method is also capable of producing very good results as far as sunshine duration (SD) measurements are concerned. The mean overestimation in daily SD derived from this method, when compared to the value from a pyrheliometer by using the standard DSI threshold of $120\,W\,m^{-2}$, is less than 5 %, which is very close to the accepted uncertainty of the traditional manual screening of cards. Differences could be reduced if advice proposed by WMO is taken regarding the rounded ends of the burned areas and the cases of intermittent burning. In addition, the thresholds applied in the image processing could be defined differently for each card type. These improvements, however, will be the subject of future research.

The two CSSRs studied here give almost the same SD despite different geometries and cardboard types, but CSSR1 is slightly more sensitive than the CSSR2, therefore producing systematically longer SD: mean bias of 0.1–0.3 h depending on the method (manual or automatic). This difference is very small and of the same order as the instrumental uncertainty. Again, the differing sensitivity of CSSR1 and CSSR2 is shown by the threshold values that should be applied to DSI for an exact match of CSSR SD values: 55 and $110\,W\,m^{-2}$, respectively. The different sensitivity of the instruments is also reflected in the burn width measurements: the mean relative difference is about 30 % (CSSR1 burns wider than CSSR2 burns). Thus, it is important to know which instrument has been used if we want to estimate other magnitudes such as DSI.

Hourly DSI can be satisfactorily estimated from the burn width measurements. The estimation based on burn width by using a logistic fit is better than that based on counting the time length of the burn (i.e., hourly SD). For example, in the case of CSSR1 we have obtained a mean bias error of $-0.6\,W\,m^{-2}$, a coefficient of determination of 0.81, and a relative root mean squared error of 24 % when comparing hourly DSI estimations from burn width with actual measurements (the corresponding values for CSSR2 are $6.6\,W\,m^{-2}$, 0.75, and 27 %). These indexes are notably better than those obtained when SD is used ($-30\,W\,m^{-2}$, 0.70, and 30 % approximately for the two instruments). This result shows that the burn width gives more information about the variation of DSI on high-resolution temporal data, while the other method is useful only for estimations for longer periods (e.g., daily). The parameters of the equation that relates burn width with DSI depend on the sensitivity of the instrument and the maximum DSI of the region.

Future research may consider taking into account the perforated part of the burn (Roberts, 2012) as well as the burn width (perforated plus scorch parts); a priori, this would help in the estimation of DSI and help to explain the reason for the tail in the scatter plot of burn width versus DSI. Another point to consider is the use of different threshold values when defining the burned regions depending on the type of CSSR and cards; the thresholding process is the most important factor if we are interested in relating the burn width with DSI, so it is crucial to define it correctly for each CSSR and card. Moreover, further analysis needs to consider the shift between the burn and DSI measurements if analyses with less than a few minutes resolution are to be performed.

Once this method is implemented, other magnitudes can be introduced. Since DSI is affected by atmospheric turbidity, especially at times near sunrise and sunset because of the longer optical path, having an estimation of DSI from the burn width may be used to estimate turbidity, i.e., CSSR records can become a proxy measurement for turbidity and atmospheric aerosol loading. This possibility has been proposed before by Jaenicke and Kasten (1978), Helmes and Jaenicke (1984, 1985, 1986), and more recently by Horseman et al. (2008, 2013) as reviewed by Sanchez-Romero et al. (2014).

Thus, the measurements of burn width in CSSR cards obtained from the semiautomatic method described in this study can provide a way to create time series of solar irradiance in less than daily resolution, a fact that was proposed before by some authors (Wright, 1935; Wood and Harrison, 2011; Horseman et al., 2013). We have shown that it is possible to use readily available technology to mine the archives of a simple, reliable, and widely used meteorological instrument to provide a proxy record of DSI. In our case, the process of scanning and manual intervention can last only 1.5–2 min for each card; i.e., it is possible to process a year of exposed cards in about 12 h of work. In fact, the reconstruction can reach as much as around 120 years from the present day because long-term records of SD cards are available at some historical meteorological stations such as, for example, the Blue Hill Observatory in Milton, Massachusetts (US), where CSSR cards have been stored since the late 1880s (Magee et al., 2014; Mike Iacono, personal communication, 2014).

Acknowledgements. This research was supported by the Spanish Ministry of Science and Innovation project NUCLIERSOL (CGL2010-18546). The first author enjoys a grant from the FPU program (FPU AP2010-0917) of the Spanish Ministry of Education. A. Sanchez-Lorenzo was supported by the "Secretaria per a Universitats i Recerca del Departament d'Economia i Coneixement, de la Generalitat de Catalunya i del programa Cofund de les Accions Marie Curie del 7è Programa marc d'R+D de la Unió Europea" (2011 BP-B 00078) and the postdoctoral fellowship JCI-2012-12508. The Negretti & Zambra Campbell–Stokes recorder was provided by the Pyrenean Institute of Ecology (CSIC). We thank Juan Jose Vizcaino Morillas (AEMET) for providing sunshine duration cards of the model Mod.98.

Edited by: F. Pope

References

Aguilar, E., Auer, I., Brunet, M., Peterson, T. C., and Wieringa, J.: Guidelines on climate metadata and homogenization, Malaga, 2003.

Angell, J. K., Korshover, J., and Cotton, G. F.: Variation in United States cloudiness and sunshine, 1950–82, J. Clim. Appl. Meteorol., 23, 752–760, 1984.

Ångström, A.: Solar and terrestrial radiation, Q. J. Roy. Meteor. Soc., 50, 121–125, 1924.

Baker, D. G. and Haines, D. A.: Solar radiation and sunshine duration relationships in the north-central region and Alaska, Agricultural Experiment Station, University of Minnesota., 1969.

Bakirci, K.: Correlations for estimation of daily global solar radiation with hours of bright sunshine in Turkey, Energy, 34, 485–501, doi:10.1016/j.energy.2009.02.005, 2009.

Benson, R. B. B., Paris, M. V. V, Sherry, J. E. E., and Justus, C. G. G.: Estimation of daily and monthly direct, diffuse and global solar radiation from sunshine duration measurements, Sol. Energy, 32, 523–535, doi:10.1016/0038-092X(84)90267-6, 1984.

Bider, M.: Uber die Genauigkeit der Registrierungen des Sonnenscheinautographen Campbell-Stokes, Arch. Meteor. Geophy. B, 9, 1922–230, 1958.

Brázdil, R., Flocas, A. A., and Sahsamanoglou, H. S.: Fluctuation of sunshine duration in central and south-eastern Europe, Int. J. Climatol., 14, 1017–1034, doi:10.1002/joc.3370140907, 1994.

Cerveny, R. and Balling, R. C.: Inhomogeneities in the long-term United States' sunshine record, J. Climate, 3, 1045–1048, 1990.

Döll, P., Kaspar, F., and Lehner, B.: A global hydrological model for deriving water availability indicators: model tuning and validation, J. Hydrol., 270, 105–134, doi:10.1016/S0022-1694(02)00283-4, 2003.

Fan, Q. and Zhang, Y.: A scorch extraction method for the Campbell-Stokes sunshine recorder based on multivariable thresholding, in International Conference on Imaging Systems and Techniques, Beijing, 2013.

Galindo Estrada, I. and Fournier D'Albe, E.: The use of the Campbell-Stokes sunshine recorder as an integrating actinometer, Q. J. Roy. Meteor. Soc., 86, 270–272, 1960.

Gilgen, H., Wild, M., and Ohmura, A.: Means and trends of shortwave irradiance at the surface estimated from global energy balance archive data, J. Climate, 11, 2042–2061, 1998.

Good, E.: Estimating daily sunshine duration over the UK from geostationary satellite data, Weather, 65, 324–328, 2010.

Gueymard, C.: Analysis of monthly average solar radiation and bright sunshine for different thresholds at Cape Canaveral, Florida, Sol. Energy, 51, 139–145, 1993.

Guijarro, J. A.: Cambios en la medida de las horas de insolación: Análisis de su impacto en dos observatorios de las Islas Baleares (España), Rev. Climatol., 7, 27–32, 2007.

Helmes, L. and Jaenicke, R.: Experimental verification of the determination of atmospheric turbidity from sunshine recorders, J. Clim. Appl. Meteorol., 23, 1350–1353, 1984.

Helmes, L. and Jaenicke, R.: Hidden information within series of measurements – Four examples from atmospheric sciences, J. Atmos. Chem., 3, 171–185, 1985.

Helmes, L. and Jaenicke, R.: Atmospheric turbidity determined from sunshine records, J. Aerosol Sci., 17, 261–263, 1986.

Hinssen, Y. B. L.: Comparison of different methods for the determination of sunshine duration, KNMI Scientific Rep. WR-2006-06, 72 pp., De Bilt, Netherlands, 2006.

Hinssen, Y. B. L. and Knap, W. H.: Comparison of pyranometric and pyrheliometric methods for the determination of sunshine duration, J. Atmos. Ocean. Tech., 24, 835–846, doi:10.1175/JTECH2013.1, 2007.

Horseman, A. M., Mackenzie, A. R., and Timmis, R.: Using bright sunshine at low-elevation angles to compile an historical record of the effect of aerosol on incoming solar radiation, Atmos. Environ., 42, 7600, doi:10.1016/j.atmosenv.2008.06.033, 2008.

Horseman, A. M., Richardson, T., Boardman, A. T., Tych, W., Timmis, R., and MacKenzie, A. R.: Calibrated digital images of Campbell–Stokes recorder card archives for direct solar irradiance studies, Atmos. Meas. Tech., 6, 1371–1379, doi:10.5194/amt-6-1371-2013, 2013.

Jaenicke, R. and Kasten, F.: Estimation of atmospheric turbidity from the burned traces of the Campbell-Stokes sunshine recorder, Appl. Optics, 17, 2617–2621, 1978.

Katsuyama, M.: On comparison between rotating mirror sunshine recorders and Jordan sunshine recorders, Weather Serv. Bull., 54, 169–183, 1987.

Kerr, A. and Tabony, R.: Comparison of sunshine recorded by Campbell–Stokes and automatic sensors, Weather, 59, 90–95, doi:10.1256/wea.99.03, 2004.

Kothe, S., Good, E., Obregón, A., Ahrens, B., and Nitsche, H.: Satellite-Based Sunshine Duration for Europe, Remote Sens., 5, 2943–2972, doi:10.3390/rs5062943, 2013.

Lally, M. J.: Solar beam intensity inferred from Campbell-Stokes Sunshine Records, 24 pp., University of Reading, UK., 2008.

Louche, A., Notton, G., Poggi, P., and Simonnot, G.: Correlations for direct normal and global horizontal irradiation on a French Mediterranean site, Sol. Energy, 46, 261–266, doi:10.1016/0038-092X(91)90072-5, 1991.

Maduekwe, A. A. L. and Chendo, M. A. C.: Predicting the components of the total hemispherical solar radiation from sunshine duration measurements in Lagos, Nigeria, Renew. Energ., 6, 807–812, 1995.

Magee, N. B., Melaas, E., Finocchio, P. M., Jardel, M., Noonan, A., and Iacono, M. J.: Blue Hill Observatory Sunshine – Assesment of climate signals in the longest continuous meteorological record in North America, B. Am. Meteorol. Soc., online first, doi:10.1175/BAMS-D-12-00206.1, 2014.

Martínez-Lozano, J. A., Tena, F., Onrubia, J. E., and De La Rubia, J.: The historical evolution of the Ångström formula and its modifications: Review and bibliography, Agr. Forest Meteorol., 33, 109–128, 1984.

Matuszko, D.: A comparison of sunshine duration records from the Campbell-Stokes sunshine recorder and CSD3 sunshine duration sensor, Theor. Appl. Climatol., 6 pp., doi:10.1007/s00704-014-1125-z, 2014.

Michalsky, J. J.: Comparison of a National Weather Service Foster Sunshine Recorder and the World Meteorological Organization Standard for sunshine duration, Sol. Energy, 48, 133–141, 1992.

Middleton, W. E. K.: Invention of the meteorological instruments, 362 pp., The Johns Hopkins University Press, 1969.

Monteith, J. L.: Climate and the efficiency of crop production in Britain, Philos. T. R. Soc. Lond., 281, 277–294, 1977.

Nfaoui, H. and Buret, J.: Estimation of daily and monthly direct, diffuse and global solar radiation in Rabat (Morocco), Renew. Energ., 3, 923–930, 1993.

Olseth, J. A. and Skartveit, A.: Solar irradiance, sunshine duration and daylight illuminance derived from METEOSAT data for some European sites, Theor. Appl. Climatol., 69, 239–252, 2001.

Painter, H. E.: The performance of a Campbell-Stokes sunshine recorder compared with a simultaneous record of a normal incidence irradiance, Meteorol. Mag., 110, 102–109, 1981.

Powell, G. L.: Instrumental effects on the sunshine record, J. Clim. Appl. Meteorol., 22, 962–963, 1983.

Power, H. C.: Estimating clear-sky beam irradiation from sunshine duration, Sol. Energy, 71, 217–224, 2001.

Prescott, J.: Evaporation from a water surface in relation to solar radiation, Trans. R. Soc. Sci. Aust., 64, 114–125, 1940.

Roberts, C. J.: Recovering Direct Solar Downwelling Flux Time Series from the Campbell-Stokes Sunshine Card Archive, MSc dissertation, 221 pp., University of Birmingham, UK, 2012.

Roldán, A., Righini, R., and Grossi Gallegos, H.: Preliminary analysis of tghe Behaviour of the cards used in the Argentine sunshine recorders, Av. en Energías Renov. y Medio Ambient., 9, 9–13, 2005.

Sanchez-Lorenzo, A. and Wild, M.: Decadal variations in estimated surface solar radiation over Switzerland since the late 19th century, Atmos. Chem. Phys., 12, 8635–8644, doi:10.5194/acp-12-8635-2012, 2012.

Sanchez-Lorenzo, A., Brunetti, M., Calbó, J., and Martin-Vide, J.: Recent spatial and temporal variability and trends of sunshine duration over the Iberian Peninsula from a homogenized data set, J. Geophys. Res., 112, D20115, doi:10.1029/2007JD008677, 2007.

Sanchez-Lorenzo, A., Calbó, J., Wild, M., Azorin-Molina, C., and Sanchez-Romero, A.: New insights into the history of the Campbell-Stokes sunshine recorder, Weather, 68, 327–331, doi:10.1002/wea.2156, 2013.

Sanchez-Romero, A., Sanchez-Lorenzo, A., Calbó, J., González, J. A., and Azorin-Molina, C.: The signal of aerosol-induced changes in sunshine duration records: A review of the evidence, J. Geophys. Res.-Atmos., 119, 4657–4673, 2014.

Sears, R. D., Flocchini, R. G., and Hatfield, J. L.: Correlations of total, diffuse and direct solar radiation with the percentage of possible sunshine for Davis, California, Sol. Energy, 27, 357–360, doi:10.1016/0038-092X(81)90070-0, 1981.

Stanhill, G.: Estimation of direct solar beam irradiance from measurements of the duration of bright sunshine, Int. J. Climatol., 18, 347–354, 10.1002/(SICI)1097-0088(19980315)18:3<347::AID-JOC239>3.0.CO;2-O, 1998.

Stanhill, G.: Through a glass brightly: Some new light on the Campbell-Stokes sunshine recorder, Weather, 58, 3–11, 2003.

Stanhill, G. and Cohen, S.: Global dimming: a review of the evidence for a widespread and significant reduction in global radiation with discussion of its probable causes and possible agricultural consequences, Agr. Forest Meteorol., 107, 255–278, doi:10.1016/S0168-1923(00)00241-0, 2001.

Stanhill, G. and Cohen, S.: Solar radiation changes in the United States during the Twentieth Century: Evidence from sunshine duration measurements, J. Climate, 18, 1503–1512, 2005.

Stanhill, G. and Cohen, S.: Solar radiation changes in Japan during the 20th Century: Evidence from sunshine duration measurements, J. Meteorol. Soc. Jpn., 86, 57–67, 2008.

Steurer, P. M. and Karl, T. R.: Historical sunshine and cloud data in the United States, Oak Ridge National Laboratory, Pennsylvania State University, United States, 1991.

Suehrcke, H.: On the relationship between duration of sunshine and solar radiation on the Earth's surface: Ångström's equation revisited, Sol. Energy, 68, 417–425, 2000.

Vignola, F., Michalsky, J., and Stoffel, T.: Solar and Infrared Radiation Measurements, Boca Raton, Florida, CRC Press, 2012.

Vuerich, E., Morel, J. P., Mevel, S., and Oliviéri, J.: Updating and development of methods for worldwide accurate measurements of sunshine duration, Paper presented at the WMO Technical Conference of CIMO (TECO – 2012), Brussels, Belgium, 16–18 October 2012.

WMO: Abridged Final Report of the Third Season of the Commission for Instruments and Methods of Observations, WMO-No. 116, Geneva, Switzerland, 1962.

WMO: Guide to Meteorological Instruments and Methods of Observation, 7th Edition, World Meteorological Organisation, Secretariat of the World Meteorological Organisation, Geneva, Switzerland, 2008.

Wood, R. C. and Harrison, R. G.: Scorch marks from the sky, Weather, 66, 39–41, doi:10.1002/wea.664, 2011.

Wood, J., Muneer, T., and Kubie, J.: Evaluation of a new photodiode sensor for measuring global and diffuse irradiance, and sunshine duration, J. Sol. Energ.-T ASME, 125, 43–48, doi:10.1115/1.1531149, 2003.

Wright, H. L.: Some notes on readings at Kew Observatory of the Gorczynski pyrheliometer the sunshine recorder and the black bulb thermometer, Professional Notes, No. 68, Volume V, Met Office, HMSO, London, 7 pp., 1935.

Xia, X.: Spatiotemporal changes in sunshine duration and cloud amount as well as their relationship in China during 1954–2005, J. Geophys. Res., 115, D00K06, doi:10.1029/2009JD012879, 2010.

9

Correction of raindrop size distributions measured by Parsivel disdrometers, using a two-dimensional video disdrometer as a reference

T. H. Raupach and A. Berne

Environmental Remote Sensing Laboratory, School of Architecture, Civil, and Environmental Engineering, École Polytechnique Fédérale de Lausanne (EPFL), 1015 Lausanne, Switzerland

Correspondence to: A. Berne (alexis.berne@epfl.ch)

Abstract. The raindrop size distribution (DSD) quantifies the microstructure of rainfall and is critical to studying precipitation processes. We present a method to improve the accuracy of DSD measurements from Parsivel (particle size and velocity) disdrometers, using a two-dimensional video disdrometer (2DVD) as a reference instrument. Parsivel disdrometers bin raindrops into velocity and equivolume diameter classes, but may misestimate the number of drops per class. In our correction method, drop velocities are corrected with reference to theoretical models of terminal drop velocity. We define a filter for raw disdrometer measurements to remove particles that are unlikely to be plausible raindrops. Drop concentrations are corrected such that on average the Parsivel concentrations match those recorded by a 2DVD. The correction can be trained on and applied to data from both generations of OTT Parsivel disdrometers, and indeed any disdrometer in general. The method was applied to data collected during field campaigns in Mediterranean France for a network of first- and second-generation Parsivel disdrometers, and on a first-generation Parsivel in Payerne, Switzerland. We compared the moments of the resulting DSDs to those of a collocated 2DVD, and the resulting DSD-derived rain rates to collocated rain gauges. The correction improved the accuracy of the moments of the Parsivel DSDs, and in the majority of cases the rain rate match with collocated rain gauges was improved. In addition, the correction was shown to be similar for two different climatologies, suggesting its general applicability.

1 Introduction

The raindrop size distribution (DSD) quantifies the microstructure of rainfall. The DSD describes the statistical distribution of falling drop sizes: it is the number of drops with a given equivolume diameter per unit volume of air. The DSD plays a fundamental role in the analysis of rainfall. Interception of precipitation by vegetation canopies or city environments, erosion of soil through raindrop impact, and pollutant dispersal both on the ground and in the atmosphere are all fields in which the DSD is important (e.g. Uijlenhoet and Sempere Torres, 2006). Knowledge of the DSD is required in order to study the behaviour of electromagnetic waves in the atmosphere, so it is highly relevant to rainfall remote sensing and telecommunication links (Olsen et al., 1978; Jameson and Kostinski, 2001; Uijlenhoet and Sempere Torres, 2006). Moreover, all bulk rainfall variables of interest can be derived as weighted moments of the DSD (e.g. Ulbrich, 1983; Testud et al., 2001). In order to study rainfall microstructure effectively, we require accurate measurements of the DSD.

Disdrometers are instruments that measure the DSD at a point location. There are various types, each with advantages and disadvantages. In this paper we are concerned with the OTT Hydromet particle size and velocity (Parsivel) disdrometer, and the two-dimensional video disdrometer (2DVD) from Joanneum Research. The original Parsivel was by PM Tech Inc. OTT Hydromet purchased the rights to the instrument and redesigned it in 2005; the result was the first-generation Parsivel. The second-generation Parsivel^2 was introduced in 2011, and provided improvements over the first-generation model (Tokay et al., 2014). The Parsivel is a laser optical disdrometer that uses a sheet of light through which drops fall. The diameter and velocity of a drop is then determined by sensing the shadow it casts and for how long it casts it (Löffler-Mang and Joss, 2000). Parsivels bin drops into classes of velocity and diameter and record the number of drops measured per class over an integration time. Parsivel disdrometers have been shown to be susceptible to errors in the recorded drop concentrations, particularly for small and large drops (Krajewski et al., 2006; Tokay et al., 2013). The Parsivel measurement technique assumes properties of the precipitation that are far more appropriate for rain than for solid precipitation; for example, that particles will be spheroidal, have a horizontal orientation of their major axis, and that only one particle will be in the beam at once (Yuter et al., 2006; Battaglia et al., 2010). The Parsivel is, however, a low cost, durable, and reliable instrument that makes it particularly well-suited for deployment in networks to study the small-scale variability of the DSD (e.g. Tapiador et al., 2010; Jaffrain et al., 2011).

The 2DVD [1] uses two perpendicular high-speed line-scan cameras, each with an opposing light source, to measure particles from orthogonal angles and thus record their shape (e.g. Thurai and Bringi, 2005; Thurai et al., 2007) as well as their size and velocity (Kruger and Krajewski, 2002; Schönhuber et al., 2008). Information on each individual particle that falls through the measurement area of the 2DVD is recorded. A particle's fall speed is determined by the difference in time between its detection in the two camera planes, which are offset vertically by 6.2–7 mm. Thus, the 2DVD uses no literature-derived estimates for raindrop shape or velocity; these parameters are measured directly (Schönhuber et al., 2008). Some drawbacks of the 2DVD have been noted. In particular, drops with diameters smaller than 0.2 mm have been found to be unreliably measured (Tokay et al., 2001); Tokay et al. (2013) recommend taking 0.3 mm as a minimum measured diameter in 2DVD data due to underestimation of drop counts below this diameter. In earlier designs of the instrument, the reliability of measurements decreased with increasing wind speed (Nešpor et al., 2000). This has subsequently been addressed through design improvements (Schönhuber et al., 2007).

Several comparisons between 2DVD and Parsivel disdrometers have been reported on in the literature. In experimental trials the 2DVD has been found to produce better matches to rain gauges than Joss and Waldvogel (Tokay et al., 2001) and Parsivel (Thurai et al., 2011; Tokay et al., 2013) disdrometers. Krajewski et al. (2006) showed that PM Tech Parsivel disdrometers measured higher numbers of small drops (0.2 to 0.4 mm) than the 2DVD and generally reported higher rain rates. In a study in Alabama, USA, using first-generation Parsivels, Tokay et al. (2013) found that Parsivel disdrometers were less sensitive to small drops than the 2DVD, and that they overestimated the numbers of drops over 2.44 mm in diameter, while underestimating the numbers of drops under 0.76 mm in diameter. Furthermore, they found that Parsivels measured fall velocities lower than the expected terminal fall speeds for drops larger than 2.44 mm in diameter. Tokay et al. (2013) concluded that inhomogeneous laser beams in first-generation Parsivel disdrometers were the cause of the misestimation of drop counts. Thurai et al. (2011) found that first-generation Parsivels recorded higher mass-weighted mean diameter and rain rate than 2DVD, mostly when the rain rate exceeded $20\,\mathrm{mm\,h^{-1}}$.

Disdrometers can record erroneous measurements due to wind turbulence, splashing, mismatching between cameras (in the case of the 2DVD), multiple drops appearing at the same time, margin-fallers, or external interference from, for example, insects or spiderwebs. Minimal data treatment for disdrometer measurements usually involves removing outlier points by reference to expected terminal fall velocity (e.g. Tokay et al., 2001; Kruger and Krajewski, 2002; Thurai and Bringi, 2005). For example, Tokay et al. (2013) removed drops exceeding $\pm 50\,\%$ of the expected terminal fall speed, while Jaffrain and Berne (2011) used a threshold of $\pm 60\,\%$ of the expected fall speed. This existing approach removes particles that are obviously erroneous, but it has some shortcomings. By only removing measurements, it does not allow for the fact that the disdrometer may underestimate the number of drops falling. Most importantly, the treatment is based solely on bulk variables such as rain rate, and does not test whether the resulting DSDs after the correction are physically viable.

In this paper we present a correction method for DSD measurements provided by Parsivel disdrometers, using a 2DVD as a reference instrument. The correction is designed to ensure that the DSDs recorded by Parsivel disdrometers are accurate, in terms of both the raw DSD and its moments. The correction method adjusts two properties of the recorded DSDs. First, drop velocities per diameter class are shifted such that the mean velocity per diameter class aligns with the theoretical terminal drop velocity for raindrops of that diameter; these raw measurements can then be screened for implausible measurements. Second, per-diameter-class volumetric drop concentrations are scaled such that they match, in a statistical way, the concentrations measured by a collocated 2DVD.

The rest of this paper is organised as follows: the DSD is introduced in detail in Sect. 2. The data used are described in Sect. 3. Measurement of the DSD and the instruments we are concerned with in this work are discussed in Sect. 4. The correction is introduced in Sect. 5. The results of the correction applied to the data are shown in Sect. 6 for first-generation Parsivels, and in Sect. 7 for Parsivel^2. The application of

[1] The 2DVD was called the two-dimensional video distrometer by Schönhuber et al. (2008), to emphasise that the instrument collects information on the distribution of particles. To avoid confusion we use the standard spelling of disdrometer.

the technique to another climatology is addressed in Sect. 8. Concluding remarks are made in Sect. 9.

2 The drop size distribution of precipitation

On average, during precipitation, 1 m^3 of air contains about 10^3 raindrops, with many more small drops than large ones (Uijlenhoet and Sempere Torres, 2006). Small raindrops are close to spherical, but in drops larger than about 1 mm in diameter, the bottom of the drop flattens out progressively with drop size (Beard and Chuang, 1987; Andsager et al., 1999; Pruppacher and Klett, 2000; Thurai and Bringi, 2005; Thurai et al., 2007). For this reason the size of a raindrop is generally characterised by its equivolume diameter, which is the diameter of a sphere containing the same volume of water as the drop. Raindrops are primarily between 0.1 and 6 mm in equivolume diameter, and they fall at speeds from 0.1 to greater than 9 $\mathrm{m\,s^{-1}}$ (Uijlenhoet and Sempere Torres, 2006; Roe, 2005), with the terminal fall speed of a drop dependent on its size, plus the atmospheric temperature, relative humidity, and altitude above sea level (Beard, 1976). The volumetric DSD is written $N(D)$, and is the number of raindrops with equivolume diameter D per unit volume of air (Jameson and Kostinski, 2001).

The drop size distribution $N(D)$ ($\mathrm{m^{-3}\,mm^{-1}}$) can be described as the total drop concentration multiplied by a probability density function $f(D)$, such that

$$N(D) = N_t f(D). \quad (1)$$

The total drop concentration N_t ($\mathrm{m^{-3}}$) is the total number of drops falling per cubic metre of air. It is the zeroth moment of the DSD, such that

$$N_t = \int_{D_{min}}^{D_{max}} N(D)\mathrm{d}D. \quad (2)$$

The great power of the DSD comes from the fact that, because the shape and fall velocity of a raindrop can be reliably described once its equivolume diameter is known, all integral rainfall parameters of interest can be derived as weighted moments of the DSD. These are also known as bulk rainfall parameters. Any bulk rainfall parameter P can be written as

$$P = a_P \int_{D_{min}}^{D_{max}} w_P D^p N(D)\mathrm{d}D, \quad (3)$$

where a_P and p are constants (Ulbrich, 1985) and w_P is a weight that possibly depends upon D. For example, the rain rate R ($\mathrm{mm\,h^{-1}}$) is defined from the DSD as

$$R = 6\pi\, 10^{-4} \int_{D_{min}}^{D_{max}} N(D)v(D)D^3\mathrm{d}D, \quad (4)$$

where $v(D)$ ($\mathrm{m\ s^{-1}}$) is the terminal fall velocity for drops with equivolume diameter D (mm). Terminal fall velocities in still air can be calculated from the equivolume diameter; popular formulas include those of Atlas et al. (1973), Beard (1976), and Brandes et al. (2002).

The definitions given in this section assume a continuous DSD function of which the integral can be taken. When measured by an instrument, however, the DSD is usually provided as the concentration of drops per discrete class of equivolume diameter. In this case the above equations are modified, such that the integration becomes a sum over all classes, $N(D)$ becomes N_i ($\mathrm{mm^{-1}\,m^{-3}}$), the drop concentration for the ith class, and $\mathrm{d}D$ becomes ΔD_i (mm), the width of the ith class. When the diameter D of drops in a class is required, for example in Eq. (4), we use the centre of the ith diameter class, which we call D_i (mm).

3 Data

The Parsivel DSD correction was developed and tested on first-generation Parsivel data collected during two consecutive autumns in Ardèche, France, as part of the Hydrological Cycle in the Mediterranean Experiment (HyMeX[2], Drobinski et al., 2014). The method was then also tested on second-generation Parsivel (Parsivel^2 hereafter) data collected in the same region in autumn 2013. To test the method in a different climatology and region, we used data from Payerne, Switzerland, using a first-generation Parsivel and the same 2DVD used in the HyMeX campaign. In this section these data sets are briefly described.

3.1 HyMeX SOPs 2012 and 2013

Two autumn campaigns in the same region in Ardèche, France, provided the primary data used in this work. The campaigns were special observation periods (SOPs) run between September and November in both 2012 (SOP2012) and 2013 (SOP2013). The field site was a roughly $5 \times 5\,\mathrm{km^2}$ area in the Cévennes region; see the map in Fig. 1. Cévennes has a Köppen–Geiger Cfa climate type, which indicates that it has a temperate climate with no dry season and a hot summer (Peel et al., 2007). The town of Montélimar, about 18 km from the SOP2012 and SOP2013 field area, records an average annual rainfall of 905 mm, with 77 rainy days per year on average (MeteoFrance, 2014). Cévennes experiences Mediterranean rainfall and has a well-defined precipitation maximum in October (Frei and Schär, 1998). It is subject to heavy precipitation events that can produce large rainfall totals (greater than 150 mm) in a day (Ricard et al., 2012).

In 2012, seven first-generation Parsivel disdrometers (two of which were collocated) and a 2DVD were deployed. In 2013, the same network was deployed with the addition of two more first-generation Parsivel disdrometers. The 2DVD

[2] See http://www.hymex.org

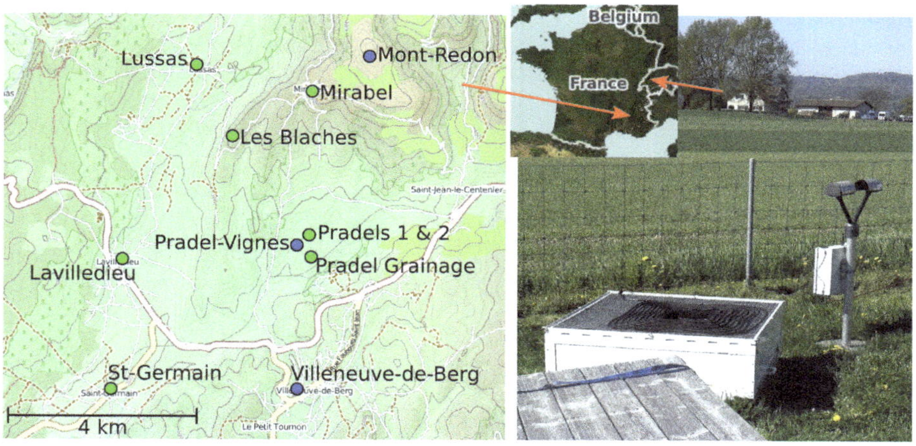

Figure 1. The setup of the field campaigns. For HyMeX, in Ardèche, on the left: Parsivel stations (green) and Parsivel^2 stations (blue). Montbrun and Pradel Grainage first-generation stations were deployed only in 2013. Pradel was the location of collocated Parsivel first-generation stations. Pradel Grainage was the location for both first and second-generation instruments in 2013. Two Parsivel^2 stations were collocated in Villeneuve-de-Berg. The 2DVD was located at Pradel Grainage. The inset map shows the location of the two field areas. On the right is a picture of the setup for Payerne 2014. Maps © Thunderforest (CC BY-SA, http://www.thunderforest.org/), map data © OpenStreetMap (ODbL, http://www.openstreetmap.org/copyright).

was collocated with a Parsivel and a tipping-bucket rain gauge in 2013. Collocated rain gauge measurements were available for all disdrometers, with the exception of Montbrun in 2013. Furthermore, we used data from a network of five Parsivel^2 disdrometers that was deployed in the same region during the 2013 campaign. Station information is summarised for all the disdrometers in Table 1. In this paper we refer to the two campaigns as SOP2012 and SOP2013. For our purposes, the main difference between the setup of the two campaigns is that in SOP2013, there was a Parsivel, Parsivel^2, and rain gauge collocated with the 2DVD at the Pradel Grainage site. In SOP2012, the closest Parsivel and rain gauge to the 2DVD were at the site of Pradel 1 and Pradel 2, about 480 m away. For some analyses we combined data from SOP2012 and SOP2013 into a single data set, which we refer to as the "combined SOPs" data set.

Due to a clock error with the 2DVD, a variable clock drift was present in the 2DVD data. During the campaign, Parsivel clocks were synchronised using inbuilt global positioning system (GPS) receivers and were thus more reliable than the 2DVD clock. Adjustments were made to the 2DVD data for SOP2013 in order to synchronise the clocks of the instruments, for events where it was possible to do so. This synchronisation was done manually, by comparing time series of the rain rate from the 2DVD and a collocated Parsivel. The 2DVD time series was shifted forward in time to match the Parsivel time series as closely as possible, at 30 s temporal resolution. The adjustment was then applied to the series of individual 2DVD drops. Table A1 shows the adjustments made per event in SOP2013, which are between 30 and 60 s. For SOP2012, no adjustments were made because the 2DVD was not collocated with any instruments.

3.2 Payerne 2014

To test the method on data collected in a different region and a different climatology, we used data collected in Payerne, Switzerland. Payerne has a Köppen–Geiger Cfb climate type, which indicates that it has a temperate climate, without a dry season, and with a warm summer. It has an average annual rainfall of 891 mm, with an average of 114 rainy days per year (MeteoSwiss, 2013), meaning that the precipitation in Payerne is more evenly spread across the year than in Ardèche. In contrast to the drier summers of Ardèche, Payerne's rainfall is higher during the summer months (MeteoSwiss, 2013). A first-generation Parsivel disdrometer was collocated with a 2DVD in Payerne between April and June 2014. Table 2 shows the station information. In this paper, we describe the development of the Parsivel correction technique by focusing on data from the HyMeX SOP2012 and SOP2013 campaigns. We then discuss the application of the technique to data from Payerne, to examine its applicability to a different climatological region.

4 Processing of disdrometer measurements

Disdrometer measurements must be processed to convert raw measurements into more useful forms. In this section we describe the processing of data from the Parsivel disdrometer and 2DVD.

4.1 Parsivel

Parsivel disdrometers bin measured particles into particle counts per velocity and diameter class. There are 32 velocity

Table 1. Disdrometer station information for the HyMeX campaigns, showing the instrument (P1 – first-generation Parsivel, P2 – Parsivel^2, 2DVD – two-dimensional video disdrometer), the WSG84 (World Geodetic System 1984) coordinates of each station, its altitude (m) above sea level, the number of hours it recorded liquid precipitation ($R > 0.01\,\mathrm{mm\,h^{-1}}$) with all quality control flags positive during 2012 (H12) and 2013 (H13), and the total amount (mm) it recorded for those times in 2012 (A12) and in 2013 (A13). Note that hours and amounts are calculated using Parsivel data after the correction has been applied. The station at St-Germain had technical problems in 2013, which accounts for its lower hours and total amount.

Inst	Name	Lat (°N)	Long (°E)	Alt	H12	H13	A12	A13
P1	Lavilledieu	44.5772	4.4532	227	169	111	252	223
P1	Les Blaches	44.6008	4.4810	429	148	107	258	210
P1	Lussas	44.6123	4.4706	289	117	102	226	181
P1	Mirabel	44.6069	4.4987	496	168	125	260	242
P1	Pradel 1	44.5829	4.4987	278	145	117	259	203
P1	Pradel 2	44.5829	4.4987	278	149	106	286	157
P1	Pradel Grainage	44.5790	4.5011	271		105		197
P1	St-Germain	44.5551	4.4497	204	158	89	283	97
P2	Mont-Redon	44.6141	4.5148	636		134		219
P2	Pradel Grainage	44.5790	4.5011	271		118		196
P2	Pradel-Vignes	44.5801	4.4950	256		47		128
P2	Villeneuve-de-Berg	44.5547	4.4954	301		111		189
P2	Villeneuve-de-Berg 2	44.5547	4.4954	301		113		198
2DVD	2DVD	44.5790	4.5011	271	129	96	230	199

Table 2. Disdrometer station information for the Payerne 2014 campaign, showing the instrument (P1 – first-generation Parsivel, 2DVD – two-dimensional video disdrometer), the WSG84 coordinates of each station, its altitude (m) above sea level, the number of hours it recorded liquid precipitation ($R > 0.01\,\mathrm{mm\,h^{-1}}$) with all quality control flags positive, and the total amount (mm) it recorded for those times. SSP stands for Station SwissMetNet Payerne. Note that hours and amounts are calculated using Parsivel data after the correction has been applied. The 2DVD was deployed later than the Parsivel.

Inst	Name	Lat (°N)	Long (°E)	Alt	H14	A14
P1	SSP	46.8115	6.9424	490	208	263
2DVD	2DVD	46.8115	6.9424	490	130	84

classes and 32 diameter classes, with varying widths. Parsivels also determine the rainfall intensity (or rain rate), and two status flags: one provides an indication of the type of precipitation being observed (liquid or solid, for example), and another provides information on the quality of the measurement. For example, if the glass in front of the Parsivel's laser beam is dirty and reliable measurements are no longer possible, that will be indicated by a quality flag with value of 2. Value 0 indicates normal operation, while value 1 indicates dirty glass but that measurements are still possible. Value 3 indicates that the laser is damaged. We make use of these flags to restrict our analysis to high-quality measurements.

The effective sampling area of the Parsivel disdrometer is about 54 cm^2, but is different for different diameter classes, due to the fact that the whole drop diameter must be included in the sampling area for the drop to be counted. So-called "margin-fallers" are automatically removed, which reduces the effective sampling area. For the ith class, the sampling area is (Löffler-Mang and Joss, 2000; Battaglia et al., 2010)

$$S_i^{\mathrm{Pars}} = 10^{-6} \times L\left(B - \frac{D_i}{2}\right), \tag{5}$$

where S_i^{Pars} (m^2) is the effective sampling area, D_i (mm) is the class-centre equivolume drop diameter for the ith diameter class, L (mm) is the length of the Parsivel beam (180 mm), and B (mm) is the width of the beam (30 mm).

Let $C_{v,i}$ (–) be the raw number of particles recorded by the Parsivel for the vth velocity class and the ith equivolume drop diameter class. Let Δt (s) be the measurement integration time, V_v (ms^{-1}) the class-centre velocity of the vth velocity class, and ΔD_i (mm) the width of the ith diameter class. Then we can convert the raw number of particles into a per-diameter-class volumetric drop concentration N_i^{Pars} (m^{-3} mm^{-1}) using

$$N_i^{\mathrm{Pars}} = \frac{1}{S_i^{\mathrm{Pars}} \Delta D_i \Delta t} \sum_{v=1}^{32} \frac{C_{v,i}}{V_v}. \tag{6}$$

It is worth noting that the Parsivel instrument itself calculates and provides an estimate of the rain intensity. In this paper we always refer to the estimate of rain rate provided by the Parsivel as the "Parsivel-derived intensity", to avoid confusion with the DSD-derived rain rate R, which is defined by Eq. (4). The values of these two variables are usually very similar, but they are not exactly the same; differences are possibly due to peculiarities of the implemented Parsivel processing algorithm that is not public.

4.2 Two-dimensional video disdrometer

The 2DVD records details of individual drops, including the diameter and velocity of each and the effective sampling area of the instrument at the moment the drop was recorded. For our purposes it is practical to bin the drops into diameter classes. Let M be the number of drops that were recorded within one integration time of length Δt, and let S_j^{2DVD} (m^2) and V_j ($m\,s^{-1}$) be respectively the effective sampling area and fall velocity for the jth recorded particle. Then the ith equivolume diameter class, where the class width is ΔD_i (mm), will have a drop concentration N_i^{2DVD} ($m^{-3}\,mm^{-1}$) of

$$N_i^{2DVD} = \frac{1}{\Delta D_i \Delta t} \sum_{j=1}^{M} \frac{1}{S_j^{2DVD} V_j}. \tag{7}$$

While most 2DVD-derived bulk rainfall variables are calculated using this N_i^{2DVD}, the rainfall rate R ($m\,h^{-1}$) for a given time step can be calculated directly from the individual drop measurements without binning the drops into classes. The rain rate is given by

$$R^{2DVD} = \frac{6\pi \times 10^{-4}}{\Delta t} \sum_{j=1}^{M} \frac{D_j^3}{S_j^{2DVD}}, \tag{8}$$

where D_j (mm) is the equivolume diameter of the jth recorded drop. The difference between the drop-wise rain rate and the rain rate calculated from a binned DSD is very small; in the 2DVD data used in this paper, the mean relative difference between DSD-derived rain rate and rain rate calculated drop-wise was less than 0.5 %.

While the classes for the Parsivel disdrometer are predefined, we can choose any class definition for the 2DVD data. For comparisons of drop concentrations with the Parsivel records, we used Parsivel diameter classes for the 2DVD. For computation of the other bulk parameters from 2DVD data we used diameter classes with a constant width of 0.2 mm, corresponding to the resolution of the 2DVD.

4.3 Criteria for suspicious particles

Before converting our raw drop counts into per-diameter-class volumetric drop counts, we perform some data processing, the aim of which is to filter out particles recorded by the Parsivels and the 2DVD that are very unlikely to be raindrops. These measurements are assumed to be caused by external interferences such as insects, or droplets of water caught in spiderwebs inside the measurement area. We use simple thresholds to exclude classes of velocity and diameter which are unfeasible. To decide on the values for the thresholds, the 2DVD was used as the reference because it is not as easily affected by these external factors as Parsivel disdrometers.

Drops can only reach a certain size (about 10 mm) before they break up into smaller drops due to aerodynamic forces

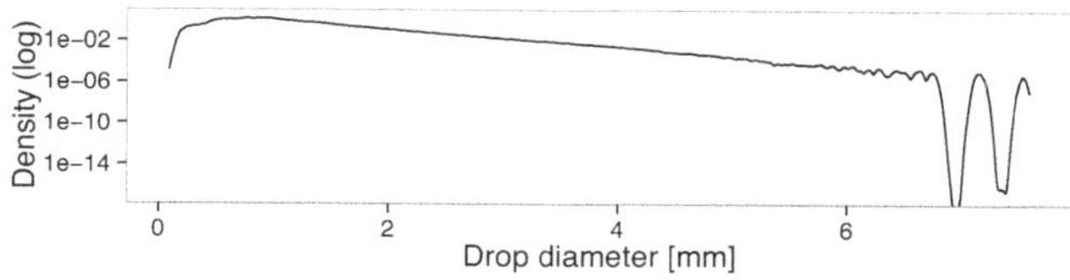

Figure 2. Distribution of drop diameters recorded by the 2DVD in SOP2012 and SOP2013 events, with the y axis on a log scale.

(Pruppacher and Klett, 2000). Figure 2 shows the distribution of drop sizes recorded in rain events from HyMeX (2012 and 2013) by the 2DVD. Table A2 shows the number of drops per diameter class for larger drops. Based on this information, and by looking at the velocity/diameter combinations that the 2DVD hardly ever recorded, we chose a filter that removes a drop with diameter D (mm) and velocity V ($m\,s^{-1}$) if any of the following conditions are true:

$$D > 7.5, \tag{9}$$

$$V > v(D) + 4, \tag{10}$$

$$V < v(D) - 3, \tag{11}$$

where $v(D)$ is the terminal velocity for a drop of equivolume diameter D as defined by Beard, 1976. Figure 3 shows the occurrence of velocity/diameter combinations recorded by the 2DVD during the combined SOPs. Figure 5 shows similar plots for sums of drop counts per Parsivel diameter and velocity class, for both the 2DVD and Parsivel. In these figures, the grey area is the region in which drops will be removed. Over the combined SOPs data set, the filter removed 0.2 % of the drops recorded by the 2DVD. This filtering of suspicious records was applied to both Parsivel and 2DVD data before resampling to any different time resolutions. To resample Parsivel records, the mean DSD was found over each new time period and bulk rainfall variables were then calculated from each mean DSD. The Parsivel precipitation type flag was resampled to give an indication of the proportion of the time period for which solid precipitation was recorded. Note that here, solid precipitation refers to any precipitation that does not fit into the Parsivel instrument's criteria for liquid precipitation, which is based on the velocity and size of the particle (see Löffler-Mang and Joss, 2000). The worst quality flag was kept for each resampled time step, to give an indication of whether any low-quality flags were raised during the resampled integration time.

4.4 2DVD as reference instrument

Given that the 2DVD has previously been shown to produce better matches to independent rain rate measurements than Parsivel (e.g. Tokay et al., 2001; Krajewski et al., 2006), and that it provides higher-resolution DSD measurements than Parsivel, both temporally and in the drop sizes it can discern, we used the 2DVD as the reference instrument for this work. To test the reliability of the 2DVD we compared the 2DVD

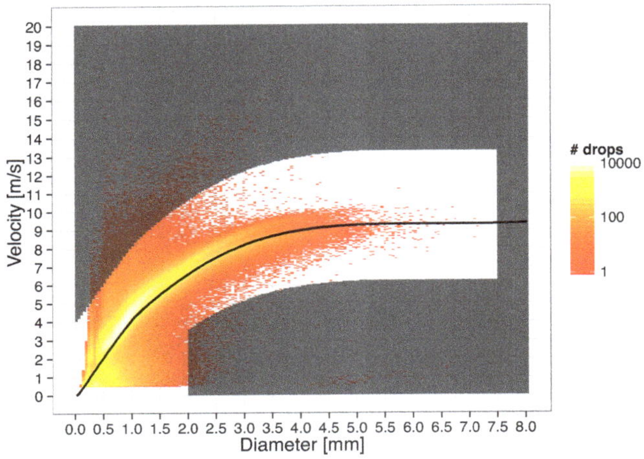

Figure 3. Occurrence of velocity/diameter combinations, with drop counts on a log scale, recorded by 2DVD during the HyMeX campaigns in the autumns of 2012 and 2013. The physical-drop filter is overlaid in grey. The black line indicates the Beard (1976) expected terminal drop velocity.

measurements to collocated rain gauges for the HyMeX SOP2013 campaign. Two separate instruments were collocated with the 2DVD during SOP2013: a Vaisala weather station equipped with a rain cap, and a tipping-bucket rain gauge. We compared the rain rate derived from the 2DVD drop data (Eq. 8) to rain gauge records. To remove solid particles we considered only time steps for which the collocated Parsivel recorded at least 90 % liquid precipitation, and for which the 2DVD and rain gauge both recorded a rain rate greater than or equal to 0.1 mm h^{-1}. One outlier time step, for which the 2DVD was only partially working (4 October 2013, 18:00 UTC – Coordinated Universal Time), was removed. The comparisons are shown in Fig. 6.

It is worth noting here the performance statistics we use in this work. In all scatterplots in this paper, the one-to-one line is shown in red dashes, while the blue line indicates the line of best fit found using linear least squares regression, with standard error shaded in grey. The reference instrument is always on the x axis. The regression slope (reg. slope) is the slope of the regression line. For a given time t, let the reference value be R_t and the observed value be O_t. Let the total number of time steps be T. Then the mean ratio is defined as the reference mean divided by the observed mean:

$$\text{Mean ratio} = \langle R\rangle/\langle O\rangle. \tag{12}$$

Let E_t be the difference for the tth time step, defined as $E_t = O_t - R_t$. RMSE is the root mean squared error,

$$\text{RMSE} = \sqrt{\frac{\sum_{t=1}^{T} E_t^2}{T}}. \tag{13}$$

r^2 is the squared Pearson correlation coefficient between reference and observed data sets. Bias is the mean of the differences, $\langle E\rangle$. Relative bias (rel. bias) is the median of the relative errors, a percentage defined as

$$\text{Rel. bias} = \text{median}\{(O_t - R_t)/R_t \times 100\}. \tag{14}$$

We are only concerned with liquid precipitation in this paper, so we subset time steps to those in which the Parsivel recorded no solid precipitation (for 5 min resolution) or at most 10 % solid precipitation (for 1 h resolution), and for which the Parsivels recorded no non-zero quality status flags. Furthermore, we only compared time steps for which both instruments being compared recorded non-zero rain amounts. We take 0.01 mm h^{-1} as the minimum rain rate the Parsivel can record in one 30 s integration time. Thus, we use a non-zero rain rate threshold of 0.001 mm h^{-1} at 5 min resolution and of 8.3×10^{-5} mm h^{-1} at 1 h resolution. Because each tip of the tipping-bucket rain gauges indicates 0.1 mm of accumulated precipitation, the minimum rain rate that a rain gauge can measure in 5 min is 1.2 mm h^{-1}, and in 1 h the minimum is 0.1 mm h^{-1}. When comparing to rain gauges, the non-zero rain rate threshold therefore becomes 1.2 mm h^{-1} for 5 min resolution and 0.1 mm h^{-1} for 1 h resolution. Because our correction affects the DSD-derived rain rates from the Parsivels, we use the Parsivel-derived intensity when applying the non-zero threshold to Parsivel data. We refer to time steps that satisfy these criteria as those with non-zero liquid DSDs.

The 2DVD showed excellent agreement with the tipping-bucket rain gauge and Vaisala weather station, with high correlation coefficients (r^2 at least 0.98) and low bias amounts for both comparisons (absolute bias less than or equal to 0.2 mm h^{-1}). In both cases the 2DVD tended to slightly underestimate the rain amount given by the other gauge. We conclude, however, that the 2DVD provided reliable measurements of the rain rate. Note that the relative bias between 2DVD and gauge was −14 %, and between 2DVD and Vaisala it was 9 %. The difference in these relative biases can be explained largely by differences in measurements of very small rain rates. This is equivalent to the relative bias we observed between two collocated Parsivels (Pradel 1 and Pradel 2) using the same constraints to choose comparison time steps, also at 1 h resolution, after filtering for unfeasible records (but before any other correction was applied). Using Pradel 1 as reference, the relative bias was −8 % (15 %) in 2013 (2012). Using Pradel 2 as reference, the relative bias was 9 % (−13 %) in 2013 (2012). This means that when we compare Parsivel rain rates to rain gauges or to the 2DVD, we cannot distinguish the level of agreement when the relative bias is less than about 10 %, due to instrumental uncertainty.

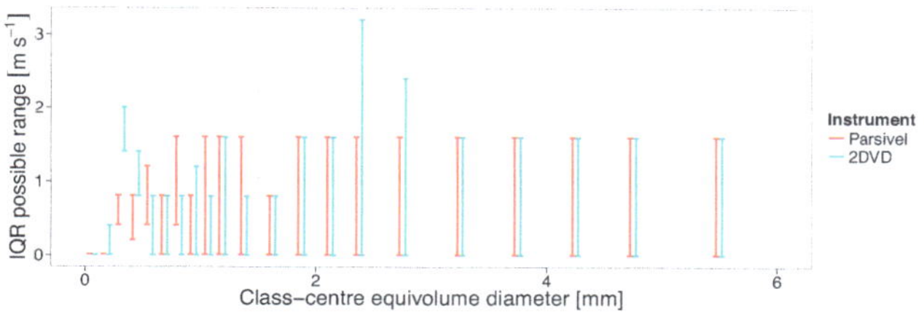

Figure 4. Velocity IQR possible ranges, by Parsivel diameter and velocity class, for mean drop counts for 2DVD and the collocated Parsivel in HyMeX 2013. Only time steps for which both instruments recorded a value and the Parsivel-recorded liquid rain were included. The grey vertical bars indicate the Parsivel diameter class boundaries. Above the 21st diameter class (drops above 6 mm), there were not enough drops to meaningfully calculate a velocity range; for this reason the plot is truncated to 6 mm.

5 Correction of Parsivel DSDs

The correction of Parsivel DSDs is made in two steps. The two steps were chosen so that both the velocity and diameter measurements made by Parsivel disdrometers are addressed. First, the raw Parsivel data is corrected so that per-diameter-class mean velocities match the expected terminal velocity for each class. At this point the raw data can be screened for unfeasible measurements as described in Sect. 4.3. Second, a per-diameter-class adjustment factor is applied to Parsivel classes, in order to make the drop size distribution match, in a statistical way, that recorded by a 2DVD. This adjustment of drop concentrations in effect changes the drop equivolume diameters measured by the Parsivel. In this section we address each correction in turn.

5.1 Correction of per-diameter-class drop velocities

Figures 3 and 5 show the density of particles recorded at each diameter/velocity combination, by the 2DVD and Parsivel disdrometers. Both the 2DVD and Parsivel record drops at a range of velocities for a given equivolume diameter or diameter class. In these plots, the black line is the expected terminal velocity per drop diameter, calculated using the method of Beard (1976). The 2DVD recorded the highest concentrations of drops on and very near the expected terminal velocities. Indeed in SOP2013, for time steps for which the nearest Parsivel recorded liquid precipitation, the bias between expected terminal velocity and velocity recorded by the 2DVD was 0.04 $m\,s^{-1}$ and the relative bias was 2 % (over the combined SOPs the bias was 0.2 $m\,s^{-1}$ and relative bias was 6 %). We hence consider the terminal fall velocity from Beard (1976) as the reference value for fall velocity. The Parsivel tends to overestimate the velocities of small drops.

To correct the velocities in the Parsivel data, we take the set of recorded velocities for each drop diameter class, and shift the values such that the mean velocity is equal to the expected terminal velocity as calculated by the algorithm of

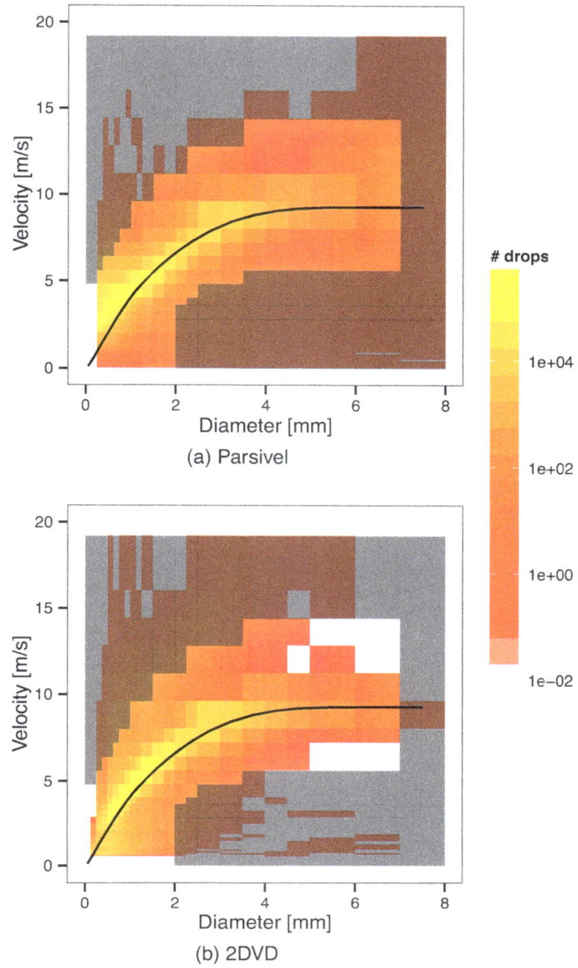

Figure 5. Sum of raw drop occurrences per Parsivel class, for the 2012 and 2013 campaigns. Parsivel counts are summed at stations Pradel 1 (for 2012) and Pradel Grainage (for 2013). The filtered areas are overlaid in grey. The black line is the expected terminal drop velocity calculated by Beard (1976). Drop counts are specified by colour on a log scale.

Beard (1976). Because the velocity classes do not have constant width, the classes are first subsampled into classes of width 0.1 $m\,s^{-1}$, then shifted and regrouped into the original class sizes. Except when some drops were counted in very low-velocity classes and are shifted out of the valid velocity range, the number of drops per diameter class remains the same before and after the velocity shift. An example plot of drop counts per velocity and diameter class before and after the velocity shift is shown in Fig. 7. The velocity shift is equivalent to shifting each column up or down such that the mean velocity for each column (which is usually close to the brightest point) aligns with the line that indicates the expected terminal velocity. As an example, for the average drop counts per velocity and diameter class for SOP2013, using the Parsivel at Pradel Grainage, the mean shift required

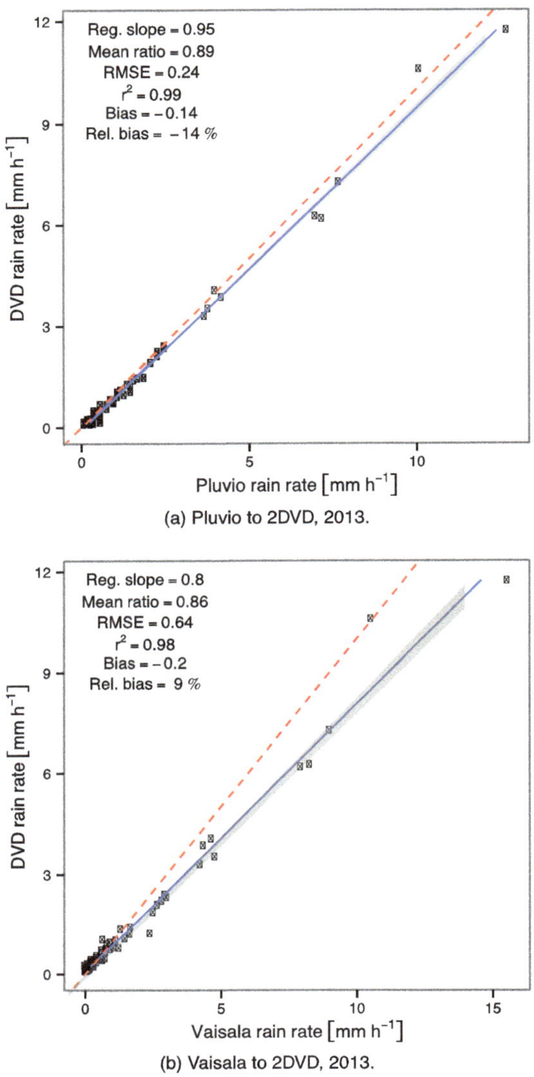

Figure 6. Scatterplots showing the comparison between the 2DVD and **(a)** a collocated tipping-bucket rain gauge (Pluvio), and **(b)** a collocated Vaisala weather station. Time steps compared are at 1 h resolution from HyMeX SOP2013, and include only those times for which the collocated Parsivel recorded a Parsivel-derived rain rate $\geq 0.1\,\mathrm{mm\,h^{-1}}$, $\leq 10\,\%$ of the time step was marked as solid precipitation, and for which both the 2DVD and gauge recorded a rain rate $\geq 0.1\,\mathrm{mm\,h^{-1}}$.

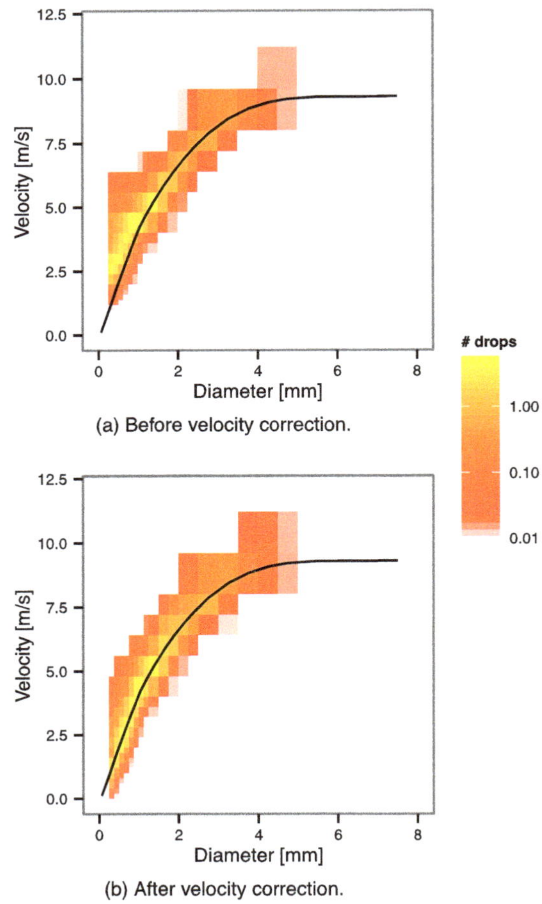

Figure 7. An example of the velocity correction. Average drop counts (on a log scale) for liquid rain from the Parsivel at Pradel Grainage for SOP2013, shown **(a)** before the velocity correction and **(b)** afterwards.

per diameter class from 0 to 5 mm was $-0.29\,\mathrm{m\,s^{-1}}$. Once the velocities are corrected in the raw Parsivel data, any suspicious particles are removed using the criteria shown in Sect. 4.3, and the volumetric drop concentrations per diameter class are found using Eq. (6). This correction and filtering was applied before resampling to any lower time resolutions.

Figure 4 shows a comparison between the interquantile ranges (IQRs) of the recorded velocities, by diameter class. We calculated the mean drop counts per Parsivel velocity and diameter class for SOP2013, for 2DVD and Parsivel. We then applied the velocity correction to the Parsivel drop counts, and calculated the velocity class quantiles, weighted by the drop counts per diameter class, for each set. Given that the velocity classes can be quite large, it is possible that the quantiles both fall within one velocity class; in this case, the IQR could be between 0 and the width of the velocity class. We calculated the minimum and maximum possible IQR range for each diameter class. The plot shows that when binned into the Parsivel velocity classes, the spread of velocities was of the same order of magnitude between the two instruments. We conclude that to correct the velocities measured by Parsivel, it is sufficient to shift the mean velocity closer to the expected terminal velocity as described above, then remove suspicious particles (see Eqs. 9, 10, and 11).

5.2 Correction of diameter-class concentrations

We now turn to correcting the drop concentrations per diameter class with reference to the 2DVD. Let $P(i)$ be the ratio

Table 3. Calibrated first-generation Parsivel correction factors for Parsivel-derived intensity classes for the SOP2013 campaign. Each row contains the class number, the centre equivolume diameter for the class (D_i), and the calibrated factors $P(i)$ for each class of Parsivel-derived intensity. Intensity class boundaries are provided in millimetres per hour ($\mathrm{mm\,h^{-1}}$).

Class (i)	D_i (mm)	$[0, 0.5)$	$[0.5, 1)$	$[1, 2)$	$[2, 200)$
3	0.31	0.05	0.06	0.09	0.12
4	0.44	0.12	0.15	0.24	0.28
5	0.56	0.38	0.44	0.63	0.66
6	0.69	0.48	0.54	0.71	0.85
7	0.81	0.70	0.77	0.95	1.13
8	0.94	0.73	0.74	0.97	1.09
9	1.06	0.84	0.84	1.03	1.26
10	1.19	0.90	0.84	1.04	1.27
11	1.38	0.84	0.81	1.00	1.21
12	1.62	0.75	0.71	0.88	1.03
13	1.88	0.74	0.57	0.77	0.96
14	2.12	0.66	0.54	0.71	0.88
15	2.38	0.51	0.56	0.63	0.83
16	2.75	0.47	0.45	0.47	0.77
17	3.25	0.42	0.46	0.39	0.71
18	3.75	0.47		0.46	0.53
19	4.25				0.43
20	4.75				0.20
21	5.50				0.42

of 2DVD drop concentration to Parsivel drop concentration, defined such that for the ith equivolume diameter class with centre-diameter D_i (mm), at any given time step,

$$P(i) = \frac{N_i^{\mathrm{2DVD}}}{N_i^{\mathrm{Pars}}}. \tag{15}$$

$P(i)$ is thus the correction factor for that time step: when the Parsivel drop concentration for class i is multiplied by $P(i)$ it will match the 2DVD drop concentration for class i. To "train" the correction for a given data set, we find median values of $P(i)$ per class of Parsivel-derived rain intensity. Parsivel-derived intensity is used as it is a measurement of the rain intensity that is always available with Parsivel disdrometers, and is independent of our DSD correction. It is hence easily accessible to all potential users. The result is a collection of correction factors for each Parsivel-derived intensity class. When Parsivel records are multiplied by these correction factors, the per-diameter-class drop counts are scaled to match the corresponding 2DVD drop counts.

To explain the correction in more detail, we take as an example the HyMeX 2012 and 2013 SOPs and show each step of the correction calibration. We used data from SOP2013 to train the correction, because there was a Parsivel collocated with the 2DVD at Pradel Grainage in that campaign. We used a time resolution of 1 h, in order to increase the chance of a time step sampling large drops, and in order to smooth outliers. Assuming the obtained correction is not dependent on the temporal resolution, it will be applied at resolutions higher than 1 h in order to have reliable Parsivel DSD measurements for studies of small-scale DSD variability. A strict set of criteria was used to choose which time steps the comparison should be performed on. We used time steps for which the 2DVD and the collocated Parsivel recorded a non-zero liquid DSD. For all of SOP2013 there were 234 such time steps, corresponding to 234 h of rainfall over which we trained the correction factors. For each valid time step, we compared the mean DSD recorded by the 2DVD and collocated Parsivel.

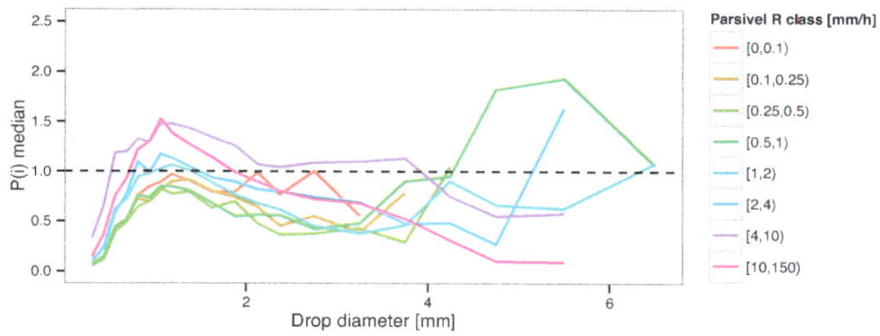

Figure 8. Median $P(i)$ values classed by Parsivel-derived intensity.

Values of $P(i)$ were calculated for each time step in the training set, by comparing the Parsivel DSD to the 2DVD DSD. The result is a distribution of P values for each drop diameter class. To investigate the effect of rain intensity on the values of $P(i)$, we divided the time steps into classes of intensity, using the Parsivel-derived intensity modelled by the sensor. The median $P(i)$ values of each intensity and diameter class are shown in Fig. 8. There is clearly a dependency between the values in the $P(i)$ curve and the rain intensity.

The most notable feature of Fig. 8 is that the numbers of small drops (under about 0.7 mm) were overestimated by the Parsivel. For these classes, the values of $P(i)$ are low, indicating that the Parsivel drop counts need to be scaled down to match the corresponding 2DVD drop counts. For low rain rates, below $1\,\mathrm{mm\,h^{-1}}$, the Parsivel overestimated drop counts in all classes up to 4 mm. Note that large drops are very rare in these rain rate classes and, as we will see, the values of $P(i)$ are more reliable for smaller drop sizes. We identified groups of behaviour of $P(i)$ by ranges of Parsivel-derived intensity, and thus divided the intensities into four classes for ranges $[0, 0.5)$, $[0.5, 1)$, $[1, 2)$ and $[2, 200)\,\mathrm{mm\,h^{-1}}$. Using these ranges as the class definitions for Parsivel-derived intensity, we obtained distributions of $P(i)$ per drop diameter and Parsivel-derived intensity class that are shown in Fig. 9. The distributions are over all time steps and they get larger as the drop diameter increases, which shows that there was much more uncertainty in the correction factors for large diameters than for small diameters.

Across these rain rate classes there was a tendency for the Parsivel to overestimate the numbers of drops smaller than 0.81 mm in diameter and greater than 1.88 mm in diameter,

Table 4. Time series statistics per moment, comparing Parsivel data (at Pradel Grainage) before (bef.) and after (aft.) the correction is applied to the 2DVD, at 5 min resolution, for event times. The 2DVD is taken as the reference. Units of bias and RMSE are per cubic metres per millimetre p (m^{-3} mm^{p}) where p is the moment order. R.b. stands for relative bias.

Moment	Bias bef.	Bias aft.	R.b. bef.	R.b. aft.	RMSE bef.	RMSE aft.	r^2 bef.	r^2 aft.
0	113.21	11.01	139.53	15.70	197.46	43.82	0.57	0.91
1	51.00	9.58	87.08	18.18	89.59	36.83	0.77	0.93
2	30.07	9.59	67.49	21.05	54.46	41.06	0.93	0.94
3	32.21	12.27	63.90	26.31	69.92	62.65	0.96	0.95
4	69.43	22.52	66.18	32.88	213.84	131.01	0.95	0.95
5	218.98	62.88	79.25	41.28	831.12	374.18	0.93	0.93
6	818.32	245.64	97.61	49.48	3481.05	1446.36	0.90	0.89
7	3402.98	1176.15	123.17	61.57	15604.67	7069.59	0.85	0.82

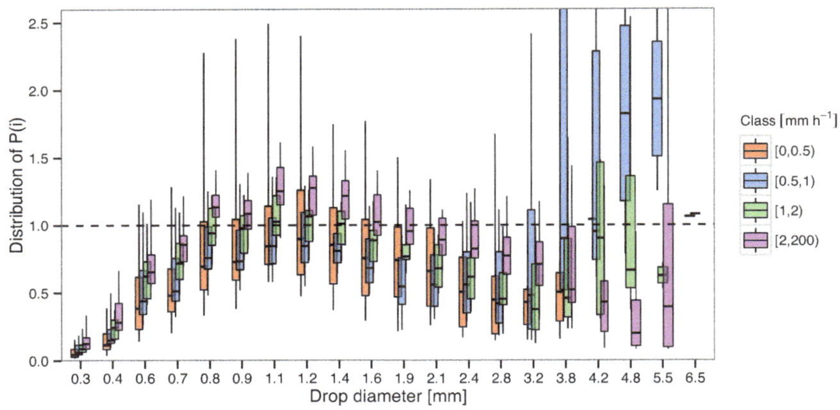

Figure 9. Distributions of $P(i)$ values classed by Parsivel-derived intensity. The correction factors used are the medians of the distributions. The boxes show the interquartile ranges while lines show the 10–90 % quantiles of each distribution. The y axis is cut at 2.6.

with the best performance occurring in the 1–2 mm drop diameter range. For rain rates above 2 mm h^{-1}, the Parsivel more closely matched the 2DVD and indeed underestimates the numbers of drops between about 0.8 and 1.6 mm in diameter.

To train the correction factors, we randomly selected sets of 80 % of the valid training time steps. To determine the impact of sampling effect, we reran the calibration 100 times with different randomly chosen calibration time steps, taking the median of the per-class $P(i)$ distribution each time, and recording the range of resulting values. These ranges are shown per Parsivel-derived intensity class in Fig. 10. We see that the sampling effect for small drops was very small, but that it was larger for larger drop size classes. To ensure a more robust correction, we want to only apply the correction to drop diameter classes for which the training sampling effect (the spread) on $P(i)$ is small. However, in order for the correction to affect all moments of the DSD it is important that it is applied to larger drops as well as smaller ones. We decided to apply a threshold on the spread of the sampling effect. The correction was kept for increasing drop diameters until the sampling effect first surpassed this threshold.

There are hence two threshold values that must be chosen to train correction factors. The first is the minimum-allowed volumetric drop concentration for which 2DVD and Parsivel classes will be compared; let this threshold be Q (mm^{-1} m^{-3}). The second threshold is the maximum-allowed spread in values of $P(i)$ over 100 training iterations of the filter; let this threshold be A. Q was set to 1×10^{-5} and acted simply to stop diminishingly small drop concentrations from adversely affecting the correction calibration. A was set to 0.7. The spread was also required to be larger than 1×10^{-6}, to ensure enough samples were available to give a representative calibration for each diameter class. A sensitivity analysis showed that the values of Q and A did not affect greatly the outcome of the calibration, so long as Q was low enough and A was large enough to allow for sampling and therefore correction of larger drops sizes.

To derive the final correction factors we iterated over 100 sets of training time steps, selecting randomly 80 % of the available times for each iteration. The per-diameter and per-intensity class correction factor is the mean value of $P(i)$ medians for each class over all iterations. The calibrated correction factors for SOP2013 are shown in Table 3. The largest drop diameter class affected by the correction was the 21st

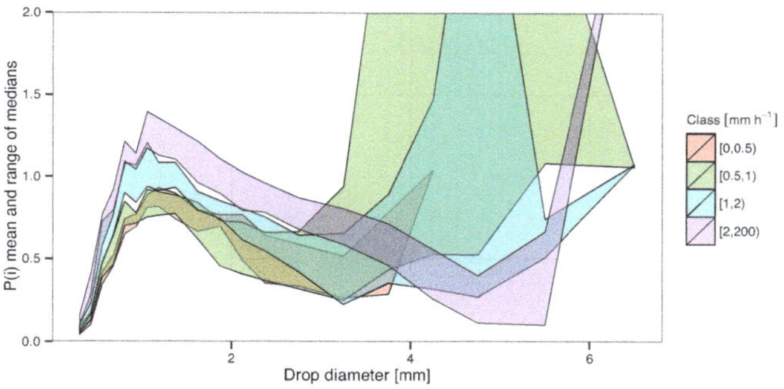

Figure 10. Sampling effect per diameter class, for different classes of Parsivel-derived intensity. The coloured regions represent the minimum and maximum median $P(i)$ per equivolume drop diameter class observed over 100 iterations. The y axis is cut at 2.0.

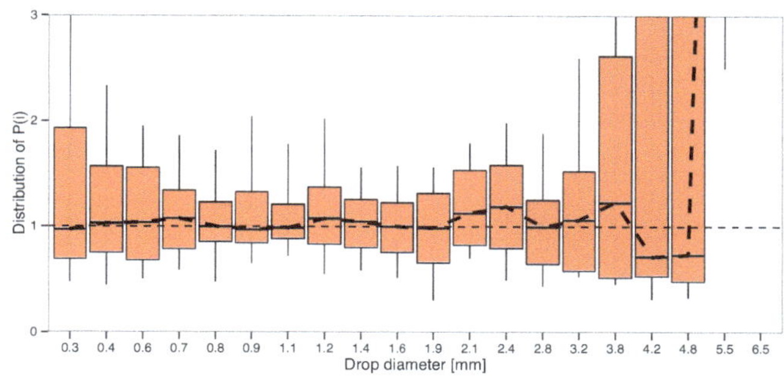

Figure 11. The distributions of $P(i)$ values for the corrected DSD, on an example set of validation time steps from SOP2013, and for all classes of Parsivel-derived intensity, for the Parsivel collocated with the 2DVD. The y axis is cut at 3.

Parsivel class, with a centre size of 5.5 mm. Drops up to the 17th class (class-centre diameter 3.25 mm) are corrected no matter the Parsivel-derived intensity. To apply the correction, each time step was taken separately and, depending on the Parsivel-derived intensity of the time step, the appropriate scaling factors were applied to each equivolume drop diameter amount.

The correction ensures that the corrected DSD more closely matches the DSD recorded by the 2DVD. For example, for the HyMeX SOP2013 data, Fig. 11 shows the distributions of $P(i)$ after the correction, for one example validation set of 20 % of the 1 h time steps in SOP2013. After the correction the DSD much more closely matched that of the 2DVD, especially for small drop diameter classes. For larger drops of greater than about 3 mm the match was not as close, but note this is 20 % of the data and sampling effect changes large drop comparisons much more than small ones. The fact that the large drops differed from the 1 : 1 line reflects the difficulty in training a correction for classes in which there are not many drops to use as a training set, and demonstrates why we chose to train on 1 h time steps and to use the mean $P(i)$ values over many iterations.

6 Drop concentration correction results

In this section we explore the effect of the correction on the moments of the DSD, including the derived rain rate. Our goal in this work is to have reliable DSD measurements from networks of Parsivel disdrometers, in order to be able to study the small-scale variability of the DSD in space and time. We are therefore interested in higher time resolutions than the 1 h resolution we used to train the correction factors. Recall that the choice of 1 h resolution for the training set was made to increase the numbers of sampled large drops, but that we aim to have a correction that is independent of the time resolution. We thus applied the trained correction to 5 min time resolution data to evaluate its effects, for all first-generation Parsivels in the SOP2013 campaign. We also applied the correction to data from SOP2012, as an independent validation data set, and to the combined SOPs. Recall that because we are only interested in liquid precipitation, we subset the available time steps for each Parsivel station to those that contained no Parsivel warning flags regarding data quality, and no solid precipitation markers, and we only compared time steps for which both instruments being compared measured non-zero rain rates.

6.1 DSD moments

To demonstrate the effect of the Parsivel DSD correction on the moments of the DSD, we compare the first seven moments of the DSD recorded by the 2DVD, to the same moments derived from Parsivel DSDs before and after the correction is applied. For these comparisons we used HyMeX SOP2013 event time steps at 5 min resolution, and the Parsivel collocated with the 2DVD at Pradel Grainage. Comparisons of moments of orders 0, 1, 4, and 6 are displayed in Fig. 12, Q–Q plots for these moments are shown in Fig. 13, and time series statistics are shown in Table 4. We see from the densities and Q–Q plots that the correction shifted the distributions of all the moments towards those of the 2DVD. The statistics show an improvement in the relative bias of all moments, by a maximum of 124 % for moment zero and a minimum of 33 % for moment four. RMSE was improved for all moments. r^2 was improved for moments of orders 0, 1, and 2, and remained very similar for higher moments. These results demonstrate that the correction improves Parsivel DSDs at high temporal resolution even when it is trained from 1 h DSD spectra.

6.2 Effect on rain rates

Having confirmed that the correction shifts the densities of the DSD moments towards those of the 2DVD, we used independent instruments – collocated tipping-bucket rain gauges – to test the effect of the correction on the rain rates produced

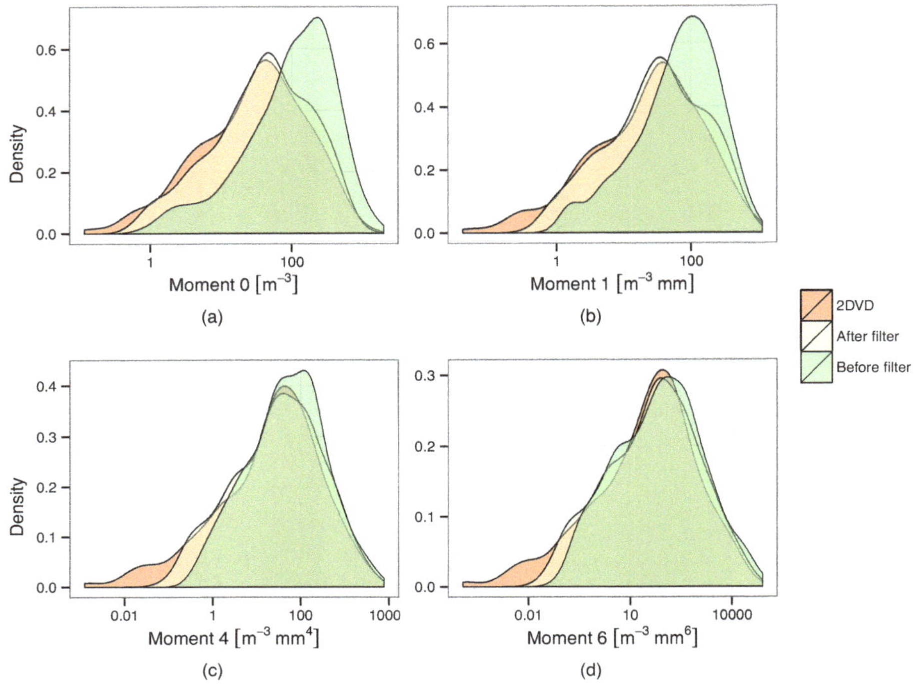

Figure 12. The effect of the correction on DSD moments **(a)** 0 and **(b)** 1, **(c)** 4, and **(d)** 6, showing the densities of the Parsivel-derived DSD moments before and after the correction was applied, and the 2DVD moments, for HyMeX SOP2013 event time steps at Pradel Grainage. The x axis has a log scale.

by Parsivel DSDs. Two of the rain gauges provided measurements that we considered to be suspicious. The station at Mirabel-Pradel-Ferme-2, which is physically closest to our Parsivels Pradel 1 and Pradel 2, produced a marked overestimate of the rain amounts compared to those Parsivels, the 2DVD, and the rain gauge at Mirabel-Pradel-Ferme-1. For this reason we used Mirabel-Pradel-Ferme-1 as the reference gauge at this location. Mirabel-Pradel-Ferme-1 was located approximately 12 m away from Mirabel-Pradel-Ferme-2. Similarly, the rain gauge at Lavilledieu-Ecole-2 was physically closest to our Parsivel at Lavilledieu but, for a period of 1.5 h on 18 September 2012, this rain gauge produced rain rates that were markedly smaller than the rain rates produced by our Parsivel and the nearby rain gauge Lavilledieu-Ecole-1. This gauge, which was approximately 12 m away, provided measurements that more closely matched the Parsivel during this time. We thus used Lavilledieu-Ecole-1 as the reference rain gauge for this station.

We compiled performance statistics for each of the first-generation Parsivel stations, before and after the correction was applied, for a 5 min time resolution. As an example, Fig. 14 shows a scatterplot of rain rates compared to a collocated rain gauge for Pradel 1, the Parsivel that was closest to the 2DVD and deployed in both 2013 and 2012, for 5 min time resolution across both campaigns. The statistics for this station show that the correction produced a clear improvement in the rain rate; the relative bias was reduced by 12 %, the mean ratio and regression slope were both closer to 1 and the RMSE was reduced.

Given that the correction was trained only on SOP2013 data, it makes sense to look at the results from SOP2012 and SOP2013 separately as well as together. For SOP2012 only, the performance effects per statistic are shown in Table 5. For SOP2013 only, the performance effects are shown in Table 6. The differences shown are between the performance statistics after the corrections (velocity and concentration) had both been made, minus the statistics when no correction had been made. The before and after sets were both screened to remove implausible measurements. For SOP2012, the correction improved the RMSE, bias, and relative bias at four of the seven stations. At two of the stations (Pradel 1 and Mirabel) the performance was hardly affected by the DSD correction. At the remaining two stations (Lussas and Lavilledieu) the relative bias was degraded, leaving the final relative bias at these stations as −10 and −9 % respectively; both these relative biases are close to the instrumental variability we observed in Sect. 4.4. Recall that the 2DVD slightly underestimated the rain rate with respect to collocated gauges. For SOP2013, the RMSE and bias were improved at six of eight stations, relative bias was improved at seven of eight stations, and r^2 was hardly changed. The remaining station (Mirabel) showed a degradation of relative bias to an after-correction relative bias of −20 %. At Mirabel, the Parsivel was placed on the edge of

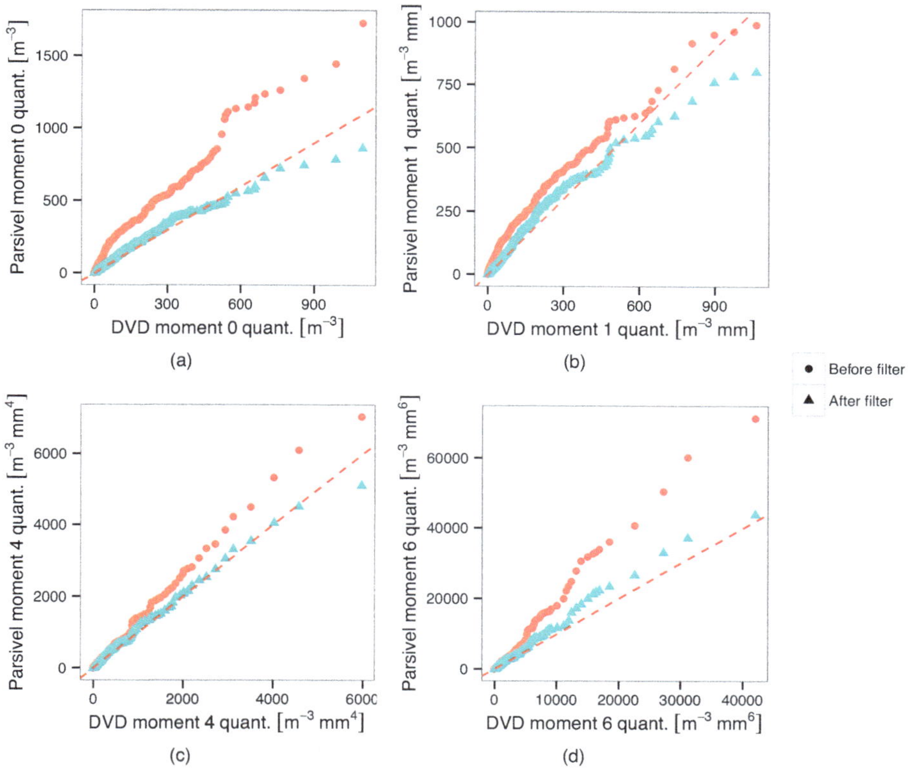

Figure 13. Quantile-to-quantile plots showing the effect of the correction on Parsivel DSD moments **(a)** 0, **(b)** 1, **(c)** 4, and **(d)** 6, by comparing to the 2DVD moments, for HyMeX SOP2013 event time steps at Pradel Grainage.

Table 5. Performance effects of the proposed correction on Parsivel data, and stations on which comparisons were performed, for SOP2012 only at 5 min time resolution. N is the number of time steps on which comparison was possible (high quality, liquid precipitation only). ΔRMSE and Δ|bias| are in units of millimetres per hour, while Δ|r.bias| is a percentage.

Parsivel	Pluvio	ΔRMSE	Δ|bias|	Δ|r.bias|	Δr^2	N
Mirabel	Mirabel-Mairie	0.660	0.532	1.032	0.003	271
Lussas	Lussas-Salle-Polyvalente	0.362	0.454	9.980	−0.002	291
St-Germain	Saint-Germain-Ecole	−0.092	−0.252	−7.427	−0.011	655
Lavilledieu	Lavilledieu-Ecole-2	0.316	0.378	5.760	−0.005	641
Les Blaches	Mirabel-Les-Blaches	−0.212	−0.512	−10.610	−0.006	299
Pradel 1	Mirabel-Pradel-Ferme-1	−0.165	−0.229	−1.949	−0.010	301
Pradel 2	Mirabel-Pradel-Ferme-1	−0.785	−0.548	−12.484	−0.007	327

a retaining wall, which may have introduced turbulence and affected the Parsivel measurements.

For the combined SOPs data set, the Parsivel performance statistics before any correction are shown in Table A3, after both velocity and concentration corrections in Table A4, and the changes made to the performance by the DSD correction are shown in Table 7. Again, all data sets were screened for implausible particles. From these data we can see that RMSE, bias, and relative bias were all improved at six of the eight stations. At the other two stations (Lavilledieu and Mirabel) there was a degradation of performance in terms of rain rate, by about 5 % in terms of relative bias. The relative bias at Lavilledieu after the correction was applied was −6 %, which is within the instrumental error limits. Mirabel may have suffered from turbulence effects; its relative bias across the combined SOPs was already −15 % before any correction was performed. Despite degradations in R bias that were limited to two disdrometers, this analysis of the influence of the correction on the combined SOPs data set confirms its overall benefit to the DSD recorded by Parsivel disdrometers, even at high temporal resolution.

Table 6. Performance effects of the proposed correction on Parsivel data, and stations on which comparisons were performed, for SOP2013 only at 5 min time resolution. N is the number of time steps on which comparison was possible (high quality, liquid precipitation only). ΔRMSE and Δ|bias| are in units of millimetres per hour, while Δ|r.bias| is a percentage.

Parsivel	Pluvio	ΔRMSE	Δ|bias|	Δ|r.bias|	Δr^2	N
Mirabel	Mirabel-Mairie	1.953	1.365	8.332	−0.000	133
Lussas	Lussas-Salle-Polyvalente	−0.874	−0.710	−11.928	−0.006	375
St-Germain	Saint-Germain-Ecole	0.554	0.091	−4.466	−0.017	204
Lavilledieu	Lavilledieu-Ecole-2	−0.597	−0.645	−14.260	−0.006	387
Pradel Grainage	Pradel Grainage	−0.943	−0.773	−15.389	−0.009	377
Les Blaches	Mirabel-Les-Blaches	−0.350	−0.245	−3.536	−0.006	194
Pradel 1	Mirabel-Pradel-Ferme-1	−0.899	−0.722	−16.377	−0.004	218
Pradel 2	Mirabel-Pradel-Ferme-1	−0.946	−0.707	−16.812	−0.004	167

Table 7. Performance effects of the proposed correction on Parsivel data, and stations on which comparisons were performed, for combined SOPs at 5 min time resolution. N is the number of time steps on which comparison was possible. ΔRMSE and Δ|bias| are in units of millimetres per hour, while Δ|r.bias| is a percentage.

Parsivel station	Rain gauge	ΔRMSE	Δ|bias|	Δ|r.bias|	Δr^2	N
Mirabel	Mirabel-Mairie	1.192	0.813	4.902	0.002	404
Lussas	Lussas-Salle-Polyvalente	−0.408	−0.408	−11.340	−0.006	666
St-Germain	Saint-Germain-Ecole	0.053	−0.173	−7.387	−0.025	859
Lavilledieu	Lavilledieu-Ecole-2	−0.135	0.192	2.351	−0.002	1028
Pradel Grainage	Pradel Grainage	−0.943	−0.773	−15.389	−0.018	377
Les Blaches	Mirabel-Les-Blaches	−0.269	−0.404	−10.363	−0.012	493
Pradel 1	Mirabel-Pradel-Ferme-1	−0.478	−0.618	−12.473	−0.018	519
Pradel 2	Mirabel-Pradel-Ferme-1	−0.838	−0.603	−13.473	−0.012	494

6.3 Results at lower temporal resolution

To further test the effects of the correction on Parsivel DSD-derived rain rates compared to collocated rain gauges, and to test the applicability of the filter to different time resolutions, we performed the same analysis as in the previous section but for 1 h temporal resolution on the combined SOPs data set. The differences made by the correction to the DSD moments at 1 h time resolution are shown in Table 8. At 1 h time resolution, the correction improved the bias and relative bias on all moment orders, while RMSE was improved for all orders except the third, for which it was hardly changed. r^2 between moment orders before and after the correction was improved for moments of order 1–3, and maintained at the same level for the other moments. The differences made to the rain rate to gauge comparisons at 1 h resolution are shown in Table 9. RMSE was improved at five of the eight stations, and bias at three of the eight. Relative bias was degraded in all but two cases. This degradation is attributable to very small rain rates; indeed, when we selected time steps for which the rain rate was greater than 1.2 mm h^{-1}, the relative bias was improved at five of the eight stations, and only the station at Mirabel had an after-correction relative bias that was greater than the instrumental variability. For rain rates between 0.1 and 1.2 mm h^{-1}, the after-correction bias was negative at all stations and the per-station mean bias was −0.12 mm h^{-1}. This bias is similar to the bias of the 2DVD compared to the gauge for the same rain rates (−0.08 mm h^{-1}). We conclude that our correction procedures result in Parsivel measurements that better match those of the 2DVD, which itself underestimated rain rate for low rain rates when compared to a collocated gauge. We recommend that care is taken with the application of this correction to rain rates below 1.2 mm h^{-1}.

7 Application to Parsivel^2

We applied our method to second-generation Parsivels (Parsivel^2) that were also deployed in the HyMeX 2013 campaign. To train the correction for Parsivel^2 we followed the same method of comparing Parsivel records for the station at Pradel Grainage to the collocated 2DVD to train the correction factors per Parsivel-derived rain intensity class. The only difference was that, due to changes between the first- and second-generation Parsivels, the curves of $P(i)$ per Parsivel-derived intensity class showed different and more complex behaviour to those of the first-generation Parsivel. The classes we used were $[0, 0.1)$, $[0.1, 0.25)$, $[0.25, 0.5)$, $[0.5, 1)$, $[1, 2)$ and $[2, 200)$ mm h^{-1}.

Apart from the different Parsivel-derived rain intensity class definitions, the training process was identical to that

Table 8. Time series statistics per moment, comparing Parsivel data (at Pradel Grainage) before (bef.) and after (aft.) the correction is applied to the 2DVD, at 1 h resolution, for event times in SOP2013. The 2DVD is taken as the reference. Units of bias and RMSE are m^{-3} mm^p where p is the moment order. R.b. stands for relative bias.

Moment	Bias bef.	Bias aft.	R.b. bef.	R.b. aft.	RMSE bef.	RMSE aft.	r^2 bef.	r^2 aft.
0	70.73	−0.29	107.59	−2.07	132.63	44.10	0.45	0.76
1	30.41	0.56	60.87	0.44	65.45	38.02	0.66	0.81
2	16.30	−0.23	42.24	1.10	51.92	46.51	0.82	0.83
3	16.07	−3.29	38.72	7.10	80.21	80.20	0.84	0.82
4	35.78	−12.85	46.74	10.10	201.90	190.73	0.80	0.78
5	120.92	−41.83	63.10	18.83	662.64	582.25	0.72	0.71
6	481.63	−119.21	82.49	22.79	2526.18	2107.27	0.61	0.62
7	2127.64	−232.92	102.07	40.65	10688.27	8598.09	0.50	0.52

Table 9. Performance effects of the proposed correction on Parsivel data, and stations on which comparisons were performed, at 1 h time resolution, for the combined SOPs. N is the number of time steps on which comparison was possible. ΔRMSE and Δ|bias| are in units of millimetres per hour, while Δ|r.bias| is a percentage.

Parsivel station	Rain gauge	ΔRMSE	Δ|bias|	Δ|r.bias|	Δr^2	N
Mirabel	Mirabel-Mairie	0.486	0.332	16.265	0.004	122
Lussas	Lussas-Salle-Polyvalente	−0.088	0.136	14.326	0.006	183
St-Germain	Saint-Germain-Ecole	0.018	0.059	14.730	−0.011	223
Lavilledieu	Lavilledieu-Ecole-2	0.146	0.292	23.862	−0.005	277
Pradel Grainage	Pradel Grainage	−0.108	−0.096	−14.164	−0.020	131
Les Blaches	Mirabel-Les-Blaches	−0.089	0.031	12.701	−0.006	117
Pradel 1	Mirabel-Pradel-Ferme-1	−0.141	−0.034	12.195	−0.011	130
Pradel 2	Mirabel-Pradel-Ferme-1	−0.220	−0.156	−4.583	−0.014	118

shown in Sect. 5. The resulting correction factors are shown for the HyMeX Parsivel^2 data set in Table 10. A comparison of the corrections for first- and second-generation Parsivel is shown in Fig. 15 and shows significant differences. Differences are expected, given that at a minimum the laser is different between the two instruments. Both filters were similar for drops up to about 1 mm in diameter, in that they both show the Parsivel overestimating drops in comparison to the 2DVD. Parsivel^2 is shown to underestimate the numbers of drops between 1.38 and 3.25 mm diameter. Drops larger than 3.5 mm were overestimated by both generations of Parsivel, but less so by Parsivel^2.

After training the correction factors we applied them to Parsivel^2 data for all available stations. Due to small differences in clock times between the rain gauges and Parsivel^2 stations we used 1 h time resolution. We first compared the moments to the 2DVD moments for event time steps only; these results are shown in Table 12. The bias was improved for moments of order 0–3, 6, and 7, but was degraded for moments of order 4 and 5. Moments of orders 4 and 5 had the two lowest biases before the correction. In contrast, the relative bias was improved for all moment orders except the sixth, where it was maintained essentially at the same level, and the seventh. This indicates that the distribution of differences for moment of orders 4 and 5 may have included outliers which affected the bias. RMSE and r^2 were improved for all moment orders.

We compared the rain rates after the correction of Parsivel^2 to those recorded by collocated rain gauges, for all available time steps. Due to timing errors with the Parsivel^2 network, we applied the correction to 1 h time steps. The results are shown in Table 11. Absolute and relative bias were improved at one station, but degraded at the others. Again, there appears to be an effect of low rain rate on these performance statistics. When we counted only time steps with rain rates at or above 1.2 mm h^{-1}, the worst degradation in relative bias dropped from 20 to 14 %. There were many outliers in these data sets, and work is ongoing to further refine the correction on these Parsivel^2 data. Despite this degradation, the correction improved the DSD moments compared to the 2DVD. We hypothesise that training the Parsivel^2 correction factors using more data and therefore a lower time resolution, plus fixing potential clock issues in this data set, would improve the performance of the correction on Parsivel^2 data.

8 Application to another climatology

Finally, we applied our technique to data collected in a different region and climatology (see Sect. 3.2). In the Payerne 2014 campaign, a 2DVD and first-generation Parsivel

Table 10. Calibrated Parsivel^2 correction factors for Parsivel-derived intensity classes for the HyMeX 2013 campaigns. Each row contains the class number, the centre equivolume diameter for the class (D_i), and the calibrated factors P(i) for each class of Parsivel-derived intensity. Intensity class boundaries are provided in millimetres per hour.

Class (i)	D_i (mm)	[0, 0.1)	[0.1, 0.25)	[0.25, 0.5)	[0.5, 1)	[1, 2)	[2, 200)
3	0.31	0.02	0.04	0.04	0.05	0.06	0.07
4	0.44	0.03	0.05	0.05	0.07	0.11	0.16
5	0.56	0.11	0.16	0.19	0.22	0.30	0.36
6	0.69	0.20	0.26	0.29	0.36	0.45	0.54
7	0.81	0.36	0.47	0.52	0.53	0.71	0.78
8	0.94	0.55	0.55	0.67	0.67	0.80	0.86
9	1.06	0.86	0.85	0.94	0.89	1.01	1.03
10	1.19	0.74	0.84	1.08	0.90	1.17	1.03
11	1.38	1.04	1.13	1.22	1.12	1.36	1.12
12	1.62	1.10	1.20	1.35	1.19	1.37	1.10
13	1.88	1.14	0.97	1.34	1.17	1.41	1.04
14	2.12			1.25	1.17	1.22	0.97
15	2.38			1.29	1.17	1.43	1.06
16	2.75			1.43		1.37	1.07
17	3.25			0.51		1.31	1.02
18	3.75						0.97
19	4.25						0.73
20	4.75						0.58
21	5.50						0.45
22	6.50						0.32

Table 11. Performance effects of the proposed correction on Parsivel^2 network data, and stations on which comparisons were performed. N is the number of time steps on which comparison was possible. ΔRMSE and Δ|bias| are in units of millimetres per hour, while Δ|r.bias| is a percentage.

Parsivel	Pluvio	ΔRMSE	Δ|bias|	Δ|r.bias|	Δr^2	N [h]
Villeneuve-de-Berg	Villeneuve-de-Berg-2	0.05	0.11	7.34	−0.01	129
Mont-Redon	Mirabel-Mont-Redon	0.03	−0.08	−19.68	−0.00	128
Pradel-Vignes	Mirabel-Pradel-Vignes	0.07	0.17	0.94	0.00	58
Pradel Grainage	Pradel Grainage	0.29	0.34	9.65	−0.01	154
Villeneuve-de-Berg 2	Villeneuve-de-Berg-2	0.17	0.33	17.49	−0.01	132

were collocated. When compared to a collocated rain gauge at 1 h resolution, the 2DVD recorded rain rates with a performance very similar to that shown in the SOP2013 campaign (see Sect. 4.4), with a mean ratio of 0.84, an RMSE of 0.24 mm h^{-1}, an r^2 of 0.98, bias of −0.15 mm h^{-1}, and a relative bias of −19 %.

We used the same technique as described in Sect. 5, using data with 1 h temporal resolution for training, and 10 min resolution (the resolution of the reference rain gauge) for testing. We used the same Parsivel-derived intensity classes as for SOP2013. The resulting correction factors are shown in Table 13. The correction is compared to the correction found in SOP2013, per Parsivel-derived rain rate class, in Fig. 16. These plots show that the trained correction factors were very similar across climatologies, for classes of low rain rates (0–1 mm h^{-1}). For larger rain rates, there were some differences between the correction factors, but the general shape remained the same. It should be noted that in the Payerne 2014 data set, the precipitation sampled had much lower intensity than that found in SOP2013, and we hypothesise that the differences are due to these sampling effects. We tested the results using both 1 h and 10 min resolution data sets.

For consistency, we kept the threshold for the maximum-allowed spread in $P(i)$ at 0.7. In the Payerne 2014 data set, there appears to be greater sensitivity to this threshold, indicating that the sample size available for training was smaller than in SOP2013. For this reason we also tested the results when the correction trained using SOP2013 was applied to the Payerne 2014 data set. The effect on Parsivel rain rate performances are summarised in Table 14. The comparisons of moments with those of the 2DVD are included in the appendix, in Tables A5, A6, and A7.

The results show that when the Payerne data set was used to train the correction factors, there was a slight improvement

Table 12. Time series statistics per moment, comparing Parsivel^2 data (at Pradel Grainage) before (bef.) and after (aft.) the correction is applied, to the 2DVD, at 1 h resolution, for event times. The 2DVD is taken as the reference. Units of bias and RMSE are per cubic metres per millimetre (p m^{-3} mm^p) where p is the moment order. R.b stands for relative bias.

Moment	Bias bef.	Bias aft.	R.b. bef.	R.b. aft.	RMSE bef.	RMSE aft.	r^2 bef.	r^2 aft.
0	166.10	−3.20	268.90	−0.46	252.12	44.46	0.39	0.78
1	74.22	−1.25	172.75	0.65	113.59	38.17	0.55	0.82
2	34.10	−1.51	90.83	0.78	65.85	45.19	0.77	0.85
3	13.52	−3.78	49.64	0.12	73.42	70.83	0.87	0.88
4	1.56	−11.12	29.01	−0.18	144.25	144.89	0.89	0.90
5	9.54	−35.47	14.18	−3.64	399.39	381.95	0.89	0.91
6	150.37	−123.07	4.64	−4.66	1553.18	1261.94	0.88	0.92
7	1141.04	−462.26	−2.21	−7.28	7617.83	4977.42	0.87	0.91

Table 13. Calibrated Parsivel correction factors for Parsivel-derived intensity classes for the Payerne 2014 campaign. Each row contains the class number, the centre equivolume diameter for the class (D_i), and the calibrated factors $P(i)$ for each class of Parsivel-derived intensity. Intensity class boundaries are provided in millimetres per hour.

Class (i)	D_i (mm)	[0, 0.5)	[0.5, 1)	[1, 2)	[2, 200)
3	0.31	0.04	0.04	0.10	0.08
4	0.44	0.08	0.13	0.22	0.21
5	0.56	0.28	0.37	0.44	0.50
6	0.69	0.40	0.44	0.51	0.65
7	0.81	0.58	0.61	0.70	0.86
8	0.94	0.61	0.65	0.66	0.82
9	1.06	0.78	0.77	0.74	0.94
10	1.19	0.68	0.75	0.71	0.96
11	1.38	0.72	0.73	0.70	0.93
12	1.62	0.70	0.62	0.68	0.98
13	1.88	0.69	0.50	0.54	0.96
14	2.12	0.50	0.49	0.55	0.88
15	2.38	0.59	0.52	0.34	0.74
16	2.75	0.40	0.58	0.47	0.62
17	3.25	0.50	0.61	0.27	0.68
18	3.75		0.66	0.53	0.52
19	4.25		0.32		0.51
20	4.75				0.19
21	5.50				0.46

in the Parsivel's rain rate estimation at 10 min resolution. At 1 h resolution, the absolute bias was maintained but the relative bias was degraded. Correlations were maintained by the correction. When the HyMeX-trained correction was applied to the Payerne data set, the performance was improved again. This indicates again that the sample size of the Payerne data set may have been smaller than required for a representative set of correction factors to be trained. Whether the Payerne-trained or HyMeX-trained correction factors were used, there was an improvement in the match between Parsivel and the 2DVD at Payerne for all moments. This suggests that the correction is robust and can be applied as such in different climatic regions.

Table 14. Summary of performance effects of the Parsivel correction, for Payerne 2014. Set indicates which data set was used to train the correction factors using one resolution (Pay – Payerne 2014, S13 – SOP2013), Res. is the temporal resolution to which the corrections were applied, and N is the number of time steps to which the correction was applied. ΔRMSE and Δ|bias| are in units of millimetres per hour, while Δ|r.b.| (relative bias) is a percentage.

Set	Res.	ΔRMSE	Δ|bias|	Δ|r.b.|	Δr^2	N
Pay	10 m	−0.108	−0.048	−4.937	−0.002	435
Pay	1 h	−0.015	−0.003	3.879	−0.004	164
S13	10 m	−0.131	−0.131	−9.570	−0.003	435

9 Conclusions

We have developed a method to correct raindrop size distributions recorded by Parsivel disdrometers, using a two-dimensional video disdrometer as a reference instrument. The correction is made in two steps. First, raw Parsivel drop counts binned by velocity and diameter are shifted so that per-diameter-class mean velocities align with expected terminal velocities. The raw data can then be screened for particles that are unlikely to be raindrops, and per-diameter-class volumetric drop concentrations can then be calculated. Second, these volumetric drop concentrations are adjusted by factors trained by reference to the 2DVD. The adjustment causes the drop concentrations to match those of the 2DVD in a statistical way.

The correction was applied to Parsivel and Parsivel^2 data from two autumn field campaigns in Ardèche, France. The results showed an improvement in the accuracy of moments of the DSD, when compared to the 2DVD as the reference instrument. Comparison of rain rate with collocated rain gauges showed changes that are acceptable, given the overall improvement in the accuracy of the DSD afforded by the

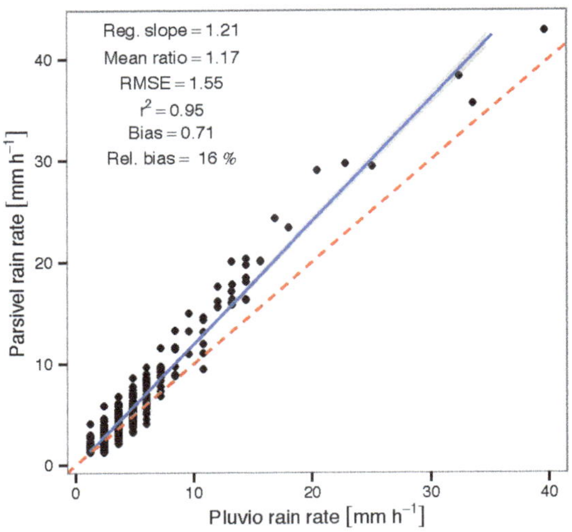

(a) Physical drops only.

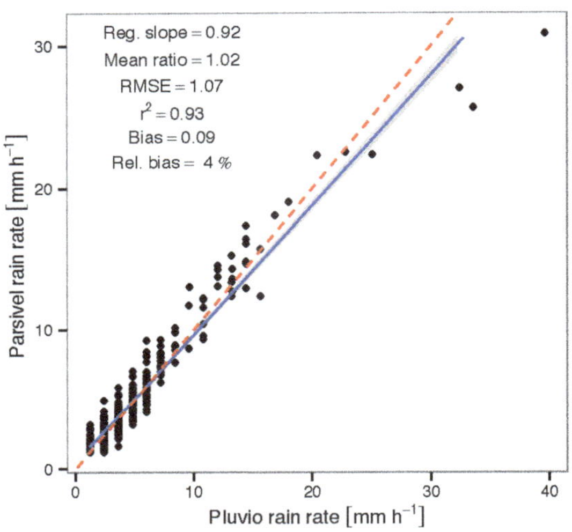

(b) Concentrations corrected.

Figure 14. Scatterplots showing the effect of the proposed correction, for the combined SOPs, with liquid precipitation only and rain rates over 1.2 mm h^{-1}, for Pradel 1, the closest station to the 2DVD that was present in both 2012 and 2013.

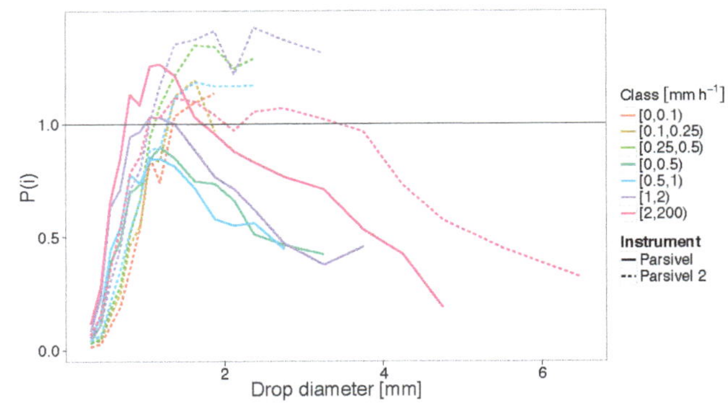

Figure 15. Comparison between correction factors for different generations of Parsivel disdrometers in SOP2013.

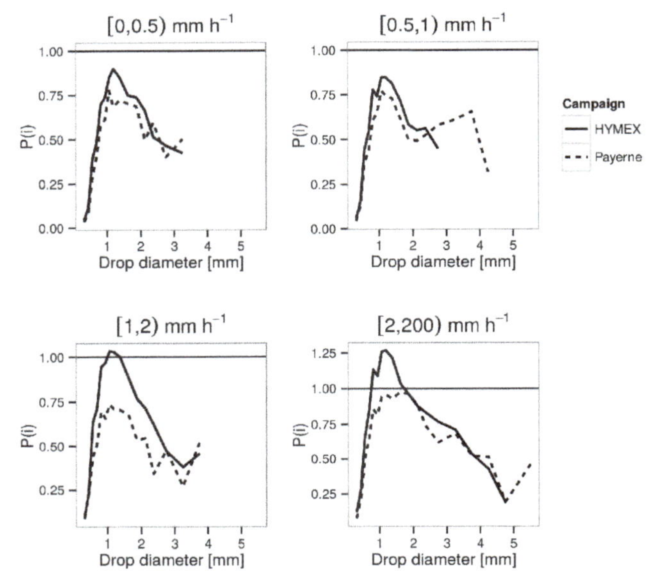

Figure 16. Comparison between correction factors for different campaigns, Payerne 2014 and SOP2013. Both sets were trained on data at 1 h time resolution.

correction. It must be noted that because the 2DVD is used as the reference instrument, the adjusted Parsivel drop size distributions will be, at best, as accurate as the measurements obtained by the 2DVD. If a better reference becomes available, exactly the same approach could be applied to correct the Parsivel (or indeed any other disdrometer) and to improve the agreement with the reference.

The correction was shown to be timescale-independent through application to both 5 min and 1 h Parsivel records. While in this case the correction was trained on data sets containing mainly light to intermediate rain rates (mostly below 20 mm h^{-1}), the method is flexible because it is conditioned on the Parsivel-derived rain intensity, and could be trained for higher rain rate classes as required. The method does not involve changing the hardware or software of the instrument, and it can be applied retrospectively to existing data sets. The correction offers the ability to improve the accuracy of the DSDs recorded by Parsivel disdrometers, which are instruments that are especially suitable for deployment in networks. High-quality DSD measurements from networks of Parsivel disdrometers can be used in valuable work on topics such as the small-scale variability of the drop size distribution. The correction has been demonstrated to work across two different climatologies in Europe, and work is ongoing to further test the transferability of the method.

Appendix A: Extra data tables

Table A1. Clock adjustments (A) for 2DVD events in HyMeX SOP2013.

Event	From (UTC)	To (UTC)	A (s)
24,25	20-10-2013, 00:00:00	24-10-2013, 00:00:00	60
26	27-10-2013, 00:00:00	28-10-2013, 00:00:00	30
27	02-11-2013, 00:00:00	03-11-2013, 00:00:00	60
28	04-11-2013, 00:00:00	05-11-2013, 00:00:00	60
29	05-11-2013, 00:00:00	06-11-2013, 00:00:00	30
30	18-11-2013, 00:00:00	19-11-2013, 00:00:00	30

Table A2. Numbers of large drops recorded by the 2DVD during the combined SOP event times.

Diameter class (mm)	Number of drops	% total drops
(5, 5.5]	273	0.00531
(5.5, 6]	97	0.00189
(6, 6.5]	36	0.00070
(6.5, 7]	10	0.00019
(7, 7.5]	3	0.00006
(7.5, 8]	1	0.00002

Table A3. Performance statistics for rain rate per Parsivel station for the combined SOPs at 5 min resolution, before the DSD correction is applied. RMSE (E) and bias are in units of millimetres per hour; relative bias (R.b.) is a percentage. F stands for fit slope, M stands for mean ratio, and Pradel Grain. stands for Pradel Grainage.

Station	E	Bias	r^2	R.b.	F	M
Mirabel	1.49	−0.65	0.98	−15.18	1.03	0.90
Lussas	1.55	0.51	0.94	11.99	1.18	1.12
St-Germain	1.06	0.35	0.96	7.83	1.12	1.08
Lavilledieu	1.24	0.22	0.96	3.75	1.14	1.05
Pradel Grain.	2.04	1.12	0.97	26.22	1.29	1.28
Les Blaches	1.35	0.50	0.95	10.46	1.17	1.13
Pradel 1	1.55	0.71	0.95	16.12	1.21	1.17
Pradel 2	1.69	0.88	0.97	20.40	1.29	1.22

Table A4. Performance statistics for rain rate per Parsivel station for the combined SOPs at 5 min resolution, after the DSD correction is applied. RMSE (E) and bias are in units of $mm\ h^{-1}$; relative bias (R.b.) is a percentage. F stands for fit slope, M stands for mean ratio, and Pradel Grain. stands for Pradel Grainage.

Station	E	Bias	r^2	R.b.	F	M
Mirabel	2.69	−1.46	0.98	−20.08	0.76	0.79
Lussas	1.14	−0.10	0.94	0.65	0.90	0.98
St-Germain	1.11	−0.18	0.94	−0.45	0.85	0.96
Lavilledieu	1.11	−0.41	0.96	−6.11	0.88	0.92
Pradel Grain.	1.09	0.35	0.96	10.83	0.94	1.08
Les Blaches	1.08	−0.10	0.94	−0.10	0.87	0.98
Pradel 1	1.07	0.09	0.93	3.64	0.92	1.02
Pradel 2	0.85	0.27	0.96	6.93	0.97	1.06

Table A5. Time series statistics per moment, comparing Parsivel data for Payerne 2014 before (bef.) and after (aft.) the correction is applied, to the 2DVD, at 10 min resolution. The 2DVD is taken as the reference. Units of bias and RMSE are per cubic metre per millimetre p (m^{-3} mm^p) where p is the moment order. R.b stands for relative bias.

Moment	Bias bef.	Bias aft.	R.b. bef.	R.b. aft.	RMSE bef.	RMSE aft.	r^2 bef.	r^2 aft.
0	133.26	2.01	231.25	3.60	196.68	24.64	0.67	0.94
1	64.71	2.02	154.15	4.73	94.61	17.40	0.79	0.96
2	38.07	1.87	109.95	6.75	55.90	16.22	0.90	0.97
3	29.06	1.71	88.31	8.39	44.92	20.72	0.95	0.97
4	31.66	1.53	78.41	12.34	60.23	37.56	0.97	0.96
5	53.86	2.16	80.14	17.03	153.13	98.52	0.97	0.94
6	141.20	11.87	87.77	22.58	567.75	356.38	0.95	0.89
7	505.55	91.94	102.59	31.68	2536.41	1660.16	0.90	0.77

Table A6. Time series statistics per moment, comparing Parsivel data for Payerne 2014 before (bef.) and after (aft.) the correction is applied, to the 2DVD, at 1 h resolution. The 2DVD is taken as the reference. Units of bias and RMSE are per cubic metre per millimetre p (m^{-3} mm^p) where p is the moment order. R.b stands for relative bias.

Moment	Bias bef.	Bias aft.	R.b. bef.	R.b. aft.	RMSE bef.	RMSE aft.	r^2 bef.	r^2 aft.
0	81.23	0.65	266.11	0.92	129.14	11.97	0.77	0.96
1	39.42	0.69	172.77	3.22	63.03	8.62	0.85	0.97
2	23.08	0.56	120.47	1.65	37.52	7.72	0.91	0.98
3	17.26	0.37	80.41	3.04	29.04	8.99	0.95	0.98
4	17.74	0.18	66.01	3.28	32.94	14.15	0.98	0.98
5	27.28	0.73	69.61	7.50	66.29	30.07	0.99	0.98
6	64.91	8.04	74.55	12.27	216.27	92.68	0.99	0.96
7	221.52	62.86	85.83	20.23	898.23	439.25	0.97	0.89

Table A7. Time series statistics per moment, comparing Parsivel data for Payerne 2014 before (bef.) and after (aft.) the correction is applied, to the 2DVD, at 10 min resolution. In this case the SOP2013 correction is applied to the Payerne 2014 data set. The 2DVD is taken as the reference. Units of bias and RMSE are m^{-3} mm^p where p is the moment order. R.b stands for relative bias.

Moment	Bias bef.	Bias aft.	R.b. bef.	R.b. aft.	RMSE bef.	RMSE aft.	r^2 bef.	r^2 aft.
0	133.26	22.61	231.25	31.53	196.68	43.57	0.67	0.94
1	64.71	18.33	154.15	32.52	94.61	35.98	0.79	0.95
2	38.07	16.34	109.95	33.84	55.90	34.20	0.90	0.96
3	29.06	16.07	88.31	34.31	44.92	37.72	0.95	0.95
4	31.66	17.58	78.41	36.23	60.23	51.95	0.97	0.95
5	53.86	22.61	80.14	39.17	153.13	108.50	0.97	0.93
6	141.20	41.75	87.77	45.30	567.75	366.53	0.95	0.89
7	505.55	140.46	102.59	53.95	2536.41	1687.96	0.90	0.77

Acknowledgements. The authors thank the maintainers of the HPicoNet rain gauge and Parsivel^2 network: Brice Boudevillain, Gilles Molinié, and Simon Gérard from the Laboratoire d'étude des Transferts en Hydrologie et Environnement (LTHE) at Grenoble University. Rain gauge and Parsivel^2 data were obtained from the HyMeX program, sponsored by grants MISTRALS/HyMeX, ANR-2011-BS56-027 FLOODSCALE project and OHMCV. The authors thank Jacopo Grazioli, Marc Schleiss, Joël Jaffrain, and Daniel Wolfensberger, for their work on the Parsivel and 2DVD network deployed in HyMeX SOPs 2012 and 2013. The authors thank Joël Jaffrain and Meteoswiss for their work on installation and maintenance of instruments in the Payerne 2014 campaign. Finally, the authors acknowledge the financial support of the Swiss National Science Foundation under grant 2000021_140669.

Edited by: G. Vulpiani

References

Andsager, K., Beard, K. V., and Laird, N. F.: Laboratory measurements of axis ratios for large rain drops, J. Atmos. Sci., 56, 2673–2683, 1999.

Atlas, D., Srivastava, R., and Sekhon, R. S.: Doppler radar characteristics of precipitation at vertical incidence, Rev. Geophys., 11, 1–35, 1973.

Battaglia, A., Rustemeier, E., Tokay, A., Blahak, U., and Simmer, C.: PARSIVEL snow observations: a critical assessment, J. Atmos. Oceanic Technol., 27, 333–344, doi:10.1175/2009JTECHA1332.1, 2010.

Beard, K. V.: Terminal velocity and shape of cloud and precipitation drops aloft, J. Atmos. Sci., 33, 851–864, 1976.

Beard, K. V. and Chuang, C.: A new model for the equilibrium shape of raindrops, J. Atmos. Sci., 44, 1509–1524, 1987.

Brandes, E., Zhang, G., and Vivekanandan, J.: Experiments in rainfall estimation with a polarimetric radar in a subtropical environment, J. Appl. Meteor., 41, 674–685, 2002.

Drobinski, P., Ducrocq, V., Alpert, P., Anagnostou, E., Béranger, K., Borga, M., Braud, I., Chanzy, A., Davolio, S., Delrieu, G., Estournel, C., Filali Boubrahmi, N., Font, J., Grubisic, V., Gualdi, S., Homar, V., Ivancan-Picek, B., Kottmeier, C., Kotroni, V., Lagouvardos, K., Lionello, P., Llasat, M. C., Ludwig, W., Lutoff, C., Mariotti, A., Richard, E., Romero, R., Rotunno, R., Roussot, O., Ruin, I., Somot, S., Taupier-Letage, I., Tintore, J., Uijlenhoet, R., and Wernli, H.: HyMeX, a 10-year multidisciplinary program on the Mediterranean water cycle, Bull. Amer. Meteor. Soc., 95, 1063–1082, doi:10.1175/BAMS-D-12-00242.1, 2014.

Frei, C. and Schär, C.: A precipitation climatology of the Alps from high-resolution rain-gauge observations, Int. J. Climatol., 18, 873–900, 1998.

Jaffrain, J. and Berne, A.: Experimental quantification of the sampling uncertainty associated with measurements from Parsivel disdrometers, J. Hydrometeor., 12, 352–370, doi:10.1175/2010JHM1244.1, 2011.

Jaffrain, J., Studzinski, A., and Berne, A.: A network of disdrometers to quantify the small-scale variability of the raindrop size distribution, Water Resour. Res., 47, W00H06, doi:10.1029/2010WR009872, 2011.

Jameson, A. R. and Kostinski, A. B.: What is a raindrop size distribution?, Bull. Amer. Meteor. Soc., 82, 1169–1177, 2001.

Krajewski, W. F., Kruger, A., Caracciolo, C., Golé, P., Barthes, L., Creutin, J.-D., Delahaye, J.-Y., Nikolopoulos, E. I., Ogden, F., and Vinson, J.-P.: DEVEX-disdrometer evaluation experiment: basic results and implications for hydrologic studies, Adv. Water Resour., 29, 807–814, doi:10.1016/j.advwatres.2005.03.018, 2006.

Kruger, A. and Krajewski, W. F.: Two-dimensional video disdrometer: a description, J. Atmos. Oceanic Technol., 19, 602–617, 2002.

Löffler-Mang, M. and Joss, J.: An optical disdrometer for measuring size and velocity of hydrometeors, J. Atmos. Oceanic Technol., 17, 130–139, 2000.

MeteoFrance: Données climatiques de la station de Montélimar, http://www.meteofrance.com/climat/france/station/26198001/normales (last access: 20 November 2014), 2014.

MeteoSwiss: Federal Office of Meteorology and Climatology MeteoSwiss, Climate normals Payerne, Reference period 1981–2010, http://meteosuisse.ch/files/kd/climsheet/en/PAY_norm8110.pdf (last access: 20 November 2014), 2013.

Nešpor, V., Krajewski, W. F., and Kruger, A.: Wind-induced error of raindrop size distribution measurement using a two-dimensional video disdrometer, J. Atmos. Ocean. Technol., 17, 1483–1492, 2000.

Olsen, R. L., Rogers, D. V., and Hodge, D. B.: The aR^b relation in the calculation of rain attenuation, IEEE T. Antenn. Propag., 26, 318–329, 1978.

Peel, M. C., Finlayson, B. L., and McMahon, T. A.: Updated world map of the Köppen-Geiger climate classification, Hydrol. Earth Syst. Sci., 11, 1633–1644, doi:10.5194/hess-11-1633-2007, 2007.

Pruppacher, H. R. and Klett, J. D.: Microphysics of clouds and precipitation, Kluwer Academic Publishers, Dordrecht, 2nd rev. and enl. Edn., with an introduction to cloud chemistry and cloud electricity Edn., 2000.

Ricard, D., Ducrocq, V., and Auger, L.: A climatology of the mesoscale environment associated with heavily precipitating events over a northwestern Mediterranean area, J. Appl. Meteorol. Climatol., 51, 468–488, 2012.

Roe, G. H.: Orographic precipitation, Ann. Rev. Earth Planet Sci., 33, 645–671, 2005.

Schönhuber, M., Lammer, G., and Randeu, W. L.: One decade of imaging precipitation measurement by 2D-video-distrometer, Adv. Geosci., 10, 85–90, doi:10.5194/adgeo-10-85-2007, 2007.

Schönhuber, M., Lammer, G., and Randeu, W. L.: The 2D-video-distrometer, in: Precipitation: Advances in measurement, estimation and prediction, 3–31, Springer, 2008.

Tapiador, F., Checa, R., and De Castro, M.: An experiment to measure the spatial variability of rain drop size distribution using sixteen laser disdrometers, Geophys. Res. Lett., 37, L16803, doi:10.1029/2010GL044120, 2010.

Testud, J., Oury, S., Black, R. A., Amayenc, P., and Dou, X.: The concept of "normalized" distribution to describe raindrop spectra: a tool for cloud physics and cloud remote sensing, J. Appl. Meteor., 40, 1118–1140, 2001.

Thurai, M. and Bringi, V. N.: Drop axis ratios from a 2D video disdrometer, J. Atmos. Oceanic Technol., 22, 966–978, doi:10.1175/JTECH1767.1, 2005.

Thurai, M., Huang, G., Bringi, V., Randeu, W., and Schönhuber, M.: Drop shapes, model comparisons, and calculations of polarimetric radar parameters in rain, J. Atmos. Ocean. Technol., 24, 1019–1032, 2007.

Thurai, M., Petersen, W. A., Tokay, A., Schultz, C., and Gatlin, P.: Drop size distribution comparisons between Parsivel and 2-D video disdrometers, Adv. Geosci., 30, 3–9, doi:10.5194/adgeo-30-3-2011, 2011.

Tokay, A., Kruger, A., and Krajewski, W.: Comparison of drop size distribution measurements by impact and optical disdrometers, J. Appl. Meteor., 40, 2083–2097, 2001.

Tokay, A., Petersen, W. A., Gatlin, P., and Wingo, M.: Comparison of raindrop size distribution measurements by collocated disdrometers, J. Atmos. Ocean. Technol., 30, 1672–1690, doi:10.1175/JTECH-D-12-00163.1, 2013.

Tokay, A., Wolff, D. B., and Petersen, W. A.: Evaluation of the New Version of the Laser-Optical Disdrometer, OTT Parsivel2, J. Atmos. Ocean. Technol., 31, 1276–1288, doi:10.1175/JTECH-D-13-00174.1, 2014.

Uijlenhoet, R. and Sempere Torres, D.: Measurement and parameterization of rainfall microstructure, J. Hydrol., 328, 1–7, doi:10.1016/j.jhydrol.2005.11.038, 2006.

Ulbrich, C. W.: Natural variations in the analytical form of the raindrop-size distribution, J. Climate Appl. Meteor., 22, 1764–1775, 1983.

Ulbrich, C. W.: The Effects of Drop Size Distribution Truncation on Rainfall Integral Parameters and Empirical Relations, J. Climate Appl. Meteor., 24, 580–590, 1985.

Yuter, S. E., Kingsmill, D. E., Nance, L. B., and Löffler-Mang, M.: Observations of precipitation size and fall speed characteristics within coexisting rain and wet snow, J. Appl. Meteorol. Climatol., 45, 1450–1464, doi:10.1175/JAM2406.1, 2006.

The Orbiting Carbon Observatory (OCO-2): spectrometer performance evaluation using pre-launch direct sun measurements

C. Frankenberg[1], R. Pollock[1], R. A. M. Lee[1], R. Rosenberg[1], J.-F. Blavier[1], D. Crisp[1], C. W. O'Dell[2], G. B. Osterman[1], C. Roehl[3], P. O. Wennberg[3], and D. Wunch[3]

[1]Jet Propulsion Laboratory, California Institute of Technology, Pasadena, CA, USA
[2]Colorado State University, Fort Collins, CO, USA
[3]California Institute of Technology, Pasadena, CA, USA

Correspondence to: C. Frankenberg (christian.frankenberg@jpl.nasa.gov)

Abstract. The Orbiting Carbon Observatory-2 (OCO-2), launched on 2 July 2014, is a NASA mission designed to measure the column-averaged CO_2 dry air mole fraction, X_{CO_2}. Towards that goal, it will collect spectra of reflected sunlight in narrow spectral ranges centered at 0.76, 1.6 and 2.0 μm with a resolving power ($\lambda/\Delta\lambda$) of 20 000. These spectra will be used in an optimal estimation framework to retrieve X_{CO_2}. About 100 000 cloud free soundings of X_{CO_2} each day will allow estimates of net CO_2 fluxes on regional to continental scales to be determined. Here, we evaluate the OCO-2 spectrometer performance using pre-launch data acquired during instrument thermal vacuum tests in April 2012. A heliostat and a diffuser plate were used to feed direct sunlight into the OCO-2 instrument and spectra were recorded. These spectra were compared to those collected concurrently from a nearby high-resolution Fourier Transform Spectrometer that was part of the Total Carbon Column Observing Network (TCCON). Using the launch-ready OCO-2 calibration and spectroscopic parameters, we performed total column scaling fits to all spectral bands and compared these to TCCON results. On 20 April, we detected a CO_2 plume from the Los Angeles basin at the JPL site with strongly enhanced short-term variability on the order of 1 % (3–4 ppm). We also found good (<0.5 ppm) inter-footprint consistency in retrieved X_{CO_2}. The variations in spectral fitting residuals are consistent with signal-to-noise estimates from instrument calibration, while average residuals are systematic and mostly attributable to remaining errors in our knowledge of the CO_2 and O_2 spectroscopic parameters. A few remaining inconsistencies observed during the tests may be attributable to the specific instrument setup on the ground and will be re-evaluated with in-orbit data.

1 Introduction

After the launch failure of OCO (Crisp et al., 2004) in 2009, the National Aeronautics and Space Administration (NASA) authorized the development of OCO-2, which was successfully launched on 2 July 2014 at 2:56 a.m. Pacific daylight time (PDT) from Vandenberg Air Force Base on the California coast. To achieve its mission goal, OCO-2 spectra must enable retrievals of column averaged atmospheric CO_2 (denoted as X_{CO_2}) with an accuracy of 1 ppm or better. A typical algorithm for the retrieval of X_{CO_2} (Bösch et al., 2006; Butz et al., 2009; O'Dell et al., 2012) concurrently employs three spectral bands, centered around 0.76 μm (O_2 A-band), 1.61 μm (weak CO_2 band) and 2.06 μm (strong CO_2 band). By using this multi-channel approach, X_{CO_2}, surface albedos, as well as aerosol properties can be retrieved concurrently.

2 OCO-2 instrument overview

This paper describes spectral fitting results from direct sun observations collected with the OCO-2 instrument during thermal vacuum tests (TVACs) in April 2012 at the Jet Propulsion Laboratory (JPL) in Pasadena, California. We compare the instrument performance with measurements

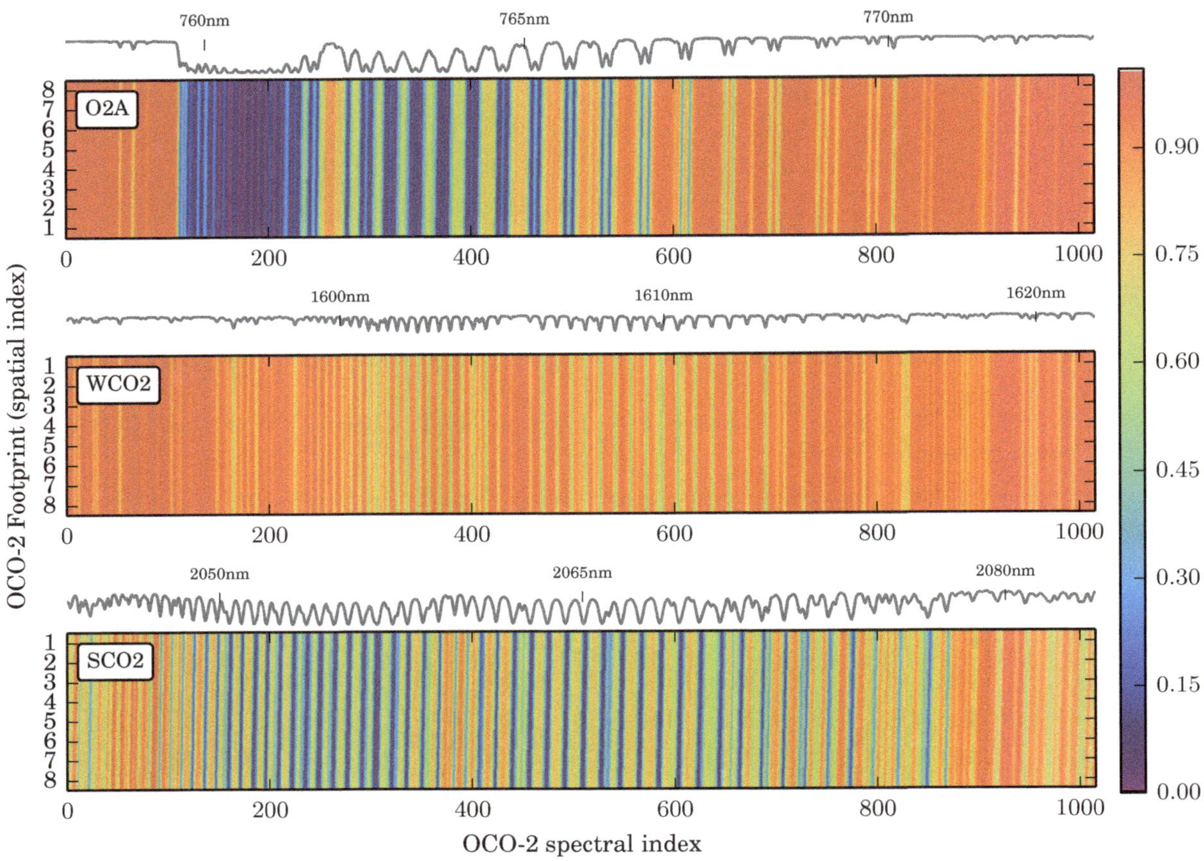

Figure 1. OCO-2 transmission spectra from TVAC direct-sun measurements at an air mass factor (AMF) of ≈ 2.5. Transmission levels from 0 to 1 are color-coded for the O2A, WCO2 and SCO2 bands, from top to bottom. The x axis represents the spectral dimension (1016 elements) of the OCO-2 focal plane and the y axis the spatial (160 elements co-added to eight individual footprints). As a guide for the eye, a spectral cross section of the center footprint is shown on top of each image, including wavelength indicators at the top. The detectors are slightly tilted with respect to the slit orientation for the O_2 A-band in particular, but also for the strong CO_2 band, causing a stronger footprint dependence of the dispersion coefficients.

collected by a high-resolution Fourier transform spectrometer (FTS) from the Total Carbon Column Observing Network (TCCON) (Washenfelder et al., 2006; Wunch et al., 2011) that was located at the same altitude, ≈ 200 m from the TVAC facilities at JPL. By using actual retrievals from the OCO-2 data, we can evaluate retrieval-relevant instrument properties that will help guide users of the OCO-2 spectra. While retrievals using direct sun observations are simpler than retrievals using reflected sunlight from space, these measurements from OCO-2 provide an end-to-end description of the combined impact of the OCO-2 instrument calibration as well as other aspects of the retrieval, such as the spectrally dependent gas absorption cross sections. Detailed descriptions of the instrument calibration can be found in Rosenberg et al. (2015); O'Dell et al. (2011) for radiometric calibration and Lee et al. (2015); Day et al. (2011) for instrument line-shape (ILS) and spectral characterization. This paper is organized as follows: Sect. 2 provides a succinct overview of the OCO-2 instrument. Sect. 3 discusses the retrievals based on TVAC data acquired by the OCO-2. Section 4 focuses on the fortuitous detection of a CO_2-rich plume associated with the Los Angeles urban dome during the TVAC tests and Sect. 5 summarizes the overall work.

The OCO-2 instrument is a three-channel grating spectrometer. It records spectra of the O_2 A-band (0.757–0.775 μm, full width at half maximum (FWHM) = 0.042 nm, denoted as O2A), a weak CO_2 band (1.594–1.627 μm, FWHM = 0.076 nm, denoted as WCO2) and a strong CO_2 band (2.043–2.087 μm, FWHM = 0.097 nm, denoted as SCO2) with eight independent along-slit focal plane array readouts, denoted as footprints (1–8). A common telescope feeds all three spectrometers through a series of beam splitters and re-imagers, with a linear polarizer selecting only the polarization vector perpendicular to the entrance slit. Each spectrometer works in first order of the holographic grating. The system is optically fast (F/1.8) and yields a high signal-to-noise ratio (SNR). See Crisp et al. (2004), Pollock et al.

(2010) and Haring et al. (2008) for a more general description of the OCO instrument and mission.

At each spectrometer's focus, a 1024 × 1024 imaging array collects the spectrum, such that one dimension measures field angles along the slit, while the other dimension measures wavelengths. The O_2 A-band detector is a silicon (HyViSI) Hawaii-1RG, and the two CO_2 detectors are HgCdTe Hawaii-1RG; all were manufactured by Teledyne Scientific and Imaging, LLC. Only 160 of the 1024 pixels in the spatial dimension are used, and sets of ≈ 20 are averaged onboard to constitute the eight spatial along-slit footprints. In the spectral dimension, four reference pixels are blacked out on each end of the array, leaving 1016 pixels (or channels) per band and footprint. Figure 1 shows an example of the focal plane array (1016 × 160, with block averages of 20 rows in the spatial dimension) readout of all three bands using a direct sun measurement, corresponding to an air mass equivalent to a typical in-space viewing geometry (solar zenith angle of 67°). For retrieval purposes, it is important to note that each spectral band (three), footprint (eight) and spectral pixel (1016) has its own characterization; this is in contrast to the FTS instrument onboard GOSAT (Hamazaki et al., 2005; Kuze et al., 2009). The ILS, for instance, is given for each of the 3 × 8 × 1016 detector elements independently, resulting in a 4-dimensional array with the dimensions 3 × 8 × 1016 × 200 (the ILS is defined for 200 spectral points around the center point).

The OCO-2 instrument was designed to have a spectral sampling of approximately 2.5 detector elements per full width at half maximum (FWHM), in each band, and a spectral resolving power $\lambda/\Delta\lambda$ of approximately 20 000 in the CO_2 channels and 17 000 in the O_2 A channel. The shape of the OCO-2 ILS function is determined by the slit width, pixel pitch, optical aberrations, diffraction, and detector crosstalk. Details of the ILS as well as spectral and radiometric characterization that are used in this manuscript can be found in Rosenberg et al. (2015), Lee et al. (2015), Day et al. (2011) and O'Dell et al. (2011). Figure 2 shows an example of the OCO-2 ILS for footprint 4 in each band at four different detector array spectral positions. The ILS variation in the spatial direction (footprint) is not as strong as in the spectral direction but still needs to be taken into account for accurate retrievals. As mentioned above, ILS functions are provided in tables that can be interpolated because conventional line-shape functions (e.g., Gaussian, Voigt) cannot fit the shapes well enough for accurate X_{CO_2} retrievals.

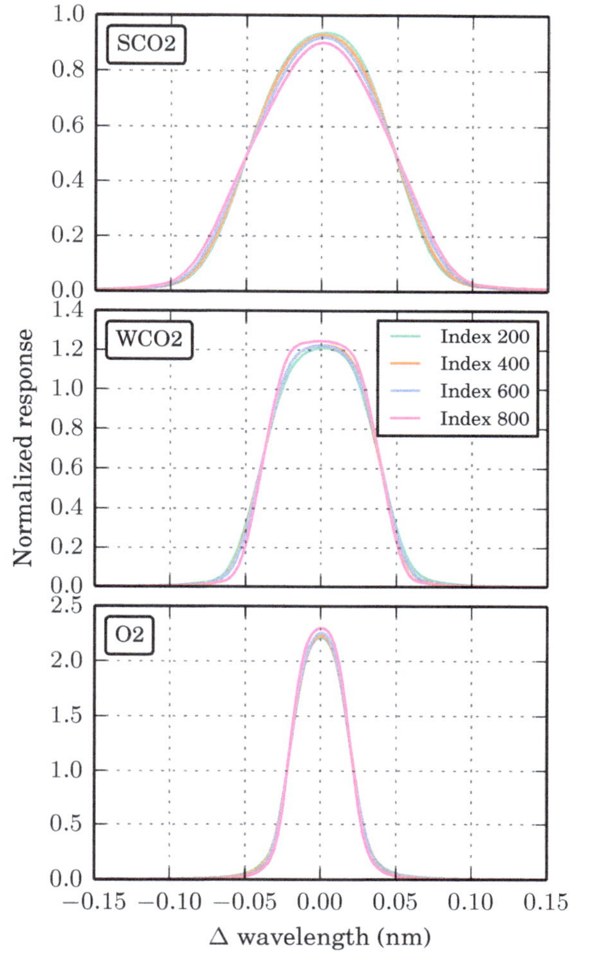

Figure 2. OCO-2 ILSs for a center footprint (#4) at four different spectral positions for each band independently. FWHM values are about 0.04, 0.075 and 0.1 nm for the O2, WCO2 and SCO2 bands, respectively.

3 Thermal Vacuum Test (TVAC) results

3.1 Experimental setup

During the characterization and calibration of the OCO-2 flight instrument at the JPL, a heliostat was used to direct sunlight onto a diffuser that was viewed by the flight instrument through a window in the TVAC chamber. Simultaneously, a TCCON instrument located at the same altitude about 200 m away recorded high-resolution solar spectra through essentially the same atmospheric column. The FTS spectra were acquired at an unapodized resolution of 0.013 cm^{-1} (45 cm optical path difference), which is approximately 20 times higher than the spectral resolution of the OCO-2 instrument. In addition, the FTS spectra had high SNR and were characterized by a single, well-determined ILS for the entire spectral range (Wunch et al., 2011). Apart from the spectral resolution, there are a few other differences between the two measurements. First, the integration time of the OCO-2 instrument was 0.333 s (a continuous 3 Hz measurement) while the FTS required 79 s to complete a scan. Second, the OCO-2 instrument had a larger field of view (FOV) than the FTS, observing the full solar disk and some of the surrounding sky, while the FTS observed the center

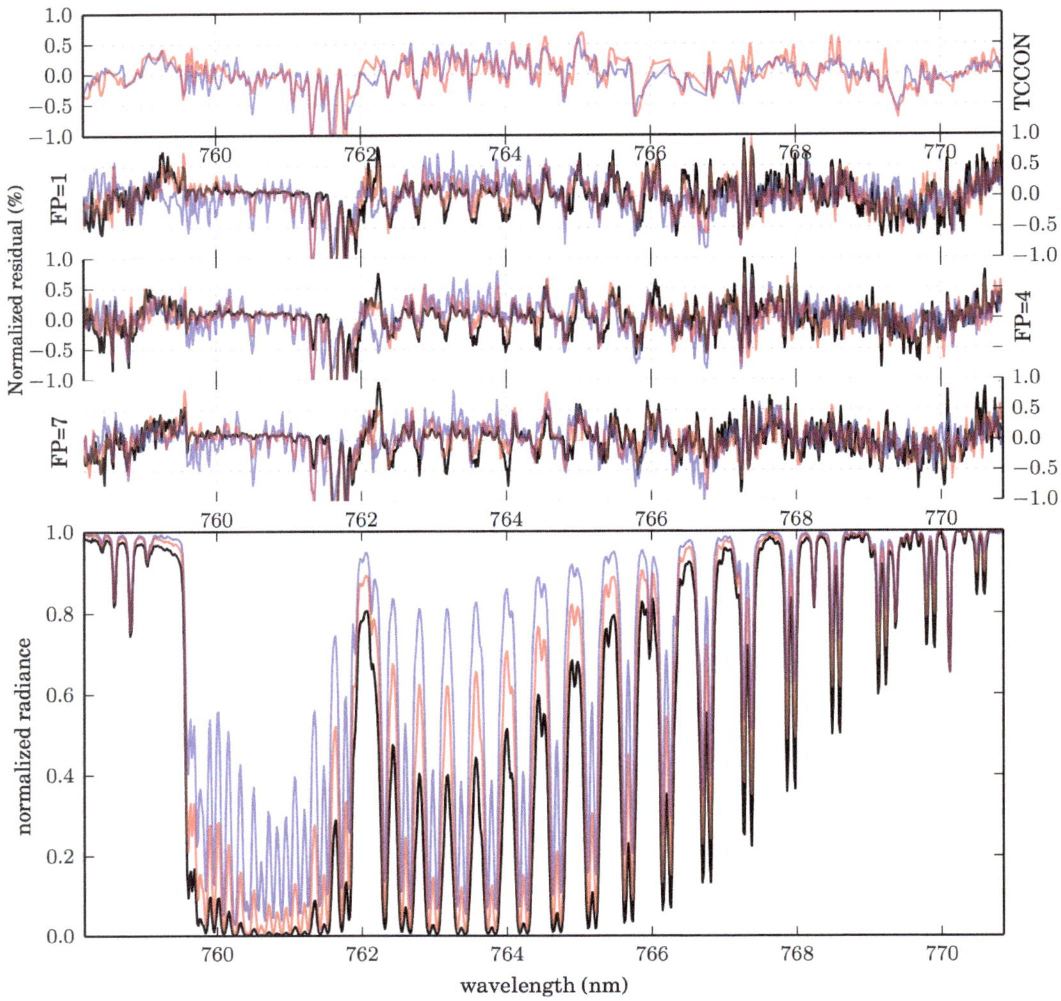

Figure 3. Spectral fits to direct-sun measurements in the O2A band of OCO-2 and TCCON. Normalized radiances from OCO-2 are shown in the bottom panel for three different air masses (low (1.1), typical (2.5) and high (5.1) in blue, red and black, respectively). The middle panels show residuals of fits to normalized spectra, for three OCO-2 footprints, and the top panel shows fit residuals to a high-resolution FTS spectrum of identical air mass, which was convolved and resampled to OCO-2 spectral resolution prior to fitting.

of the solar disk. The solar lines observed by OCO-2 were therefore broadened relative to those observed by the FTS due to Doppler shifts caused by the sun's rotation. For this reason, solar line pixels in the FTS fits were excluded in the following analysis, as we aimed at using exactly the same retrieval setup as for the OCO-2 fits.

3.2 Spectral fits

For the spectral fits, we performed total column retrievals of trace gases for each band (O2A, WCO2, SCO2) independently. For this purpose, we used the fast IMAP-DOAS preprocessor in a special up-looking retrieval mode (Frankenberg et al., 2005) using a single temperature and pressure a priori profile for the entire day, which was extracted from the National Centers for Environmental Prediction (NCEP) data, but replacing the surface pressure with the one locally measured at JPL. In this up-looking mode, the IMAP-DOAS algorithm merely fits spectral dispersion, solar shift and continuum baseline as a third-order polynomial, as well as a total column scaling of a pre-defined O_2, H_2O and CO_2 profile shape. Neither surface pressure nor a temperature scaling was retrieved.

In the following, we will show spectral fits for a low (1.1), typical (2.5) and high (5.1) air mass factor (AMF) in blue, red and black, respectively. The "typical" AMF will be very close to the lower end of AMFs encountered during flight, as the AMF in perfect nadir viewing simplifies to $1 + 1 / \cos(\mathrm{SZA})$ with SZA denoting the solar zenith angle, and the instrument is required to return data at solar zenith angles between ~ 21.5 and $85°$. To compare the OCO-2 spectral fit results (i.e., trace gas columns and fit residuals)

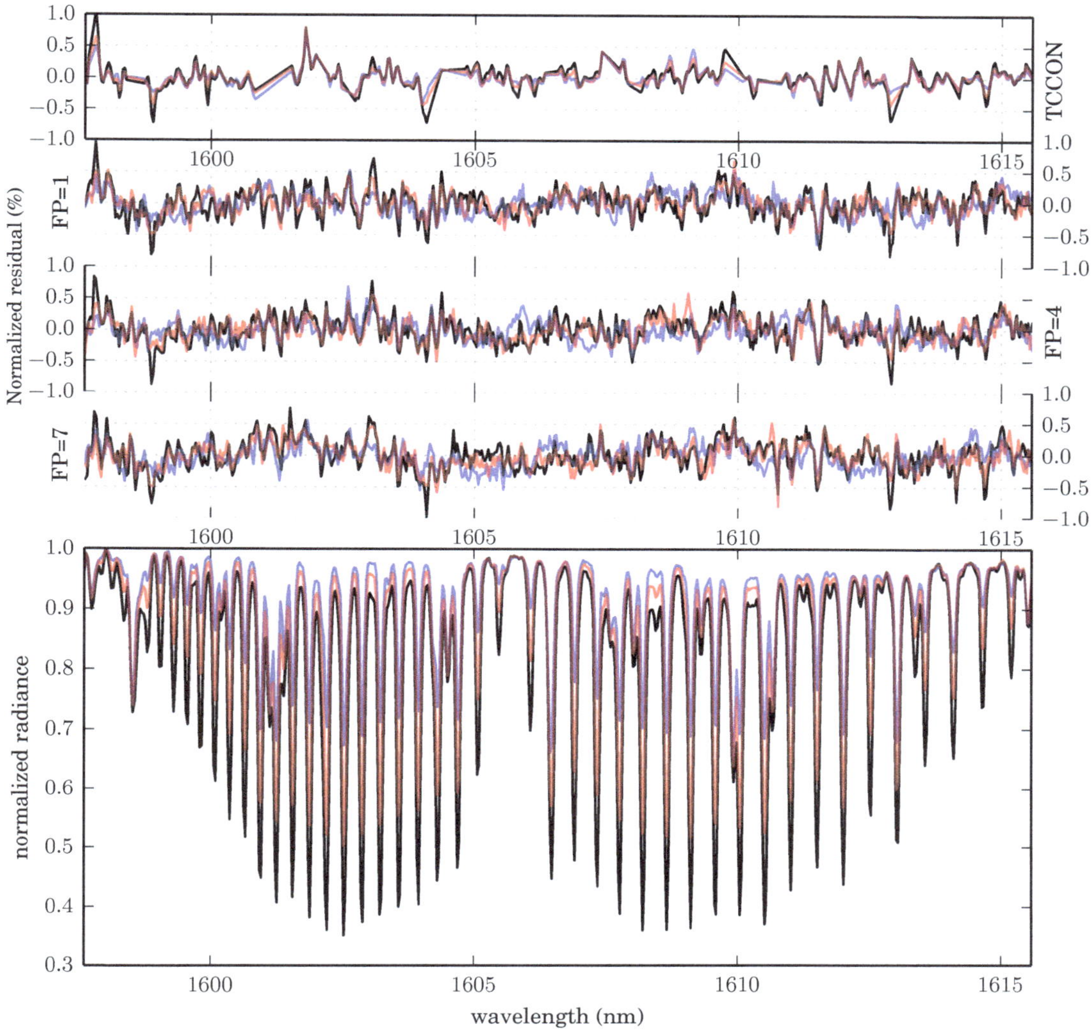

Figure 4. Spectral fits to direct-sun measurements in the WCO2 band of OCO-2 and TCCON (analogous to Fig. 3).

to a high-quality reference, we used collocated FTS spectra and convolved them with the OCO-2 ILS, and performed fits on those spectra using exactly the same fitting routines as for the OCO-2 data. This ensures that differences in the fits are caused neither by differences in spectral resolution nor by specific software setups. Also, convolved spectral residuals of fits to high-resolution (e.g., FTS) data are not equal to residuals of fits to convolved spectra. In addition, solar lines were masked out for the FTS fits since, as noted above, the OCO-2 instrument, in TVAC mode, records full-disk solar spectra while the FTS records narrow FOV disk-centered spectra. In the following, we will display fit residuals in percent, normalized by a representative continuum level radiance, chosen as the maximum signal within each band. For the comparison, we chose OCO-2 and TCCON spectra where the zero-path difference (ZPD) was within a few seconds of the OCO-2 spectra acquisition. For the sake of clarity, we show residuals from only three of the eight footprints. However, we did verify that all footprints behave similarly. For all tests, we used the absorption coefficient (ABSCO) v4.2 spectroscopy database developed for OCO-2 (Thompson et al., 2012).

Spectral fits to the oxygen A-band are shown in Fig. 3. The top panels show normalized residuals of the convolved FTS fits, followed by spectral residuals for 3 OCO-2 footprints (footprints 1, 4 and 7). The bottom panel shows normalized radiances for each air mass. At higher air masses, the R-branch from 760 to 762 nm saturates almost entirely even at OCO-2 spectral resolution, with near-zero transmission across 1–2 nm in wavelength range.

In general, fit residuals of the convolved TCCON spectra strongly resemble those of the OCO-2 flight instrument. Given the high SNRs for individual OCO-2 spectra in TVAC mode (600–1200 for all bands), purely noise-driven residu-

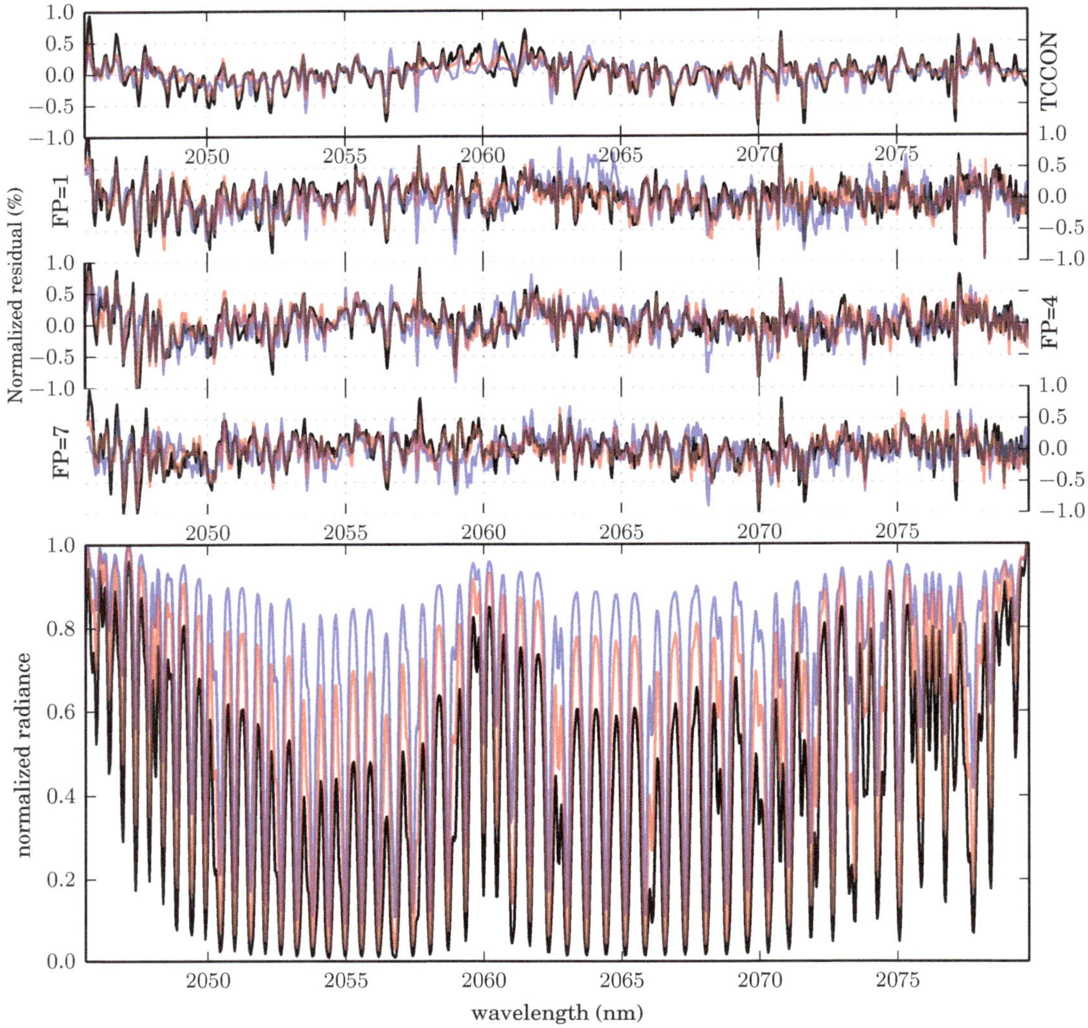

Figure 5. Spectral fits to direct-sun measurements in the SCO2 band of OCO-2 and TCCON (analogous to Fig. 3).

als should be on the order of 0.1–0.2 %. Systematic features appear in both TCCON and OCO-2 residuals with maxima of up to 1 %. Larger deviations in the R-branch are most likely caused by the choice of the a priori temperature and pressure profile. Given that the two instruments show very similar residuals, we conclude that they are still dominated by errors in the ABSCO tables of the O_2 A-band absorption cross sections, specifically relating to line mixing, line-shape and collision-induced absorption. This can also be seen in the air mass dependence of systematic residuals. At higher air masses, the far line-wing shape becomes more important as line centers are entirely saturated. The periodic structures in the P branch residuals point to the importance of the line shape as residuals are smaller at lower air mass, at which the far wing line shape becomes less important.

Weak CO_2 band (WCO2) fits are shown in Fig. 4. At OCO-2 spectral resolution this band is not saturated even at high air masses. Most of the information content is thus in the line centers as measured by OCO-2, and the continuum can be isolated much more easily from well-separated CO_2 lines. Again, we observe residuals that strongly resemble those from TCCON fits (with many solar features masked), pointing to small remaining errors in CO_2 absorption cross sections. There is a strong similarity between footprints but, in general, most features are below 0.5 %, similar to measurements from the GOSAT satellite as well as TCCON. Some residual features, such as the slope at the short-wavelength edge of the fit window, can also be caused by broad spectral variations in the solar spectrum that are not accurately represented here.

SCO2 fits are shown in Fig. 5. Line centers at OCO-2 resolution now gradually saturate at higher air mass and the continuum level radiances are harder to define than in the other bands because of overlapping line wings. In general, conclusions for this band are similar to the other bands, with spectral residuals roughly resembling the TCCON residuals. The only exception may be at the short-wavelength edge where OCO-2 residuals are somewhat higher, potentially caused by either dispersion or ILS changes at the band edge, where the instrument calibration is less well characterized.

In summary, spectral fits to OCO-2 data yield expected fitting results but, as with all other current instruments measuring this spectral range, systematic residuals still appear. Given the high SNR of the OCO-2 instrument, this would result in unrealistically high χ^2 values for spectral fits, which is why we will implement a residual fitting technique based on empirical orthogonal functions derived from systematic features as is currently done for data from GOSAT. Using these basis functions, as additional fit parameters, we expect fit residuals to be mostly noise driven, resulting in realistic χ^2 values that will help the fit converge, allow us to use χ^2 as quality criteria, and result in more realistic a posteriori uncertainty estimates, as the true OCO-2 measurement noise can be provided to the weighted least squares fit.

To determine whether the variability in spectral residuals can be explained by OCO-2 detector noise, we computed the standard deviation of spectral residuals using 2000 OCO-2 fits and define this as a surrogate for empirically derived noise in the OCO-2 data. Using the mean of the 2000 radiance spectra as signal level surrogate, we then derive a set of SNR points at varying signal levels within a spectrum, mostly caused by CO_2 and O_2 absorptions. Figure 6 shows the empirically derived SNR points as well as a typical SNR relationship based on the launch-ready OCO-2 calibration (Rosenberg et al., 2015). There is excellent agreement between calibration curves and empirically derived SNR based on spectral fits. The spread of the individual data points for each band is due to the behavior of the detector, with varying gain coefficients and shot noise in each pixel. This is reflected in the OCO-2 launch-ready calibration curves as well but, for clarity, the theoretical SNR curve of just a typical pixel is shown in transparent lines, and corresponds very well with the empirical noise estimates. After accounting for the systematic residual features, expected χ^2 distributions of the residuals are thus attained.

Two advantages of the OCO-2 grating spectrometer are its high dynamic range and relatively low noise at low signal levels. This results in much lower noise levels within deep absorption lines of the O2A and the SCO2 bands than would be possible with space-based FTS data, which exhibit a constant absolute noise level across the entire spectrum (which would yield a straight line through the origin in the SNR vs. signal level curve when plotting individual spectral points within a single acquisition).

3.3 Footprint dependencies

The OCO-2 spectrometer records spectra separately in the spatial domain of the focal plane array. To increase SNR, photons collected in 20 adjacent spatial rows are averaged to define a footprint, hence the 160 spatial detector rows produce eight spatial footprints (see Fig. 1). Given that each footprint spans a considerable fraction of the focal plane array, instrument calibration (most importantly dispersion and ILS) varies from footprint to footprint. With the unprecedented accuracy requirements in the sub 1 % range for atmospheric CO_2 measurements, inter-footprint difference in retrieved quantities for imaging spectrometers will cause challenges in using this data.

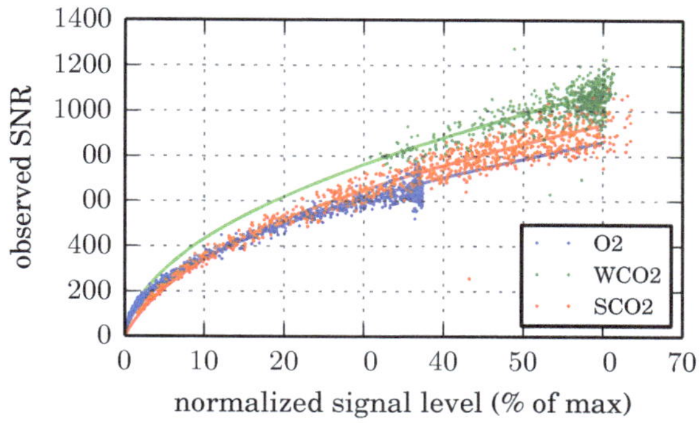

Figure 6. Empirical SNRs of each OCO-2 band derived from the standard deviation of spectral residuals from 2000 separate fits. Solid transparent lines represent the expected OCO-2 SNR for a typical detector pixel, which agrees well with the observed SNR. For each band (O2A, WCO2, SCO2), the maximum observable signal is defined as 7.0×10^{20}, 2.45×10^{20} and 1.25×10^{20} photons $s^{-1}\,m^{-2}\,sr^{-1}\,\mu m^{-1}$, respectively, where it is taken into account that OCO-2 will only measure one polarization direction (i.e., the maximum signal level in terms of full intensity would be twice as high as provided here).

In Fig. 7, we show differences between the total column CO_2 estimates from each footprint and the all-footprint average, for the weak and strong CO_2 bands. Two main features are apparent: (1) there is a very small inter-footprint variation but at least six to seven of the footprints agree to within ±0.5 ppm; (2) the inter-footprint variations are somewhat unstable and appear to be influenced by how the heliostat is illuminating the OCO-2 instrument, as can be seen by some jumps around 16:00, when the heliostat was realigned. In the test setup, the heliostat was not large enough to fill the entire OCO-2 pupil, which can slightly affect the characteristics (Liebe et al., 2009). This effect, which will not happen in orbit, might change when the heliostat is realigned. Hence, it is not yet clear what fraction of the variation is merely caused by the experimental setup and what is caused by potential small calibration inconsistencies. In orbit, the inter-footprint differences are not expected to vary in time and the variability of ±0.5 ppm can be characterized using in-orbit data and corrected for using constant correction factors. It should be mentioned that the OCO-2 pupil during calibrations using the integrating sphere was fully illuminated.

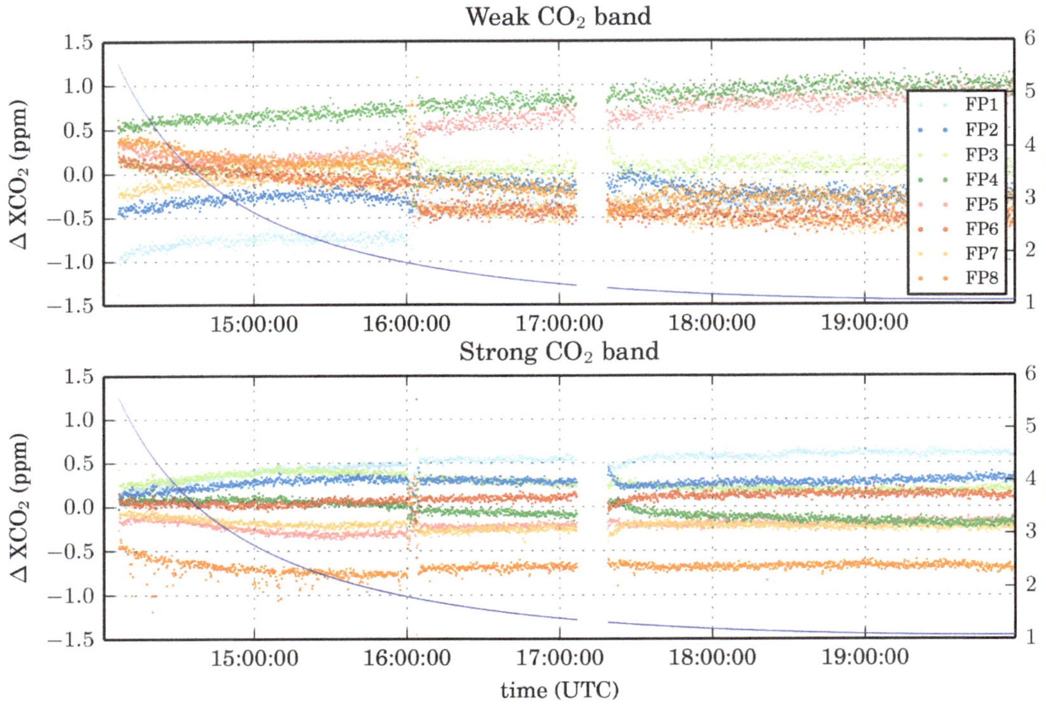

Figure 7. Footprint differences in retrieved CO_2 columns using the strong and weak bands from OCO-2 TVAC data acquired on 21 April 2012.

For the pre-launch OCO-2 calibration, two sets of instrument line shapes were tested (Lee et al., 2015), of which one will eventually be implemented for official OCO-2 retrievals. One set is derived from tunable diode laser measurements and a second, enhanced set (see Lee et al., 2015) using direct comparisons against the high-resolution FTS spectra, where the ILS was optimized to best match the FTS data. In Fig. 8, the inter-footprint variations averaged over a longer time-period are shown independently for the FTS enhanced ILS (solid lines) and the original laser-based ILS (dashed lines). The impact is not large and mostly affects footprint #1, at the edge of the FOV. For this particular footprint, both CO_2 bands would behave somewhat abnormally but deviate in opposing directions (positive or negative). This could have implications for a joint three-band retrieval. The enhanced ILS largely mitigates this problem to the best possible degree. Once OCO-2 is collecting in-orbit data, both ILS models will be evaluated in order to decide whether the FTS-enhanced ILS mitigated only an issue associated with the TVAC test setup or a real calibration issue. In any case, most footprints agree very well with each other for both choices of ILS and provide confidence, especially given the small fitting residuals discussed earlier. However, even this small footprint dependence will have to be scrutinized since the accuracy requirements of better than 1 ppm are very stringent. This also has implications for future imaging spectrometers with hundreds of spatial pixels, for which a careful per-footprint calibration as for OCO-2 might not be feasible.

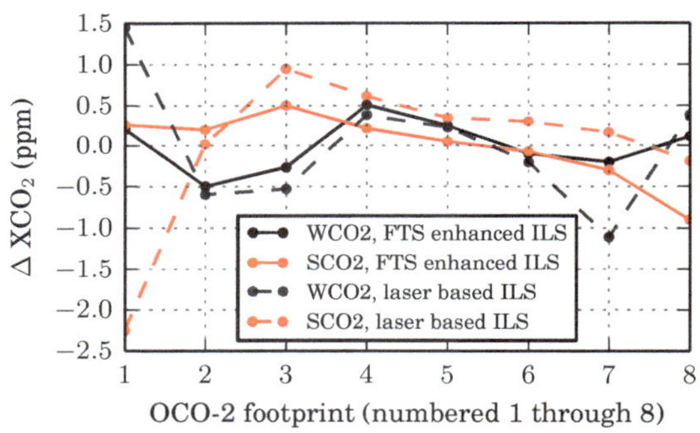

Figure 8. Average of inter-footprint differences in retrieved CO_2 columns from OCO-2 using two different sets of instrument line shapes, namely the laser-based ILS and the FTS-enhanced ILS estimate.

3.4 The Matador test

The OCO-2 retrieval strategy imposes stringent requirements on knowledge about the detector linearity and dark current. Any imperfections can slightly modify the fractional depths of absorption lines depending on the detector fill level, which can create regional and/or seasonal biases in retrieved X_{CO_2}. For OCO, the so-called "Matador test" was devised to detect any apparent non-linearities or dark offsets (O'Dell et al., 2011).

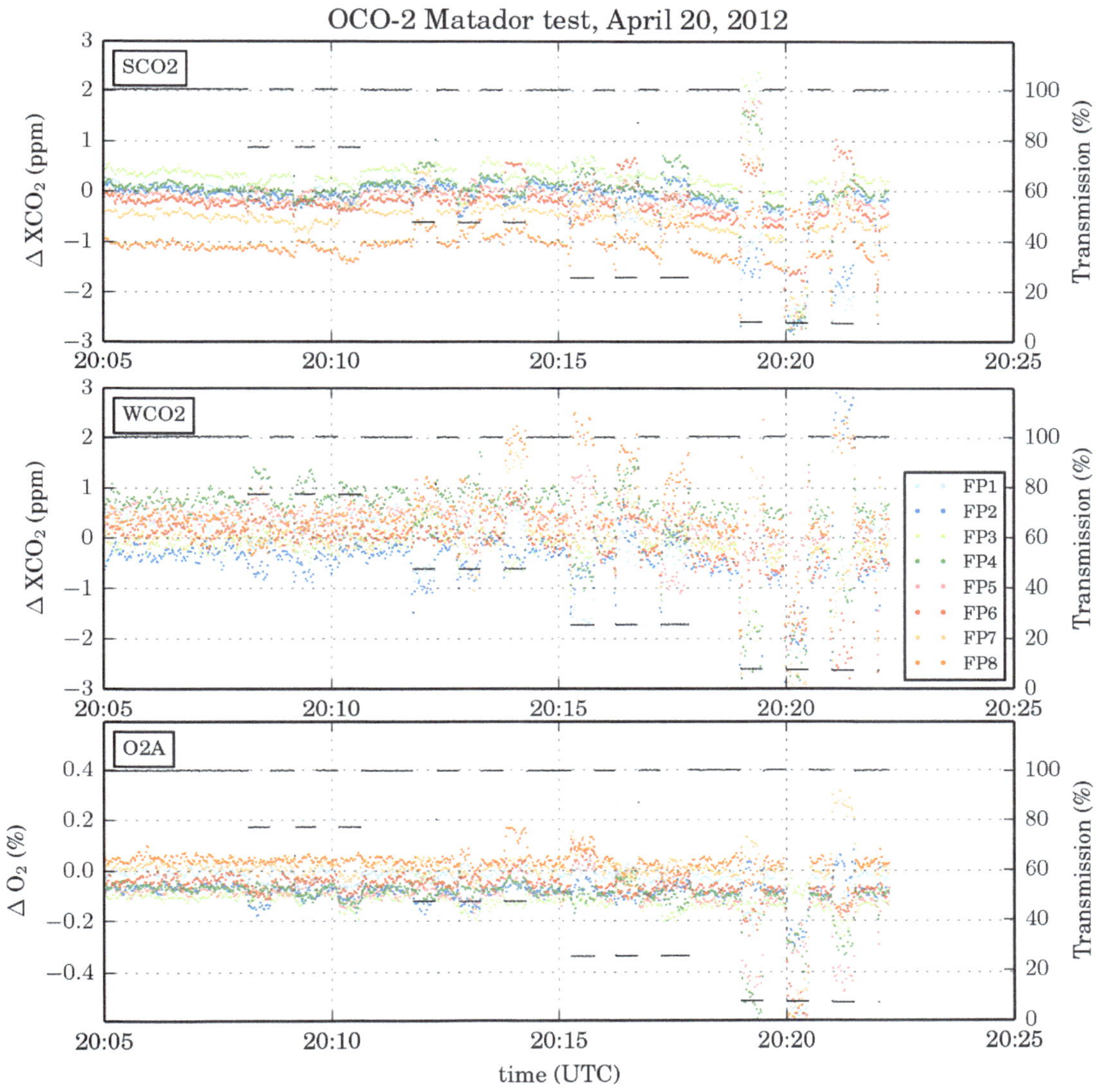

Figure 9. X_{CO_2} (in ppm) and O_2 (in %) differences from a common average in the OCO-2 Matador test, which changed signal levels through inserting aluminum sheets with variable hole density into the light beam. Gray shades indicate the time periods that the aluminum sheets were in the beam and the black horizontal lines indicate the transmission level for each of these tests. Each footprint for both CO_2 bands and the O_2 band is displayed separately.

The general TVAC setup was used with direct sunlight illuminating a diffuser viewed by the OCO-2 spectrometers via a series of mirrors. For the Matador test, special sheets with varying transmission were optionally inserted into the optical path. To create a spectrally uniform reduction in the solar intensity, aluminum sheets with small holes were fabricated with different hole-densities in a hexagonally packed pattern to create effective transmissions of approximately 75, 50, 25 and 10 %. For the test, the sheets were rapidly (and manually) inserted into and removed from the heliostat optical path, hence the name “Matador test” due to the visual analogy to a matador’s cape. The test was performed near local noon on 20 April 2012.

Figure 9 shows results for both CO_2 bands and the O_2 A-band independently, for each footprint, expressed as deviations in X_{CO_2} or O_2 column from the average across all footprints at the initial 100 % transmission level (i.e., before the sheets were inserted). The periods with reduced transmission are indicated with a gray background, as each sheet, in descending order of transmission, was inserted and removed three times.

For the 75, 50 and 25 % transmission levels, changes on the order of much less than 0.5 ppm occur. More importantly, deviations are neither consistent across footprints nor consistent across all three independent Matador tests per transmission level. This behavior hints at subtle changes in how the OCO-2 pupil was illuminated for each of these tests, which

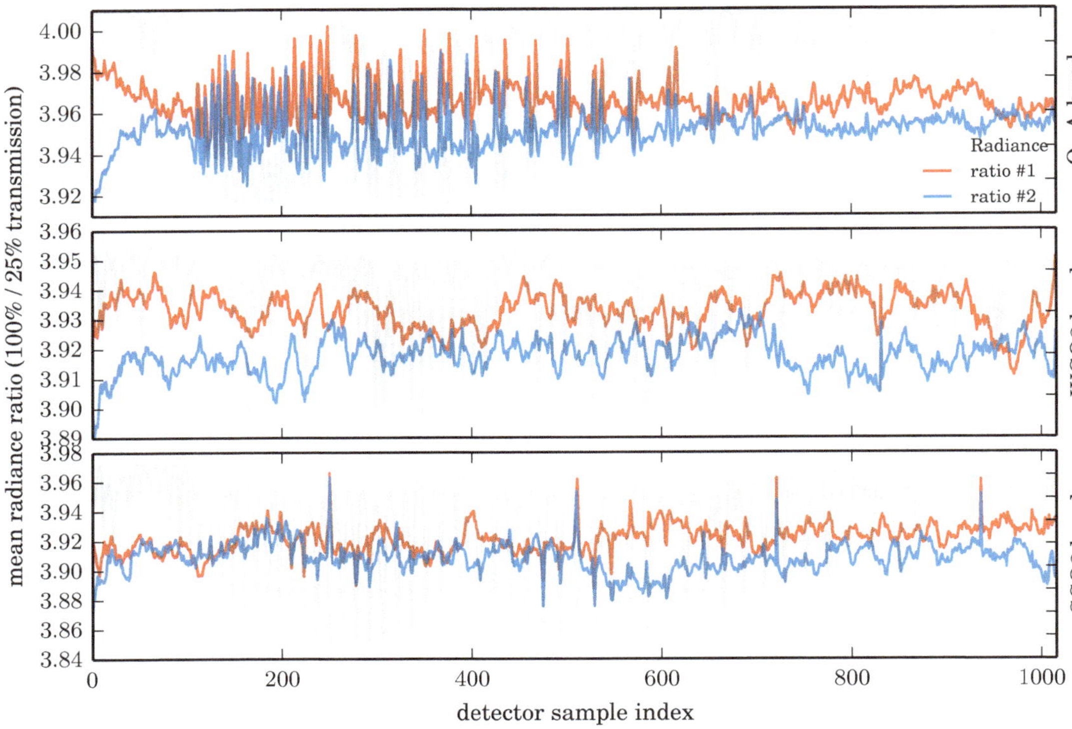

Figure 10. Radiance ratios derived from the Matador tests for footprint #4. We used the first two 25 % transmission tests recorded between 20:15 and 20:17 UTC and computed three average spectra: one each for for the two periods with the sheet included and one for the full illumination in between the tests. As a visual guide for where the absorption lines are located, a scaled spectrum is plotted in a light gray for each band.

can also cause minor variations in the ILS and, hence, X_{CO_2}. This is even more obvious at the < 10 % transmission levels, in which jumps in retrieved CO_2 are considerably higher but without any consistent pattern across all three independent tests.

For O_2 columns, all footprints agree to about 0.05 %, which would translate to better than 0.5 hPa or 0.2 ppm if errors propagate linearly into X_{CO_2}. In general, the O_2 A-band appeared to be somewhat more consistent than the CO_2 bands. It is, however, not clear whether calibration parameters are the cause of the very small variations in the O_2 fits or whether these can be merely attributed to the shape of the oxygen absorptions itself, where a lot of information is also coming from the deep Q-R-branch, which is less sensitive to ILS variations as no individual lines are sampled.

As an additional test, we derived radiance ratios of the full illumination level and two separate Matador tests for the 25 % transmission level, as shown in Fig. 10. No absorption features can be detected in the CO_2 band ratios, which confirms an accurate OCO-2 calibration in terms of linearity and dark subtraction. However, the two separate tests exhibit ratios with some broad-band features that differ between tests and might be causing slight changes in X_{CO_2} between tests, potentially due to changes in the pupil illumination. Simi-

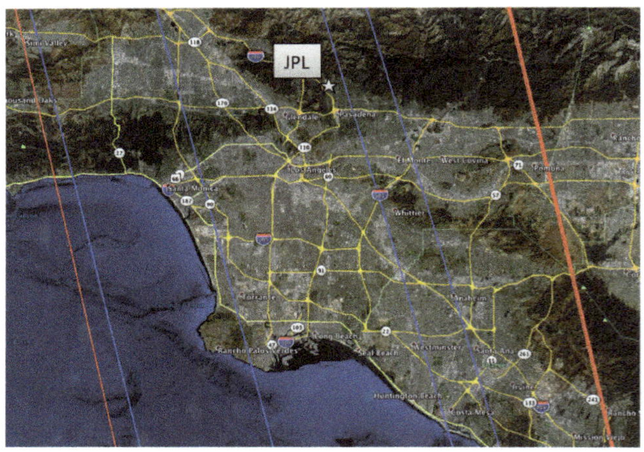

Figure 11. OCO-2 Orbit track simulations for Nadir (red) and Glint (blue), in each season, in the larger Los Angeles area. The nominal target tracks are shown here. These may change slightly, depending on the exact orbit achieved. Nadir repeat cycles of 16 days are close repeats but Glint viewing orbits shift during the year owing to changes in sun–earth geometry.

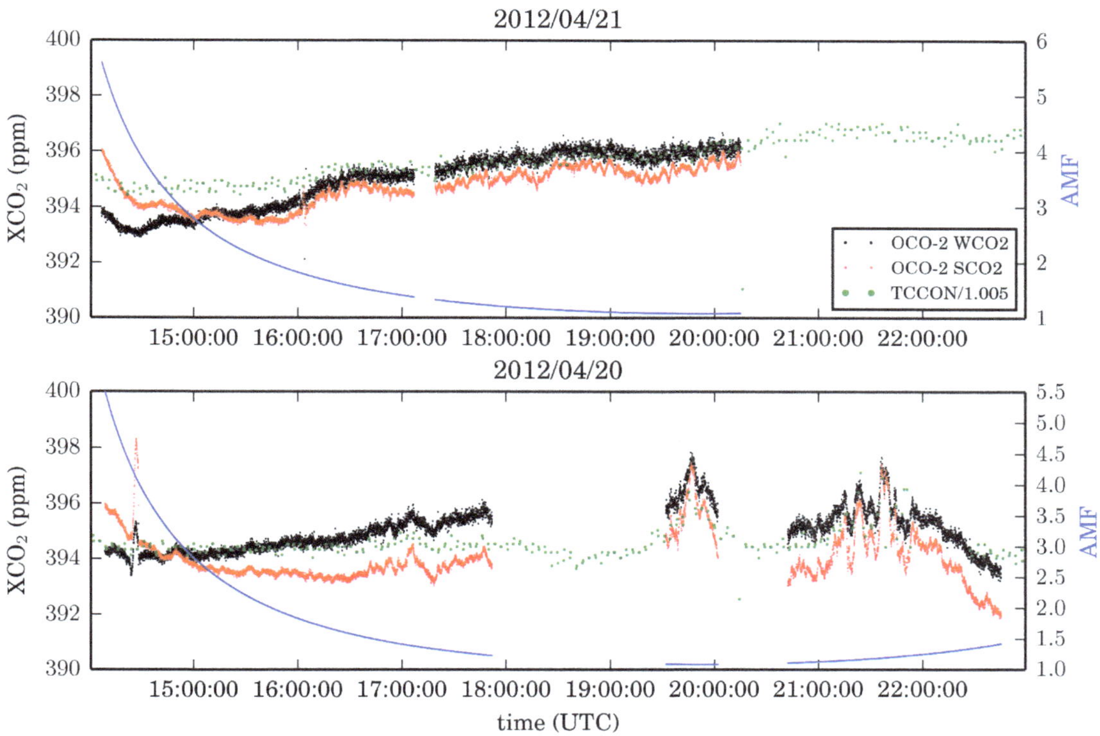

Figure 12. X_{CO_2} time series on 20/21 April 2012 acquired at JPL near Pasadena, CA. Air mass is depicted in blue, SCO2 and WCO2 results in red and black, respectively, and scaled official TCCON data in green. Only every 10th OCO-2 data point is plotted but all eight footprints are averaged.

larly, there appears to be a small change in the ILS for the O_2 A-band, most likely also related to the illumination changes.

In summary, all of the three spectral bands show very good consistency, but achieving a near perfect consistency is currently not possible and small calibration factors for each footprint may need to be included for OCO-2. At this stage, we have gained confidence that the OCO-2 behaves well at variable illumination levels but also that we have obtained the maximum amount of information possible from the TVAC setup. For these pre-launch tests, it is hard to reach the stability in terms of instrument thermal control or illumination that will be achieved in orbit.

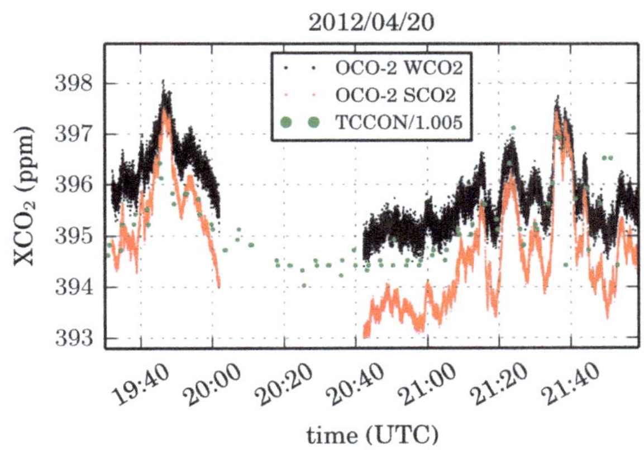

Figure 13. Zoom on the CO_2 column time series on 20 April 2012 acquired at JPL. Every OCO-2 data point is plotted and all eight footprints are averaged.

4 Observing the Los Angeles urban dome with the OCO-2 instrument during the TVAC tests

Apart from the primary mission objective to better understand regional-scale biospheric fluxes, the OCO-2 data can also help us to estimate carbon fluxes from megacities (Kort et al., 2012). Even though its narrow swath will only rarely cross these localized sources, it will have unprecedented spatial resolution and X_{CO_2} sensitivity along the orbit track. Also, the glint repeat cycles will have shifted orbit paths, increasing the chances of obtaining direct cross sections of the urban dome at least once in the OCO-2 mission. An example of typical OCO-2 orbit paths nearby a large metropolitan area is shown in Fig. 11 for the case of Los Angeles.

In the Los Angeles basin, ground-based studies observed elevated amounts of CO_2 and CH_4 from total column measurements (Wunch et al., 2009). For CO_2, column enhancements measured with a TCCON FTS in 2007/2008 at JPL revealed diurnal changes in X_{CO_2} on the order of up to 4 ppm, caused by the Los Angeles urban dome approaching the foothills at JPL around midday.

Here, we report on the first urban plume observed from the OCO-2 even before launch. Figure 12 shows time series for 20/21 April of CO_2 column retrievals using the WCO2 and SCO2 bands of OCO-2 as well as the TCCON FTS retrievals located nearby (Wunch et al., 2011). For TCCON retrievals, we extracted the official X_{CO_2} values retrieved using the latest GGG2012 data version (https://tccon-wiki.caltech.edu/Network_Policy/Data_Use_Policy/Data_Description). These data are based on two weak CO_2 bands, one of which is identical with the OCO-2 WCO2 band. CO_2 columns are scaled by retrievals of dry air, obtained from fits to the oxygen band at 1.27 μm. The official TCCON data applies both an AMF correction and a scaling factor to bring the measured X_{CO_2} onto the NOAA standard CO_2 scale as used for the in situ networks.

For OCO-2, all eight footprints have been averaged but no further smoothing has been applied. Owing to the strengths of absorption features in the SCO2 band, retrieval scatter in the strong band is substantially lower than for the weak band. TCCON data was divided by 1.005, indicating a small but consistent offset between the official TCCON results and OCO-2. The X_{CO_2} observed on these two TVAC days exhibits substantially different behavior. On 21 April, a relatively smooth diurnal cycle, with maxima around solar noon, was observed with an overall amplitude of about 2 ppm in X_{CO_2}. Some short-term features, such as the increase at 16:00 or the swings around 18:30 and 19:30, can be clearly observed in both TCCON and OCO-2, while the latter allows for a much faster sampling in time due to the 3 Hz readout rate. This also allows us to sample the very erratic behavior seen on 20 April. This day, interrupted by Matador tests around local noon, shows very high-frequency variations of observed CO_2 starting around 20:00 UTC. Figure 13 shows a zoom plot into this specific anomalous time-period with high-frequency variability unresolved by the TCCON sampling, but similar general features observed by both instruments. This behavior suggests the edge of the urban dome moving in and out of the observation site. Swings can be up 3–4 ppm within a 5–10 min period, as is observed at about 21:40 UTC. Assuming typical wind-speeds of 5–10 m s^{-1}, these temporal scales correspond to spatial scales of 1.5–3 km, which will be matched by the in-orbit OCO-2. In other words, spatial gradients of about 2-4 ppm within a few kilometers can be expected in strongly emitting areas such as Los Angeles. In addition, cross sections of the urban dome can be mapped with the OCO-2 with the caveat of a more complex retrieval from space, where retrievals are based on backscattered radiation and have to take aerosols into account. However, 1 % changes in X_{CO_2} as observed here are substantial and should be differentiable from light-path modifications within the urban dome.

Another feature is apparent in the time series: the amplitude of the variation in the weak band is smaller than in the strong band. This could be caused by differences in the averaging kernels (AK) of both channels in total column scaling mode. The AK of the strong band, with more saturation, typically peaks at much higher values near the surface if a simple total column fit is applied. Thus, CO_2 variability at the surface will be more amplified in the SCO2 band retrievals, which is well reflected in the OCO-2 measurements and is another hint at boundary-layer CO_2 enhancements.

5 Conclusions

In this work, we performed spectral fits on calibrated OCO-2 data obtained during thermal vacuum tests at the JPL in April 2012. Direct sunlight was fed to the OCO-2 instrument via a heliostat and data could be compared against a high-quality reference taken from an FTS located in close proximity. OCO-2 spectra were calibrated with the launch-ready calibration version from the OCO-2 instrument team. We performed the first CO_2 and O_2 total column retrievals from the OCO-2, providing a high-level evaluation and reference for the expected OCO-2 instrument performance. We find that spectral residuals are still mostly dominated by systematic features but that these are also apparent in other measurements, including high-resolution FTS data. Hence, most of the systematic features can be attributed to uncertainties in the spectroscopy. We also find that OCO-2 footprints agree very well with each other: better than 1 ppm. There is a small time-dependence on the footprint dependency, which is probably related to the specific test setup and would thus not occur in space.

A strong CO_2 plume from the Los Angeles metropolitan area (3–4 ppm enhancement in the total column average) was observed on 20 April 2012, when the plume caused very high-frequency (5–10 min) swings of CO_2 at the JPL site. These are indicative of strong spatial gradients in X_{CO_2}, which will be very helpful in characterizing localized carbon emissions when OCO-2 orbit tracks passes over such regions.

In summary, the OCO-2 instrument performed very well on the ground and none of the spectral fits calculated in this work showed anything unexpected. The team is now eagerly waiting for in-space OCO-2 data.

Acknowledgements. The research described in this paper was carried out by the Jet Propulsion Laboratory, California Institute of Technology, under a contract with the National Aeronautics and Space Administration. We would like to acknowledge all the JPL employees who worked tirelessly to acquire the OCO-2 TVAC data. The TCCON instrument was built by the California Institute of Technology with support from NASA's OCO-2 project. TCCON data used in this analysis are available at https://tccon-wiki.caltech.edu/Network_Policy/Data_Use_Policy. The TCCON instrument used here is currently operating at the NASA Armstrong Flight Research Center in Edward, California. Government sponsorship acknowledged.

Edited by: D. Griffith

References

Bösch, H., Toon, G., Sen, B., Washenfelder, R., Wennberg, P., Buchwitz, M., de Beek, R., Burrows, J., Crisp, D., and Christi, M.: Space-based near-infrared CO_2 measurements: Testing the Orbiting Carbon Observatory retrieval algorithm and validation concept using SCIAMACHY observations over Park Falls, Wisconsin, J. Geophys. Res., 111, 0148–0227, 2006.

Butz, A., Hasekamp, O. P., Frankenberg, C., and Aben, I.: Retrievals of atmospheric CO_2 from simulated space-borne measurements of backscattered near-infrared sunlight: accounting for aerosol effects, Appl. Optics, 48, 3322–3336, 2009.

Crisp, D., Atlas, R. M., Breon, F.-M., Brown, L. R., Burrows, J. P., Ciais, P., Connor, B. J., Doney, S. C., Fung, I. Y., Jacob, D. J., Miller, C. E., O'Brien, D., Pawson, S., Randerson, J. T., Rayner, P., Salawitch, R. J., Sander, S. P., Sen, B., Stephens, G. L., Tans, P. P., Toon, G. C., Wennberg, P. O., Wofsy, S. C., Yung, Y. L., Kuang, Z., Chudasama, B., Sprague, G., Weiss, B., Pollock, R., Kenyon, D., and Schroll, S.: The orbiting carbon observatory (OCO) mission, Adv. Space Res., 34, 700–709, 2004.

Day, J. O., O'Dell, C. W., Pollock, R., Bruegge, C. J., Rider, D., Crisp, D., and Miller, C. E.: Preflight spectral calibration of the Orbiting Carbon Observatory, IEEE T. Geosci. Remote Sens., 49, 2793–2801, 2011.

Frankenberg, C., Platt, U., and Wagner, T.: Iterative maximum a posteriori (IMAP)-DOAS for retrieval of strongly absorbing trace gases: Model studies for CH4 and CO2 retrieval from near infrared spectra of SCIAMACHY onboard ENVISAT, Atmos. Chem. Phys., 5, 9–22, doi:10.5194/acp-5-9-2005, 2005.

Hamazaki, T., Kaneko, Y., Kuze, A., and Kondo, K.: Fourier transform spectrometer for greenhouse gases observing satellite (GOSAT), in: Proceedings of SPIE, Vol. 5659, p. 73, 2005.

Haring, R. E., Pollock, R., Sutin, B. M., Blakley, R., Scherr, L. M., and Crisp, D.: Fabrication and assembly integration of the orbiting carbon observatory instrument, Proc. SPIE, 7082, 708213, doi:10.1117/12.796289, 2008.

Kort, E. A., Frankenberg, C., Miller, C. E., and Oda, T.: Space-based observations of megacity carbon dioxide, Geophys. Res. Lett., 39, L17806, doi:10.1029/2012GL052738, 2012.

Kuze, A., Suto, H., Nakajima, M., and Hamazaki, T.: Thermal and near infrared sensor for carbon observation Fourier-transform spectrometer on the Greenhouse Gases Observing Satellite for greenhouse gases monitoring, Appl. Optics, 48, 6716–6733, 2009.

Lee, R. A. M., et al.: Preflight Spectral Calibration of the Orbiting Carbon Observatory-2, in preparation, 2015.

Liebe, C. C., Pollock, R., Hannah, B., Bartman, R., Radulescu, C., Rud, M., and Esposito, J. A.: System for establishing best focus for the Orbiting Carbon Observatory instrument, Opt. Eng., 48, 073605, doi:10.1117/1.3180867, 2009.

O'Dell, C. W., Day, J. O., Pollock, R., Bruegge, C. J., O'Brien, D. M., Castano, R., Tkatcheva, I., Miller, C. E., and Crisp, D.: Preflight radiometric calibration of the orbiting carbon observatory, IEEE T. Geosci. Remote Sens., 49, 2438–2447, 2011.

O'Dell, C. W., Connor, B., Bösch, H., O'Brien, D., Frankenberg, C., Castano, R., Christi, M., Eldering, D., Fisher, B., Gunson, M., McDuffie, J., Miller, C. E., Natraj, V., Oyafuso, F., Polonsky, I., Smyth, M., Taylor, T., Toon, G. C., Wennberg, P. O., and Wunch, D.: The ACOS CO_2 retrieval algorithm – Part 1: Description and validation against synthetic observations, Atmos. Meas. Tech., 5, 99–121, doi:10.5194/amt-5-99-2012, 2012.

Pollock, R., Haring, R. E., Holden, J. R., Johnson, D. L., Kapitanoff, A., Mohlman, D., Phillips, C., Randall, D., Rechsteiner, D., Rivera, J., Rodriguez, J. I., Schwochert, M. A., and Sutin, B. M.: The Orbiting Carbon Observatory instrument: performance of the OCO instrument and plans for the OCO-2 instrument, Proc. SPIE, 7826, 78260W, doi:10.1117/12.865243, 2010.

Rosenberg, R., Maxwell, S., Johnson, B., and Pollock, R.: Preflight Radiometric Calibration of the Orbiting Carbon Observatory-2, in preparation, 2015.

Thompson, D. R., Benner, D. C., Brown, L. R., Crisp, D., Devi, V. M., Jiang, Y., Natraj, V., Oyafuso, F., Sung, K., Wunch, D., Castaño, R., and Miller, C. E.: Atmospheric validation of high accuracy CO_2 absorption coefficients for the OCO-2 mission, J. Quant. Spectrosc. Ra., 113, 2265 – 2276, doi:10.1016/j.jqsrt.2012.05.021, 2012.

Washenfelder, R. A., Toon, G. C., Blavier, J.-F., Yang, Z., Allen, N. T., Wennberg, P. O., Vay, S. A., Matross, D. M., and Daube, B. C.: Carbon dioxide column abundances at the Wisconsin Tall Tower site, J. Geophys. Res.-Atmos., 111, D22305, doi:10.1029/2006JD007154, 2006.

Wunch, D., Wennberg, P. O., Toon, G. C., Keppel-Aleks, G., and Yavin, Y. G.: Emissions of greenhouse gases from a North American megacity, Geophys. Res. Lett., 36, L15810, doi:10.1029/2009GL039825, 2009.

Wunch, D., Toon, G. C., Blavier, J.-F. L., Washenfelder, R. A., Notholt, J., Connor, B. J., Griffith, D. W. T., Sherlock, V., and Wennberg, P. O.: The Total Carbon Column Observing Network, Philos. T. Roy. Soc. A, 369, 2087–2112, doi:10.1098/rsta.2010.0240, 2011.

11

An airborne perfluorocarbon tracer system and its first application for a Lagrangian experiment

Y. Ren, R. Baumann, and H. Schlager

Deutsches Zentrum für Luft- und Raumfahrt (DLR), Institut für Physik der Atmosphäre, Oberpfaffenhofen, Germany

Correspondence to: Y. Ren (yu.ren@dlr.de)

Abstract. A perfluorocarbon tracer system (PERTRAS), specifically designed for Lagrangian aircraft experiments, has been developed by the Deutsches Zentrum für Luft- und Raumfahrt (German Aerospace Center, DLR). It consists of three main parts: a tracer release unit (RU), an adsorption tube sampler (ATS), and a tracer analytical system. The RU was designed for airborne tracer release experiments; meanwhile, it can be used on various platforms for different experimental purposes (here research vessel). PERTRAS was for the first time applied in the field campaign Stratospheric ozone: Halogen Impacts in a Varying Atmosphere (SHIVA) in November 2011. An amount of 8.8 kg perfluoromethylcyclopentane (PMCP) was released aboard the research vessel *Sonne* (RV *Sonne*) near the operational site of this campaign, Miri, Malaysia, on 21 November. The tracer samples collected using the ATS onboard the DLR research aircraft Falcon were analyzed in the laboratory using a thermal desorber–gas chromatography–mass spectrometry (TD–GC–MS) system. Guided by forecasts calculated with the Lagrangian model Hybrid Single Particle Lagrangian Integrated Trajectory (HYSPLIT), 64 tracer samples were collected onboard the Falcon approximately 5 and 25 h after the release, mostly with a time resolution of 1 min. Enhanced PMCP concentrations relative to ambient PMCP background values (mean: 6.62 fmol mol^{-1}) were detected during three intersects of the fresh tracer plume (age 5 h), with a maximum value of 301.33 fmol mol^{-1}. This indicates that the fresh tracer plume was successfully intercepted at the forecast position. During the second flight, 25 h after the release, the center of tracer plume was not detected by the sampling system due to a faster advection of the plume than forecast. The newly developed PERTRAS system has been successfully deployed for the first time. The instrumental setup and comparisons between the measurements and HYSPLIT simulations are presented in this study.

1 Introduction

Lagrangian experiments are a powerful tool in atmospheric investigations of the evolution of aerosols, clouds, and reactive chemical species during transport (Huebert and Lenschow, 1999). The approach in Lagrangian experiments is to measure the change of the composition of a single air mass as it moves using successive aircraft flights. A number of successful airborne Lagrangian experiments in the free troposphere have been performed using ambient tracers, including ASTEX/MAGE (Clarke et al., 1996), ACE-1 (Bates et al., 1998), COBRA (Lin et al., 2004), and ICARTT (Fehsenfeld et al., 2006; Methven et al., 2006). However, it was recognized during these experiments that the ability to quantify the microphysical and chemical processes during transport could be greatly enhanced with the use of an artificial tracer for non-ambiguous air mass identification and information on mixing processes.

Perfluoroalkylcycloalkanes (hereafter PFCs) are regarded as ideal tracer compounds for the investigation of atmospheric transport and dispersion because they are non-toxic, non-reactive, non-scavenging, and non-depositing, and they contribute neither to stratospheric ozone depletion nor to tropospheric air pollution. Furthermore, PFCs have also been used for many other applications, including detection of fluid leaks in underground pipes, testing of effectiveness of underground storage tanks and covered waste sites, and determination of indoor air exchange rates. Furthermore, PFCs have a limited industrial use and thus have a low atmospheric background concentration, which is important for long-range Lagrangian experiments. The observed background concentrations of most PFCs used are at mixing ratios of typically a few parts per 10^{15} (fmol mol^{-1}) (Piringer et al., 1997; Straume et al., 1998; Kim et al., 2002; Simmonds et al., 2002; Watson et al., 2007). They are potent greenhouse gases with a high global warming potential (GWP of about 7000). However, due to the very small atmospheric burden their global warming impact is insignificant compared to other greenhouse gases (Watson et al., 2007). Watson et al. (2007) estimated that the present total PFC radiative forcing (RF) is approximately 2×10^{-5} W m^{-2}. A PFC release into the atmosphere of 10 kg during a tracer experiment represents a negligible greenhouse contribution ($\sim 5 \times 10^{-11}$ W m^{-2}) compared to the combined anthropogenic RF (1.6 W m^{-2}) estimated by the 2007 IPCC (International Panel on Climate Change) Fourth Assessment Report.

PFCs can be detected down to background concentrations by enrichment on a solid adsorbent, separation by gas chromatography (GC), and detection using electron capture detection (ECD) (DeBortoli and Pecchio, 1985; Dietz, 1986; Lagomarsino, 1996; Straume et al., 1998). In recent years, due to the excellent selectivity and sensitivity and simple manipulation process, the analytical method with negative ion chemical ionization mass spectrometry (NICI-MS) for PFCs has increasingly been used (Begley et al., 1988; Galdiga and Greibrokk, 2000; Cooke et al., 2001; Simmonds et al., 2002).

The major PFC tracers – such as PMCP, perfluoromethylcyclohexane (PMCH), and perfluorodimethylcyclohexanes (PDCHs) – are liquids with the appreciable vapor pressure and relatively low boiling point at atmospheric temperature and pressure. They can be released into the atmosphere by either atomization or vaporization followed by dilution to prevent condensation until the vapors are sufficiently diluted below the ambient dew point concentration (Dietz, 1986). In previous experiments, the PFCs were released by being directly introduced into the fast-moving air stream (Ferber et al., 1981) or through the evaporation after increasing the temperature above the tracers' boiling point with a heater (Draxler, 1985; Draxler et al., 1991). Afterwards, adsorbent samplers, as the most popular established technique for sampling PFCs, were used to collect the tracer gas in the air (Ferber et al., 1981; Dietz, 1986; Draxler et al., 1991). So far, mainly ground-based tracer experiments using PFCs have been performed in North America and Europe at continental and regional scales (Ferber et al., 1986; Draxler and Heffter, 1989; Ambrosetti et al., 1998; Nodop et al., 1998; Green, 1999; Johnson et al., 2000; Green et al., 2003). During these experiments the PFCs have been released from ground sites and ships and subsequently measured at different ground stations. Although airborne measurements (Ferber et al., 1986; Ambrosetti et al., 1998; Nodop et al., 1998; Johnson et al., 2000; Green et al., 2003) have also been conducted for studies of the vertical mixing and dispersion of PFC-tagged air mass in some campaigns, these experiments were focused on the tracer transport near the ground. So far, no PFC tracer release and sampling system suitable for a deployment onboard research aircraft in the free troposphere (FT) and the upper troposphere–lower stratosphere (UT–LS) have been reported.

In this paper, we present a new perfluorocarbon tracer system, PERTRAS, which was developed especially for airborne tracer release and detection. Two newly constructed automatic modules are parts of the system: a PFC release unit (RU) and an adsorption tube sampler (ATS). Combined with the well-established analytical technique of thermal desorber (TD)–GC–NICI-MS, the PERTRAS system was applied for the first time in the field campaign SHIVA. Section 2 gives a detailed description of the system. The tracer experiment during SHIVA is described in Sect. 3. In Sect. 4, the results of the first of the PERTRAS applications are presented. Finally, a summary and conclusion are given in Sect. 5.

2 Instrumental setup

A very first prototype of an airborne PFC release and sampling system was built in cooperation with the Brookhaven National Laboratory and used for a pilot tracer experiment. However, leakage problems and aircraft certification issue required the construction of new RU and ATS modules, specifically designed for airborne operation. Here, stainless-steel adsorption tubes used for the sampler are completely compatible with most commercially available thermal desorber systems. In the next subsections, the instrumental setup of the PFC RU and ATS and the post analysis process in the laboratory are described in more detail.

2.1 PFC release unit (RU)

The basic principle of the PFC RU is to release the tracer liquid by spraying aerosol into a gas flow in the atmosphere through the process of high-pressure atomization using a spray nozzle. Figure 1 shows a schematic diagram of the design of the PFC RU (left) and the aircraft module (right). It consists of three 6.6 L ($\sim$ 10 kg liquid PFCs each) stainless-steel tanks. Depending on the planned experiments, up to 30 kg in total and three different PFCs can be released. The release rate can be set between 0.1 and 500 mL min^{-1} with an uncertainty of 2 % using a high-pressure liquid chromatography (HPLC) pump (Knauer, Germany, type 1800). The pressure in the tube to the PFC outlet is adjustable between 40 and 100 bar. A 500 mL stainless-steel cylinder of nitrogen (grade 5.0, Praxair) is installed ahead of the PFC tanks, and during release nitrogen is used for equalization of the pressure in the tanks. The operation of the RU is achieved by using an onboard PC running LabVIEW through a touch screen control panel, or by remote controlling from a laptop through virtual network computing (VNC). The different PFC tanks can be selected by switching a three-way valve through con-

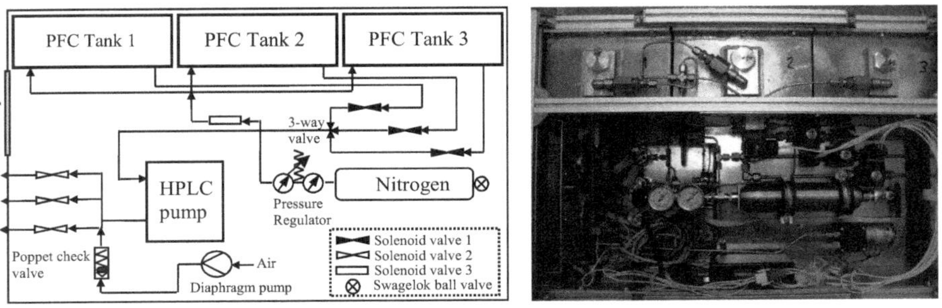

Figure 1. PFC release unit: schematic of gas flow (left); aircraft release module (right).

trolling the solenoid valves (Fig. 1 (left), solenoid valve 1). Then the release can start automatically by pressing the button on the controlling panel after a release program is set up. Three spray nozzles are used to disperse PFC droplets with diameters less than 20 μm into the fast-moving airstream. They are selected by controlling the solenoid valves (Fig. 1 (left), solenoid valve 2), which are connected to the outlet tube of the HPLC pump. The HPLC pump maintains the pressure necessary for the nozzles to operate in the range where the droplets will be within this size range needed to avoid sedimentation of the droplets before evaporation. The RU has a total weight of 47 kg.

During the release operation, the small PFC droplets are dispersed into the wake of the aircraft or ship or the air flow at ground level. When the release is performed aboard an aircraft, the RU is installed in a rack in the aircraft cabin and the PFC release is through a special outlet mounted into the fuselage. For security reason, an additional container is integrated in the rack capable of housing the total amount of PFCs in case of a leak during flight. The PFC RU can both be used for ground-based and ship-based releases as well, depending on the different objectives of the experiments. During SHIVA the latter release platform was chosen, and the PFC RU (65.0 cm × 48.5 cm × 35.5 cm), shown in Fig. 1 (right), was installed on the deck of the stern of the research vessel (RV) *Sonne*. The sprayer was fixed at the pillar of the gunwale about 10 m above the sea surface.

2.2 Aircraft adsorption tube sampler (ATS)

The components of the aircraft ATS are shown schematically in Fig. 2. This sampler was constructed specifically for the purpose of airborne sampling. It consists of four 16-position multiport valves (Valco Instruments Co. Inc.). The sampler can collect a maximum of 64 samples with the standard adsorption tubes (outer diameter: 6.35 mm; inner diameter: 5 mm; length: 89 mm; material: stainless steel) continuously or selectively according to the experimental requirements. The four multiport valves are incorporated into a modified metal bellow pump (Ansyco, Germany). The flow rate is controlled by a metal mass flow controller (MFC, Bronkhorst, USA). The selection of the sampling tube is operated by the

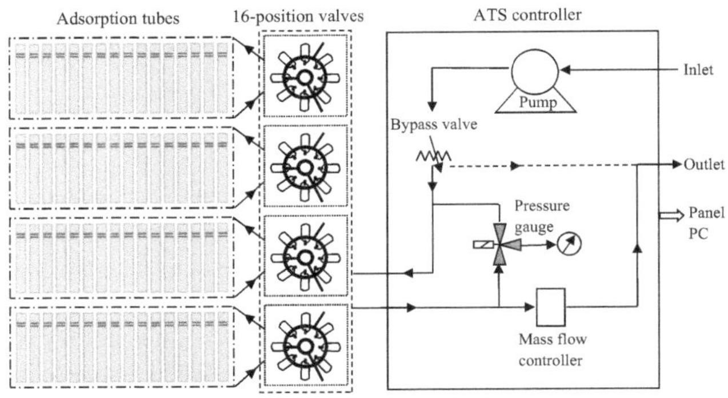

Figure 2. Schematic of the ATS. In standby or waiting mode, gas flow occurs through the dashed lines to the outlet. In sampling mode, gas flow is through the solid lines into the adsorption tubes.

touch screen of an integrated aircraft ATS controller. Depending on different experimental purposes, the sampling flow, sampling time and the interval between two samples, etc., can be controlled through the touch screen as well. After a sampling sequence is programmed, the system runs automatically.

The gas flow has two different modes: standby/waiting and sampling. When the sampler is in the mode of standby (ready for sampling) or waiting for sampling (during the time intervals between two samples), ambient air is drawn into the system by the metal bellow pump (Fig. 2, through the dashed lines) directly without flowing through the adsorption tubes. When the ATS starts to sample, ambient air is drawn into selected adsorption tubes and the sampling flow is controlled by a MFC. Different amounts of sample air can pass through the tubes at optional intervals, and perfluorocarbon tracer molecules are trapped and concentrated in the tubes onto adsorbent material. The volume and the time of air passed through the tubes are recorded. After one sampling sequence is finished, the ATS is switched back to standby mode automatically. Before installing the ATS into an aircraft, the test of airtightness for each tube is conducted at the ground with high-purity, PFC-free nitrogen. During the field experiment SHIVA, the ATS (53.0 cm × 48.5 cm × 16.0 cm, 23.6 kg) (Fig. 3) was installed in the DLR research aircraft

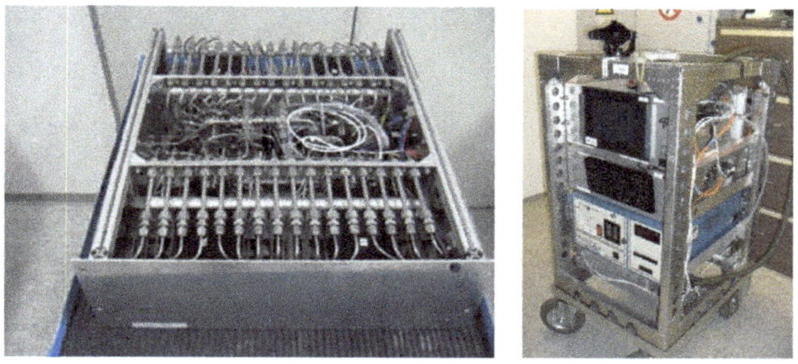

Figure 3. Adsorption tube setup in the sampler ATS (left) and aircraft instrument rack including PERTRAS and ozone and CO instruments (right).

Falcon. The inner part of the ATS module is depicted in Fig. 3 (left). Figure 3 (right) shows the ATS integrated in the aircraft rack together with ozone and CO instruments.

Stainless-steel adsorption tubes (PerkinElmer, USA) are used to collect PFCs, filled with about 200 mg Carboxen 569 (20–45 mesh, Supelco, USA). After filling with Carboxen 569, all sampling tubes are conditioned at 260 °C for at least 5 h under a constant flow of high-purity helium (Alphagaz™ He, 6.0, Air Liquide). The tubes are periodically checked by GC–MS to ensure that they are free of PFCs before sampling.

Using adsorption tubes for the collection of gas samples, the suitable sampling volume related to the breakthrough of the analytes on the adsorbent material needs to be considered. There are many different definitions of breakthrough volume (BTV). It is usually defined as the gas volume which passes through the adsorbent bed before a given compound begins to be eluted from the sorbent. Different sample volumes were tested to ensure that no PFC tracer had broken through the adsorbent bed during sampling. For this, the first blank adsorption tube of Carboxen 569 was loaded with a certain amount of gas PFC standard. This tube was connected in series with another blank tube to retain the PFCs eluted from the front to the back sorbent tube (Aragón et al., 2013). The results showed that there was no PFC penetration with 40 L PFC-free nitrogen at the flow rate of 200 mL min^{-1} flushing through the adsorption tubes at 20 °C. The theoretical BTV of these sample tubes with Carboxen 569 should be higher than 40 L for most PFCs. In addition, according to the results of Yang et al. (2005), the penetration volume of perfluorodimethylcyclobutane (PDCB) is almost independent of the flow rate ($\leq$ 1.5 L min^{-1}) of the mobile phase. Other PFCs have stronger adsorption ability than PDCB on Carboxen 569. Therefore, several liters of PFCs with less than 1.5 L min^{-1} flow rate can be trapped onto the sampling tube without exceeding the theoretical BTV.

2.3 Analytical instruments

The identification of the PFCs and quantification of their concentrations is accomplished offline by using an analytical system consisting of a GC (TRACE GC Ultra) connected with a pre-concentrated thermal desorber and a mass spectrometer in the laboratory. Here only a brief description of the analytical process is given. More details are given by Ren et al. (2014). Thermal desorption was carried out with a thermal desorber ATD650 (PerkinElmer, USA). The ATD650 was connected to the injection port of the GC through a temperature-controlled interface. With a two-step desorption, the sample which has been enriched into the cold trap, maintained at a temperature of −30 °C, can be desorbed in ∼ 3 s onto the analytical column in the GC (Tracer GC Ultra) oven in split mode. The cold trap was filled with about 100 mg Carboxen 569 (20–45 mesh, Supelco, USA) and conditioned by heating at 260 °C with carrier gas for at least 5 h under high-purity helium to clean prior to the normal analysis. The temperature of both the interface and the valve was set at 200 °C.

The separation of the tracer was carried out on a 30 m × 0.25 mm × 5.0 μm Al_2O_3-PLOT-S capillary column (HP-PLOT Al_2O_3 S, Agilent, USA). A particle trap 5 m × 0.25 mm deactivated fused silica column (Agilent, USA) was connected to the Al_2O_3-PLOT-S to avoid particles shedding from the stationary phase layer of PLOT capillary column entering the MS detector. The following oven temperature program was set up: 100 °C (0.5 min), 30 °C min^{-1} to 130 °C (6.5 min), 30 °C min^{-1} to 140 °C (23 min), then 20 °C min^{-1} to 190 °C (5 min). High-quality helium (Alphagaz™ He, 6.0, Air Liquide) was used as carrier gas with a flow rate of 1 mL min^{-1}. The mass spectrometer was operated in negative ion chemical ionization (NICI) mode at 70 eV and in selected ion monitoring (SIM) mode. Methane (N55 quality, Air Liquide) was used as reagent gas. The GC–MS interface and the source temperature were set to 190 °C. It has been demonstrated by Ren et al. (2014) that this detection system can be used for routine laboratory analysis of several PFCs with good sensitivity and selectivity including PMCP, PMCH, PDCH, PDCB, and PECH (perfluoroethylcyclohexane).

Samples were calibrated using a gas standard prepared at the University of Bristol (Simmonds et al., 2002). The limit of detection (LOD) of the method for PMCP was 1.2 ppq (fmol mol^{-1}), detected with a 3 times signal-to-noise ratio. The precision of PMCP during the analytical process was ±4.6 %. The calibration appeared to be linear at least 3 orders of magnitude for the concentrations that were studied.

3 Field deployment of PERTRAS

3.1 SHIVA – measurement campaign

SHIVA is a European Union (EU)-funded research project (http://shiva.iup.uni-heidelberg.de/) with the main objective of reducing uncertainties in present and future stratospheric halogen loading and ozone depletion. Of particular relevance are studies of short- and very short-lived substances (VSLS) with climate-sensitive natural emissions. A core field campaign was conducted during November–December 2011 in the tropical western Pacific region with measurements along the coastline of the Peninsular Malaysia and Borneo using ships and aircraft. The campaign base of the DLR research aircraft Falcon was Miri on Borneo, Malaysia. One measurement objective was to study the transport and dispersion of VSLS halocarbons produced from seaweed along the coast of Borneo in the marine boundary layer and the entrainment of these emissions in the FT (Aschmann et al., 2009; Tegtmeier et al., 2012, 2013). SHIVA provided the opportunity to deploy the PERTRAS system for the first time in the framework of a Lagrangian experiment. In the following, the experimental details of the tracer release and collection are given.

3.2 Tracer release

The PFC tracer release was conducted from the German RV *Sonne*. Therefore the release unit with PMCP was set up on the deck of RV *Sonne*. The time of the tracer release was selected based on HYSPLIT forecasts. An amount of 8.8 kg pure PMCP was released on 21 November 2011 from 02:24 UTC (112.96° E, 4.51° N) to 03:03 UTC (113.01° E, 4.56° N). The release position along the *Sonne* route is depicted in Fig. 4 as site C4.

3.3 HYSPLIT simulation and the flight patterns

Version 4 of HYSPLIT (Draxler and Hess, 1997, 1998; http://www.arl.noaa.gov/ready/hysplit4.html) was employed to forecast and simulate the transport and dispersion of the PFC tracer and to support flight planning. Data from the Global Forecast System (GFS, for forecast) and Global Data Assimilation System (GDAS, for analysis) with 1° × 1° horizontal resolution and 3 h time intervals were used as meteorological input for HYSPLIT. Data were given on pressure levels with a resolution of 25 hPa within the lower atmosphere. HYSPLIT was operated in particle dispersion mode, and all parameterizations were set to the standard values. The release species was treated as a passive gas tracer without deposition. The simulation takes into account the movement of the ship during the release; that is, the release is split into 39 point sources (one for each minute) shifted in space and time according to the ship course. The HYSPLIT forecast was regularly updated every 6 h based on the latest available GFS data.

Figure 5 shows HYSPLIT forecasts of tracer concentrations after PMCP was released from the RV *Sonne* on 21 November 2011 over a dispersion period of 5 and 25 h. The forecasted maximum concentration in the PMCP tracer plume is depicted in dark red. Based on these forecasts, the corresponding flight plan of the Falcon on these 2 days was made, and the flight tracks for the two flights are also illustrated in Fig. 5.

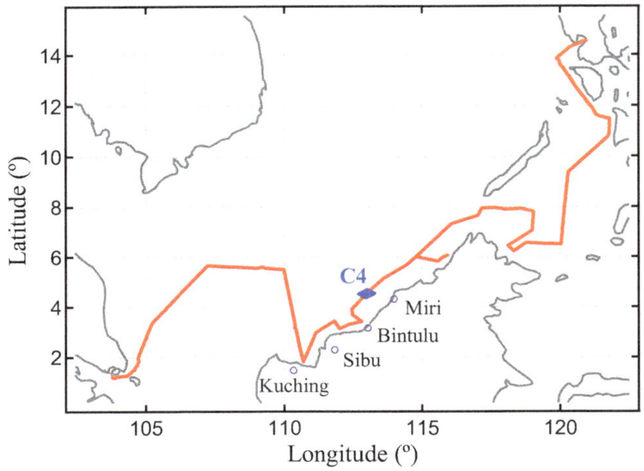

Figure 4. Route of the RV *Sonne* during SHIVA field campaign (red line) and the position (C4) of the tracer release: PMCP was released on the deck of *Sonne* from 02:24 to 03:03 UTC on 21 November 2011.

3.4 Tracer sampling

The ATS was successfully deployed onboard the Falcon during two flights conducted on 21 and 22 November 2011. The aim was to test and verify the operation of the whole analytical system and to compare the tracer measurements with the HYSPLIT simulation. Therefore, multiple Falcon transects were planned through the center of the tracer plume based on the HYSPLIT forecasts, as shown in Fig. 5. In general, the collection of samples was started several minutes before the expected entering of the tracer plume according to the forecast, and sampling was terminated a few minutes after the expected exit of the plume. On 21 November, the Falcon performed two intersections of the plume rotated by 90°, at two altitudes (170 and 500 m), about 75 km away from the release point. Before flying back to Miri, the aircraft ascended to about 1000 m, trying to catch the highest plume part according to the vertical cross-section map. On 22 November, the Falcon flew through the elongated plume at about 800 and 1200 m, about 150 km away from the release point. All samples were collected at 1 L min^{-1} for 1–1.5 min. There were 0.5 to 2 min intervals between two samples in a sampling sequence, and the intervals were adjusted due to the actual sampling situation. During the sampling periods, the aircraft speed ranged from 101 to 112 m s^{-1} (21 November) and from 123 to 140 m s^{-1} (22 November); thus 1–1.5 min collection times result in a horizontal grid spacing of about 6–10 and 7–13 km, respectively. In accordance with the flight plan, in total 64 samples were obtained on these two flights, including four blanks for quality control to check whether any contamination of the sampling tubes had taken place during transport, sampling, or analysis. To avoid contamination, the sampling tubes had been separated strictly from the tracer

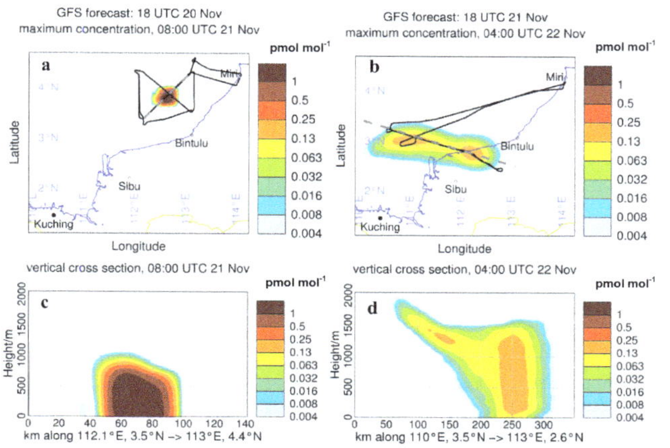

Figure 5. HYSPLIT forecasts of the horizontal and vertical distribution of the PFC concentration at the northern coast of Borneo, 5 h (**a** and **c**) and 25 h (**b** and **d**) after the release of the tracer. The flight tracks of the two Falcon sampling flights are also shown (black lines in **a** and **b**). The thick dashed lines (grey) indicate the location of vertical cross sections (**c** and **d**).

storage tank and the tracer releaser system during the whole campaign.

In addition, a commercial sequential tube sampler (STS 25, PerkinElmer, USA) was used to collect background samples at a rural site (112.978° E, 4.398° N) near the airport of Miri, about 9 km south of the center of Miri. Using a sampling flow rate of 200 mL min^{-1}, air samples were trapped onto adsorption tubes for 40 min. A total of 37 samples were collected between 20 November and 5 December 2011, including 8 blanks, which were also used for quality control purposes. Since the plume of the released PMCP tracer was not advected to Miri, background samples could still be collected at the ground in Miri after the release experiment. Before and after the ground-based sampling, the flow through every sampling orifice of the ground sampler was measured with a Gillian Galibrator-2 (Scientific Instrument Services, Inc™, USA). After the sampling period, the tubes were sealed immediately at both ends with 1/4 in. brass storage caps with 1/4 in. PTFE ferrules and stored until analysis.

4 Results and discussion

The analysis of all blank tubes showed that no contamination occurred during the transport of the tubes to and from the campaign site, nor during sampling, storage, or analysis. The measured background concentrations of PMCP at the ground near Miri range between 5.18 and 8.95 fmol mol^{-1}, with a mean value of 6.62 fmol mol^{-1} and a standard deviation of 1.02 fmol mol^{-1} (6.62 $\pm$ 1.02 fmol mol^{-1}). This is comparable to PMCP levels reported from previous field experiments (Kim et al., 2002; Simmonds et al., 2002; Watson et al., 2007; Ren et al., 2014). In the following subsections, the analytical results of the tracer measurements with PERTRAS are presented and comparisons between the measurement and HYSPLIT simulation are discussed.

4.1 Time series of PMCP measurements

Figure 6a shows a time series of PMCP tracer concentrations (red bars) along with altitude (in black) for the flight on 21 November 2011. In total, 33 air samples were collected during five crossings at three different flight levels (170, 500, and 1000 m). The measured PMCP concentrations range from 9.36 to 301.33 fmol mol^{-1}, largely exceeding the mean PMCP background concentration measured at the ground (6.62 fmol mol^{-1}) for most samples. Three time periods of clearly elevated PMCP concentrations are observed for the first three plume crossings during 07:00–07:25, 07:45–08:05, and 08:25–08:45 UTC (about 5 h after the release), at altitudes of 170, 170, and 500 m, respectively. This demonstrates that the Falcon successfully sampled the tracer plume during these periods.

During the first sampling period at 170 m from northeast (NE) to southwest (SW), three samples collected between 07:06 and 07:15 UTC showed enhanced tracer concentrations relative to background concentrations. The maximum detected PMCP concentration (301.33 fmol mol^{-1}) at 07:07 UTC is about 45 times higher than the mean background at the ground. Along both sides of the maximum, the detected PMCP concentrations decrease, reaching values close to the background level. At around 07:10 UTC there is a 5 min sampling gap, due to a short period of malfunction of the sampling pump. During this time interval, two samples could not be taken. During the second sampling period (07:45–08:05 UTC), when the Falcon flew from northwest (NW) to southeast (SE), most samples were clearly higher than the background, with a maximum value of 203.39 fmol mol^{-1}. The expansion of the tracer plume in NW–SE direction is found to be broader than along the perpendicular direction (NE–SW), and the concentration variation of PMCP are comparatively moderate. During the third sampling period at around 08:25–08:45 UTC, when the Falcon flew along the NE–SW direction again but at 500 m, the PMCP concentrations show a quasi-symmetrical distribution. Very likely the Falcon flew through a narrow region of the tracer plume or just touched the fringe of the tracer plume. During the last sampling period from 09:05 to 09:25 UTC, two of the samples have slightly higher concentrations than the background. The samples collected at 1000 m (09:45–10:00 UTC) have PMCP concentrations close to background values. Probably the Falcon flew above the tracer plume here.

For the tracer sampling flight on 22 November, about 25 h after the release, a total of 27 samples, all except one, were successfully analyzed. The corresponding time series of PMCP concentrations is shown in Fig. 6b (purple bars). Most PMCP values detected (maximum concentration

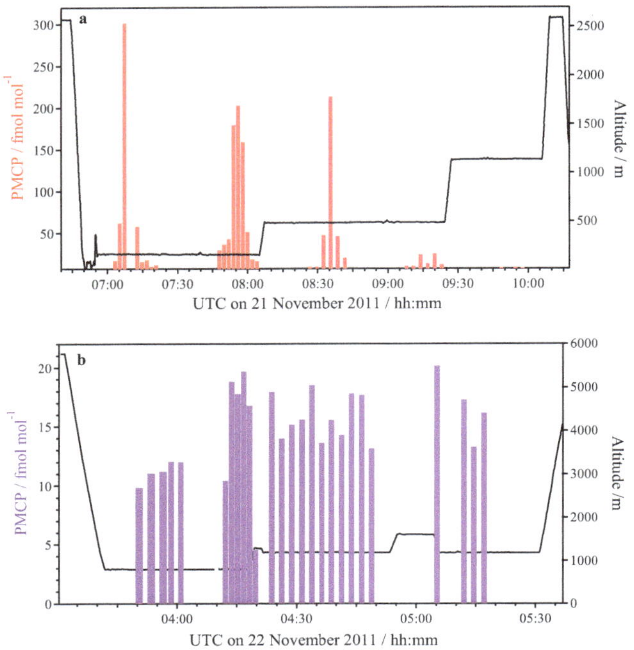

Figure 6. Time series of PMCP concentrations on 21 November **(a)** and 22 November 2011 **(b)**. Each bar represents the averaged value during each sampling interval.

of 20.14 fmol mol^{-1}) are higher than background. Unfortunately, the tracer plume was advected further south than forecast and partly over land. Therefore the lowest level of the Falcon possible was 1000 m, and the sampling occurred only across the outer edge of the tracer plume.

4.2 Comparison between the measurement and HYSPLIT simulations

The comparisons between the PMCP measurements and the HYSPLIT simulations based on GDAS data for the flight on 21 November show in general a reasonably good agreement as illustrated in Fig. 7.

During the first sampling period (marked as "1" in Fig. 7) at around 07:10 UTC, the peak PMCP concentration simulated by HYSPLIT (6190.74 fmol mol^{-1}) is much higher than the maximum PMCP concentrations observed (301.33 fmol mol^{-1}) (07:06–07:07 UTC); however, the HYSPLIT tracer maximum is calculated for the period where two samples could not be taken due to the short pump failure. The average PMCP concentration (279.31 fmol mol^{-1}) calculated with HYSPLIT for the period 07:06–07:07 UTC is similar to the measurements.

During SHIVA, the University of Frankfurt operated also an in situ gas chromatograph, GHOST (Gas cHromatograph for the Observation of Stratospheric Tracers – coupled with a mass spectrometer), onboard the Falcon for measurements of halogen compounds (Bujok et al., 2001; Sala et al., 2014). GHOST was not prepared or calibrated to measure PFCs;

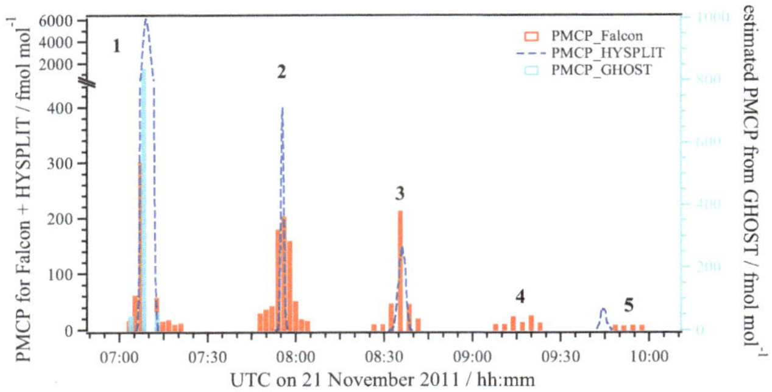

Figure 7. Time series of measured PMCP concentrations (red bars) on 21 November and HYSPLIT simulations using analyzed meteorological fields (blue dashed lines). Bluish-green bars represent the PMCP concentrations estimated from the signals of the GHOST instrument.

however it detected signals of PMCP during the tracer plume sampling.

Interestingly, during the sampling period 07:00–07:15 UTC, three high signals of the PMCP tracer were detected by GHOST, one being extraordinarily high. One of the GHOST signals corresponds well in time with one of the PERTRAS samples during the period 07:12–07:13 UTC (the third bluish-green bar, Fig. 7). Based on the ratios of both signals to the calibration gas used for GHOST, the PMCP concentrations of the two other GHOST signals can be estimated collected from 07:03 to 07:04 UTC, between the first and the second sample of ATS, and from 07:07 to 07:08 UTC, starting just after the sample from ATS being finished at around 07:07 UTC. The GHOST estimates for the PMCP concentrations amount to 46.8 and 840.6 fmol mol^{-1}, respectively, and are also shown in Fig. 7 as bluish-green bars (the first and the second one). Overall the PMCP measurements and the HYSPLIT simulations indicate that the Falcon was traversing the center of the tracer plume during the first sampling period.

For further comparisons of the PMCP measurements and HYSPLIT simulations, time snapshots of the tracer plume for each 10 min time interval for the sampling flight on 21 and 22 November are shown in Figs. 8 and 9. Here, the horizontal model layers are depicted which match best with the Falcon flight altitude during each 10 min time window from $t-5$ min to $t+5$ min for the corresponding samples. For those cases where the flight altitude varied by more than 100 m during the corresponding time window, the contours show, at each horizontal grid point, the highest volume mixing ratio (VMR) of any vertical model layer which the aircraft crossed. Those height ranges are indicated in the headings of the plots.

As shown in panels a and b of Fig. 8, which correspond to the sampling period of "1" in Fig. 7, the flight track almost passed through the center of the tracer plume as planned. In panel b, the samples collected at 07:07 UTC (yellow oval)

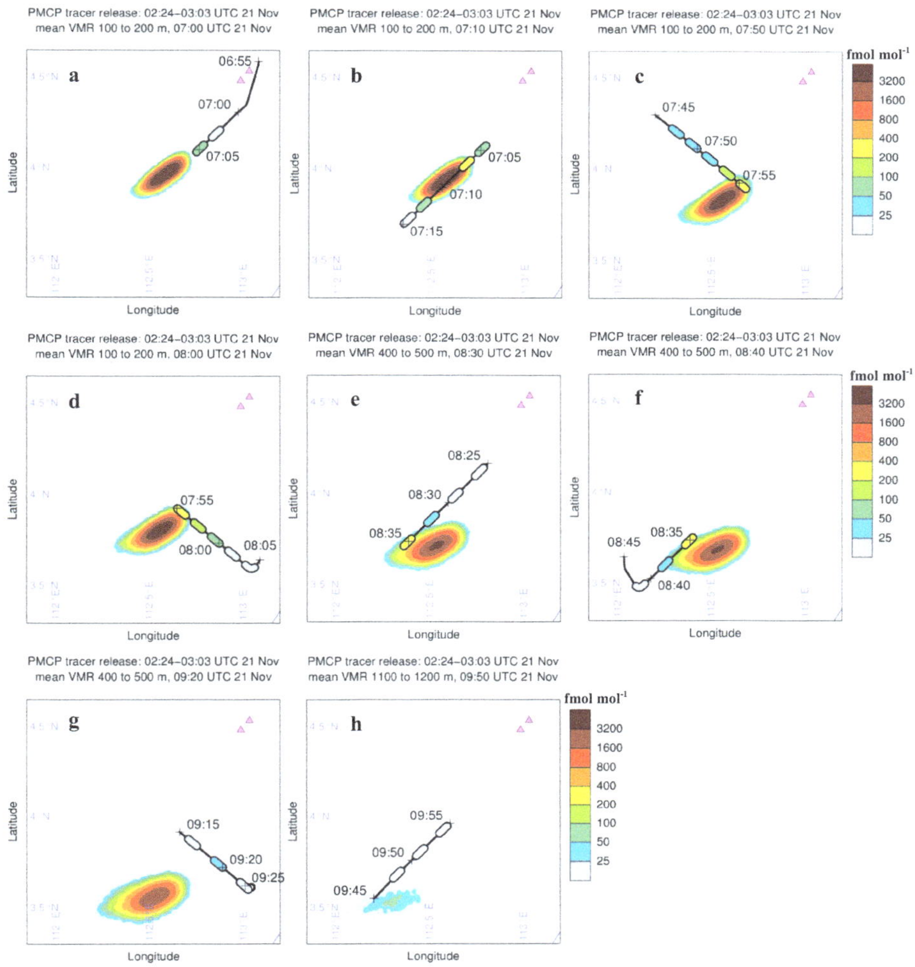

Figure 8. Time snapshots of the simulated tracer plume with HYSPLIT using analyzed meteorological fields (big plume) on 21 November, and the corresponding sampling tracks (black line). **(a** and **b)** The plume at $t = 07{:}00$ and 07:10 UTC, **(c** and **d)** the plume at $t = 07{:}50$ and 08:00 UTC, **(e** and **f)** the plume at $t = 08{:}30$ and 08:40 UTC, **(g)** the plume at $t = 09{:}20$ UTC, and **(h)** the plume at $t = 09{:}50$ UTC. The volume mixing ratio (VMR) is calculated as mean value within vertical layers of 100 m thickness, for **(a–d)** 100–200 m, for **(e–g)** 400–500 m, and for **(h)** 1100–1200 m. Small colored ovals represent the samples collected along the flight track (same color scale as used for the HYSPLIT data). Two purple triangles indicate the positions of the start and the end of the tracer release.

and 07:12 UTC (green oval), with PMCP concentrations of 301.33 and 58.54 fmol mol^{-1}, respectively, are quantitatively similar to the simulation results. However, one sample (another green oval in Fig. 8a and b) with a PMCP concentration of 62.58 fmol mol^{-1}, very well above background values, was collected out of the main forecasted plume. This indicates that the simulation is underestimating the size of the plume in NE–SW direction at least by 10 km. However, this difference can be regarded as reasonable since the model cannot reproduce such small variations in the measured tracer concentrations because it is below the resolution of the meteorological input data (Stohl et al., 1998).

For the other two sampling periods during 07:45–08:05 UTC (Fig. 7, case 2) and 08:25–08:45 UTC (Fig. 7, case 3), the simulated PMCP concentrations are similar in magnitude to the measurements. These samples were mainly collected in the boundary region of the plume. As can be seen from Fig. 8c, during 07:45–08:05 UTC, two samples, illustrated as light-green and yellow ovals, with PMCP concentrations of 179.66 and 203.39 fmol mol^{-1}, respectively, show reasonable agreement with the model but are located in the boundary of the plume in the northeastern direction. The plume was transported a little bit further to the south compared to the forecast used for flight planning. Two other samples illustrated as light-green (159.36 fmol mol^{-1}) and green ovals (51.96 fmol mol^{-1}) (Fig. 8d) are located outside of the simulated plume, which indicates that the plume in this area probably dispersed more (at least 12 km) in the southeastern direction than simulated by the model. From the reanalysis for the time period 08:30–08:40 UTC (Fig. 8e and f), the

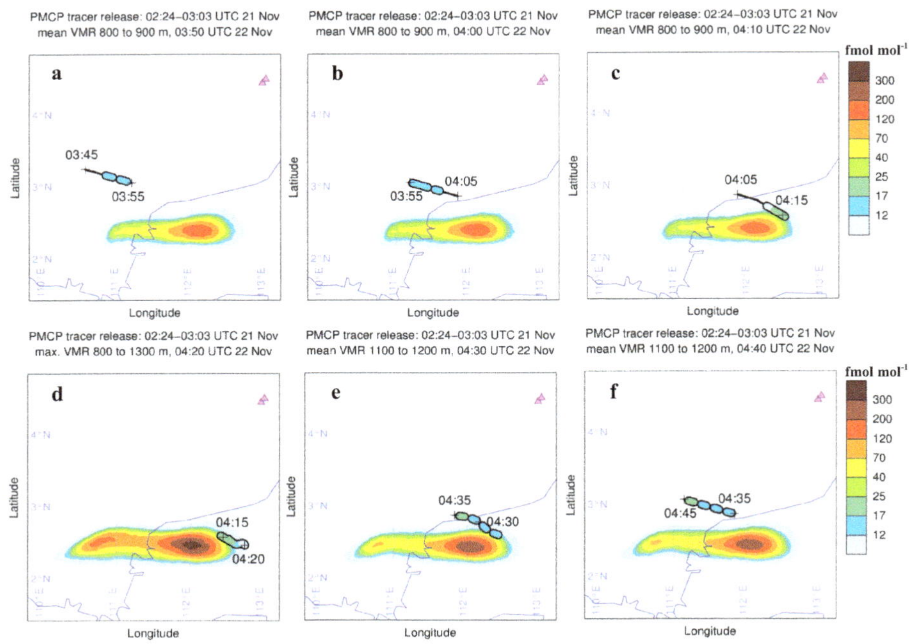

Figure 9. Time snapshots of the simulated tracer plume with HYSPLIT using analyzed meteorological fields (big plume) on 22 November, and the corresponding sampling tracks (black line). **(a–f)** The plume at $t = $ 03:50, 04:00, 04:10, 04:20, 04:30, and 04:40 UTC, respectively. VMR is calculated as mean value within vertical layers from 800 to 900 m **(a–c)** and 1100 to 1200 m **(e** and **f)**. Panel **(d)** picks, at any location, the maximum out of five vertical layers between 800 and 1300 m, because the flight altitude changed during this section. Small colored ovals represent the samples collected along the flight track (same color scale as used for the HYSPLIT data). Two purple triangles indicate the positions of the start and the end of the tracer release.

flight track seems to have touched the southwestern edge of the plume as speculated in Sect. 4.1. This implies that the tracer plume in reality was advected further to the southeast. Although the Falcon mostly did not fly through the plume center as planned, the results from the measurements (Fig. 7) coincide with the HYSPLIT simulations quite well.

The results of samples collected during the fourth and fifth sampling period (marked as "4" and "5" in Fig. 7) at around 09:15 and 09:55 UTC are not captured by the simulations. This suggests that the tracer plume moved away already from the sampling area, which can be seen more clearly in Fig. 8g and h.

The plume simulation at 5 h after the release underestimates the along-wind size by some 20 % (Fig. 8b) and the cross-wind size by a factor of roughly 1.5–2 (Fig. 8c/d) compared with the size inferred from the locations of the samples taken with the ATS. This is consistent with an overestimation of maximum concentrations in the dispersion simulation compared with the results from PERTRAS for the crossings at 07:10 and 07:55 UTC, implying an underestimation of dispersion and mixing by HYSPLIT. However, at this time the size of the plume is significantly smaller than the grid spacing of the driving wind field data, which might result in an underestimation of the stretching and folding of the plume by local sub-grid scale wind variations. In addition, an underestimation of the turbulent velocity components for that specific meteorological condition cannot be ruled out. As a consequence, preparation and execution of future tracer release experiments would strongly benefit from the use of higher-resolution meteorological forecast data, if available.

For the sampling flight on 22 November, if the analysis of HYSPLIT is correct, the Falcon sampled north of the main plume, barely touching the northeastern edge from 04:10 to 04:30 (shown in Fig. 9). According to the HYSPLIT forecast, which was used for the flight planning, the tracer plume should have been located further north. Another difficulty for the flight plan was that the plume was located partly over land and a minimum possible flight altitude of the Falcon was required to be 1000 m. The latest updated HYSPLIT forecasts for the flight on 22 November after the takeoff of the Falcon already showed that the tracer plume center was moving faster and more to the south; however, the flight track of the Falcon could not be changed due to telecommunication loss with the Falcon because of the low flight altitude. Therefore the Falcon collected samples according to the flight plan made before takeoff. The results from the measurement basically confirm the post-flight simulation of HYSPLIT using analyzed meteorological data.

5 Summary and conclusion

We have presented a newly developed perfluorocarbon tracer system, PERTRAS, including a tracer release unit, an adsorption tube sampler, and an analyzing system (TD–GC–NICI-MS). The tracer RU is highly portable, can be automatically operated, and is suitable for the deployment on different platforms (ground station, ship, and aircraft) depending on the objective of the tracer experiment, although it was designed for the application aboard aircraft. The tracer release rate can be adjusted between 0.1 and 500 mL min^{-1} with an uncertainty of 2 % using a HPLC pump, and up to a total of 30 kg and three different PFCs can be released. Up to 64 tracer samples can be taken during a flight with the tracer sampler ATS. The sampling procedure can be programmed including sequences with standby/waiting and sampling modes. All sample tubes are analyzed with a high selective and sensitive TD–GC–NICI-MS system in the laboratory with a limit of detection of 1.2 fmol mol^{-1}.

During the SHIVA field campaign in November 2011, the new PERTRAS system was applied for the first time. The tracer release was conducted on the deck of RV *Sonne*, and 8.8 kg PMCP was released during a period of 39 min (02:24 to 03:03 UTC). The tracer was sampled by the DLR research aircraft Falcon using the ATS during two flights on 21 and 22 November, approximately 5 and 25 h after the release, respectively. In total, 63 samples including four blanks were collected, with a time resolution of 1 min, and analyzed after the campaign in the DLR laboratory using a TD–GC–NICI-MS system. The result of all blank samples shows that no contamination occurred during the transport of the sampling tubes to the campaign site and back, as well as during sampling and analysis.

During the sampling flight on 21 November the tracer plume was intercepted successfully several times and a reasonable agreement was found between the tracer measurements and post-flight HYSPLIT simulations using analyzed meteorological fields. It could be demonstrated that the entire PERTRAS system worked reliably. The plume simulated by HYSPLIT is somewhat smaller than the one inferred from the measurement for the first three sampling periods for the 21 November flight, implying a slight underestimation of dispersion and mixing by HYSPLIT.

The tracer plume was largely missed during the flight on 22 November for several reasons. Firstly, there was a restriction of the minimum flight altitude to 1000 m since the tracer plume was advected partly over land. Secondly, a needed adjustment of the flight pattern after the Falcon takeoff based on the latest HYSPLIT forecasts could not be transmitted to the Falcon crew due to a loss of telecommunication because of the low flight altitude. These problems should not occur for tracer experiments at higher altitudes.

The tracer experiment during SHIVA demonstrated the capability of the new tracer system PERTRAS to unambiguously identify and trace an air mass tagged with a PFC. This proof-of-concept study provides the practical basis for the application of the PFC tracer system in future aircraft experiments in the FT and UT–LS to quantify small-scale diffusion parameters and synoptic-scale flow distortion in this altitude region. Furthermore, the release of two different PFC tracers will be helpful to understand the mixing between distinct air masses, in particular near the tropopause. Moreover, the tagging of air masses with PFC tracers can significantly support Lagrangian experiments to study chemical process and aerosol transformations. In addition, investigations of transport and multiphase processes in convective clouds can be facilitated by tagging the air with a PFC tracer in the inflow region and sampling the tracer in the outflow region.

After the successful deployment of PERTRAS during the SHIVA field campaign, further tracer experiments are planned to study transport and transformation processes in deep convection in the tropics and in large scale pollution plumes of megacities in Asia. Recently, the PERTRAS system has been applied successfully during the campaign T-NAWDEX-Falcon in a warm conveyor belt over Europe. Analysis of this data is ongoing, and results will be presented in a forthcoming paper.

Acknowledgements. This work was funded by the Deutsche Forschungsgemeinschaft (DFG) under Priority Program HALO SPP 1244 (SCHL 1857/1-2) and by the EU project SHIVA (EU-226224). We especially thank Michael Lichtenstern (DLR) for his excellent work of constructing PERTRAS instrumentation. We thank Russel N. Dietz and Thomas B. Watson of the Brookhaven National Laboratory for very helpful advice in all aspects of the PFC technology. We also thank Dominik Schäuble and Tina Jurkat (DLR) for the tracer release from RV *Sonne* and Klaus Pfeilsticker and Marcel Dorf (University of Heidelberg) for coordinating the SHIVA campaign. Stephan Sala and Andreas Engel (Goethe University Frankfurt) are acknowledged for providing the GHOST data. Thanks are also due to Damien Martin (University of Bristol) for providing the gas standard. We acknowledge the support of the pilots and staff of the DLR Flight Department and the help of Anke Roiger, Anja Reiter, and Heinfried Aufmhoff (DLR) during the campaign. We are thankful to the NOAA Air Resources Laboratory for providing the HYSPLIT model.

The service charges for this open access publication have been covered by a Research Centre of the Helmholtz Association.

Edited by: W. T. Sturges

References

Ambrosetti, P., Anfossi, D., Cieslik, S., Graziani, G., Lamprecht, R., Marzorati, A., Nodop, K., Sandroni, S., Stingele, A., and Zimmermann, H.; Mesoscale transport of atmospheric trace constituents across the central Alps: Transalp tracer experiments, Atmos. Environ., 32, 1257–1272, 1998.

Aragón, M., Borrull, F., and Marcé, R. M.: Thermal desorption-gas chromatography-mass spectrometry method to determine phthalate and organophosphate esters from air samples, J. Chromatogr. A, 1301, 76–82, 2013.

Aschmann, J., Sinnhuber, B.-M., Atlas, E. L., and Schauffler, S. M.: Modeling the transport of very short-lived substances into the tropical upper troposphere and lower stratosphere, Atmos. Chem. Phys., 9, 9237–9247, doi:10.5194/acp-9-9237-2009, 2009.

Bates, T. S., Huebert, B. J., Gras, J. L., Griffiths, F. B., and Durkee, P. A.: International Global Atmospheric Chemistry (IGAC) Project's First Aerosol Characterization Experiment (ACE1): Overview, J. Geophy. Res., 103, 16297–16318, doi:10.1029/97JD03741, 1998.

Begley, P., Foulger, B., and Simmonds, P.: Femtogram detection of perfluorocarbon tracers using capillary gas chromatography-electron-capture negative ion chemical ionization mass spectrometry, J. Chromatogr., 445, 119–128, 1998.

Bujok, Q., Tan, V., Klein, E., Nopper, R., Bauer, R., Engel, A., Gerhards, M., Afchine, A., Mckenna, D. S., Schmidt, U., Wiedhold, F. G., and Fischer, H.: GHOST – A novel airborne gas chromatograph for in situ measurements of long-lived tracers in the lower stratosphere: method and applications, J. Atmos. Chem., 39, 37–64, 2001.

Clarke, A. D., Uehara, T., and Porter, J. N.: Lagrangian evolution of an aerosol column during the Atlantic Stratocumulus Transition Experiment, J. Geophy. Res., 101, 4351–4362, doi:10.1029/95JD02612, 1996.

Cooke, K. M., Simmonds, P. G., Nickless, G., and Makepeace, A. P. W.: Use of capillary gas chromatography with negative ion-chemical ionization mass spectrometry for the determination of perfluorocarbon tracers in the atmosphere, Anal. Chem., 73, 4295–4300, 2001.

DeBortoli, M. and Pecchio, E.: Sensitive determination of perfluorocarbon tracers in air by intermediate trapping and GC-ECD Analysis, J. High Res. Chromotog., 8, 422–425, 1985.

Dietz, R. N.: Perfluorocarbon tracer technology, Regional and Long-Range Transport of Air Pollution, edited by: Sandroni, S., Elsevier, Amsterdam, 215–247, 1986.

Draxler, R. R.: Metropolitan Tracer Experiment (METREX), NOAA Tech. Memo. ERL, ARL-140, 1985.

Draxler, R. R. and Heffter, J. L.: Across North America Tracer Experiment (ANATEX), Volume 1: description, ground-level sampling at primary sites, NOAA Tech. Memo., ERL, ARL-167, 1989.

Draxler, R. R. and Hess, G. D.: Description of the HYSPLIT_4 modeling system, NOAA Tech. Memo. ERL, ARL-224. 1997.

Draxler, R. R. and Hess, G. D.: An overview of the HYSPLIT_4 modeling system of trajectories, dispersion and deposition, Aust. Meteorol. Mag., 47, 295–308, 1998.

Draxler, R. R., Dietz, R. N., Lagomarsino, R. J., and Start, G.: Across North America Tracer Experiment (ANATEX): sampling and analysis, Atmos. Environ., 25A, 2815–2836, 1991.

Fehsenfeld, F. C., Ancellet, G., Bates, T. S., Goldstein, A. H., Hardesty, R. M., Honrath, R., Law, K. S., Lewis, A. C., Leaitch, R., McKeen, S., Meagher, J., Parrish, D. D., Pszenny, A. A. P., Russel, P. B., Schlager, H., Seinfeld, J., Talbot, R., and Zbinden, R.: International Consortium for Atmospheric Research on Transport and Transformation (ICARTT): North America to Europe-Overview of the 2004 summer field study, J. Geophy. Res., 111, D23S01, doi:10.1029/2006JD007829, 2006.

Ferber, G. J., Telegadas, K., Heffter J. L., Dickson, C. R., Dietz, R. N., and Philip, W. K.: Demonstration of a long-range atmospheric tracer system using perfluorocarbons, Final Report, EPA-600, 1981.

Ferber, G. J., Heffter, J. L., Draxler, R. R., Lagomarsion, R. J., Thomas, F. L., Dietz, R. N., and Benkovitz, C. M.: Cross-Appalachian Tracer Experiment (CAPTEX 83), Final Report, NOAA Tech. Memo., ERL, ARL-142, 1986.

Galdiga, C. U. and Greibrokk, T.: Ultra trace detection of perfluorocarbon tracers in reservoir gases by adsorption/thermal desorption in combination with NICI-GC/MS, Fresen, J. Anal. Chem., 367, 43–50, 2000.

Green, M., Kuhns H., Pitchford, M., Dietz, R., Ashbaugh, L., and Watson, T.: Application of the tracer-aerosol gradient interpretive technique (TAGIT) to sulfur attribution for the Big Bend Regional Aerosol and Visibility Observational (BRAVO) study, J. Air Waste Manage., 53, 586–595, 2003.

Green, M. C.: The project MOHAVE tracer study: study design, data quality, and overview of results, Atmos. Environ., 33, 1955–1968, 1999.

Huebert, B. J. and Lenschow, D. H.: What have Lagrangian experiments accomplished?, IGACtivity Newsletter, 17, 4–8, 1999.

Johnson, D. W., Osborne, S., Wood, R., Suhre, K., Johnson, R., Businger, S., Quinn, P. K., Wiedensohler, A., Durkee, P. A., Russell, L. M., Andree, M. O., O'Dowd, C., Noone, K. J., Bandy, B., Rudolph, J., and Rapsomanikis, S.: An overview of the Lagrangian experiments undertaken during the North Atlantic regional Aerosol Characterization Experiment (ACE-2), Tellus, 52B, 290–320, 2000.

Kim, H. K., Yea, S. K., Ro, C. U., Lee, C. B., Jang, M., Lee, G., Yoo, E., and Han, J. S.: Determination of atmospheric perfluorocarbon background concentrations of fl/L Range at the western Coastal Area of Korea, B. Korean Chem. Soc., 23, 301–308, 2002.

Lagomarsino, R. J.: An improved gas chromatographic method for the determination of perfluorocarbon tracers in the atmosphere, J. Chromatogr. Sci., 34, 405–412, 1996.

Lin, J. C., Gerbig, C., Wofsy, S. C., Andrews, A. E., Daube, B. C., Grainger, C. A., Stephens, B. B., Bakwin, P. S., and Hollinger, D. Y.: Measuring fluxes of trace gases at regional scales by Lagrangian observations: Application to the CO_2 Budget and Rectification Airborne (COBRA) study, J. Geophy. Res., 109, D15304, doi:10.1029/2004JD004754, 2004.

Methven, J., Arnold, S. R., Stohl, A., Evans, M. J., Avery, M., Law, K., Lewis, A. C., Monks, P. S., Parrish, D., Reeves, C., Schlager, H., Atlas, E., Blake, D., Coe, H., Cohen, R. C., Crosier, J., Flocke, F., Holloway, J. S., Hopkins, J. R., Hübler, G., McQuaid, J., Purvis, R., Rappenglück, B., Ryerson, T. B., Sachse, G. W., Singh, H., Watson, N., Whalley, L. K., and Williams, P.: Establishing Lagrangian connections between observations within air masses crossing the Atlantic during the ICARTT experiment, J. Geophy. Res., 111, D23S62, doi:10.1029/2006JD007540, 2006.

Nodop, K., Connolly, R., and Girardi, F.: The field campaigns of the European Tracer Experiment (ETEX): Overview and results, Atmos. Environ., 32, 4095–4108, 1998.

Piringer, M., Baumann, K., Rötzer, H., Riesing, J., and Nodop, K.: Results on perfluorocarbon background concentrations in Austria, Atmos. Environ., 31, 515–527, 1997.

Ren, Y., Schlager, H., and Martin, D.: The application of TD/GC/NICI-MS with an Al_2O_3-PLOT-S column for the determination of perfluoroalkylcycloalkanes in the atmosphere, Chromatogr., 77, 309–316, 2014.

Sala, S., Bönisch, H., Keber, T., Oram, D. E., Mills, G., and Engel, A.: Deriving an atmospheric budget of total organic bromine using airborne in situ measurements from the western Pacific area during SHIVA, Atmos. Chem. Phys., 14, 6903–6923, doi:10.5194/acp-14-6903-2014, 2014.

Simmonds, P. G., Greally, B. R., Olivier, S., Nickless, G., Cooke, K. M., and Dietz, R. N.: The background atmospheric concentrations of cyclic perfluorocarbon tracers determined by negative ion-chemical ionization mass spectrometry, Atmos. Environ., 36, 2147–2156, 2002.

Straume, A. G., Dietz, R. N., Koffi, E. N., and Nodop, K.: Perfluorocarbon background concentrations in Europe, Atmos. Environ., 32, 4109–4122, 1998.

Stohl, A., Hittenberger, M., and Wotawa, G.: Validation of the Lagrangian particle dispersion model FLEXPART against large-scale tracer experiment data, Atmos. Environ., 32, 4245–4264, 1998.

Tegtmeier, S., Krüger, K., Quack, B., Atlas, E. L., Pisso, I., Stohl, A., and Yang, X.: Emission and transport of bromocarbons: from the West Pacific ocean into the stratosphere, Atmos. Chem. Phys., 12, 10633–10648, doi:10.5194/acp-12-10633-2012, 2012.

Tegtmeier, S., Krüger, K., Quack, B., Atlas, E., Blake, D. R., Boenisch, H., Engel, A., Hepach, H., Hossaini, R., Navarro, M. A., Raimund, S., Sala, S., Shi, Q., and Ziska, F.: The contribution of oceanic methyl iodide to stratospheric iodine, Atmos. Chem. Phys., 13, 11869–11886, doi:10.5194/acp-13-11869-2013, 2013.

Watson, T. B., Wilke, R., Dietz, R. N., Heiser, J., and Kalb, P.: The atmospheric background of perfluorocarbon compounds used as tracers, Environ. Sci. Technol., 41, 6909–6913, 2007.

Yang, Y. Q., He, J. B., Ye, S. Y., and Yang, T. Z.: Penetration capacity of trace perfluorodimethylcyclobutane through Carboxen 569 Adsorbent, Atomic Energy Science Technology, 39, 404–408, 2005.

A new airborne laser-induced fluorescence instrument for in situ detection of formaldehyde throughout the troposphere and lower stratosphere

M. Cazorla[1,*], G. M. Wolfe[1,2], S. A. Bailey[1], A. K. Swanson[1,3], H. L. Arkinson[4], and T. F. Hanisco[1]

[1]Atmospheric Chemistry and Dynamics Lab, NASA Goddard Space Flight Center, Greenbelt, MD, USA
[2]Joint Center for Earth Systems Technology, University of Maryland Baltimore County, Baltimore, MD, USA
[3]Goddard Earth Sciences Technology and Research, University of Maryland Baltimore County, Baltimore, MD, USA
[4]Department of Atmospheric and Oceanic Science, University of Maryland, College Park, MD, USA
[*]now at: Universidad San Francisco de Quito, Instituto de Investigaciones Atmosféricas, Colegio de Ciencias e Ingeniería, Diego de Robles y Vía Interoceánica, Quito, Ecuador

Correspondence to: T. F. Hanisco (thomas.hanisco@nasa.gov)

Abstract. The NASA In Situ Airborne Formaldehyde (ISAF) instrument is a high-performance laser-based detector for gas-phase formaldehyde (HCHO). ISAF uses rotational-state specific laser excitation at 353 nm for laser-induced fluorescence (LIF) detection of HCHO. A number of features make ISAF ideal for airborne deployment, including (1) a compact, low-maintenance fiber laser, (2) a single-pass design for stable signal response, (3) a straightforward inlet design, and (4) a stand-alone data acquisition system. A full description of the instrument design is given, along with detailed performance characteristics. The accuracy of reported mixing ratios is ±10 % based on calibration against IR and UV absorption of a primary HCHO standard. Precision at 1 Hz is typically better than 20 % above 100 pptv, with uncertainty in the signal background contributing most to variability at low mixing ratios. The 1 Hz detection limit for a signal / noise ratio of 2 is 36 pptv for 10 mW of laser power, and the *e* fold time response at typical sample flow rates is 0.19 s. ISAF has already flown on several field missions and platforms with excellent results.

1 Introduction

Formaldehyde (HCHO) is a ubiquitous constituent of the Earth's atmosphere. It is primarily produced during hydrocarbon oxidation, with an estimated source strength of $250 \pm 54\,\mathrm{Tg\,yr^{-1}}$ (Fortems-Cheiney et al., 2012). In the background troposphere, the oxidation of methane (CH_4) by hydroxyl radical (OH) is the main source. Photochemical degradation of anthropogenic and biogenic hydrocarbons greatly enhances HCHO production in the lower troposphere, with the most significant precursor being biogenic hydrocarbon isoprene. HCHO is also emitted through fuel combustion (Olaguer et al., 2009; Luecken et al., 2012), biomass burning (Yokelson et al., 2013) and vegetation (DiGangi et al., 2011). These sources are generally minor globally compared to secondary production, but they may be significant locally. HCHO is lost via photolysis and reaction with OH, with a typical daytime lifetime of 2–3 h. Wet and dry deposition are minor sinks globally, but can be significant locally, such as during precipitation. Typical mixing ratios range from tens of parts per trillion by volume (pptv) in pristine air to several hundred parts per billion by volume (ppbv) near combustion sources.

HCHO is a tracer for, and active participant in, multiple atmospheric processes. As a byproduct of hydrocarbon oxidation, HCHO can provide quantitative constraints on the photochemical link between primary emissions and secondary pollutants (Chatfield et al., 2010; Duncan et al., 2010). A number of studies have employed satellite-based HCHO observations to better constrain isoprene emission inventories (Palmer et al., 2003, 2006; Millet et al., 2008; Marais et al., 2012). Because of its short lifetime, the mixing ratio of HCHO in the upper troposphere is typically 10 to 100 times less than that near continental surfaces. Convection can pump surface air to the upper troposphere very rapidly, and HCHO is a valuable marker for the efficiency of this process (Barth et al., 2007; Fried et al., 2008) and a participant in ensuing chemistry (Apel et al., 2012). Furthermore, photolysis of HCHO is a potent source of radicals and thus helps to propagate chemical cycles (Edwards et al., 2011, 2013).

A handful of techniques are currently available to measure HCHO in the atmosphere; a review of these methods and inter-comparisons can be found elsewhere (Gilpin, et al., 1997; Cardenas et al., 2000; Hak et al., 2005; Wisthaler et al., 2008; Kaiser et al., 2014). Laser-based and spectroscopic methods are generally employed when high sensitivity and time response are required. The most common among these include tunable diode laser-based absorption spectroscopy (Weibring et al., 2007; Fried et al., 2008), differential optical absorption spectroscopy (DOAS) (Baidar et al., 2013) and laser-induced fluorescence (LIF) (Hottle et al., 2009). Diode laser spectroscopy and DOAS rely on molecular absorption in the IR and UV regions, respectively, while LIF utilizes UV-stimulated emission. DOAS can be applied for in situ or remote (column-averaged) observations, and a similar principle underlies satellite-based HCHO observations (Chance et al., 2000). Indeed, HCHO is one of only a few hydrocarbons observable from space, and high-quality in situ observations are critical for retrieval validation.

Instrumentation for airborne observations is held to stringent engineering standards. An ideal instrument is small, lightweight, low power, and able to withstand vibrations and large temperature variations. Low maintenance, calibration stability, and fast time response (≥ 1 Hz) are also desirable. For HCHO, a substantial dynamic range – roughly 4 orders of magnitude – is also necessary to adequately sample mixing ratios throughout the troposphere and lower stratosphere. Here, we present the NASA In Situ Airborne Formaldehyde (ISAF) instrument, a new LIF-based HCHO detector that meets or exceeds all of the above requirements. After a brief synopsis of the LIF method in Sect. 2, we provide a detailed description of instrument construction (Sect. 3) and performance characteristics (Sect. 4). Section 5 gives an example of observations taken during a recent field campaign.

2 Principle of operation

LIF is a well-established spectroscopic method for selective measurement of trace gases. LIF detection of HCHO was first reported by Becker et al. (1975) and refined by Möhlmann (1985). Hottle et al. (2009) applied the technique to in situ atmospheric measurements. In these approaches a tunable laser is used to ensure specificity to HCHO and a pulsed operation is used to minimize background noise. Figure 1 illustrates the key processes involved. A narrow-band laser, tuned to a specific rovibronic absorption line in the 353–355 nm region of the A–X electronic transition, promotes a fraction of the HCHO molecules to the first excited electronic state. The majority of these excited HCHO molecules are quenched back to the ground electronic state by collisions with air. A small fraction ($\sim 1\,\%$) undergo vibrational relaxation and subsequently relax to the ground electronic state through fluorescence, emitting a structured band of photons in the range of 400–470 nm (Becker et al., 1975; Shibuya et al., 1979; Möhlmann, 1985). Fluorescence occurs with an exponential decay over several hundred nanoseconds (Fig. 1, right). The number of fluorescence photons per unit of laser power is proportional to the mixing ratio of HCHO in the sample gas. This relationship is readily quantified with a known standard.

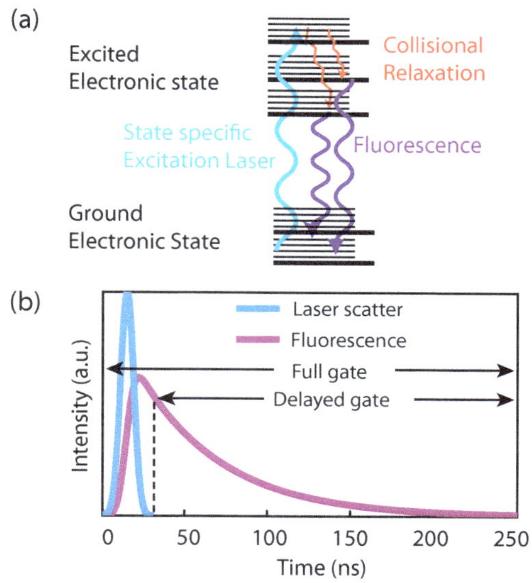

Figure 1. The spectral **(a)** and temporal **(b)** features of laser induced fluorescence detection. In **(a)** the laser excites a single rotational transition at 353.16 nm. The fluorescence occurs after collisional relaxation and appears at longer wavelengths. In **(b)** time gating of fluorescence detection provides additional discrimination between prompt and longer timescale processes. Laser excitation and elastic scattering processes that include Raman, Rayleigh, and chamber reflections have timescales that match the laser pulse (25 ns full-width). Fluorescence has a lifetime of tens of nanoseconds and the signal extends for more than 100 ns beyond the end of the laser pulse and elastic scattering. The delayed gate captures HCHO fluorescence while the full gate captures both fluorescence and scattered light.

For ISAF, the chosen absorption feature is centered at 353.163 nm, which has a cross-section of 3.77×10^{-19} cm^2 at 150 Torr (Co et al., 2005). While this cross-section is

~7 % lower than the peak value in this band, it is chosen because of the presence of several unique adjacent peaks, notably a triplet feature (see Fig. 6). This area of the spectrum serves as a "fingerprint" for HCHO, simplifying the wavelength-tracking algorithm. The offline position is only 0.005 nm away from the peak. This wavelength difference provides a high differential cross-section for HCHO, but the differential is negligible for species that have non-structured absorptions in this region. This allows for precise and selective measurements of HCHO. To date, we have not identified any interfering species that has a measurable differential absorption and fluorescence between the online and offline positions.

3 Instrument description

ISAF is comprised of four main subsystems: a UV fiber laser, an optical bench comprising the detection and reference cells, gas handling components (pressure controller, inlet system, tubing, valves and pump), and the data acquisition computer. Each of these subsystems is described in greater detail below. Aside from the vacuum pump, all components are contained in a single custom-built chassis with dimensions of 60 cm length × 43 cm width × 35 cm height. The layout of the chassis is shown in Fig. 2. This chassis is designed to run on 28 V DC, which is supplied directly from aircraft power or through an external AC/DC interface unit. The instrument itself weighs 25 kg and draws 200 W. An external pump, computer, and control box are used for operation on larger aircraft with on-board operators. Including all peripherals, the total instrument package weighs 60 kg and draws a peak power of 600 W.

3.1 Laser

Pulsed, tunable UV laser radiation is provided by a NovaWave (now Thermo Scientific) TFL 3000 fiber laser. The laser uses a master oscillator power amplifier architecture to provide narrow wavelength, tunable, pulsed light at 353 nm. A single distributed feedback laser operating at 1060 nm is chopped with an acousto-optic modulator and amplified in two stages to 2 W with a series of diode-pumped, semiconductor-doped, polarization-maintaining optical fibers. The 1060 nm output is fiber-coupled into a harmonic generation module containing two periodically poled nonlinear crystals and collimating optics. The harmonic generation housing is mounted on the optical bench (Fig. 2) and coupled with free-space optics. The repetition rate is 300 kHz, with a pulse width of 25 ns and a typical output power of 10–40 mW. The output wavelength is tunable to ~0.0001 nm, permitting resolution of individual rovibronic absorption features. Computer control is provided via RS232. The entire unit weighs ~12 kg and draws < 60 W,

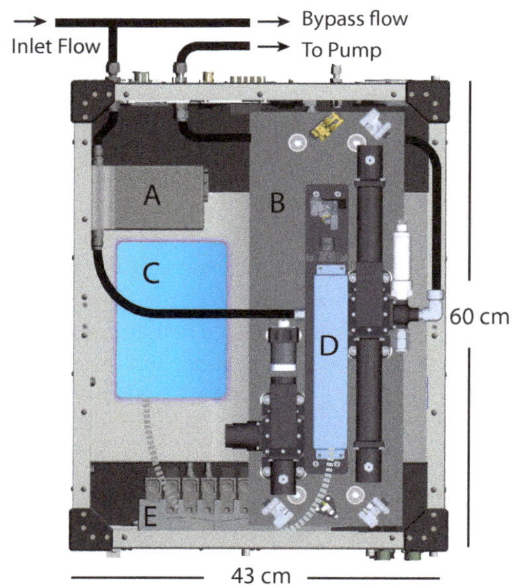

Figure 2. A top view of the ISAF instrument layout. The instrument chassis is 60 × 43 × 35 cm and 25 kg. The major subsystems are labeled: (A) pressure controller, (B) optical bench, (C) tunable fiber laser, (D) harmonic generation housing, and (E) data acquisition system.

making it an attractive alternative to larger laser systems typically used for LIF.

3.2 Optical bench

The optical bench is shown in greater detail in Figs. 3 and 4. The bench is an aluminum sandwich panel with 1.25 cm thick honeycomb core and with 0.1 cm cladding. The bench is mounted to the instrument chassis with vibration damping mounts (Sorbothane) to minimize vibrations. Figure 3 shows the optical bench and the path of the laser and gas flow. Figure 4 shows a cross-sectional view of the detection cell. The simplicity and single-laser pass design of this system is a major advantage for field applications; alignment rarely degrades throughout flight campaigns or during shipping. Likewise, the instrument shows no effects in high vibration environments or in aircraft maneuvers. Both cells are heated to a constant temperature of 35 °C, and capability exists to heat other parts of the instrument if needed (i.e., for high-altitude applications).

UV laser light is collimated at 2 mm diameter at the output of the harmonic stage with a lens (Thor LA4600-UV, $f = 100$ mm) and directed to the detection cell via four dielectric-coated mirrors (Edmund 47-323). Two of these mirrors, along with the two between the detection and reference cells, utilize piezo-driven mounts (Newport Agilis) for remote alignment. The laser beam enters and exits the reduced-pressure detection cell (Fig. 4) through V-type antireflection-coated fused silica windows (Edmund 48-207)

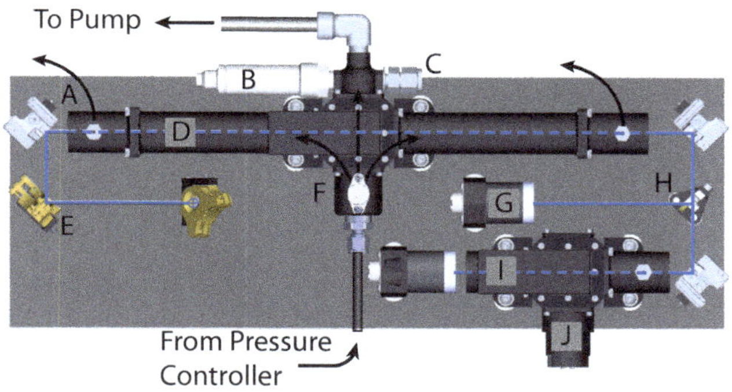

Figure 3. This diagram shows a top view of the optical bench with the harmonic generation housing and platform removed. The laser propagation is indicated by the solid (open to air) and dashed (inside the cells) blue lines. The sample flow is indicated by the black arrows. The components are (A) piezo actuated mirror (3); (B) pressure sensor; (C) thermistor; (D) sample detection cell; (E) fixed mirror; (F) sample heater; (G) reference cell (2); (H) beam splitter; (I) reference cell; (J) formaldehyde source.

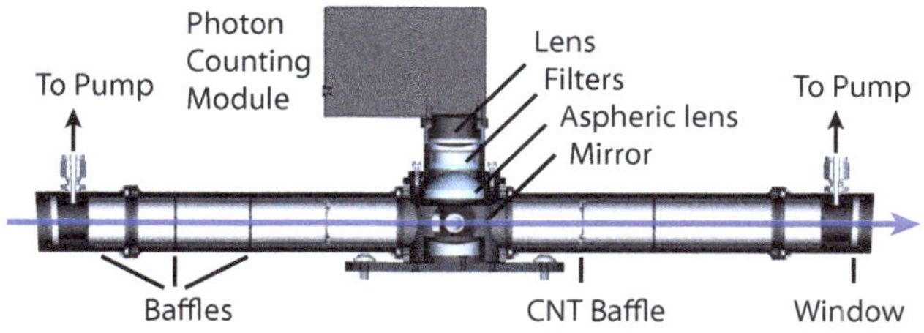

Figure 4. This diagram shows a side-view of the cross-section of the detection cell. The blue line indicates the laser path and propagation direction.

tilted at 5° to redirect reflections from the window surfaces out of the beam propagation path. The arms of the detection cell contain a series of circular baffles with 2.5–3.5 mm apertures to eliminate stray light. Baffles and cell walls are coated with a combination of Avian black paint (Avian Technologies), black anodize, and molybdenum oxide treated aluminum (Insta-Black 380, EPI). The innermost baffles are coated with an ultra-black custom carbon nanotube treatment (Hagopian, 2011). The laser beam, gas flow, and focus of the collection lens intersect orthogonally in the center of the cell. Along this detection axis, photons are imaged with an aspheric lens (Edmund 6-321) through a 370 nm long pass absorption filter (Edmund 66-049) and a 400 nm long pass interference filter (Omega 3RD400LP). The image is partially focused with another lens (Thor LA1608, $f = 75$ mm) onto a photomultiplier tube (PMT) (Hamamatsu 7360). A spherical mirror (Thor CM254, $f = 25$ mm) is located opposite the aspheric lens to improve collection efficiency. Since this is a fluorescence measurement, only the sample inside the imaging volume of the collection lens (the center of the cell) is detected.

After the detection cell, a beam sampler (Thor BSF05-UV) diverts $\sim 4\,\%$ of the laser to a power monitor comprised of a diffuser (Thor DGUV10-600), filter (Thor FGUV11), and amplified photodiode (OSI 555-UV). The photodiode has a large active area (1 cm^2) to ensure that the entire laser beam is detected. The diffuser is needed to spread the laser over enough of the photodiode to avoid inhomogeneities in the photodiode responsivity. The main laser beam is directed into a reference cell (Fig. 3, lower right), which is used to monitor the laser wavelength. This cell is nearly identical to the sample cell shown in Fig. 4; the main differences are shorter arms and a neutral density filter (Thor NE30A) to limit fluorescence signal intensity. A heated cartridge filled with powdered urea formaldehyde glue (DAP Weldwood) provides a large concentration of HCHO (~ 1 ppmv). This cell is static (no gas flow) but is maintained at the same pressure as the sample cell by coupling the arms to the sample cell exhaust. This is critical for maintaining the same absorption line width in both cells, as the reference cell signal is used to correct sample cell data (see Sect. 3.5). A normally closed solenoid valve and an activated charcoal scrubber are placed inline to limit sample cell contamination. The laser beam is ultimately dumped into a second power monitor, which is primarily used to gauge laser alignment.

3.3 Gas handling

The primary components of the air sampling system include the aircraft inlet, pressure controller, and vacuum pump. Some care must be taken with the inlet system to minimize adsorption and production/loss of HCHO on surfaces (Wert et al., 2002). Outgassing of HCHO from tubing walls, physisorption and desorption of HCHO, and the chemical or catalytic conversion of oxygenated hydrocarbons on metal surfaces to HCHO are primary concerns in the design of the sampling system. Fast flows, low surface area fluorocarbon or fluorocarbon-coated surfaces, and minimal dead space are the design features. For example, there are no valves or sensors between the tip of the inlet and the instrument; the sampling path is continuous tubing with VCO (Swagelok) fittings. Furthermore, all metal surfaces are coated with fluorocarbon (FluoroPel, Cytonix), all tubing is PTFE or FEP, and pressure, temperature, and flow are measured after the detection cell.

Figure 5 depicts the inlet design used on the NASA DC-8 aircraft. The inlet tube is a 45 cm length of 0.48 cm ID Silcosteel (Restek) coated with FluoroPel mounted at 90° to the airflow to reject particles. The tip of the inlet is chamfered at 15° to enhance the particle rejection and is heated to 30–60 °C inside the winglet to prevent condensation and sticking. A tapered aluminum nacelle oriented parallel to the direction of flight serves as a flow straightener and generates a ram pressure equal to ~ 1.15 times that of ambient at the inlet tip, driving an air flow of 10–25 sLm through the inlet. The corresponding volumetric flow at typical pressures and tem-

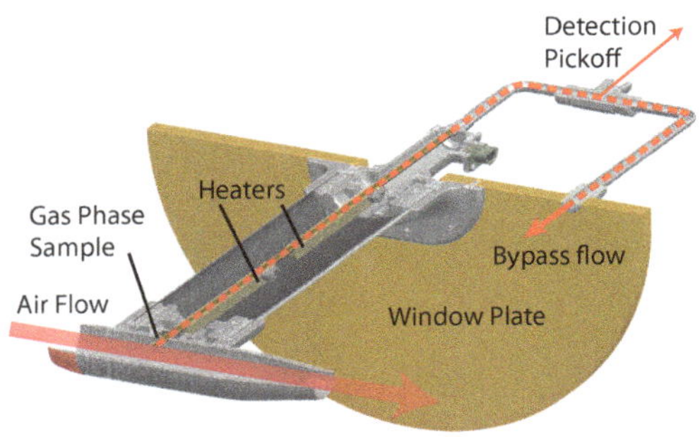

Figure 5. This diagram shows a cross-section view of the particle rejection inlet mounted on a window plate. A ram air nacelle mounted on an airfoil 30 cm from the plate generates ram pressure that directs the ambient air through the heated tube mounted at 90° to the flow. A small fraction (2 sLm) of the heated air enters the instrument and the remainder (~ 20 sLm) returns to ambient. The plumbing on the cabin side of the window plate is not drawn to scale. There is 1 m of tubing between the inlet and detection pick-off.

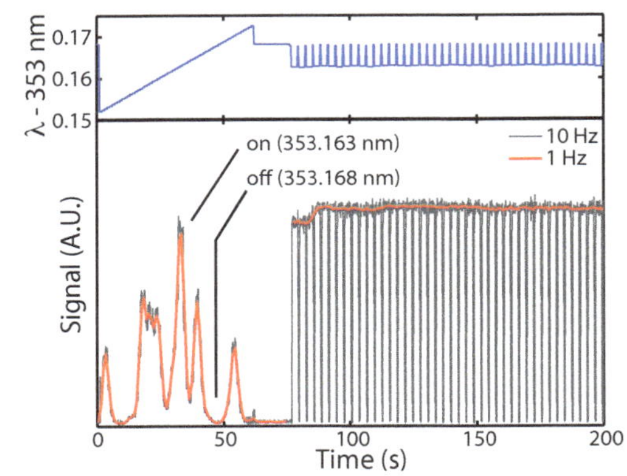

Figure 6. These plots show the wavelength (top) and fluorescence signal (bottom) during a scan of the laser, followed by the standard data acquisition mode. The computer algorithm uses the reference cell signal to maintain the laser on the large rotational feature at 353.163 nm. The contributions of species other than HCHO to the fluorescence signal are removed by subtracting the signal at the offline position. The difference between the signal at the online position and the offline position is proportional to [HCHO].

peratures is 25–65 L min^{-1}. Inside the aircraft, air is carried via 1 m of 0.7 cm ID PTFE tubing to the instrument chassis and is sub-sampled at 2–4 sLm. The excess flow passes through a flow meter (Honeywell) and exits behind the inlet. The residence time in the heated Silcosteel portion of the tube is 8–20 ms, with an additional 35–100 ms in the transfer tubing.

Inside the chassis, sample air passes through a pressure controller (MKS 640A) heated to 35 °C, followed by a ~ 20 cm length of 0.4 cm ID PFA tubing connected to the detection cell. The residence time in the detection cell is ~ 115 ms. Sample pressure, temperature, and flow rate are monitored immediately downstream of the cell to minimize surface area. A small amount of sample air (~ 100 $cm^3\ s^{-1}$) is drawn through the laser arms to reduce contamination from painted surfaces or dead space. Sample air is exhausted through the vacuum pump (Agilent IDP-3 or Vacuubrand MD-1). During instrument shutdown, contamination is minimized by backfilling the cell with ambient air scrubbed by a Drierite/molecular sieve cartridge, which reduces HCHO to < 100 pptv.

3.4 Data acquisition

All hardware interfacing is handled with an all-solid-state CompactRIO (National Instruments) with realtime operating system and a field programmable gate array (FPGA). The FPGA and acquisition software are configured for fast photon counting and autonomous operation. Six 200 MHz counters are used to integrate PMT signals at a nominal rate of 10 Hz. Pulses from the sample cell PMT are sampled by three of these counters, one non-gated (continuous) and two each with a separate gate (Fig. 1, right). The "full" gate samples both scattered laser light and HCHO fluorescence, while the "delayed" gate excludes the laser. The reference cell gate is the same width as the full gate. Laser power is also monitored at 10 Hz. Other diagnostics (laser parameters, temperature, pressure, flow, GPS, and aircraft data) are recorded at 1 Hz.

Figure 6 illustrates the standard data acquisition routine. Online counting is done for 2.5 s (25 points), followed by 0.5 s (5 points) offline. Laser drift (mainly due to temperature of the seed laser diode) requires an active line-locking algorithm. The online position is alternately shifted to a slightly (±0.0001 nm) lower or higher wavelength on adjacent online cycles, and the reference cell is monitored for maximum signal. When the laser is tuned precisely to the online position the reference cell is at the maximum signal. When the laser is tuned 0.0001 nm (or 0.0002 nm) off of the peak position the signal drops by about ~ 5 % (or ~ 10 %). The resulting small changes in the reference cell signal are used to maintain the laser on the peak of the line. The signal is also used to correct the sample signal by the ~ 5 % (or ~ 10 %) needed to normalize the signal to the signal at the peak position.

Once per hour, a full scan over the HCHO fingerprint is recorded. This is primarily used to optimize the harmonic generation crystal temperatures and as a performance diagnostic. For example, the presence of interfering species can be detected in the spectrum. To date, we have not detected any features other than those shown in Fig. 6. This is not surprising. To be detected, the potential species must absorb at 353.163 nm, fluoresce in the 400–470 nm band, and have a significant fluorescence lifetime.

3.5 Data reduction

Both the full and delay gate signals can be used to derive HCHO mixing ratios. The delay gate excludes $\sim 53\,\%$ of the fluorescence signal, but it is also relatively insensitive to laser scatter (from Raman, Rayleigh, or particulates) and gives superior signal / noise (S / N) ratios below ~ 550 pptv HCHO. Thus, the delay gate signal is typically used, and all results discussed below stem from this signal. The full gate data provides a useful metric for scatter and other potential laser-related issues.

The detected signal is proportional to the abundance of HCHO and the laser power. To the first order, we can determine HCHO from the ratio of the laser power-normalized difference between the online and offline signals and the calibration factor. To minimize systematic noise, however, we use additional steps to convert raw counts to mixing ratios. These analysis steps are (1) normalization by laser power; (2) background subtraction, determined by a smoothed interpolation of offline signals; (3) normalization by the reference cell signal to account for deviations from the maximum absorption wavelength due to the line-locking algorithm; and (4) application of a pressure-dependent calibration factor (see Sect. 4.1).

4 Performance

4.1 Sensitivity

Detection sensitivity (signal response per unit of HCHO) is a function of laser power, collection efficiency, PMT response, and counter gating. As mentioned above, signals are normalized by laser power. Because of the single-pass design, laser alignment does not affect sensitivity except in extreme cases when the laser is partially blocked. In practice, the laser maintains alignment for entire field campaigns and usually during shipments to and from deployment locations. During flight, the ratio of the two power monitors (at the exits of the detection and reference cells) is constant to $\pm 3\,\%$.

Collection efficiency depends on the physical characteristics of the sample cell, including the types of collection optics (lenses and filters) and their relative spacing. We do not expect the collection efficiency to change under normal operation. However, catastrophic events, such as the degradation of optics from contamination could reduce the sensitivity. The responsivity of the PMT should be stable in the short term (months to years), but it can degrade over the lifetime of the PMT. The gated counter timing is a fixed parameter and is optimized for signal / noise. Changes in the gate timing relative to the laser pulse could affect the sensitivity. These changes are easily diagnosed in data reduction and can be corrected by using the non-delayed gates.

In practice, instrument sensitivity is determined via calibration against a primary standard, consisting of a 500–700 ppbv mixture of HCHO in N_2 (Scott Gas/Air Liquide). This mixture is calibrated every few months (typically before and after a field deployment) with an MKS Multigas 2031 Fourier transform infrared spectrometer using manufacturer-provided IR cross-sections tied to gravimetric permeation tube calibrations. The 1σ uncertainty in mixing ratio from any single determination is typically 2–3 %. HCHO mixtures are also stable over time; for example, five calibrations performed on a single tank over the course of a year give a mixing ratio of 611 ± 8 ppbv and show no trend.

To better constrain the calibration to literature standards, a series of experiments were performed to compare concentrations derived from Fourier transform infrared spectroscopy (FTIR) and long-path UV absorption. The UV absorption cell was a custom-built, 200 cm long, 1.1 cm diameter stainless-steel tube fitted with windows, gas ports, and a pressure gauge. Visible light from a Cathodeon deuterium lamp was filtered (Edmund U-330) and focused through the cell and measured with an Ocean Optics Maya 2000 spectrometer. The cell pressure was held above atmosphere (1110–1150 Torr) with a gas flow rate of 6 sLm. Absorption measurements were carried out simultaneously in both the IR and UV systems using a high concentration (nominally 11 ppmv) HCHO standard. Literature UV cross-sections (Meller and Moortgat, 2000) in the 285–370 nm range were fit to the measured absorption spectra (nominal resolution ~ 1 nm) to obtain HCHO concentrations and mixing ratios. Based on 11 such experiments, the mixing ratios from UV and IR are 10.49 ± 0.20 ppmv and 10.96 ± 0.02 ppmv, respectively. The ratio of UV / IR $= 0.9571$ is applied to all IR-derived values; for example, the IR-derived average mixing ratio of 611 ppbv is corrected to 585 ppbv. By making this adjustment, our calibration is tied to the Meller and Moortgat (2000) cross-section rather than the MKS spectrometer.

Calibrations of ISAF are achieved via standard additions to zero air. Zero air is delivered either from an ultra zero air cylinder or from a zero air generator (Sabio 2020) with an additional Drierite/molecular sieve scrubber to reduce HCHO mixing ratios to 30 pptv or less. Small flows of the primary HCHO standard (1–50 $cm^3\,s^{-1}$) are diluted into a 5 sLm flow of zero air to yield HCHO concentrations in the range of 0–5 ppbv. Signal is plotted against the diluted HCHO mixing ratio, and the slope of this relationship gives the instrument sensitivity (see inset of Fig. 7). This sensitivity varies with sample cell pressure due to the combined effects of number density, absorption line broadening, and quenching. Figure 7 shows this pressure dependence as determined from a number of standard addition experiments. Peak sensitivity occurs at a cell pressure of 200 mbar. In flight, the system is configured to maintain a constant pressure that is dictated by the ceiling of the aircraft. For example, on the NASA DC-8, the cell pressure is typically held at 100 mbar, and the sensitivity is 75 counts $s^{-1}\,mW^{-1}\,ppbv^{-1}$. Propagation of uncertainties in the HCHO concentration, determined by both the absolute

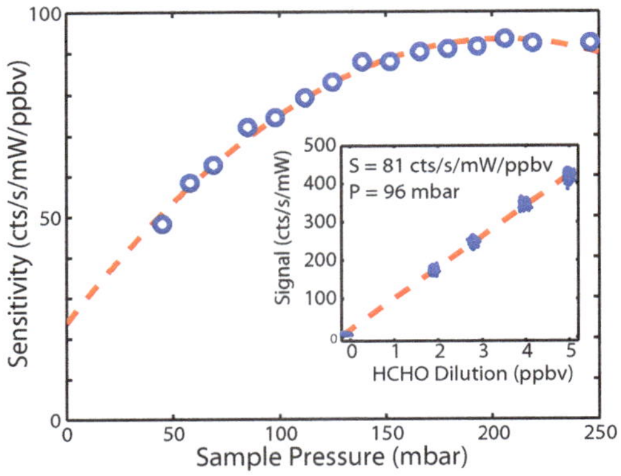

Figure 7. This plot shows the sensitivity of the instrument to HCHO versus pressure. The inset shows an example of a calibration using an HCHO standard addition to a flow of air. Several calibrations at one pressure are used to determine the pressure dependence.

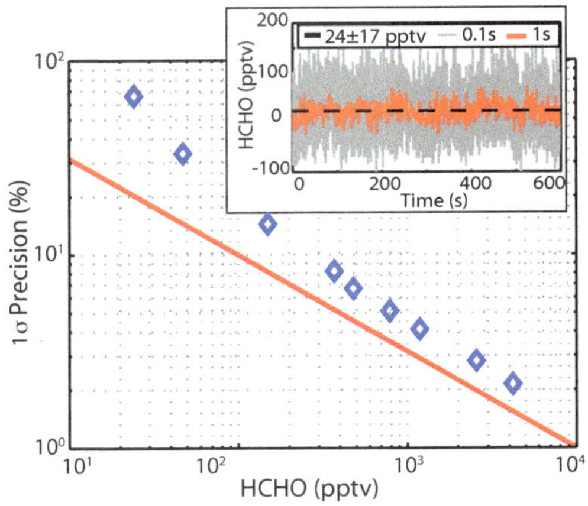

Figure 8. This plot shows the measured and theoretical 1σ precision as a percentage of HCHO concentration. The inset shows an example of the mean and standard deviation for constant HCHO. The average value (24 pptv) and the standard deviation (17 pptv) are used to determine the measured precision (70 % in this case) shown by the blue diamonds. The solid line is the theoretical maximum precision based on the counting statistics of the signal level.

UV absorption measurement and dilution flow rates, gives an estimated accuracy of $\pm 10\,\%$.

Calibrations are generally carried out in the laboratory both before and after field missions. Because HCHO adheres to surfaces in tubing, fittings, regulators and flow meters, several days are needed to ensure consistent results, especially with new PTFE or FEP tubing. Calibrations can be performed in the field during campaigns, if needed. In practice, these calibrations are used to monitor instrument performance. We have never changed the instrument sensitivity calibration numbers during a campaign. Based on four field campaigns spanning 2 years, sensitivity varies by less than 10 % over a single mission and by less than 30 % between missions. Given the difficulty of adding a known amount of HCHO, it is likely that the sensitivity of the instrument was constant over those 2 years and that the variability in the calibrations is due to differences in calibration conditions and personnel.

4.2 Precision

The instrument precision is primarily limited by processes that generate noise or affect the background (offline) signal. The largest potential sources of noise, Rayleigh and Raman scattering of laser light, are reduced with a combination of optical filters, absorbing surfaces (e.g., baffles), and gated photon counting. Laboratory tests indicate that scattering contributes less than 7 counts s^{-1} mW^{-1} to the background signal at 100 mbar.

Figure 8 shows the short-term precision estimated from the normalized standard deviation ($\sigma_x / < x >$) for data segments of 10 to 15 min at a range of constant mixing ratios. Also displayed is the inverse square root dependence expected from Poisson counting statistics. Above mixing ratios of ~ 100 pptv, the precision is better than 20 % and exhibits the expected slope but is somewhat higher than the Poisson limit. Some fraction of this offset is due to uncertainty in the background. This would also explain the reduced precision observed at the lowest mixing ratios. It is also likely that some "extra" variability is contributed by the calibration gas addition system, which would be directly proportional to the HCHO mixing ratio. We note that small changes in room temperature can cause small fluctuations in the amount of HCHO added to the calibration gas flow, presumably from increased or decreased desorption of HCHO from surfaces (i.e., regulator, valves, flow meters, and fittings). The true instrument precision thus lies between the measured and predicted values in Fig. 8.

In theory, time-averaging will improve instrument precision. Figure 9 shows the mean-normalized Allen deviation for 60 min of sampling at a constant mixing ratio of 5.1 ppbv. At short averaging times, precision improves as the inverse square root of integration time, consistent with white noise as the dominant source of variability. Averaging from the native data rate of 0.1 s to the typical reporting interval of 1 s improves precision by a factor of 3. Averaging to the "optimal" interval of 130 s improves precision by a factor of 18, although natural variability would likely dominate on this timescale in the real atmosphere.

4.3 Detection limit

For counting statistics, the signal / noise ratio is defined as

$$\frac{S}{N} = \frac{pC_f[\text{HCHO}]_t}{\sqrt{pC_f[\text{HCHO}]t + 2pBt}}, \tag{1}$$

where p is laser power, C_f is the calibration factor (sensitivity), [HCHO] is the mixing ratio, B is the background count

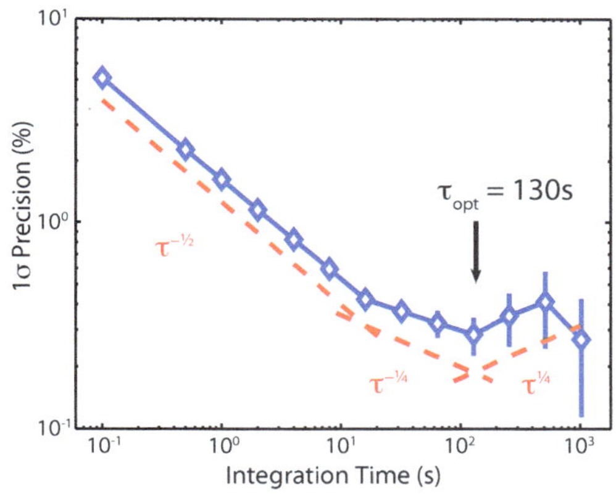

Figure 9. The mean normalized Allen deviation is shown for 60 min of sampling at a constant HCHO concentration of 5.1 ppbv. The short integration times show a $\tau^{-1/2}$ increase in precision, consistent with white noise.

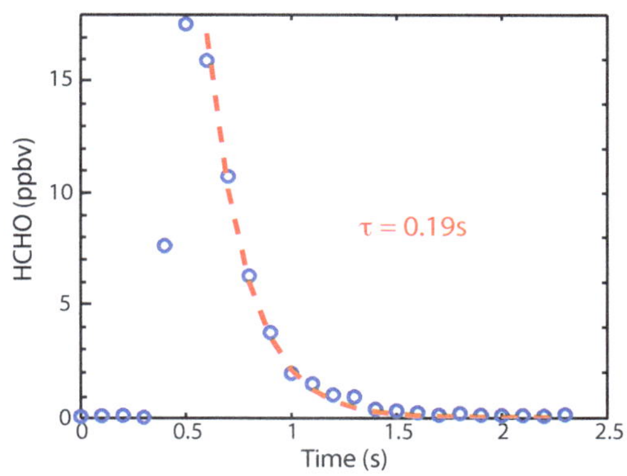

Figure 10. The time response of the instrument is demonstrated with a 10 ms pulse of HCHO added to the sample air flow in the lab. The measured HCHO shows the rapid rise expected with the flush time of ~ 115 ms. The decay of 190 ms indicates second-order effects such as wall interactions and the presence of small volumes with longer flush times.

rate, and t is the integration time. For a typical laser power of 10 mW, a sensitivity of 75 counts s^{-1} mW^{-1} $ppbv^{-1}$, a background count rate of 8 counts s^{-1} mW^{-1}, and an integration time of 1 s, the estimated detection limit is 18 pptv for S / N = 1 or 36 pptv for S / N = 2. This value agrees well with the 1σ variability of 17 pptv observed at low mixing ratios (inset of Fig. 8). Time averaging will improve the detection limit in proportion to the precision (Fig. 9). The detection limit scales as the inverse square root of laser power. The laser is capable of producing as much as 40 mW of power; thus, ISAF can potentially achieve a factor of two reduction in the detection limit by operating at higher laser power. For example, the instrument operated at 20 mW during its first deployment and achieved higher S / N. More recently, however we choose to operate at reduced power (10 mW) to extend the life of the laser diodes and nonlinear crystals.

The inset of Fig. 8 is a typical example of the signal measured with zero air added to the calibration system. Zero air and ultra zero air have 20–50 pptv HCHO, but we can obtain lower values by passing the air through a molecular sieve. With clean FEP or PFA tubing, measurements of zero or near-zero are easily obtained, even after exposure to 10 or 20 ppbv of HCHO. The ISAF instrument has no measurable chemical offset or bias of HCHO and shows little if any "memory" of exposure to high concentrations of HCHO (see Sects. 4.4 and 4.5 for more detail). However, the uncertainty shown in the inset of Fig. 8 is typical at these low concentrations. At very low fluorescence photon count rates, small changes in the laser scatter between the online and offline position can lead to an offset in the measurement. This is because we assume that the laser scatter is equal at the online and offline wavelengths. Though the wavelength difference is small (0.005 nm) there is some laser pointing change due to walk-off in the harmonic generation stage between these wavelengths, and potentially a small difference in laser scatter. This difference can result in a negative or positive offset. We assign an uncertainty of ±10 pptv to account for this potential offset.

4.4 Time response

Though the nominal sampling frequency is 10 Hz, the flush time of sampling volumes can limit the effective instrument time response. Minimizing the response time is critical for applications that require measurements of fast changes in concentration, such as eddy covariance and sampling of discrete plumes and for eliminating contamination from exposure to high concentrations of HCHO. For typical inlet pressures and flow rates, the plug-flow transit time from the tip of the inlet to the instrument chassis is 43–120 ms. This creates a delay between sampling and detection but should not cause significant "smearing" of concentration fluctuations, as the flow is typically turbulent ($Re \sim 3000$–6000). Such smearing can, however, occur in the pressure controller and sample cell. Based on the internal cell volume of 60 cm^3 and a typical volumetric flow rate of 31 L min^{-1} (2.75 sLm at 100 mbar and 35 °C), the expected $1/e$ cell flush time is 115 ms. Figure 10 shows the instantaneous response to a 10 ms pulse of HCHO, delivered via a fast switching valve (IEP series, The Lee Company). The observed decay time is 190 ms, somewhat slower than that predicted from flow characteristics. The extra delay is likely due to the pressure controller, which acts as a critical orifice and adds to the internal volume of the sampling system. A faster time response can be achieved by increasing the flow rate through the sample cell, which depends on the conductance of the pressure controller and the pumping speed.

The fast time response is an important feature of contamination-free detection at low concentrations of HCHO. HCHO continuously adsorbs and desorbs from sur-

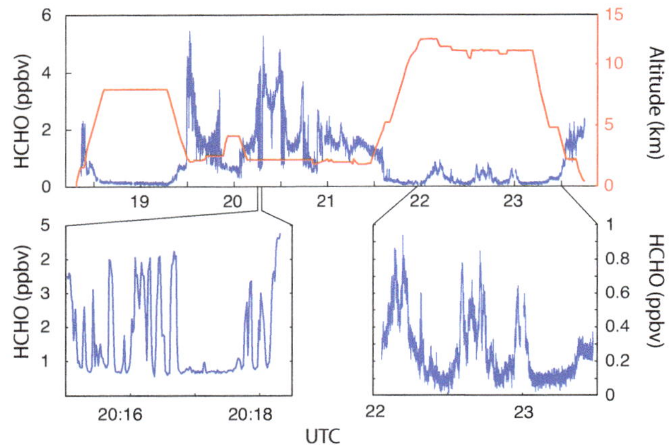

Figure 11. In situ measurements of HCHO are shown for the flight of 2 June on the NASA DC-8 during the Deep Convective Clouds and Chemistry (DC3) campaign in 2012. The top shows 1 s HCHO and GPS altitude. The lower portions are expanded views of the boundary layer (BL) leg and the upper tropospheric (UT) leg. The fast response to large changes in ambient HCHO are evident in the BL leg. The sensitivity (better than 20 pptv s^{-1}) is evident in the UT leg.

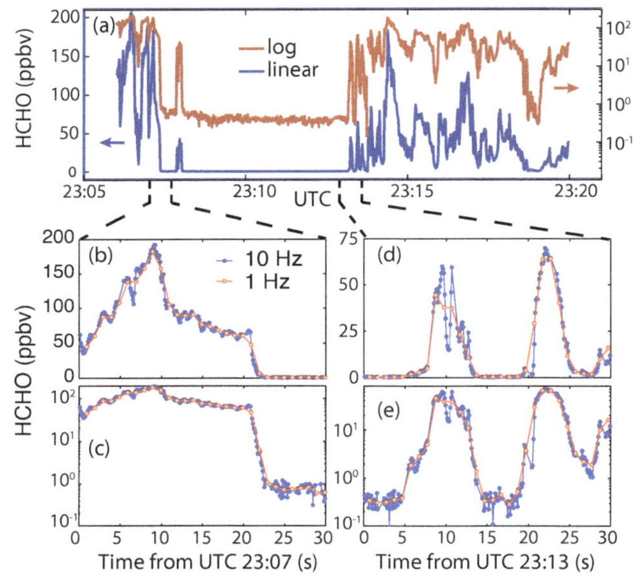

Figure 12. The time series obtained during the SEAC4RS campaign on the NASA DC-8 on 26 August 2013. The panel **(a)** shows the HCHO measured at a constant pressure altitude of 5 km as the DC-8 sampled the plume from the Rim Fire in Stanislaus National Forest as the plume traveled eastward over California and Nevada. Note the linear and logarithmic scaling. Panels **(b)** and **(c)** show a window of 30 s starting at 23:07 UTC expanded with data averaged at 10 and 1 Hz for linear and logarithmic scaling. Panels **(c)** and **(d)** show the same for the 30 s window starting at 23:13 UTC.

faces inside the instrument sampling system. This reversible process can lead to biases, especially when low concentrations are sampled after exposure of the instrument to very high concentrations. The effects of reversible adsorption and desorption of HCHO from surfaces is minimized by having a large mass flow, small volumes, and low surface area, the same features that lead to the fast time response. The measured time response includes both the effects of volumetric displacement and adsorption/desorption. The impact on in situ measurements is demonstrated in Sect. 4.5.

4.5 Field demonstration

As of March 2014, ISAF had flown successfully on four field missions spanning three platforms and a wide range of chemical conditions. For many of these flights, the instrument has flown autonomously, without an operator. Figure 11 presents observations from a single flight on the NASA DC-8 during the Deep Convective Clouds and Chemistry (DC3) experiment, which took place over the central US in May and June 2012 (Barth et al., 2015). On this flight, the DC-8 sampled at altitudes of 0–12 km, with HCHO mixing ratios ranging from 50–5500 pptv. The expanded views in Fig. 11 show a boundary layer leg and an upper tropospheric leg to illustrate the time response, dynamic range and sensitivity of ISAF. During DC3, ISAF was flown simultaneously with a well-established instrument, the University of Colorado DFGAS (Difference Frequency Generation Absorption Spectrometer) (Weibring et al., 2007). Preliminary analysis indicates excellent agreement between these two instruments, and a rigorous inter-comparison will be the subject of a future publication.

The time response of the instrument in flight is illustrated in Fig. 12. These data were obtained during the Studies of Emissions and Atmospheric Composition, Clouds and Climate Coupling Regional Surveys (SEAC4RS) campaign in 2013. During this flight on 26 August, the DC-8 sampled the smoke plume from the Rim Fire in Stanislaus National Forest as it passed from California to Nevada. In this plume, measured at a constant pressure altitude of 5 km, the HCHO abundance reached over 200 ppbv. The data is shown in linear and log scale to illustrate the several orders of magnitude change in HCHO during this segment. The observations demonstrate the ability of ISAF to measure low concentrations of HCHO even after exposure to very high concentrations for several minutes.

The expanded views provide 30 s snapshots of two regions where the abundance of HCHO is highly variable. The 10 Hz data is shown along with the 1 Hz data to demonstrate the fast response of the instrument. These 10 Hz data show much more structure than the 1 Hz and at the edges of plume filaments the transitions are much steeper. For example, the edge of the plume at time = 22 s in Fig. 12b and c drops from an abundance of 65 to 25 ppb ($1/e$) in 0.3 s. If this were limited by instrument response alone, that is if the plume were

Table 1. Summary of the ISAF instrument capability.

Size	38 × 43 × 60 cm
Weight	25 kg
Power	200 W
Precision	20 pptv s^{-1}
Time response	< 0.2 s
Accuracy	±10 %
Data rate	1 s (100 ms on request)
Operation on the DC-8 uses pump (10 kg + 300 W) and a control interface box (10 kg)	
Operation on the P3 and GV uses pump (5 kg + 200 W) and a control interface box (10 kg)	

to have an abrupt boundary, the corresponding instrument time constant would be 0.3 s. Several filaments have steeper edges, near 0.2 s, while most show broader edges that indicate more diffuse boundaries. These steep edges are consistent with the ISAF time response of around 0.2 s shown in Fig. 10. These examples – fast time response and the ability to measure highly variable concentrations – are important demonstrations of the small effect of the reversible adsorption and desorption of HCHO in the ISAF inlet and instrument. Reversible adsorption and desorption are clearly factors because the time constant is larger than the volumetric flush time, but the effects are small, on the order of 100 ms.

5 Summary

The NASA ISAF instrument is a compact, high-sensitivity, field-proven instrument for airborne observations of HCHO throughout the troposphere and lower stratosphere. The instrument capabilities are summarized in Table 1. It has already flown on the NASA DC-8, the NOAA WP-3D and the NCAR G-V aircraft on four aircraft field campaigns: DC3, Southeast Nexus (SENEX), SEAC4RS, and CONvective Transport of Active Species in the Tropics (CONTRAST). Future work will include modifications for deployment on high-altitude aircraft, such as the NASA ER-2 or WB-57, to achieve an even more complete observational data set.

Acknowledgements. This research was funded by the Goddard Internal Research and Development (IRAD) program and the NASA ROSES grant NNH10ZDA001N-SEAC4RS. The NASA Postdoctoral Program provided funding for M. Cazorla.

Edited by: E. C. Apel

References

Apel, E. C., Olson, J. R., Crawford, J. H., Hornbrook, R. S., Hills, A. J., Cantrell, C. A., Emmons, L. K., Knapp, D. J., Hall, S., Mauldin III, R. L., Weinheimer, A. J., Fried, A., Blake, D. R., Crounse, J. D., Clair, J. M. St., Wennberg, P. O., Diskin, G. S., Fuelberg, H. E., Wisthaler, A., Mikoviny, T., Brune, W., and Riemer, D. D.: Impact of the deep convection of isoprene and other reactive trace species on radicals and ozone in the upper troposphere, Atmos. Chem. Phys., 12, 1135–1150, doi:10.5194/acp-12-1135-2012, 2012.

Baidar, S., Oetjen, H., Coburn, S., Dix, B., Ortega, I., Sinreich, R., and Volkamer, R.: The CU Airborne MAX-DOAS instrument: vertical profiling of aerosol extinction and trace gases, Atmos. Meas. Tech., 6, 719–739, doi:10.5194/amt-6-719-2013, 2013.

Barth, M. C., Kim, S. W., Skamarock, W. C., Stuart, A. L., Pickering, K. E., and Ott, L. E.: Simulations of the redistribution of formaldehyde, formic acid, and peroxides in the 10 July 1996 Stratospheric-Tropospheric Experiment: Radiation, Aerosols, and Ozone deep convection storm, J. Geophys. Res.-Atmos., 112, D13310, doi:10.1029/2006jd008046, 2007.

Barth, M. C., Cantrell, C. A., Brune, W. H., Rutledge, S. A., Crawford, J. H., Huntrieser, H., Carey, L. D., MacGorman, D., Weisman, M., Pickering, K. E., Bruning, E., Anderson, B., Apel, E., Biggerstaff, M., Campos, T., Campuzano-Jost, P., Cohen, R. C., Crounse, J. D., Day, D. A., Diskin, G., Flocke, F., Fried, A., Garland, C., Heikes, B., Honomichi, S., Hornbrook, R., Huey, L. G., Jimenez, J., Lang, T., Lichtenstern, M., Mikoviny, T., O'Sullivan, D., Pan, L., Peischl, J., Pollack, I., Riemer, D., Ryerson, T., Schlager, H., St. Clair, J. M., Walega, J., Weibring, P., Weinheimer, A., Wennberg, P. O., Wisthaler, A., Wooldridge, P. J., and Ziegler, C.: The deep convective clouds and chemistry (DC3) field campaign, J. Geophys. Res., submitted, 2015.

Becker, K. H., Schurath, U., and Tatarczyk, T.: Fluorescence determination of low formaldehyde concentrations in air by dye laser excitation, Appl. Optics, 14, 310–313, doi:10.1364/ao.14.000310, 1975.

Cardenas, L. M., Brassington, D. J., Allan, B. J., Coe, H., Alicke, B., Platt, U., Wilson, K. M., Plane, J. M. C., and Penkett, S. A.: Intercomparison of formaldehyde measurements in clean and polluted atmospheres, J. Atmos. Chem., 37, 53–80, doi:10.1023/a:1006383520819, 2000.

Chance, K., Palmer, P. I., Spurr, R. J. D., Martin, R. V., Kurosu, T. P., and Jacob, D. J.: Satellite observations of formaldehyde over North America from GOME, Geophys. Res. Lett., 27, 3461–3464, doi:10.1029/2000gl011857, 2000.

Chatfield, R. B., Ren, X., Brune, W., and Schwab, J.: Controls on urban ozone production rate as indicated by formaldehyde oxidation rate an nitric oxide, Atmos. Environ., 44, 5395–5406, 2010.

Co, D. T., Hanisco, T. F., Anderson, J. G., and Keutsch, F. N.: Rotationally resolved absorption cross sections of formaldehyde in the 28100–28500 cm(-1) (351–356 nm) spectral region: Implications for in situ LIF measurements, J. Phys. Chem. A, 109, 10675–10682, doi:10.1021/jp053466i, 2005.

DiGangi, J. P., Boyle, E. S., Karl, T., Harley, P., Turnipseed, A., Kim, S., Cantrell, C., Maudlin III, R. L., Zheng, W., Flocke, F., Hall, S. R., Ullmann, K., Nakashima, Y., Paul, J. B., Wolfe, G. M., Desai, A. R., Kajii, Y., Guenther, A., and Keutsch, F. N.: First direct measurements of formaldehyde flux via eddy covariance: implications for missing in-canopy formaldehyde sources,

Atmos. Chem. Phys., 11, 10565–10578, doi:10.5194/acp-11-10565-2011, 2011.

Duncan, B., Yoshida, Y., Olson, J., Sillman, S., Martin, R., Lamsal, L., Hu, Y., Pickering, K., Retscher, C., Allen, D., and Crawford, J.: Application of OMI observations to a space-based indicator of NOx and VOC controls on surface ozone formation, Atmos. Environ., 44, 2213–2223, doi:10.1016/j.atmosenv.2010.03.010, 2010.

Edwards, P., Evans, M. J., Commane, R., Ingham, T., Stone, D., Mahajan, A. S., Oetjen, H., Dorsey, J. R., Hopkins, J. R., Lee, J. D., Moller, S. J., Leigh, R., Plane, J. M. C., Carpenter, L. J., and Heard, D. E.: Hydrogen oxide photochemistry in the northern Canadian spring time boundary layer, J. Geophys. Res.-Atmos., 116, D22306, doi:10.1029/2011jd016390, 2011.

Edwards, P. M., Young, C. J., Aikin, K., deGouw, J., Dubé, W. P., Geiger, F., Gilman, J., Helmig, D., Holloway, J. S., Kercher, J., Lerner, B., Martin, R., McLaren, R., Parrish, D. D., Peischl, J., Roberts, J. M., Ryerson, T. B., Thornton, J., Warneke, C., Williams, E. J., and Brown, S. S.: Ozone photochemistry in an oil and natural gas extraction region during winter: simulations of a snow-free season in the Uintah Basin, Utah, Atmos. Chem. Phys., 13, 8955–8971, doi:10.5194/acp-13-8955-2013, 2013.

Fortems-Cheiney, A., Chevallier, F., Pison, I., Bousquet, P., Saunois, M., Szopa, S., Cressot, C., Kurosu, T. P., Chance, K., and Fried, A.: The formaldehyde budget as seen by a global-scale multi-constraint and multi-species inversion system, Atmos. Chem. Phys., 12, 6699–6721, doi:10.5194/acp-12-6699-2012, 2012.

Fried, A., Olson, J. R., Walega, J. G., Crawford, J. H., Chen, G., Weibring, P., Richter, D., Roller, C., Tittel, F., Porter, M., Fuelberg, H., Halland, J., Bertram, T. H., Cohen, R. C., Pickering, K., Heikes, B. G., Snow, J. A., Shen, H., O'Sullivan, D. W., Brune, W. H., Ren, X., Blake, D. R., Blake, N., Sachse, G., Diskin, G. S., Podolske, J., Vay, S. A., Shetter, R. E., Hall, S. R., Anderson, B. E., Thornhill, L., Clarke, A. D., McNaughton, C. S., Singh, H. B., Avery, M. A., Huey, G., Kim, S., and Millet, D. B.: Role of convection in redistributing formaldehyde to the upper troposphere over North America and the North Atlantic during the summer 2004 INTEX campaign, J. Geophys. Res.-Atmos, 113, D17306, doi:10.1029/2007jd009760, 2008.

Gilpin, T., Apel, E., Fried, A., Wert, B., Calvert, J., Genfa, Z., Dasgupta, P. Harder, J. W., Heikes, B., Hopkins, B., Westberg, H., Kleindiens, T., Lee, Y.-N., Zhou, X., Lonneman, W., and Sewell, S.: Intercomparison of six ambient CH_2O measurement techniques, J. Geophys. Res.-Atmos., 102, 21161–21188, 1997.

Hagopian, J.: Carbon nanotubes for stray light suppression, Art. 9417, NASA Tech. Briefs, 2011.

Hak, C., Pundt, I., Trick, S., Kern, C., Platt, U., Dommen, J., Ordóñez, C., Prévôt, A. S. H., Junkermann, W., Astorga-Lloréns, C., Larsen, B. R., Mellqvist, J., Strandberg, A., Yu, Y., Galle, B., Kleffmann, J., Lörzer, J. C., Braathen, G. O., and Volkamer, R.: Intercomparison of four different in-situ techniques for ambient formaldehyde measurements in urban air, Atmos. Chem. Phys., 5, 2881–2900, doi:10.5194/acp-5-2881-2005, 2005.

Hottle, J. R., Huisman, A. J., Digangi, J. P., Kammrath, A., Galloway, M. M., Coens, K. L., and Keutsch, F. N.: A Laser Induced Fluorescence-Based Instrument for In-Situ Measurements of Atmospheric Formaldehyde, Environ. Sci. Tech., 43, 790–795, 2009.

Kaiser, J., Li, X., Tillmann, R., Acir, I., Holland, F., Rohrer, F., Wegener, R., and Keutsch, F. N.: Intercomparison of Hantzsch and fiber-laser-induced-fluorescence formaldehyde measurements, Atmos. Meas. Tech., 7, 1571–1580, doi:10.5194/amt-7-1571-2014, 2014.

Luecken, D. J., Hutzell, W. T., Strum, M. L., and Pouliot, G. A.: Regional sources of atmospheric formaldehyde and acetaldehyde, and implications for atmospheric modeling, Atmos. Environ., 47, 477–490, doi:10.1016/j.atmosenv.2011.10.005, 2012.

Marais, E. A., Jacob, D. J., Kurosu, T. P., Chance, K., Murphy, J. G., Reeves, C., Mills, G., Casadio, S., Millet, D. B., Barkley, M. P., Paulot, F., and Mao, J.: Isoprene emissions in Africa inferred from OMI observations of formaldehyde columns, Atmos. Chem. Phys., 12, 6219–6235, doi:10.5194/acp-12-6219-2012, 2012.

Meller, R. and Moortgat, G. K.: Temperature dependence of the absorption cross sections of formaldehyde between 223 and 323 K in the wavelength range 225–375 nm, J. Geophys. Res.-Atmos., 105, 7089–7101, doi:10.1029/1999jd901074, 2000.

Millet, D. B., Jacob, D. J., Boersma, K. F., Fu, T. M., Kurosu, T. P., Chance, K., Heald, C. L., and Guenther, A.: Spatial distribution of isoprene emissions from North America derived from formaldehyde column measurements by the OMI satellite sensor, J. Geophys. Res.-Atmos., 113, D02307, doi:10.1029/2007jd008950, 2008.

Möhlmann, G. R.: Formaldehyde detection in air by laser-induced fluorescence, Appl. Spectrosc., 39, 98–101, doi:10.1366/0003702854249088, 1985.

Olaguer, E. P., Rappengluck, B., Lefer, B., Stutz, J., Dibb, J., Griffin, R., Brune, W. H., Shauck, M., Buhr, M., Jeffries, H., Vizuete, W., and Pinto, J. P.: Deciphering the Role of Radical Precursors during the Second Texas Air Quality Study, J. Air Waste Manage., 59, 1258–1277, doi:10.3155/1047-3289.59.11.1258, 2009.

Palmer, P. I., Jacob, D. J., Fiore, A. M., Martin, R. V., Chance, K., and Kurosu, T. P.: Mapping isoprene emissions over North America using formaldehyde column observations from space, J. Geophys. Res.-Atmos., 108, 4180, doi:10.1029/2002jd002153, 2003.

Palmer, P. I., Abbot, D. S., Fu, T. M., Jacob, D. J., Chance, K., Kurosu, T. P., Guenther, A., Wiedinmyer, C., Stanton, J. C., Pilling, M. J., Pressley, S. N., Lamb, B., and Sumner, A. L.: Quantifying the seasonal and interannual variability of North American isoprene emissions using satellite observations of the formaldehyde column, J. Geophys. Res.-Atmos., 111, D12315, doi:10.1029/2005jd006689, 2006.

Shibuya, K., Holtermann, D. L., Peacock, J. R., and Lee, E. K. C.: Fluorescence emission-spectroscopy, radiative lifetimes, and vibrational-relaxation rates of the 40 and 41 single vibronic levels of D2CO and HDCO (A1A2), J. Phys. Chem., 83, 940–944, doi:10.1021/j100471a012, 1979.

Weibring, P., Richter, D., Walega, J. G., and Fried, A.: First demonstration of a high performance difference frequency spectrometer on airborne platforms, Opt. Express, 15, 13476–13495, doi:10.1364/oe.15.013476, 2007.

Wert, B. P., Fried, A., Henry, B., and Cartier, S.: Evaluation of inlets used for the airborne measurement of formaldehyde, J. Geophys. Res.-Atmos., 107, ACH3.1–ACH3.13, doi:10.1029/2001jd001072, 2002.

Wisthaler, A., Apel, E. C., Bossmeyer, J., Hansel, A., Junkermann, W., Koppmann, R., Meier, R., Müller, K., Solomon, S. J., Steinbrecher, R., Tillmann, R., and Brauers, T.: Technical Note: Intercomparison of formaldehyde measurements at the atmosphere simulation chamber SAPHIR, Atmos. Chem. Phys., 8, 2189–2200, doi:10.5194/acp-8-2189-2008, 2008.

Yokelson, R. J., Burling, I. R., Gilman, J. B., Warneke, C., Stockwell, C. E., de Gouw, J., Akagi, S. K., Urbanski, S. P., Veres, P., Roberts, J. M., Kuster, W. C., Reardon, J., Griffith, D. W. T., Johnson, T. J., Hosseini, S., Miller, J. W., Cocker III, D. R., Jung, H., and Weise, D. R.: Coupling field and laboratory measurements to estimate the emission factors of identified and unidentified trace gases for prescribed fires, Atmos. Chem. Phys., 13, 89–116, doi:10.5194/acp-13-89-2013, 2013.

13

Distinguishing cirrus cloud presence in autonomous lidar measurements

J. R. Campbell[1], M. A. Vaughan[2], M. Oo[3], R. E. Holz[3], J. R. Lewis[4], and E. J. Welton[5]

[1]Naval Research Laboratory, Monterey, California, USA
[2]NASA Langley Research Center, Hampton, Virginia, USA
[3]Space Sciences and Engineering Center, University of Wisconsin, Madison, Wisconsin, USA
[4]Joint Center for Earth Systems Technology, University of Maryland Baltimore County, Baltimore, Maryland, USA
[5]NASA/Goddard Space Flight Center, Greenbelt, Maryland, USA

Correspondence to: J. R. Campbell (james.campbell@nrlmry.navy.mil)

Abstract. 2012 Level-2 Cloud-Aerosol Lidar with Orthogonal Polarization (CALIOP) satellite-based cloud data sets are investigated for thresholds that distinguish the presence of cirrus clouds in autonomous lidar measurements, based on temperatures, heights, optical depth and phase. A thermal threshold, proposed by Sassen and Campbell (2001) for cloud top temperature $T_{top} \leq -37\,°C$, is evaluated versus CALIOP algorithms that identify ice-phase cloud layers using polarized backscatter measurements. Derived global mean cloud top heights (11.15 vs. 10.07 km above mean sea level; a.m.s.l.), base heights (8.76 km a.m.s.l. vs. 7.95 km a.m.s.l.), temperatures (−58.48 °C vs. −52.18 °C and −42.40 °C vs. −38.13 °C, respectively, for tops and bases) and optical depths (1.18 vs. 1.23) reflect the sensitivity to this constraint. Over 99 % of all $T_{top} \leq -37\,°C$ clouds are classified as ice by CALIOP Level-2 algorithms. Over 81 % of all ice clouds correspond with $T_{top} \leq -37\,°C$. For instruments lacking polarized measurements, and thus practical estimates of phase, $T_{top} \leq -37\,°C$ provides sufficient justification for distinguishing cirrus, as opposed to the risks of glaciated liquid-water cloud contamination occurring in a given sample from clouds identified at relatively "warm" ($T_{top} > -37\,°C$) temperatures. Although accounting for uncertainties in temperatures collocated with lidar data (i.e., model reanalyses/sondes) may justifiably relax the threshold to include warmer cases, the ambiguity of "warm" ice clouds cannot be fully reconciled with available measurements, conspicuously including phase. Cloud top heights and optical depths are investigated, and global distributions and frequencies derived, as functions of CALIOP-retrieved phase. These data provide little additional information, compared with temperature alone, and may exacerbate classification uncertainties overall.

1 Motivation

Cirrus clouds are recognized by sky gazers for their translucent and fibrous appearance, cast frequently as delicate white filaments across otherwise clear blue skies at relatively high tropospheric altitudes. To climate scientists however, cirrus clouds, which are composed almost exclusively of ice crystals, are distinct for their physical and radiative properties (e.g., Liou, 1986). As cold and optically thin counterparts to most liquid-water and mixed-phase clouds (e.g., Sassen and Cho, 1992), the net column-integrated radiative impact of cirrus cloud presence during sunlit hours varies between positive and negative, depending on the relative magnitudes of their simultaneous and offsetting contributions diurnally to tropospheric warming (infrared absorption and reemission) and cooling (solar albedo effects; Stephens et al., 1990). This attribute makes cirrus relatively unique among cloud genera. Combined with their relatively high occurrence frequencies globally (e.g., Holz et al., 2008), cirrus are significant and distinct contributors to climate overall (Sassen, 2002).

Lidars are primary remote-sensing tools used for monitoring cirrus clouds (e.g., Sassen, 1991). Two complementary NASA lidar projects are presently tasked with compiling measurements for evaluating global cirrus cloud physical and thermodynamic properties. The NASA Cloud-Aerosol Lidar and Infrared Pathfinder Satellite Observations (CALIPSO) satellite mission features the Cloud-Aerosol Lidar with Orthogonal Polarization (CALIOP[1]; Winker et al., 2010), a two-wavelength (532 and 1064 nm) instrument with linear polarization sensitivity at 532 nm. Similarly, the Micropulse Lidar Network (MPLNET[2]; Welton et al., 2002) is well into its second decade of federated ground-based observations using single-channel 523/527/532 nm (depending on the version) eye-safe elastic-scattering instruments (Campbell et al., 2002).

CALIOP and MPLNET measurements are collected autonomously, thus lacking a corresponding scene observer (e.g., a trained meteorological technician), in what has become the new *normal* for compiling large global data sets with emerging turnkey remote-sensing technologies. (This leaves aside, for the purposes of this paper, a separate debate on the representativeness of ground-based all-sky cameras and/or imaging radiometers in orbit relative to a narrow lidar profiling curtain, given the placement of CALIPSO within the NASA "A-Train" constellation.) In contrast, cloud genus, like that of cirrus, is a distinction based traditionally on visual appearance, which complies with definitions established in atlas publications and is subject to the interpretation and skill of corresponding weather observers (Lynch, 2002; Sassen, 2002). For instance, ice-phase composition, translucence (for which optical depth is an effective proxy), temperature and altitude, properties commonly referenced by researchers evaluating cirrus cloud processes (e.g., Fu et al., 1998), are all ignored as explicit components of the morphologically-based definitions for cirrus clouds (e.g., texture, color, the presence of optical phenomena) maintained by the World Meteorological Organization (WMO, 1975; Lynch, 2002). Each of these causal attributes is considered a relevant element of the WMO definitions, however, which underscores the relatively fine line between physical cloud attributes and their visual interpretation from ground level.

The remote-sensing community has now entered an age where *digital* cloud characteristics attributable to specific cloud genera are necessary for distinguishing them if we are to maintain the traditional paradigm that inventories cloud presence based on phenomenological characteristics. That is, attributes distinguishable to a ground observer must be effectively translated into a series of cloud-type-dependent parameters retrievable from or directly measurable by autonomous measurements. Without them, isolating cloud genera for potential process study is susceptible to uncertainty and, perhaps worst of all, confusion. A reading of three recent papers specifically investigating cirrus clouds with CALIOP data sets, for example, finds three very different definitions for cirrus cloud presence (Nazaryan et al., 2008; Virts et al., 2010; Thorsen et al., 2011; note that we do not question the veracity of the conclusions in these papers, or fail to recognize that previous papers served a role in motivating their cloud definitions, but are instead clarifying the depth of the challenge for the reader). The potential consequence and impact of cirrus cloud research conducted through CALIOP, MPLNET and other similar lidar projects, for instance, will depend on a consistent and robust approach to distinguishing cloud presence as interpreted from these autonomously collected data sets.

Ultimately, however, the basis for any reliable digital classification depends on the cloud-dependent information available from these relatively simple profiling instruments (simple in the sense that, as the acronym implies, lidars "detect and range"). Lidar data yield cloud boundary heights, which can be related to temperature, given a model or local radiosonde profile. Some, like CALIOP, are equipped with channels that resolve the polarization properties of particle backscatter, which can provide information on likely cloud phase and thus discriminate ice from liquid water, since the former readily depolarizes incident visible radiation (Sassen, 1991). The overwhelming majority of ground-based lidars operated presently do not, however, including those currently in MPLNET. Others (i.e., Raman and high-spectral-resolution lidars; Grund and Eloranta, 1990, Goldsmith et al., 1998) can directly measure cloud optical depth, whereas CALIOP and MPLNET instruments can be used only for estimating this parameter. Such advanced systems are a topic best considered at a future date, however. Therefore, if altitude, temperature, optical depth and phase represent the four most practical and relevant lidar-measured parameters for characterizing cirrus cloud properties and are those most relatable to phenomenological WMO definitions, the first three can be estimated (at worst) using nearly all forms of these instruments. Attributing phase, however, will depend on relative technological complexity.

Still, these four parameters in tandem do not fully relate all of the phenomenological characteristics of cirrus clouds apparent to the ground observer (which is presumably why WMO definitions fail to revert to them directly in practice). Furthermore, it is unclear whether or not a static digital classification system for cirrus can be designed based on these four variables alone and whether or not each of them tangibly contributes any practical information content. Critically, as Lynch (2002) succinctly states, "all cirrus clouds are composed of ice, but not all ice clouds are cirrus". Though the predominant nucleation mechanism responsible for their formation remains in question (e.g., Czizco et al., 2013), it is believed that, with limited exceptions (i.e., convective anvils, though a discussion of other less frequent scenarios is given by Sassen, 2002), cirrus cloud morphology begins with the freezing of submicron haze particles at temperatures below the effective threshold for homogeneous freezing of liquid water near −37 °C (Pruppacher and Klett, 1997; note that

[1]http://www-calipso.larc.nasa.gov/

[2]http://mplnet.gsfc.nasa.gov/

for consistency with discussion relating to the homogeneous freezing of liquid water in the literature, we maintain the use of degrees Celsius throughout the manuscript.). Ice phase alone, if resolvable with a given lidar, does not ensure such a distinction, however. For example, most glaciated liquid-water clouds are surely not cirrus, given fundamentally different crystal habits and ice-water paths (Sun and Shine, 1994). Further, a ground observer is likely to recognize, distinguish and report glaciation and perhaps the surrounding remnants of a supercooled liquid-water parent cloud, such as altocumulus (Wang et al., 2004). A lidar, in contrast, would only resolve relatively warm and diffuse ice-phase fallstreaks embedded among pockets of strongly scattering and signal-attenuating liquid-water droplets (e.g., Sassen, 1978). Resolving such caveats operationally within an algorithm is obviously daunting.

The results of one study help bridge the gap between visual classification and digital interpretation. Sassen and Campbell (2001; hereafter SC2001) describe a midlatitude cirrus cloud climatology developed over more than a decade using $\sim$2200 h of episodic ruby lidar (694 nm) measurements, where a trained observer specifically characterized the cloud scene at the time of profiling. They conclude that for tropospheric cloud identification "in studies lacking visual cloud observations, a minimum cloud top temperature of $-37\,°C$ be employed to 'identify' cirrus". This is based on roughly 98 % of their visibly classified cirrus cloud sample corresponding with that metric, and owing to the homogenous ice nucleation threshold that ostensibly precludes supercooled liquid-water layers from contributing significantly to what is visibly perceived as cirrus. Cloud top is the essential reference layer, then, since this is the "height at which nucleation is prevalent". SC2001 acknowledge that glaciated liquid cloud clouds (i.e., heterogenous freezing of liquid-water droplets induced by aerosol particles, such as mineral dusts (DeMott et al., 2009) or even other ice crystals (Campbell and Shiobara, 2008), as well as volcanic residues (e.g., Seifert et al., 2011)) and, more notably, sheared cirrus fallstreaks detached from their parent layer exist at relatively warmer apparent cloud top temperatures. Thus, SC2001 rightly do not rule out the existence of relatively "warm" cirrus (i.e., cloud top temperature, $T_{top} > -37\,°C$). However, and in spite of available polarization measurements, they conclude that most "warm" ice-phase clouds are not cirrus, given their likely conflicting origin. The catch, however, is that some are.

It is presumably the ambiguity in classifying a "warm" ice-phase cloud genus, combined with potential questions regarding the representativeness of a single study conducted at a single midlatitude site, which has precluded the adoption of the SC2001 threshold universally within the community with respect to autonomous lidar measurements (though some have applied it; e.g., Cadet et al., 2003). SC2001 acknowledge the latter point, declaring that "cirrus clouds are the product of weather processes such that their occurrence and macrophysical properties will vary significantly over the globe." Still, despite the increasing density of ground-based lidar profiling facilities, developing a global campaign for reconciling regional variations in cirrus cloud phenomenological characteristics with lidar measurements (i.e., height, temperatures, optical depth and phase) is clearly impractical (Dowling and Radke, 1990, notwithstanding). The community does, however, have access to the CALIOP near-global record (82° S to 82° N; 2006–current), including Level-2 retrievals for cloud altitude, base and top height temperatures, optical depth and phase. These data can be applied in a somewhat consistent fashion, by considering the variability in cloud physical and thermodynamic properties globally with the hopes of refining a set of practical digital classification metrics.

This paper thus describes such a series of tests using Level-2 CALIOP cloud height, temperature, optical depth and phase parameters to evaluate the potential of a uniform, globally applicable technique for identifying the presence of cirrus clouds in autonomous lidar measurements. Each test is designed to isolate ice-phase cloud attributes regionally and globally, in order to recognize the corresponding and relative significance of these four parameters and identify potential caveats. The goal of this work, therefore, is developing a series of dependent regional and global metrics that best characterize cirrus cloud presence, which can be adapted by researchers working within the classical phenomenological framework to specifically extract cirrus cloud observations from autonomous lidar data sets. Note that although the polarization properties of the CALIOP instrument are a primary consideration, we are particularly mindful of historical data, including those from MPLNET, for which polarization is not an option. That is, what may prove practical in distinguishing cirrus from CALIOP data sets may not ultimately satisfy applications without cloud phase estimates, and thus where only cloud heights, temperature, and/or retrievals for optical depth are available (e.g., Chew et al., 2011; Lewis et al., 2015). Therefore, a unifying digital definition should ideally be designed that is applicable to all lidar technologies.

2 CALIOP Level-2 cloud data sets

One year (2012) of CALIOP Version 3.02 Level-2 5 km cloud profile product (L2CPro-5km) data is investigated. Clouds reported in each L2_CPro-5km record have been merged so that vertically adjacent layer fragments resolved at different spatial resolutions (5, 20 and 80 km; Vaughan et al., 2009) are combined into single layers and to ensure a uniform minimum cloud separation threshold of 0.5 km consistent with SC2001. In accordance with Liu et al. (2009), a corresponding cloud aerosol discriminator (CAD Score; Liu et al., 2009) layer value between 70 and 100 (i.e., high confidence) was required of each cloud layer before merging. Note that the L2_CPro-5km product does not report broken but relatively bright liquid-water clouds occurring below 4 km (all heights a.m.s.l.). These clouds are resolved from single CALIOP signal profiles at a native 0.333 km along-track resolution, and are generally unlikely, even in polar regions, to correspond with cirrus cloud presence. The L2_CPro-5km product does, however, include those clouds resolved at coarser resolutions (i.e., 20 and 80 km) in low signal-to-noise conditions that are likely relevant to our study (i.e., optically-thin clouds like cirrus and/or secondary layers lying below optically-thick cirrus).

Any generic interpretation of sample counts and relative cloud frequencies described in the narrative for anything but the stated intention of each test applied should be considered with these caveats in mind. Further, the reader must also consider the influence of the nadir-viewing lidar geometry and how signal attenuation through optically thick clouds can limit the vertical extent of CALIOP profiling. Zhang et al. (2010), for instance, describe the synergy necessary between CALIOP and the CloudSat millimeter cloud radar (Stephens et al., 2002), also flown within the A-Train, for profiling mixed-phase liquid-layer-topped stratiform clouds. Though the CALIOP geometry is obviously far more preferable to ground-based zenith-profiling lidar application, with respect to attenuation effects, the data are not immune to some degree of sampling bias. The results below, again, are presented in a mostly relative context, except where specifically denoted.

Temperature profiles from Goddard Model Assimilation Office (GMAO) Goddard Earth Observing System Model – Version 5 (GEOS-5) products are included in the L2_CPro-5km file, and these data are collocated with reported cloud boundary heights. To suppress polar stratospheric cloud contamination, at latitudes > |60° | clouds with top height temperatures < 200 K are excluded from this analysis (Campbell and Sassen, 2008). By combining cloud phase estimates derived for each layer reported in the L2_CPro-5km files, cloud phase is determined for the merged clouds from the three available categories reported by the CALIOP phase classification algorithm: ice, liquid water and "unknown" (Hu et al., 2009). In the merged data set, ice-phase clouds correspond with a fractional cloud phase that is 100 % ice. Similarly, liquid-phase clouds are 100 % liquid water. Mixed-phase clouds are defined here as those merged layers where the ice and liquid-phase fractions are both non-zero, but where together they sum to 100 %. Unknown phase clouds represent cases where the 'unknown' parameter is anything but 0 %.

3 CALIOP cirrus cloud data set tests and discussion

3.1 Thermal thresholds vs. CALIOP Level-2 phase retrievals

The first test establishes a simple baseline comparison between apparent "bookend" scenarios for interpreting cirrus clouds from CALIOP Level-2 data sets. Here, the SC2001 threshold is compared versus all ice clouds identified with CALIOP phase retrieval algorithms. Therefore, "bookend" is defined in the sense that "warm" ice clouds are either ignored completely or considered in full within each subset. Shown in Table 1 are the total numbers of available cloud layers in the merged 2012 L2_CPro-5km product, including sample sizes and relative fractions for those clouds with $T_{top} \leq -37\,°C$ and $T_{top} > -37\,°C$ (44.66 and 55.34 %, respectively). Additionally, corresponding with each cloud and its identified phase, sizes for each sub-sample are shown for ice, liquid, mixed and unknown clouds, including relative fractions. For each of the two ice subsets ("cold" and "warm"), specific relative fractions are also shown.

Over 81 % of all clouds identified by CALIOP algorithms as ice correspond with $T_{top} \leq -37\,°C$. Accordingly, approximately 19 % of the ice-phase sample reflects warmer temperatures. However, nearly 97 % of the $T_{top} \leq -37\,°C$ cloud sample are identified as ice phase. If "unknown" cases are ignored, this increases to over 99 % of the sample. Given no other information other than simple elastic lidar backscatter (i.e., no visual cues or polarized measurements), the SC2001 threshold proves remarkably stable for distinguishing ice-phase clouds globally. If the SC2001 analysis is further taken at face value, these clouds are very likely all cirrus. SC2001 do not specifically consider or report the probability of "cold" non-cirrus ice clouds present within their sample. Presumably, this scenario occurs, albeit in likely low frequencies, given that glaciation at such temperatures would be dominated by convective processes. Anvils themselves are readily considered cirrus, however. So, it is unclear how any such clouds, if displaced from their convective core, contribute significantly to global inventories.

Figure 1 exhibits these data in a slightly different manner. For all cloud layers considered in Table 1, fractional probabilities are shown for each of the four cloud-phase types in 1 °C T_{top} intervals between −60 and 0 °C. Global results are shown, as are those from three latitudinal subsets: $\theta \leq |30°|$, $|30°| < \theta \leq |60°|$ and $\theta > |60°|$. Superimposed on each plot is the −37 °C isotherm. Corresponding sample counts with each case, as a function of phase, are shown in Fig. 2a–d. At temperatures colder than −37 °C, ice-phase clouds are dominant according to the CALIOP classification scheme, consistent with Table 1. The coldest liquid-water clouds are found in nominal amounts beginning at −39 °C, increasing rapidly for temperatures warmer than −35 °C and corresponding with a similarly rapid falloff in ice-phase frequencies. Mixed-phase clouds peak near −30 °C, and unknown cases maintain generally consistent values between −30 and 0 °C. The global profile is remarkably consistent with those derived regionally.

Shown in Table 2 are corresponding mean cloud base and top heights and temperatures and optical depths, including sample sizes, for all clouds with $T_{top} \leq -37\,°C$, all CALIOP-identified ice clouds regardless of T_{top}, and all ice clouds

Table 1. Total number of cloud layers resolved in the 2012 CALIOP V3.02 L2_CPro-5km product, number of those clouds corresponding with cloud top height temperature $\leq -37\,°C$, including sub-sample totals and relative, total and ice percentages (as denoted) as function of CALIOP-distinguished cloud phase, and those corresponding with a cloud top height temperature $> -37\,°C$, again with corresponding itemized cloud phase sample sizes and fractional percentages. Data have been filtered for polar stratospheric cloud occurrence.

Available cloud layers	CALIOP phase	$T \leq -37\,°C$	Relative %	Total %	Ice %	$T > -37\,°C$	Relative %	Total %	Ice %
34 216 819	TOTAL	15 327 653		44.80 %		18 889 166		55.20 %	
CONDITIONS	ICE	14 854 198	96.91 %	43.41 %	81.36 %	3 403 873	18.02 %	9.95 %	18.64 %
All 5 km minimum	LIQUID	12 103	0.08 %	0.04 %		11 499 175	60.88 %	33.61 %	
CAD = 70-100	MIXED	118 943	0.78 %	0.35 %		272 248	1.44 %	0.80 %	
	UNKNOWN	342 409	2.23 %	1.00 %		3 713 870	19.66 %	10.85 %	

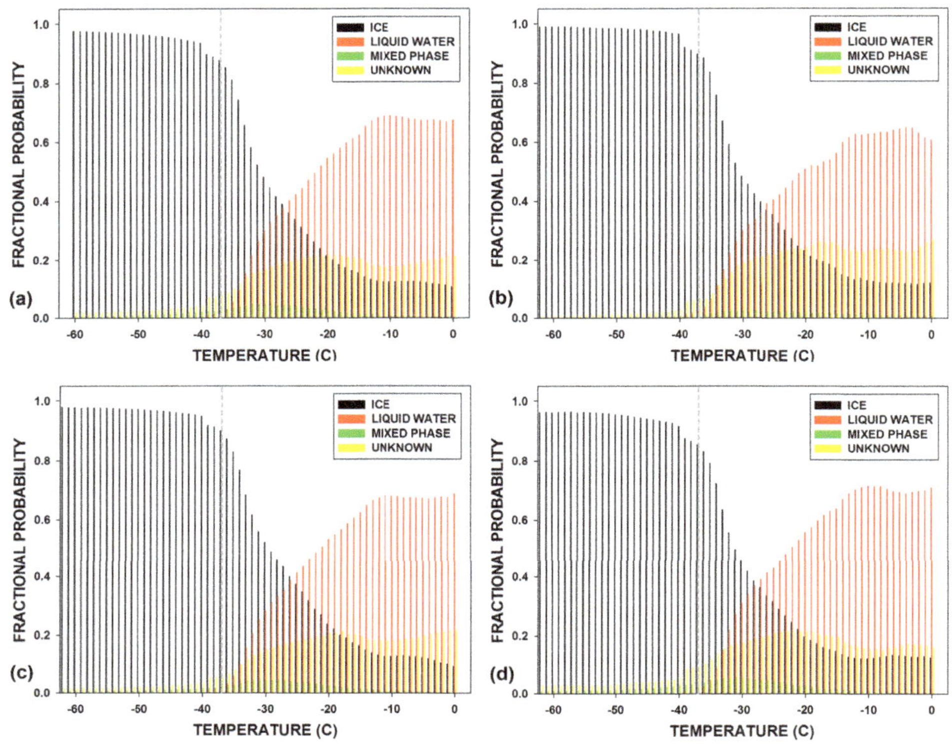

Figure 1. Fractional probabilities at 1 °C intervals between −60 and 0 °C for each of the four cloud phase types (see text) identified from the 2012 CALIOP V3.02 L2_CPro-5km, including ice (black), liquid water (red), mixed-phase (green) and unknown (yellow), for **(a)** the global scale, **(b)** the tropics ($\theta \leq |30°|$), **(c)** midlatitudes ($|30°| < \theta \leq |60°|$) and **(d)** polar regions ($\theta > |60°|$). Superimposed on these data is the −37 °C isotherm.

with $T_{top} > -37\,°C$. Results are again shown for all clouds, and the three latitudinal subsets. The impact of distinguishing cirrus cloud presence using both bookend scenarios, and thus considering the influence of signal polarization (i.e., phase), is apparent comparing the first two sets of results. Global mean cloud top heights differ by over 1 km (11.15 for ice clouds with top height temperatures $\leq -37\,°C$ vs. 10.06 km for all ice clouds). Corresponding temperatures differ by nearly 6.5 °C (−58.47 °C vs. −52.15 °C). Differences are fairly uniform in the midlatitudes and near the poles (~ 1.1 km/6.5 °C) but less so in the tropics (0.5 km/3.5 °C). Overall, however, these are two very different sets of mean thermodynamic and physical properties, with potentially significant ramifications for global climate models from effects on radiative equilibrium due to a potential parameterization basis (e.g., Stephens et al., 1990).

In Fig. 3, the occurrence frequency of the all-ice cloud subset is shown as a function of cloud top temperature in 1 °C intervals between −90 and 0 °C, including the global sample (Fig. 3a) and those for each latitudinal band (Fig. 3b–d). Superimposed on each image is the −37 °C isotherm. A primary mode is evident in the global sample centered near −55 °C. Secondary modes are evident at −70 and −80 °C. However, these latter two reflect contribution from the trop-

Table 2. Corresponding with Table 1, mean cloud base and top heights and temperatures, optical depths and sample sizes for clouds with top height $T \leq -37\,°C$, all CALIOP-identified ice clouds at all temperatures, and all such ice clouds with top height $T > -37\,°C$. For each case, values are shown for all data and those at latitudes $\theta \leq |30°|$, $|30°| < \theta \leq |60°|$ and $\theta > |60°|$.

	All $T \leq -37\,°C$ Θ = all	All $T \leq -37\,°C$ $\Theta \leq \lvert 30°\rvert$	All $T \leq -37\,°C$ $\lvert 30°\rvert < \Theta \leq \lvert 60°\rvert$	All $T \leq -37\,°C$ $\theta > \lvert 60°\rvert$	Ice; all T Θ = all	$\Theta \leq \lvert 30°\rvert$	$\lvert 30°\rvert < \Theta \leq \lvert 60°\rvert$	$> \lvert 60°\rvert$	ICE; $T > -37$ Θ = all	$\Theta \leq \lvert 30°\rvert$	$\lvert 30°\rvert < \Theta \leq \lvert 60°\rvert$	$> \lvert 60°\rvert$
BASE HEIGHT (km)	8.76	12.27	7.84	5.43	7.95	11.89	7.12	4.77	3.77	6.81	4.22	2.52
TOP HEIGHT (km)	11.15	14.50	10.25	8.00	10.07	14.02	9.22	6.89	5.02	8.14	5.51	3.71
BASE TEMP (°C)	−42.40	−51.95	−36.47	−36.69	−38.13	−49.26	−32.17	−32.54	−16.44	−13.32	−14.77	−18.66
TOP TEMP (°C)	−58.48	−67.59	−53.09	−52.77	−52.18	−64.14	−46.55	−45.45	−23.55	−22.08	−22.96	−24.44
532 nm OPTICAL DEPTH	1.18	1.02	1.43	1.16	1.23	1.05	1.46	1.20	1.51	1.56	1.63	1.40
SAMPLE	15 327 653	5 805 133	4 684 873	4 837 647	18 258 072	6 230 988	5 812 867	6 214 216	3 403 873	486 300	1 279 540	1 638 033

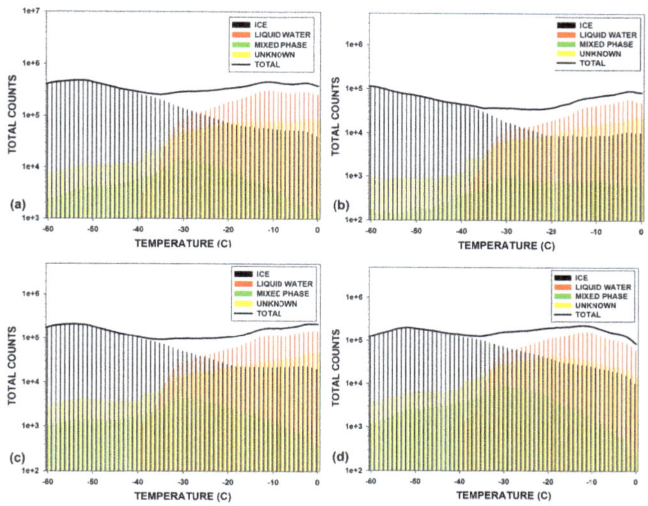

Figure 2. Corresponding with Fig. 1a–d, total counts per 1 °C interval and for each CALIOP Level-2-retrieved cloud phase and their sum.

ics only (Fig. 3b), and thus likely represent, respectively, mean convective outflow top heights and those corresponding with the tropical tropopause transition layer (Virts and Wallace, 2010; Chew et al., 2012) that are mostly confined to that latitude belt. Sample sizes drop with warming cloud top temperature, beginning at roughly −50 °C (−60 °C in the tropics), but the rate steepens in each sample, beginning near −35 °C. Interestingly, sample counts warmer than −37 °C are much lower in the tropics relative to colder temperatures than those found in the midlatitudes and poles.

As suggested above, roughly 19 % of all ice-phase clouds in this sample exhibit cloud top temperatures warmer than −37 °C. Further, beginning near −20 °C and for increasingly warmer temperatures, sample sizes flatten through 0 °C (Fig. 3a). Relative ice-phase frequencies similarly flatten for temperatures increasingly warmer than −15 °C (Fig. 1), perhaps not-so-coincidentally near the point of maximum offset in saturation vapor pressures between ice and liquid water (e.g., Bergeron–Findeisen effects that favor exclusively non-cirrus ice crystal habits, such as dendrites; Pruppacher and Klett, 1997) and the warmest practical temperatures typically associated with heterogeneous ice nucleation (DeMott et al., 1998, 2010). Less than 6 % of all ice-phase clouds correspond with $T_{top} > -20\,°C$ in the global sample. Less than 3 % correspond with $T_{top} > -10\,°C$.

Given the reasonable expectation that sample sizes should otherwise decline approaching 0 °C (e.g., as seen in Fig. 3d), this finding likely signifies the influence of a noise floor in the CALIOP products. Such an artifact will occur for any autonomously-retrieved data product, due to sensor and/or algorithm performance limitations (e.g., the misidentification of optically-thick depolarizing aerosol layers, such as dust, is one example). From Fig. 2, though total cloud counts, includ-

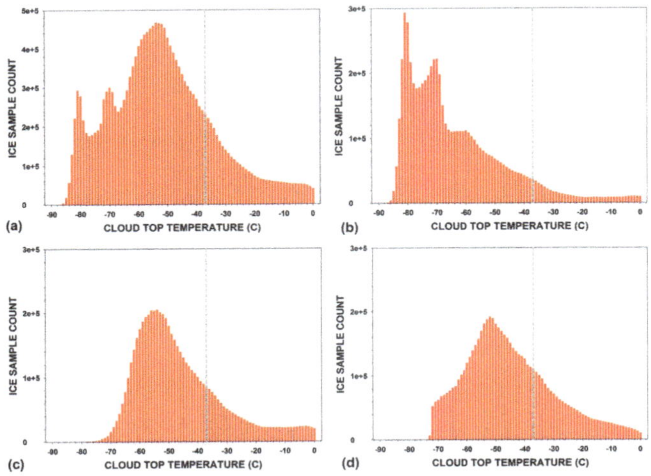

Figure 3. **(a)** Total global sample counts as a function of top height temperature for all clouds identified as ice phase from 2012 CALIOP V3.02 L2_CPro-5km data set, including overlay of −37 °C isotherm, and subset sample counts for **(b)** the tropics ($\theta \leq |30°|$), **(c)** midlatitudes ($|30°| < \theta \leq |60°|$) and **(d)** polar regions ($\theta > |60°|$). The abrupt cutoff at −73 °C in **(d)** is an artifact of the screening process imposed to eliminate polar stratospheric clouds.

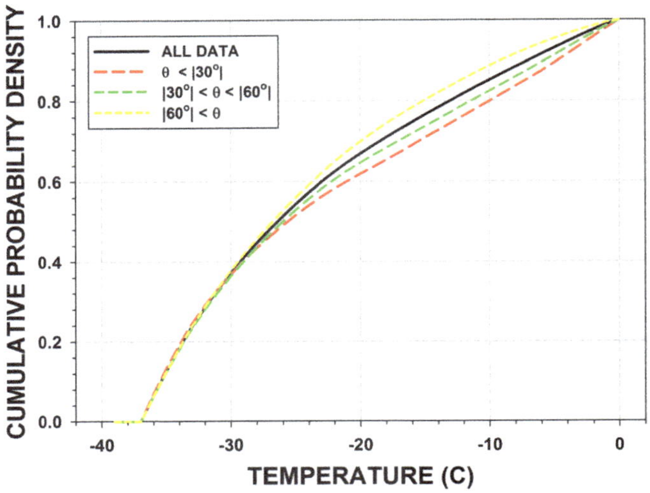

Figure 4. Cumulative probability densities between −37 and 0 °C for 2012 CALIOP V3.02 L2_CPro-5km clouds identified as ice with top height temperatures > −37 °C, including all data (black solid) and those subsets for observations found at latitudes $\theta \leq |30°|$ (red dashed), $|30°| < \theta \leq |60°|$ (green dashed) | and $|60°| < \theta$ (yellow dashed).

ing ice, are relatively low between roughly −40 and −20 °C, particularly in the tropics, they remain significant overall. Climatological characterizations of cirrus cloud properties derived from data sets that define cirrus simply as being ice clouds (e.g., according to the ice-water phase classification provided by the CALIOP Level-2 data products) are likely biased toward lower mean heights and warmer temperatures and thus do not accurately depict true cirrus cloud properties.

The third set of results in Table 2 relates to those clouds identified as ice but with cloud tops exclusively warmer than −37 °C, including total and latitudinal mean cloud base and top heights, temperatures and sample sizes. In Fig. 4, cumulative probability densities are shown for the global sub-sample between −37 and 0 °C. Mean global cloud top heights are more than 5 km lower than the $T_{top} \leq -37$ °C and all-T_{top} ice subsets and as much as 35 °C warmer. By inventorying "warm" ice clouds separately, a distinct set of clouds and corresponding physical properties are depicted in CALIOP data. However, roughly 40 % of these clouds correspond with top heights at $T_{top} \leq -30°$, which is within just a few degrees of the SC2001 threshold. From Fig. 1, and considering that liquid-phase cloud occurrence remains relatively low at temperatures below roughly −33 °C, many of the clouds right near the threshold potentially represent either "warm" cirrus or sheared fallstreaks. However, any likelihood presumably lessens with each progressively warmer 1 °C interval. It is effectively impossible to resolve one way or another, at present. However, this "grey area" and the ambiguity that it represents implies that $T_{top} \leq -37$ °C, though stable, is likely a conservative threshold, which could presumably be relaxed somewhat with sufficient justification.

In light of this latter point, one potential consideration is the uncertainty of the temperature profiles collocated with lidar-derived cloud boundaries. In deriving Table 2, uncertainties in GMAO-derived temperatures from the upper troposphere are believed to be less than 1 °C (M. Rienecker, personal communication, 2013). However, if we conservatively apply this standard, relaxing it to 2 °C, and reapply the SC2001 threshold at −35 °C, thus accounting for potential error, Tables 1 and 2 can be reconsidered. The new results are shown in Tables 3 and 4, respectively. Furthermore, in this new sample, we remove all clouds with $T_{top} \geq 0$ °C (55.37 % $T_{top} \leq -35$ °C vs. 44.63 % -35 °C $< T_{top} \leq 0$ °C).

The total ice-phase cloud partitioning fraction for $T_{top} \leq -35$ °C increases from approximately 81 % to near 84 %. Ice-phase clouds make up over 96 % of the full sample, which increases to near 99 % after removing "unknown" cases. Accordingly, mean global cloud top heights are slightly lower, falling from 11.15 to 11.01 km (over two CALIOP L2_CPro-5km bins at that altitude), and temperatures slightly warmer, rising from −58.48 to −57.75 °C. Still, the results remain significantly different than that of the corresponding all-ice sample (by offsets of approximately 1 km and 5.5 °C at cloud top). Interestingly, the all-ice sample for $T_{top} > -35$ °C exhibits lower cloud top altitudes of nearly 250 m (5.02 to 4.76 km) and higher temperatures of 2 °C (−23.55 to −21.89 °C), thus reflecting the impact of the original $T_{top} > -37$ °C sample disproportionality at temperatures just near the threshold (Fig. 3). Adjusting the SC2001

Table 3. Similar to Table 1; total number of cloud layers resolved in the 2012 CALIOP V3.02 L2_CPro-5km product; number of those clouds corresponding with cloud top height temperature $\leq -35\,°C$, including sub-sample totals and relative, total and ice percentages (as denoted) as function of CALIOP-distinguished cloud phase, and those corresponding with a cloud top height temperature ≤ 0 and $> -35\,°C$, again with corresponding itemized cloud phase sample sizes and fractional percentages. Data have been filtered for polar stratospheric cloud occurrence.

Available cloud layers	Caliop phase	$T \leq -35\,°C$	Relative %	Total %	Ice %	$0 \geq T > -35\,°C$	Relative %	Total %	Ice %
28 616 072	TOTAL	15 844 609		55.37 %		12 771 463		44.63 %	
CONDITIONS	ICE	15 284 684	96.47 %	53.41 %	83.78 %	2 959 141	23.17 %	10.34 %	16.22 %
All 5 km minimum	LIQUID	34 159	0.22 %	0.12 %		7 105 297	55.63 %	24.83 %	
CAD = 70–100	MIXED	135 266	0.85 %	0.47 %		254 875	2.00 %	0.89 %	
	UNKNOWN	390 500	2.46 %	1.36 %		2 452 150	19.20 %	8.57 %	

threshold for uncertainty, as whole, demonstrates stability consistent with applying −37 °C alone. Relaxing this value potentially captures more "warm" ice cloud cases, with still only a minor influence of supercooled liquid-water presence (Fig. 1).

3.2 Constraining cloud phase as a function of cloud top altitude

In the (increasingly unlikely) event that neither a model nor sounding profile is available for interpolating temperature to lidar-derived cloud boundary heights, the second test is designed to identify the distribution of CALIOP-identified phase frequency globally and regionally versus altitude a.m.s.l. Though this scenario is increasingly redundant, recent studies have used cloud base and top altitudes as screening metrics for distinguishing cirrus cloud presence (e.g., Nazaryan et al., 2008; Thorsen et al., 2011). Considering the direct implication of tropospheric altitude in WMO definitions for cirrus cloud classification, this constraint can become a practical means of ensuring sufficient cloud height and possibly depth, consistent with the likely expectation for cirrus cloud attributes from a corresponding ground observer.

Altitude is ultimately a proxy for temperature, however, and regional variability in that parameter would seemingly limit the application of an altitude metric toward a static global definition for cirrus presence. Global and regional fractional phase probabilities as a function of cloud top altitude, shown in Fig. 5, bear this relationship out. Though ice-phase clouds are observed at all tropospheric heights globally, except in the tropics, they are increasingly found at lower heights moving poleward. The distribution in the tropics (Fig. 5b) is particularly distinct, though, as liquid-water clouds are dominant up to near 9 km and ice is not observed below 4 km. Distributions from the midlatitudes and poles are relatively similar, accounting for the shift in increasing ice-phase frequencies to lower heights in the latter, and match the global profile well. This difference in the tropics likely reflects the prevalence of convection in forming clouds, and thus the lofting of liquid water to higher relative nucleation heights in a warmer environment overall.

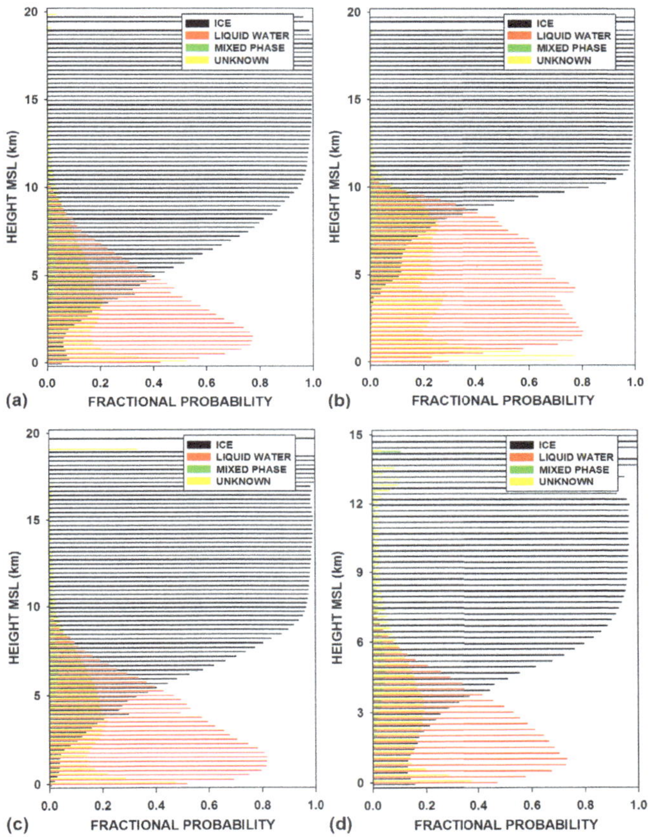

Figure 5. Fractional phase probabilities (ice, liquid water, mixed phase and unknown; see insets) in 250 m height intervals for 2012 CALIOP V3.02 L2_CPro-5km clouds from 0 to 20 km a.m.s.l for **(a)** all global cases (ice, liquid water, mixed phase and unknown), **(b)** the tropics ($\theta \leq |30°|$), **(c)** midlatitudes ($|30°| < \theta \leq |60°|$) and **(d)** from 0 to 15 km a.m.s.l. at the poles ($\theta > |60°|$).

A specific feature of these profiles that may be useful in constraining cirrus identification, particularly when depolarization measurements are not available, is represented by the crossover heights between ice and liquid-phase cloud predominance. Globally, this value is near 5 km. In the tropics, it is near 9 km. In the midlatitudes, it falls just above 5 km

Table 4. Corresponding with Table 3; mean cloud base and top heights and temperatures, optical depths and sample sizes for clouds with top height $0 > T \leq -35\,°C$; all CALIOP-identified ice clouds at all $T < 0$; and all such ice clouds with top height $0 > T > -37\,°C$. For each case, values are shown for all data and those at latitudes $\theta \leq |30°|$, $|30°| < \theta \leq |60°|$ and $\theta > |60°|$.

	All $T \leq -35\,°C$ Θ = ALL	All $T \leq -35\,°C$ $\Theta \leq \|30°\|$	All $T \leq -35\,°C$ $\|30°\| < \Theta \leq \|60°\|$	All $T \leq -35\,°C$ $\theta > \|60°\|$	Ice; $0 \geq T$ Θ = ALL	$\Theta \leq \|30°\|$	$\|30°\| < \Theta \leq \|60°\|$	$> \|60°\|$	Ice; $0 \geq T > -35$ Θ = ALL	$\Theta \leq \|30°\|$	$\|30°\| < \Theta \leq \|60°\|$	$> \|60°\|$
BASE HEIGHT (km)	8.64	12.22	7.75	5.34	7.95	11.90	7.12	4.77	3.57	6.63	4.02	2.33
TOP HEIGHT (km)	11.01	14.44	10.14	7.87	10.07	14.03	9.23	6.89	4.76	7.94	5.26	3.45
BASE TEMP (° C)	−41.83	−51.60	−35.98	−36.13	−38.17	−49.34	−32.21	−32.55	−15.28	−12.09	−13.51	−17.58
TOP TEMP (° C)	−57.75	−67.19	−52.44	−51.93	−52.22	−64.23	−46.60	−45.45	−21.89	−20.50	−21.25	−22.79
532 nm OPTICAL DEPTH	1.20	1.02	1.43	1.17	1.23	1.05	1.46	1.20	1.53	1.63	1.64	1.41
SAMPLE	15 844 609	5 879 867	4 872 132	5 092 610	18 243 824	6 223 648	5 806 984	6 213 193	2 959 141	414 586	1 114 348	1 430 207

and below 4 km at the poles. Still, these points represent only very coarse confidence levels and are in fact center points for relatively deep layers where phase ambiguity is highest among the relative distributions. Considering a 20 % probability as a simple qualitative significance threshold, globally there exists a near 3 km deep layer centered near 5 km where ice and liquid-water clouds are present in near-similar, and more importantly non-negligible, quantities. The layer is shallower in the tropics and poles, nearer to 2 km deep, and over 3 km deep in the midlatitudes. Therefore, depolarization measurements help alleviate confusion regarding phase within what are otherwise ambiguous altitude regimes that vary regionally. Still, they do not help to fully answer the question as to whether or not "warm" clouds are cirrus or glaciated liquid-water clouds in the phenomenological sense. Therefore, while that problem is better constrained, particularly if considered within specific or limited regions, it ultimately remains underdetermined.

3.3 Constraining cloud phase as a function of cloud optical depth

As described above, optical depth is a reasonable proxy for estimating cloud translucence. Cirrus clouds that are readily apparent to the ground observer typically correspond with optical depths of between 0 and approximately 3 (Sassen and Cho, 1992), whereas nearly all liquid-water cloud genera exhibit significantly greater values. This suggests that both phase and cirrus cloud identification could potentially be constrained effectively using those cloud optical depths estimated or retrieved directly from autonomous lidar measurements.

Translucence to the ground observer, which for optically-thin clouds like cirrus relates to the measure of blue sky and/or the outline of the solar disk visible through a given cloud layer, does not necessarily translate well into an optical depth collected through an autonomous lidar measurement. In particular, lidar profiles and corresponding optical-depth retrievals typically reflect processing of measurement averages and/or the integration of retrieval products derived over varying temporal periods. This is the case for CALIOP L2_CPro-5km products investigated here, for instance. Over the course of any integrating period, however, multiple cloud segments or fragments may be sampled, with backscatter then being aggregated into a single instantaneous profile average for processing (Young and Vaughan, 2009; Leahy et al., 2012). If a given cloud layer is temporally persistent and relatively homogeneous, the relationship is consistent. Otherwise, ground-based estimates for translucence and instrument-retrieved optical depth become fundamentally different parameters due to cloud gaps that are inadvertently measured during the sample period.

The third test thus examines whether or not cloud optical depth, as a proxy for translucence, can be applied practically in constraining cirrus or ice cloud presence. Shown in Fig. 6a are fractional phase probabilities for all Level-2 CALIOP cloud layers globally as a function of retrieved 532 nm optical depth. For brevity, only the global composites are shown. In Fig. 6b and c, these data are broken down into samples for $T_{top} \leq -37\,°C$ and $T_{top} > -37\,°C$. Total sample counts corresponding with Fig. 6a–c as a function of optical depth are shown in Fig. 6d–f.

At cloud optical depths below approx. 2.0, over 60 % of the bulk global sample is ice. Consistent with Table 1, for $T_{top} \leq -37\,°C$ the sample is almost exclusively ice. Of all counts, 15 % correspond with cloud optical depth ≤ 0.03, 40 % ≤ 0.30, 60 % ≤ 1.00 and 70 % ≤ 1.40. At warmer temperatures, the sample is dominated by liquid-water cloud presence, with the greatest ice probability occurring near 0.30, which corresponds with the approximate threshold for optically-thin cirrus cloud presence derived by Sassen and Cho (1992). The relative fraction for unknown cloud samples increases at the lowest values, reflecting an increased difficulty in making high-confidence phase classifications for very tenuous, weakly scattering layers. (Note that the majority of the optical depth retrievals in this analysis were derived using "unconstrained" retrievals where a priori values for the lidar extinction-to-backscatter ratio and multiple-scattering correction have been applied. These mean values discussed here, therefore, are subject to the uncertainty corresponding with the default value currently applied in Level-2 algorithms (Liu et al., 2009; Vaughan et al., 2009; Young et al., 2013).)

The abrupt changes in the optical-depth distributions that occur at an optical depth of ~2.4 are the results of an inappropriate choice of cloud lidar ratio. When reliable measurements of clear air can be made immediately above cloud top and below cloud base, estimates of the layer optical depth can be obtained directly from the CALIOP data. In these cases, the optical depths provide a constraint for the lidar equation, and the layer lidar ratios can then be retrieved as part of the solution (Young and Vaughan, 2009). However, for the 2012 CALIOP cloud data, lidar ratios could be obtained this way from only ~2.3 % of all clouds detected. Therefore again, when a measured optical depth constraint is not available, optical depth is estimated using an unconstrained retrieval that requires a priori specification of a type-dependent and layer-effective default lidar ratio. This layer-effective lidar ratio is the product of the single-scattering lidar ratio, which depends on the particle phase function and single-scatter albedo, and the effective multiple-scattering factor, which depends on both the lidar-sensing geometry (e.g., field of view and range to target) and the intrinsic and extrinsic properties of the particles being measured.

Version 3 of the CALIOP data processing uses a default lidar ratio of 25 sr for ice clouds and 19 sr for water clouds and a multiple-scattering factor of 0.6 for both cloud types. These default values are chosen to be representative of the distributions of naturally occurring lidar ratios that characterize each type. For optically-thin layers, the CALIOP default values will generate extinction solutions and optical-depth estimates with varying degrees of accuracy, depending on how closely the default lidar ratio approximates the actual layer lidar ratio (Young et al., 2013). However, as layers become more optically thick, correct specification of the lidar ratio becomes increasingly important. In particular, an overestimate of the lidar ratio can introduce a numerical instability that will cause the solution process to fail. When this condition is detected within the CALIOP retrieval scheme, the retrieval is halted and then restarted with a reduced value of the lidar ratio (Young and Vaughan, 2009). The sudden discontinuity at ~2.4 in the optical-depth distributions shown in Fig. 6 thus signifies the appearance of a family of solutions for which the initial lidar ratio had to be reduced by 5 % in order to achieve a physically meaningful solution in an optically thick layer.

Liquid-water clouds require much more frequent lidar ratio reduction than do ice clouds, which may seem paradoxical, since water cloud lidar ratios are much better known and vary much less than ice cloud lidar ratios (for water clouds see Hu et al., 2006; O'Connor et al., 2004; for ice clouds see Yorks et al., 2011; Sassen and Comstock, 2001). However, CALIOP water cloud multiple-scattering factors are currently not well-characterized, and the default multiple-scattering factor of 0.6 may be appropriate only for optically and geometrically thin water clouds. In contrast, Josset et al. (2012) use comparisons with the CALIPSO Imaging Infrared Radiometer to derive a robust empirical estimate of 0.61 for the CALIOP multiple-scattering factor for ice clouds, and this value is applicable across a broad range of optical depths and geometric thicknesses.

In summary, the use of an optical-depth threshold for constraining ice-phase cloud samples, and thus potentially cirrus, exhibits some skill at values below 2.4 in CALIOP data. Thus, consistent with the interpretations of Sassen and Cho (1992), such skill should be transferrable to all autonomous lidar data sets at optical depths less than 3. However, the ambiguity in such an analysis, due to the potential for broken liquid-water cloud layer contamination for integrated sampling averages, can limit its effectiveness and impact. Particularly for "warm" ice clouds, where some refinement would clearly be most useful, and for elastic-scattering instruments that do no measure depolarization, the information content in this parameter remains qualified and thus potentially limited if not applied under careful constraint.

4 Conclusions

The complexity in replacing visual ground-observer interpretations of classical cloud genera using autonomous remote-sensing measurements, particularly those collected with lidars, will vary with the types of clouds sampled. For cirrus clouds, the results of this study suggest that there will likely never be a single static solution for distinguishing all cirrus clouds (both "cold" and "warm" clouds relative to the

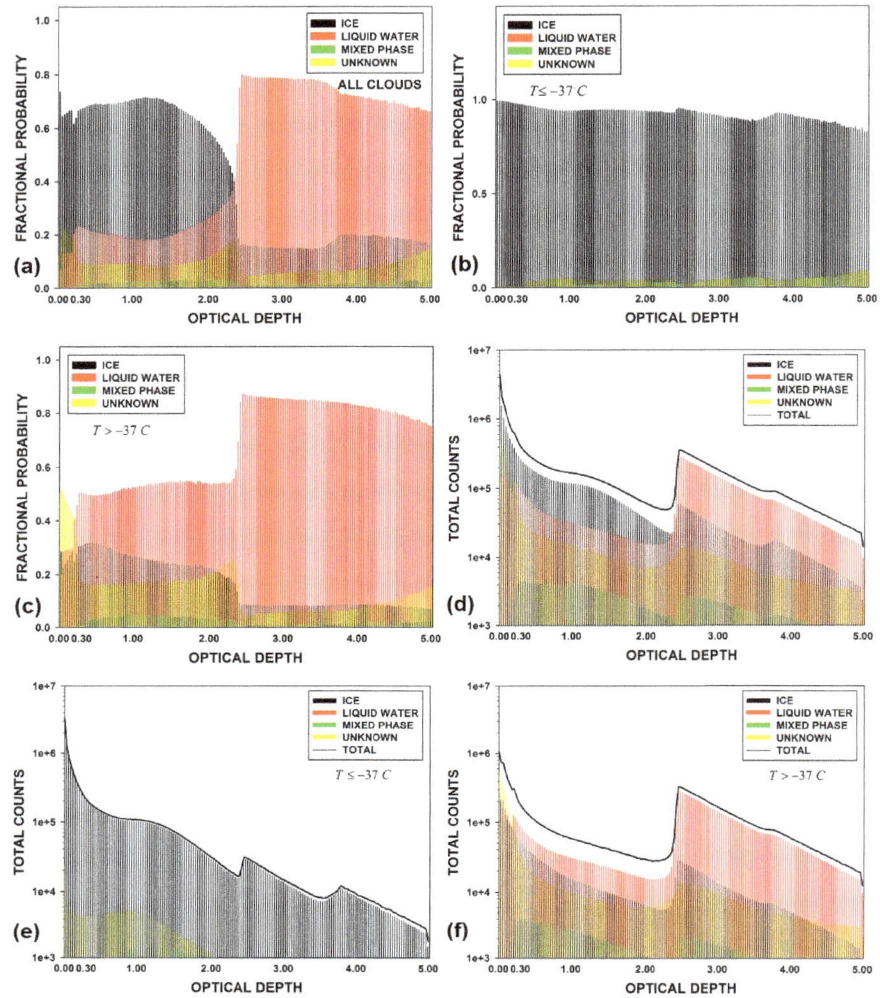

Figure 6. Fractional phase probabilities (ice, liquid water, mixed phase and unknown) in 0.03 cloud optical-depth intervals (532 nm) between 0.00 and 5.00 for 2012 CALIOP V3.02 L2_CPro-5km clouds: **(a)** for all global cases and temperatures, **(b)** for all global cases for cloud top temperature $< -37\,°C$, **(c)** for all global cases for cloud top temperature $> -37\,°C$, and **(d–f)** total counts per cloud phase type in 0.03 optical-depth intervals from the global sample corresponding with **(a–c)**.

threshold temperature for homogenous liquid water freezing) among lidar signal returns, no matter how complex the instrument (i.e., multispectral, polarized) or algorithm. The process involves both assumptions and educated guesswork, and thus some understanding of the physical nature of regional and global cirrus cloud formation and occurrence is essential. In particular, there exists a relatively broad thermal, and spatial, range where relatively warm ice-phase and liquid-water phase clouds are coincident. This ambiguity, exacerbated by some finite likelihood determining whether or not "warm" ice-phase clouds are actually cirrus or some glaciated liquid-cloud remnant (i.e., the heterogenous freezing of liquid-water droplets induced by aerosol particles), cannot be fully reconciled with autonomous measurements alone.

For cirrus, historical research has focused on their specific distinction, given that the physical and radiative properties of upper tropospheric ice-phase clouds differ fundamentally from those of lower-tropospheric glaciated liquid-water clouds. Distinguishing cirrus cloud presence then, and resolving ice cloud inventories relative to their phenomenological and radiative characteristics, has been traditionally approached as a uniquely worthy endeavor. However, considering the breadth of passive and active global satellite remote-sensing observations now available to the community, it has become timely to ask whether or not a cloud paradigm based on traditional phenomenological characteristics remains necessary going forward. For example, given that radiative transfer solutions involving ice rely on parameterizations that vary as a function of crystal habit, and thus effectively cloud top temperature, across the thermal spectrum (e.g., Gu et al., 2011), it becomes unclear whether or not the traditional distinction between ice (or any) cloud genera bear any remaining practical significance.

This study was specifically motivated by the goal of evaluating and refining dependent parameters collected and/or associated with autonomous lidar measurements for distinguishing cirrus cloud presence defined based on traditional phenomenological terms. Progress is made by documenting the difficulties surrounding the task and outlining the sensitivity to global and regional cloud properties derived under varying constraints for cirrus presence. It is ultimately hoped, however, that this paper will motivate a discussion within the community that helps resolve lingering questions that would improve such analysis further. Specifically,

1. are climatologies for cold cirrus alone sufficient for characterizing all cirrus physical and radiative properties necessary for conducting representative climate study?

2. conversely, what is the most efficient manner for inventorying ambiguously "warm" ice clouds? Are their physical and radiative characteristics sufficiently unique so as to continue distinguishing their presence separately from traditional cirrus (i.e., do they necessitate their own unique genus)? Do those elements that are actual cirrus (i.e., sheared fallstreaks) exhibit significant occurrence frequencies so as to take further steps in reducing the ambiguity in their identification?

Taking this one step further, however, and reconsidering these questions outside the framework of traditional phenomenological cloud characterization,

1. do the merits of an all-ice climatology or data sample (e.g., using phase alone in a CALIOP-like cloud data set), despite the presence of glaciated liquid-water remnants, make traditional phenomenological definitions obsolete over a significant range of climate study applications?

2. what other factors (i.e., nucleation mechanism through homogeneous vs. heterogenous freezing, supersaturation rates) drive first-order differences in ice cloud macrophysical, microphysical and radiative properties, which ultimately may require resolving in the long run to reach closure on autonomous cirrus cloud identification for climate study?

Upon resolving these questions, the results of this study will either stand on their own merit or can be refined further to make more pragmatic recommendations for the treatment of ice-phase clouds, and cirrus, in future lidar-related studies.

Summarizing then, Sassen and Campbell (2001; SC2001) recommend that, in the absence of any corresponding visual observations (a veritable luxury, considering expanding satellite data availabilities), a thermodynamic threshold of cloud top temperature $T_{top} \leq -37\,°C$ be used for identifying cirrus clouds from lidar signal returns. They acknowledge, however, the presence of sheared cirrus fallstreaks and glaciated liquid-water clouds existing at warmer apparent temperatures, though the contribution of supercooled liquid water to cirrus formation is generally considered unlikely. Hence, they also do not rule out the existence of relatively "warm" cirrus (i.e., $T_{top} > -37\,°C$). This threshold and additional tests involving cloud top heights and optical depths are evaluated primarily as a function of cloud phase using 2012 NASA CALIOP Version 3.01 Level-2 5 km Cloud Profile products. These data have been merged as a function of along-track sampling resolution to combine vertically-adjacent cloud fragments. A cloud separation threshold of 500 m is also applied to each profile, consistent with SC2001. This study considers lidar-derived cloud altitudes, optical depths and phase determinations, together with independent estimates of temperature obtained from GMAO model data. These parameters were chosen based on their relation to morphological definitions for cirrus cloud classification held currently by the World Meteorological Organization.

The SC2001 threshold proves remarkably effective when applied to the CALIOP data sets. Over 99 % of all corresponding clouds are classified as ice-phase using CALIOP algorithms. Presumably, based on their study, these clouds represent cirrus. Over 81 % of all clouds identified by CALIOP algorithms as ice globally correspond with $T_{top} \leq -37\,°C$. These findings reinforce the effectiveness of this threshold in identifying what are strongly believed to be cirrus clouds. The potential for "cold" non-cirrus clouds present in such a sample is unresolved. In particular, for instruments that do not measure signal depolarization, such as the elastic-scattering lidars used by the NASA MPLNET, the $-37\,°C$ T_{top} threshold temperature represents the only practical metric available for high-confidence identification of ice clouds, and thus presumably cirrus. Reliance solely on CALIOP Level-2 cloud phase distinction generates mean cloud base and top heights and temperatures that, when averaged globally, are 1 km lower (11.15 km vs. 10.07 km) and 6.5 °C warmer ($-58.48°$ C vs. $-52.18\,°C$) than the SC2001 threshold sample. Less than 6 % of the ice-phase CALIOP sample corresponds with $T_{top} > -20\,°C$; this group is very unlikely to represent cirrus clouds and instead likely represents a noise floor in the CALIOP products. Distinguishing the ice phase alone is not sufficient for characterizing cirrus cloud climatological properties in autonomous lidar data sets.

By considering "warm" ice cloud properties separately relative to the SC2001 threshold, a distinct set of cloud top heights and temperatures is derived (5.02 km and $-23.55\,°C$) that is over 5 km lower and 35 °C warmer than the two bulk sample means, respectively. However, nearly 40 % of the ice-phase sample for $T_{top} > -37\,°C$ are colder than $-30\,°C$, or only a few degrees warmer than that threshold. Considering and accounting for uncertainties in NASA Goddard Model Assimilation Office temperatures, for example, proves a sufficient and stable justification for relaxing the threshold slightly (conservatively reapplied here at a threshold of −35 °C), with no apparent loss of fidelity in the mean climatological statistics derived. Liquid-water cloud frequencies derived from CALIOP remain relatively low at temperatures colder than approximately −33 °C.

It is unclear how to reasonably adjust the thermal threshold any further to account for potentially "warm" cirrus and/or sheared fallstreaks. For lidar data sets that lack phase discrimination, by considering the polarization properties of backscattered signals, the fear of increasing supercooled liquid-water contamination rises with warming temperature. Tests designed for evaluating the potential gain in considering cloud top height and optical-depth information to improve the retrieval identify specific regional tendencies. When compared to the clear delineation between ice and liquid phase samples found near the SC2001 threshold alone, however, its unclear whether these tests provide any additional useful information. If anything, they likely complicate the process and exacerbate the ambiguities of "warm" ice cloud classification. Of specific note, broken liquid-water cloud presence impacts relatively low optical-depth distributions, which limits its practical use as a distinguishing attribute for ice or cirrus.

In closing, the use of polarization-sensitive instruments, and the determination of cloud phase, helps greatly in segregating ice and liquid-phase clouds. However, elastic-scattering data sets, like those presently collected by the NASA Micropulse Lidar Network, lack any other viable means for distinguishing cirrus within cloud layer products aside from a thermal threshold (e.g., Lewis et al., 2015). As suggested above, $T_{top} \leq -37$ °C proves sufficiently robust, if not practically conservative. We further demonstrate that the threshold can be justifiably relaxed through uncertainties in thermal interpolation of model and sounding data to lidar-derived cloud boundary heights. If the community decides, however, that inventorying cirrus alone going forward remains a worthy goal, further attempts to reconcile the nature of "warm" ice-phase clouds using polarization should ultimately prove consistent across all projects such as CALIOP and MPLNET. Thus, it will likely be difficult to avoid thermal constraints like that of SC2001. Otherwise, consistency and fidelity across global climatological data sets are jeopardized. Leaving aside questions raised above regarding the representativeness of ice vs. cirrus cloud inventories, the NASA investment in CALIOP and MPLNET, and the goal for robust cloud climate research overall, justifies some binding reconciliation.

Acknowledgements. This research was conducted through NASA Interagency Agreement NNG13HH10I on behalf of the NASA Micropulse Lidar Network, which itself is supported by the NASA Radiation Sciences Program (H. Maring). Authors M. Oo and R. E. Holz acknowledge support from the Oceanographer of the Navy (N2/N6E) through the Program Office at PEO C4I PMW-120.

Edited by: U. Wandinger

References

Cadet, B., Goldfarb, L., Faduilhe, D., Baldy, S., Giraud, V., Keckhut, P., and Réchou, A.: A sub-tropical cirrus clouds climatology from Reunion Island (21_S, 55_E) lidar data set, Geophys. Res. Lett., 30, 1130, doi:10.1029/2002GL016342, 2003.

Campbell, J. R. and Sassen, K.: Polar Regions stratospheric clouds at the South Pole from 5 years of continuous lidar data: macrophysical, optical, and thermodynamic properties, J. Geophys. Res., 113, D20204, doi:10.1029/2007JD009680, 2008.

Campbell, J. R. and Shiobara, M.: Glaciation of a mixed-phase boundary layer cloud at a coastal Arctic site as depicted in continuous lidar measurements, Polar Sci., 2, doi:10.1016/j.polar.2008.04.004, 2008.

Campbell, J. R., Hlavka, D. L., Welton, E. J., Flynn, C. J., Turner, D. D., Spinhirne, J. D., Scott, V. S., and Hwang, I. H.: Full-time, eye-safe cloud and aerosol lidar observation at Atmospheric Radiation Measurement program sites: Instruments and data processing, J. Atmos. Oceanic. Technol., 32, 439–452, 2002.

Chew, B. N., Campbell, J. R., Reid, J. S., Giles, D. M., Welton, E. J., Salinas, S. V., and Liew, S. C.: Tropical cirrus cloud contamination in sun photometer data, Atmos. Environ., 45, 6724–6731, doi:10.1016/j.atmosenv.2011.08.017, 2011.

Cziczo, D. J., Froyd, K. D., Hoose, C., Jensen, E. J., Diao, M., Zondlo, M. A., Smith, J. B., Twohy, C. H., and Murphy, D. M.: Clarifying the dominant sources and mechanisms of cirrus cloud formation, Science, 340, 1320–1324, doi:10.1126/science.1234145, 2013.

DeMott, P. J., Rogers, D. C., Kreidenweis, S. M., Chen, Y., Twohy, C. H., Baumgardner, D., Heymsfield, A. J., and Chan, K. R.: The role of heterogeneous freezing nucleation in upper tropospheric clouds: inferences from SUCCESS, Geophys. Res. Lett., 25, 1387–1390, 1998.

DeMott, P. J., Sassen, K., Poellot, M. R., Baumgardner, D., Rogers, D. C., Brooks, S. D., Prenni, A. J., and Kreidenweis, S. M.: Correction to "African dust aerosols as atmospheric ice nuclei," Geophys. Res. Lett., 36, L07808, doi:10.1029/2009GL037639, 2009.

DeMott, P. J., Prenni, A. J., Liu, X., Kreidenweis, S. M., Petters, M. D., Twohy, C. H., Richardson, M. S., Eidhammer, T., and Rogers, D. C.: Predicting global atmospheric ice nuclei distributions and their impacts on climate, Proc. Natl. Acad. Sci., 107, 11217–11222, 2010.

Dowling, D. R. and Radke, L. F.: A summary of the physical properties of cirrus clouds, J. Appl. Meteorol., 29, 970–978, 1990.

Fu, Q., Yang, P., and Sun, W. B.: An accurate parameterization of the infrared radiative properties of cirrus clouds for climate models, J. Climate, 11, 2223–2237, 1998.

Goldsmith, J. E. M., Blair, F. H., Bisson, S. E., and Turner, D. D.: Turn-key Raman Lidar for profiling atmospheric water vapor, clouds and aerosols, Appl. Opt., 37, 4979–4990, doi:10.1364/AO.37.004979, 1998.

Grund, C. J. and Eloranta, E. W.: The 27–28 October 1986 FIRE IFO cirrus case study: cloud optical properties determined by high spectral resolution lidar,

Mon. Weather Rev., 118, 2344–2355, doi:10.1175/1520-0493(1990)118<2344:TOFICC>2.0.CO;2, 1990.

Gu, Y., Liou, K. N., Ou, S. C., and Fovell, R.: Cirrus cloud simulations using WRF with improved radiative parameterization and increased vertical resolution, J. Geophys. Res., 116, D06119, doi:10.1029/2010JD014574, 2011.

Holz, R. E., Ackerman, S. A., Nagle, F. W., Frey, R., Dutcher, S., Kuehn, R. E., Vaughan, M. A., and Baum, B.: Global Moderate Resolution Imaging Spectroradiometer (MODIS) cloud detection and height evaluation using CALIOP, J. Geophys. Res., 113, D00A19, doi:10.1029/2008JD009837, 2008.

Hu, Y., Vaughan, M. A., Winker, D. M., Liu, Z., Noel, V., Bissonnette, L. R., Roy, G., McGill, M., and Trepte, C. R.: A simple multiple scattering-depolarization relation of water clouds and its potential applications, Proceedings of 23nd International Laser Radar Conference, Nara, Japan, 19–22, 2006.

Hu, Y., Winker, D., Vaughan, M., Lin, B., Omar, A., Trepte, C., Flittner, D., Yang, P., Sun, W., Liu, Z., Wang, Z., Young, S., Stamnes, K., Huang, J., Kuehn, R., Baum, B., and Holz, R.: CALIPSO/CALIOP Cloud Phase Discrimination Algorithm, J. Atmos. Oceanic Technol., 26, 2293–2309, doi:10.1175/2009JTECHA1280.1, 2009.

Josset, D., Pelon, J., Garnier, A., Hu, Y.-X., Vaughan, M., Zhai, P., Kuehn, R., and Lucker, P.: Cirrus optical depth and lidar ratio retrieval from combined CALIPSO-CloudSat observations using ocean surface echo, J. Geophys. Res., 117, D05207, doi:10.1029/2011JD016959, 2012.

Leahy, L. V., Wood, R., Charlson, R. J., Hostetler, C. A., Rogers, R. R., Vaughan, M. A., and Winker, D. M.: On the nature and extent of optically thin marine low clouds, J. Geophys. Res., 117, D22201, doi:10.1029/2012JD017929, 2012.

Lewis, J. R., Campbell, J. R., and Welton, E. J.: Overview and analysis of the MPLNET Version 3 cloud detection algorithm, J. Atmos. Oceanic Technol., in preparation, 2015.

Liou, K.-N.: The influence of cirrus on weather and climate processes: A global perspective, Mon. Weather Rev., 114, 1167–1199, 1986.

Liu, Z., Vaughan, M., Winker, D., Kittaka, C., Getzewich, B., Kuehn, R., Omar, A., Powell, K., Trepte, C., and Hostetler, C.: The CALIPSO lidar cloud and aerosol discrimination: version 2 algorithm and initial assessment of performance, J. Atmos. Oceanic Technol., 26, 1198–1213, 2009.

Lynch, D. K.: Cirrus: history and definition, in: Cirrus, edited by: Lynch, D. K., Sassen, K., O'C. Starr, D., and Stephens, G., Oxford University Press, New York, 480 pp., 2002.

Nazaryan, H., McCormick, M. P., and Menzel, W. P.: Global characterization of cirrus clouds using CALIPSO data, J. Geophys. Res., 113, D16211, doi:10.1029/2007JD009481, 2008.

O'Connor, E. J., Illingworth, A. J., and Hogan, R. J.: A technique for autocalibration of cloud lidar, J. Atmos. Oceanic Technol., 21, 777–786, 2004.

Pruppacher, H. R. and Klett, J. D.: Microphysics of Clouds and Precipitation, 2nd Edn., Kluwer, 954 pp., 1997.

Sassen, K.: Backscattering cross sections for hydrometeors: measurements at 6328 A°, Appl. Opt., 17, 804–806, 1978.

Sassen, K.: The polarization lidar technique for cloud research: a review and current assessment, Bull. Am. Meteorol. Soc., 72, 1848–1866, 1991.

Sassen, K.: Cirrus clouds: A modern perspective, in: Cirrus, Lynch, D., Sassen, K., O'C. Starr, D., and Stephens, G., Oxford University Press, New York, 480 pp., 2002.

Sassen, K. and Campbell, J. R.: A midlatitude cirrus cloud climatology from the Facility for Atmospheric Remote Sensing Part I: macrophysical and synoptic properties, J. Atmos. Sci., 58, 481–496, 2001.

Sassen, K. and Cho, B. S.: Subvisual-thin cirrus lidar dataset for satellite verification and climatological research, J. Appl. Meteor., 31, 1275–1285, 1992.

Sassen, K. and Comstock, J. M.: A midlatitude cirrus cloud climatology from the facility for atmospheric remote sensing, Part III: radiative properties, J. Atmos. Sci., 58, 2113–2127, 2001.

Seifert, P., Ansmann, A., Groß, S., Freudenthaler, V., Heinold, B., Hiebsch, A., Mattis, I., Schmidt, J., Schnell, F., Tesche, M., Wandinger, U., and Wiegner, M.: Ice formation in ash-influenced clouds after the eruption of the Eyjafjallajökull volcano in April 2010, J. Geophys. Res., 116, D00U04, doi:10.1029/2011JD015702, 2011.

Stephens, G. L., Tsay, S., Stackhouse Jr., P. W., and Flatau, P. J.: The relevance of the microphysical and radiative properties of cirrus clouds to climate and climate feedback, J. Atmos. Sci., 47, 1742–1753, 1990.

Stephens, G. L., Vane, D. G., Boain, R. J., Mace, G. G., Sassen, K., Wang, Z., Illingsworth, A. J., O'Connor, E. J., Rossow, W. B., Durden, S. L., Miller, S. D., Austin, R. T., Benedetti, A., and Mitrescu, C.: The Cloudsat mission and the A-Train: a new dimension of space-based observations of clouds and precipitation, Bull. Am. Meteorol. Soc., 83, 1771–1790, doi:10.1175/BAMS-83-12-1771, 2002.

Sun, Z. and Shine, K. P.: Studies of the radiative properties of ice and mixed-phase clouds, Q. J. R. Meteorol. Soc., 120, 111–137, 1994.

Thorsen, T. J., Fu, Q., and Comstock, J.: Comparison of the CALIPSO satellite and ground-based observations of cirrus clouds at the ARM TWP sites, J. Geophys. Res., 116, D21203, doi:10.1029/2011JD015970, 2011.

Vaughan, M. A., Powell, K. A., Kuehn, R. E., Young, S. A., Winker, D. M., Hostetler, C. A., Hunt, W. H., Liu, Z., McGill, M. J., and Getzewich, B. J.: Fully automated detection of cloud and aerosol layers in the CALIPSO lidar measurements, J. Atmos. Oceanic Technol., 26, 2034–2050, 2009.

Virts, K. S. and Wallace, J. M.: Annual, interannual, and intraseasonal variability of tropical tropopause transition layer cirrus, J. Atmos. Sci., 67, 3097–3112, 2010.

Virts, K. S., Wallace, J. M., Fu, Q., and Ackerman, T. P.: Tropical tropopause transition layer cirrus as represented by CALIPSO lidar observations, J. Atmos. Sci., 67, 3113–3129, 2010.

Wang, Z., Sassen, K., Whiteman, D. N., and Demoz, B. B.: Studying altocumulus with ice virga using ground-based active and passive remote sensors, J. Appl. Meteorol., 43, 449–460, 2004.

Winker, D. M., Pelon, J., Coakley Jr., J. A., Ackerman, S. A., Charlson, R. J., Colarco, P. R., Flamant, P., Fu, Q., Hoff, R., Kittaka, C., Kubar, T. L., LeTreut, H., McCormick, M. P., Megie, G., Poole, L., Powell, K., Trepte, C., Vaughan, M. A., and Wielicki, B. A.: the CALIPSO mission: A global 3D view of aerosols and clouds, Bull. Am. Meteorol. Soc., 91, 1211–1229, 2010.

World Meteorological Organization: International Cloud Atlas, vol. I, Manual on the Observation of Clouds and other Meteors, World Meteorological Organization, Geneva, 1975, updated 1995.

Yorks, J. E., Hlavka, D. L., Hart, W. D., and McGill, M. J.: Statistics of cloud optical properties from airborne lidar measurements, J. Atmos. Oceanic Technol., 28, 869–883, 2011.

Young, S. A. and Vaughan, M. A.: The retrieval of profiles of particulate extinction from Cloud–Aerosol Lidar Infrared Pathfinder Satellite Observations (CALIPSO) data: algorithm description, J. Atmos. Ocean. Technol., 26, 1105–1119, 2009.

Young, S. A., Vaughan, M. A., Kuehn, R. E., and Winker, D. M.: The retrieval of profiles of particulate extinction from Cloud-Aerosol Lidar Infrared Pathfinder Satellite Observations (CALIPSO) data: uncertainty and error sensitivity analyses, J. Atmos. Oceanic Technol., 30, 395–428, 2013.

Zhang, D., Wang, Z., and Liu, D.: A global view of midlevel liquid-layer topped stratiform cloud distribution and phase partition from CALIPSO and CloudSat measurements, J. Geophys. Res., 115, D00H13, doi:10.1029/2009JD012143, 2010.

14

Mass-specific optical absorption coefficients and imaginary part of the complex refractive indices of mineral dust components measured by a multi-wavelength photoacoustic spectrometer

N. Utry[1], T. Ajtai[2], M. Pintér[1], E. Tombácz[3], E. Illés[3], Z. Bozóki[2], and G. Szabó[2]

[1]Department of Optics and Quantum Electronics, University of Szeged, Szeged, Hungary
[2]MTA-SZTE Research Group on Photoacoustic Spectroscopy, Szeged, Hungary
[3]Department of Physical Chemistry and Materials Science, University of Szeged, Szeged, Hungary

Correspondence to: T. Ajtai (ajtai@titan.physx.u-szeged.hu)

Abstract. Mass-specific optical absorption coefficients (MACs) and the imaginary part (κ) of the refractive indices of various mineral dust components including silicate clays (illite, kaolin and bentonite), oxides (quartz, hematite and rutile), and carbonate (limestone) were determined at the wavelengths of 1064, 532, 355 and 266 nm. The MAC values were calculated from aerosol optical absorption coefficients measured by a multi-wavelength photoacoustic (PA) instrument, the mass concentration and the number size distribution of the generated aerosol samples as well as the size transfer functions of the measuring instruments. Values of κ were calculated from the measured and particle-loss-corrected data by using a Mie-theory-based retrieval algorithm. The determined values could be used for comparisons with calculated wavelength-dependent κ values typically deduced from bulk-phase measurements by using indirect measurement methods. Accordingly, the presented comparison of the measured and calculated aerosol optical absorption spectra revealed the strong need for standardized sample preparation and measurement methodology in case of bulk-phase measurements.

1 Introduction

Mineral dust (MD) in the atmosphere affects the radiative budget of the Earth both directly and indirectly. It absorbs and scatters the solar and terrestrial radiations in the near-infrared–visible–near-UV wavelength range, and its overall forcing effect is influenced by its chemical composition, temporal and spatial distribution as well as its optical properties. As a consequence of this complex dependence, current knowledge on both the sign and the magnitude of the MD-induced radiative forcing is largely uncertain (Claquin et al., 1998). Some studies estimate that the forcing effect of MD compares to or even exceeds the cumulative forcing effect of all other aerosol originating from anthropogenic sources (Sokolik and Toon, 1996). Therefore, there is a strong need for measurement methods which can characterize the optical properties of MD accurately. This is especially important in case of its anthropogenic fraction, which is estimated to make up about 30–50 % of the total atmospheric MD load (Tegen and Fung, 1995).

The wavelength-dependent optical properties of MD are represented by the complex refractive index ($Ri = n + i\kappa$), where n and κ are directly associated with scattering and absorption, respectively. Unfortunately, Ri and especially its imaginary part (κ) is notoriously difficult to be measured because unlike n, which is generally well constrained and only slightly (if at all) varies with wavelength, and is almost completely composition independent, κ has a strong chemical composition dependence, and might vary over 3 orders of magnitude in the whole climate-relevant spectral region (McConnell et al., 2010; Otto et al., 2007).

Generally speaking, light absorption is one of the most difficult aerosol parameters to be measured even when optical absorption has the same magnitude as scattering (i.e. the single-scattering albedo (SSA) is around 0.5). This is typically the case for black carbon aerosol with SSA $\approx$ 0.6. For this type of strongly absorbing aerosol the widely adopted methods for aerosol absorption measurement – i.e. either the differential method based on the simultaneous measurement of scattering and extinction, or methods based on transmission measurement on filter accumulated aerosol (Moosmüller et al., 2009) – work quite well at least in case of large aerosol concentrations. However, the former method has limited sensitivity, while the latter one has large uncertainties due to the spurious effect of filter–light interaction. Compared to this already challenging measurement task, absorption measurements on MD samples are even much more difficult due to the fact that κ is typically orders of magnitude smaller than n (i.e. the SSA is typically higher than 0.95). In this case the above-mentioned methods are no longer reliable.

Accurate optical absorption data are needed not only for modelling the effect of MD in the atmosphere by calculating radiative transfer properties such as optical thickness, single-scattering albedo, or asymmetry parameter but also for studying the variation of these parameters due to the variations of MD composition. The geographical location of the emission site was found to strongly influence the composition of the emitted MD (Claquin et al., 1998; Nickovic et al., 2012). Various efforts have already been made to calculate the optical properties of MD from the optical properties of its components (e.g. by using the effective medium approximation (Sokolik et al., 1993; Sokolik and Toon, 1999)). Most often these calculations are based on the Mie theory, and approximate the particles as being spherical and homogeneous. In these calculations the size distribution and the *Ri* of each component play a crucial role. Although refractive indexes of most abundant MD components are measured and tabulated over broad, climate-relevant wavelength ranges, these data are typically derived from bulk-phase indirect measurements, wherefore they have limited reliability. Furthermore, these measurement results are not validated by aerosol-phase measurements (i.e. the actual microphysical aerosol conditions such as size and morphology are neither taken into account nor validated). Very often *Ri* data (and especially κ) differ significantly even when measured at the same or very close wavelengths in the bulk phase due to the differences in the applied measurement methodology and sample treatment (Rao, 2002; Arakawa et al., 1997). There are serious concerns especially about the reliability of κ values determined by the most commonly applied diffuse reflectance-based techniques. Claquin et al. (1998) found that the variation of κ determined by different methodologies manifests in significant uncertainties in radiative forcing calculations, and thus not only the magnitude but also the sign of the calculated radiative forcing parameter is uncertain.

Another potential application of accurate optical absorption data in a wide wavelength range could be the so-called inverse problem (i.e. the determination of the chemical composition and thus the sources of MD (Müller et al., 2011) from its measured optical properties). This analysis is based on the fact that unlike scattering, which is mainly driven by aerosol size and morphology, absorption has rather strong sensitivity to the chemical composition too (Andreae and Gelencsér, 2006; Moosmüller et al., 2009).

The photoacoustic (PA) measurement technique has already proved its applicability under laboratory and field conditions in case of strongly absorbing aerosol components such as black carbon, brown carbon and HULIS (humic-like substances) (Andreae and Gelencsér, 2008; Cappa et al., 2008; Lack et al., 2012; Chakrabarty et al., 2010; Moosmüller et al., 2012; Ajtai et al., 2010; Utry et al., 2014). The accuracy of the method stems from both its scattering insensitivity and its filter-free sampling (measurement made on the aerosol on its natural, suspended state) (Andreae, 2001). On the other hand, application of the PA technique for the measurement of MD or MD components is a real challenge not only because of the high SSA and strong wavelength dependence of κ as described above, but also because of the relatively large size of the measured particles. Indeed, while the typical size of atmospheric black and brown carbon is in the range of a few hundred nanometres or even below, the size of the MD particles is mostly in the micron range. Consequently, the size transfer function of the detection system has to be taken into account accurately, especially for large particles, since already a few of them can considerably modify the aerosol mass concentration in the PA cell with which the PA signal is linearly proportional as described below.

Photoacoustic determination of optical absorption by aerosol is based on the absorption of periodically modulated light and the subsequent non-radiative relaxation of excited aerosol particles. Due to the high thermal conductivity and the small size of climate-relevant ambient aerosol particles, the energy absorbed by them is transferred promptly (adiabatically) to the surroundings. The subsequent thermal expansion induces periodic pressure variation, which is proportional to the concentration and the absorption cross section of light-absorbing particles, can be detected and converted into a measurable electronic signal by an acoustic transducer (typically a microphone). The optical absorption coeffitient (OAC) can be deduced from the measured PA signal by using a conversion factor determined by an independent calibration of the PA system. The major advantages of this method compared to the alternatives are selectivity, sensitivity, and accuracy. Selectivity means that PA is not only a zero background technique (i.e. in the absence of light absorption no PA signal is generated) but in addition to that the PA response is inherently unaffected by the non-absorbing aerosol particles present in the sampled aerosol stream, and also by the non-absorbing compounds of the individual aerosol particles (Moosmüller et al., 2009; Bozóki et al., 2011). The accuracy of the recently published PA instruments for aerosol measurement is in the range of a few percent, which is at least an order of magnitude better than that of the most commonly used filter-based instruments, even when their readings are corrected posteriorly after the measurement (Andreae and Gelencsér, 2006).

To the best of our knowledge, this paper presents the aerosol-phase mass-specific optical absorption coefficient (MAC) spectra of the most abundant mineralogical MD components for the first time. However it is worth noting that single-scattering albedo of several mineral dust samples was also previously determined by the combination of multi-wavelength photoacoustic spectroscopy and nephelometry (Moosmüller et al., 2012). These spectra are measured by our recently developed multi-wavelength photoacoustic spectrometer (4λ-PAS) in the UV–Vis–near-IR spectral region. Furthermore, the Mie-theory-based simulated OAC spectra are generated by using bulk-phase *Ri* values derived from published literature and the measured number size distributions corrected by the size transfer function of the measuring system. These theoretical curves are compared with measured OAC spectra, and based on these comparisons the PA method is arguably a powerful tool for verifying theoretical calculations.

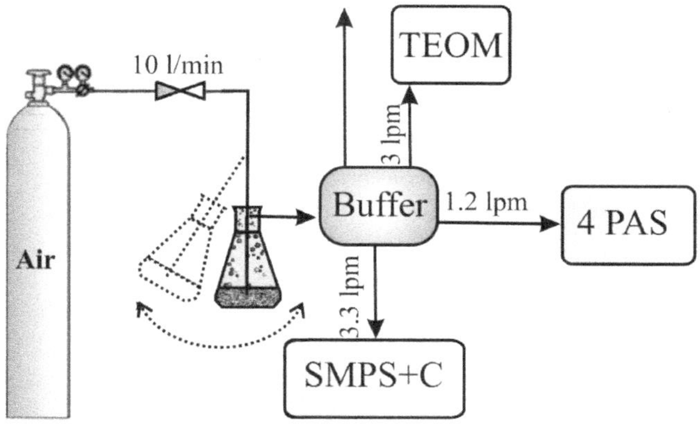

Figure 1. Schematic diagram of the experimental set-up (the vessel with dotted contour line and the arrow indicates that it is being shaken).

2 Experimental and methods

2.1 Sample preparation

In this study MD components, phyllosilicates (illite, kaolin and bentonite), oxides (quartz, hematite and rutile), and carbonate (limestone) were investigated. High-purity illite was obtained from a Hungarian source Füzérradvány and measured as received. Its mineral composition is illite 75, montmorillonite 5, quartz 14, kaolinite 1 and feldspar 4 % and its chemical composition is SiO_2 47.5, Al_2O_3 29.0, Fe_2O_3 1.2, CaO 1.8, MgO 3.0, K_2O 6.8, Na_2O 1.1, loss of ignition 8.5 and total 98.9 %. Kaolin was obtained from Zettlitz (Germany). It contains clay mineral kaolinite as the main component up to 90 %, accompanied by some micaceous minerals and small amounts of quartz. Chemical composition of kaolin is MgO 0.27, CaO 0.29, SiO_2 46.9, Al_2O_3 37.4, Fe_2O_3 0.65, K_2O 0.84, Na_2O 0.44, TiO_2 0.18, H_2O 12.95 and total 99.92 %. We used a fine-grained washed kaolin sample with a characteristic particle size of 1.2–1.4 μm. The source of montmorillonite was bentonite from Mád (Hungary). Its mineral composition is montmorillonite 86, quartz 3, feldspar 5 % and its chemical composition is SiO_2 63.8, Al_2O_3 16.0, Fe_2O_3 5.0, CaO 2.4, MgO 1.2, K_2O 0.8, Na_2O 1.1, TiO_2 0.6, loss of ignition 8.9 and total 99.8 %. The mined bentonite had a grain size of less than 90 μm, activated by 5 % Na_2CO_3 and enriched in montmorillonite. A high-purity quartz sample ($SiO_2 > 99$ %) was ground and the sieve fraction $< 200\,\mu$m was used. Hematite (Fe_2O_3) produced by Reanal Private Ltd. (Hungary) had a degree of purity > 95 %. Rutile (Bayertitan R-U-2) produced by Bayer AG contained 98 % TiO_2; it is an alumina lattice modified sample, its relative light-scattering power is 96 and its specific gravity is 4.1 g cm^{-3}. Limestone (calcium carbonate with $CaCO_3$ content 98 %) was collected in the Pilis Mountains (Hungary), and then ground to produce fine powder.

2.2 Sample generation

MD particles were re-dispersed from the prepared powder sample and added to the buffer chamber for instrument sampling in the experimental set-up depicted in Fig. 1.

Prior to the measurements, aerosol samples were desiccated in a vessel partly filled with silica gel for at least 24 h. A small amount of mineral powder was placed into the Erlenmeyer vessel and then into an automatic shaker. Synthetic air was blown through the shaker at a constant flow rate of 10 Lpm. The re-dispersed aerosol was fed into a buffer chamber. Pressure was kept at atmospheric level by an exhaust pipe implemented into the buffer chamber. Each instrument sucked the aerosol loaded air mass directly from the chamber. The relative humidity (RH) of the air stream including suspended particles in the buffer chamber was measured during the whole measurement campaign and found to be below RH $\sim$ 30 in each sample. All the presented measurement results were performed at room temperature and atmospheric pressure.

2.3 Instrumentation for aerosol characterization

The OAC was measured simultaneously at four wavelengths by a multi-wavelength photoacoustic spectrometer (4λ-PAS). It contained four measuring cells with each of them being illuminated through one of the operational wavelengths of 266, 355, 532 or 1064 nm. With the help of a magnetic valve controlled by the electronics of the system, the sample flow path through the analyser was regularly switched in a way that the purging gas with the aerosols would flow through a filter before entering the PA cells. In this way the background PA signal (i.e. the signal generated with zero aerosol concentration) was measured and subtracted from the PA signal generated by the aerosol. This repeated background subtraction was necessary because laser beams can have some pointing instability, and thus the background PA signal may vary in time. Furthermore, besides aerosol and background measurements the analysers also have a calibration mode during which the optical set-up is modified with the help of flip-up mirrors so that each cell is illuminated with the 532 nm beam and the PA signal is measured at different NO_2 concentrations. The sensitivity of each cell was calculated from the slope of the line which was fitted to the NO_2 concentrations vs. the PA signal points divided by the light power going through the cell in the calibration mode, and also by the optical absorption coefficient, which corresponded to the unit concentration of NO_2 at the measurement wavelength. Multiplying the sensitivity of a given cell with the light power going through the cell in the measurement mode gives the conversion factor with which the background-subtracted aerosol-generated PA signal has to be multiplied in order to calculate the OAC from the PA measurement. This calibration protocol ensures the wavelength- and composition-independent determination of the OAC. The accuracy of the instrument is between 4 and 10 % depending on the applied wavelength, where the decreased accuracy towards the shorter wavelengths is due to the decreased accuracy of laser power measurement. In the measurement mode a complete 15 min long measurement cycle includes a 10 min sampling and a 5 min background measurement period. Generally, eight measurement cycles were used for data evaluation. The initial transient period was not evaluated. Depending on the investigated mineral dust component and the applied wavelength, the fluctuation of the PA signal including concentration instability was found to be in between 15 and 22 %. The number concentration and size distribution of the dust minerals were measured simultaneously in the size ranges of 5 to 950 nm and 350 nm to 32 μm with a scanning mobility particle sizer (SMPS, GRIMM system Aerosol Technik, Germany, type SMPS) and an optical particle counter (OPC), respectively. The SMPS included a condensation nucleus counter (CPC Model #5.400) and a classifier "Vienna"-type long differential mobility analyzer (LDMA, Model #5.500). The LDMA separates particles according to their mobility by balancing their drag and electrical force on the equally charged aerosol stream. The sized particles are then sent to CPC for size-segregated number density measurement. The multiple charge correction defined by the ISO 15900 recommendation was used in the data evaluation during the whole measurement campaign to further increase the reliability of the measured data. Coinciding correction of the measured data was made in order to minimize the shielding effect that occurs when two or more particles arrive in the detection chamber simultaneously. The sheath and aerosol flow rate were set to 3.0 and 0.3 Lpm, respectively. In the OPC the sample air with various size particles was drawn constantly through the flat light beam produced by a focused laser diode. Each signal was then counted and classified in different size channels by an integrated pulse height analyser. The differences between the data measured by SMPS and OPC in the overlapping size domain were found to be inside the cumulative uncertainty range of the instrumentations (i.e. below 10 %); therefore, it was neglected during the data evaluation. Aerosol mass concentration was measured by a tapered element oscillating microbalance (TEOM, Rupprecht and Patashnick, Model 1400a). TEOM incorporates an inertial balance that directly measures the mass collected on an exchangeable filter cartridge by monitoring the corresponding frequency changes of a tapered element (Patashnick et al., 2002). The sample flow rate and the average time were set to be 3 Lpm and 10 min, respectively.

In addition to the online measurements, the dispersed MD components were collected on quartz filters for morphology analysis by scanning electron microscopy (SEM).

2.4 Calculations

In order to determine the number size distributions of aerosol samples within the central acoustic resonator of the PA cells where the PA signal generation occurs, as well as within each instrument connected to the aerosol generation buffer volume, the measured SMPS data were corrected with the calculated size transfer function of the experimental set-up by using the particle loss calculator (http://www.mpch-mainz.mpg.de/~drewnick/PLC; Von der Weiden et al., 2009). The model behind this code takes into account most of the physical phenomena which can cause particle losses, such as diffusion, sedimentation, turbulent deposition as a function of dimensions, flow rate, angle of bending, etc. The calculated high end (50 %) cut-off diameter in the size transfer function of the PA system is at 1.2 μm (Utry et al., 2014). The uncertainty of the particle loss calculator was determined independently in another study (von der Weiden et al., 2009). Since the experimental set-up presented here is much less complicated (i.e. devoid of double bend configuration with sharp curvature) than the one used in the validation procedure and all the applied instruments use low flow sampling in the laminar flow region where the reliability of this packet loss concealment algorithm is more adequate, we used this uncertainty in this study.

For calculating the mass-specific optical absorption coefficient (MAC) for each MD component the following quantities were used: the photoacoustically measured OAC; the mass concentration of the aerosol measured by the TEOM; and the mass concentrations calculated for the measured size distribution ratio multiplied with the aerosol mass ratio calculated from the measured size distribution with and without taking into account the particle losses. Based on these quantities MAC was calculated as follows:

$$\mathrm{MAC} = \frac{\mathrm{AOC}\left[\int_0^\infty D_\mathrm{p}^a n_N \mathrm{d}D_\mathrm{p}\right]_\mathrm{cor}}{m_\mathrm{TEOM}\left[\int_0^\infty D_\mathrm{p}^a n_N \mathrm{d}D_\mathrm{p}\right]_\mathrm{meas}}, \quad (1)$$

where m_TEOM is the mass concentration measured by TEOM, while the $\left[\int_0^\infty D_\mathrm{p}^a n_N \mathrm{d}D_\mathrm{p}\right]_\mathrm{meas}$ and the $\left[\int_0^\infty D_\mathrm{p}^a n_N \mathrm{d}D_\mathrm{p}\right]_\mathrm{cor}$ are the measured and the particle-loss-corrected volume concentrations, respectively.

For each MD component an optical absorption spectrum was calculated by using the Mie theory. This calculation also takes into account the measured and particle-loss-corrected size distribution. Furthermore, *Ri* values, which are most often applied (e.g. in effective medium approximation-based calculations), were obtained from various publications as follows:

- Bentonite: in this study we used the optical constants of montmorillonite, which is the main constituent of bentonite. Data of Arakawa et al. (1997) derived from transmission measurement on a thin film were used.
- Kaolin: both the data of Egan and Hilgeman (1979) and Arakawa et al. (1997) made on thin films by diffuse scattering and transmission measurement, respectively, were used.
- Illite: we used data from Egan and Hilgeman (1979). They applied two measurement methods: reflectance and diffuse transmittance measurements performed on a pressed pellet and on a thin section, respectively.
- Rutile: two *Ri* data sets were obtained from the SOPRA N&K database (http://www.sopra-sa.com/more/database.asp). Both of them were deduced from ellipsometric measurements conducted on thin films.
- Hematite: data from Querry (1985) which were based on diffuse reflectance from a bulk material were used. Furthermore, based on the segregated spectra for o- and e-rays, the sample anisotropy was taken into account by using the spectral averaging method. First the individual spectra for both o-ray and e-ray indices were calculated; then – since there are two equivalent axes for o-ray and only one for e-ray – a weighted average of these individual spectra in the form of [(2/3) o-ray + (1/3) e-ray] was calculated.
- As far as quartz and limestone are concerned, unfortunately no κ data were found in the literature for the studied wavelength range.

Values of κ were also calculated from the measured and particle-loss-corrected data by using a simple Mie-theory-based retrieval algorithm (Guyon et al., 2003; Hoffer et al., 2006). First, the Mie code takes the n values from the literature and the measured and particle-loss-corrected size distribution data as input parameters, and then calculates the absorption coefficient. Thereafter this value is compared to the measured absorption coefficients and following that the initial κ value is increased stepwise until the calculated and measured optical coefficients agree to within 0.5 %. The κ value associated with this agreement represents the imaginary part of the investigated sample. However, it is worth noting that although the application of Mie theory for non-spherical particles limits the reliability of the computed data, many prior works used this approach to calculate the spectral responses of dust minerals (Conant et al., 2003; DeSouza-Machado et al., 2006; Moffet and Prather, 2005; Wang et al., 2002). This is because the most widely used shape-sensitive models such as T-matrix or DDA (discrete dipole approximation) are also not using the real morphology of the investigated aerosol sample, but they are based on mathematically well characterized geometrical approximation (Kalashnikova and Sokolik, 2004; Tegen and Lacis, 1996). Kalashnikova and Sokolik demonstrated that the deviation between the spherical and non-spherical approaches becomes significant above the approximate size parameter of 5 and only in case of the asymmetry parameter and the scattering phase function, which are mainly governed by the coherent scattering physical process. However, for the absorption, which is an incoherent physical phenomenon, the deviation is much less significant even above size parameter higher than about 5.

3 Results and discussion

The MACs and the imaginary parts of the complex refractive indices deduced from the measured data are listed in Tables 1 and 2, respectively. Number size distributions can be seen in Fig. 2. Thick lines represent those measured by the OPC and SMPS, while thin lines are derived from the same OPC+SMPS measurements but corrected with the calculated size transfer function of the PA system (i.e. these are the number size distributions that are assumed to be within the PA cells). Figure 3 shows the photoacoustically measured OACs and the corresponding calculated optical absorption spectra. In some cases the results of the calculations are also presented without taking into account the particle losses in the PA system for comparison purposes. SEM pictures of the measured particles are shown in Fig. 4.

The SEM pictures are largely similar to those reported in the literature (Bohor and Hughes, 1971; Dong et al., 2014) and show that apart from rutile, which has a more or less compact and spheroid shape, all other particles show a high degree of irregularity with versatile and unclassifiable shapes.

Although all the instruments were placed as close to the buffer chamber as possible to minimize the curves and

Table 1. Mass-specific optical absorption coefficients of the measured MD components.

Mineral component	Mass-specific optical absorption coefficients [$m^2\,g^{-1}$] λ_1 266 nm	λ_2 355 nm	λ_3 532 nm	λ_4 1064 nm
Illite	$1.0\times10^{-1}\pm2\times10^{-2}$	$1.9\times10^{-2}\pm2\times10^{-3}$	$5.1\times10^{-3}\pm5\times10^{-4}$	$2.5\times10^{-3}\pm5\times10^{-4}$
Kaolin	$4.8\times10^{-1}\pm7\times10^{-2}$	$2.3\times10^{-1}\pm3\times10^{-2}$	$2.1\times10^{-2}\pm3\times10^{-3}$	$8.8\times10^{-3}\pm1\times10^{-3}$
Bentonite	$3.1\times10^{-1}\pm6\times10^{-2}$	$5.8\times10^{-2}\pm1\times10^{-2}$	$5.2\times10^{-3}\pm1\times10^{-3}$	$4.6\times10^{-3}\pm9\times10^{-4}$
Hematite	$5.3\times10^{-1}\pm1\times10^{-1}$	$5.7\times10^{-1}\pm1\times10^{-1}$	$5.4\times10^{-1}\pm1\times10^{-1}$	$3.4\times10^{-2}\pm8\times10^{-3}$
Rutile	$3.5\times10^{-1}\pm5\times10^{-2}$	$7.1\times10^{-1}\pm1\times10^{-1}$	$2.2\times10^{-3}\pm3\times10^{-4}$	$1.4\times10^{-3}\pm2\times10^{-4}$
Quartz	$1.0\times10^{-2}\pm2\times10^{-3}$	$5.8\times10^{-3}\pm1\times10^{-3}$	$2.3\times10^{-3}\pm4\times10^{-4}$	$2.0\times10^{-3}\pm4\times10^{-4}$
Limestone	$5.3\times10^{-3}\pm5\times10^{-4}$	$6\times10^{-3}\pm7\times10^{-4}$	$5.5\times10^{-4}\pm1\times10^{-4}$	$3.8\times10^{-4}\pm4\times10^{-5}$

Table 2. Imaginary parts of the complex refractive indices (κ) of the measured MD components. For the sake of clarity these data are supplemented with the n and κ values taken from the literature, where the former ones are used in the κ calculation while the latter ones are shown in this table for inter-comparison purposes.

Mineral component		λ_1 266 nm		λ_2 355 nm		λ_3 532 nm		λ_4 1064 nm	
		n	k	n	k	n	k	n	k
Illite	reference[1]	1.42	2.0×10^{-3}	1.41	1.5×10^{-3}	1.41	8.3×10^{-4}	1.39	1.2×10^{-3}
	calculated		1.1×10^{-2}		1.7×10^{-3}		7.0×10^{-4}		8×10^{-4}
Kaolin	reference[2]	1.67	4.0×10^{-2}	1.61	9.1×10^{-3}	1.57	6.3×10^{-3}	1.56	5.9×10^{-3}
	calculated		4.2×10^{-2}		6.2×10^{-3}		7.0×10^{-4}		9.0×10^{-4}
Bentonite	reference[2]	1.64	3.2×10^{-2}	1.58	2.4×10^{-2}	1.54	2.4×10^{-3}	1.52	1.1×10^{-3}
	calculated		3.1×10^{-2}		2.5×10^{-2}		2.0×10^{-3}		1.3×10^{-3}
Hematite	reference[3]	1.94	1.25×10^{0}	2.37	1.07×10^{0}	3.14	5.5×10^{-1}	2.69	2.0×10^{-2}
	calculated		1.26×10^{0}		1.1×10^{-0}		5.5×10^{-1}		2.4×10^{-2}
Rutile	reference[4]	2.31	1.2×10^{0}	3.17	8.4×10^{-1}	3.05	0	2.74	0
	calculated		1.1×10^{0}		1.0×10^{-1}		1.2×10^{-4}		1.6×10^{-4}
Quartz	reference[5]	1.60		1.57		1.55		1.54	
	calculated		6.5×10^{-3}		4.1×10^{-3}		2.3×10^{-3}		3.1×10^{-3}
Limestone	reference[5]	1.68		1.63		1.61		1.59	
	calculated		3.0×10^{-3}		2.6×10^{-3}		4.8×10^{-4}		6.0×10^{-4}

Sources of the complex refractive indices used in the calculations: (1) Egan and Hilgeman (1979); (2) Arakawa et al. (1997); (3) Querry (1985); (4) Sopra 2; (5) Ghosh (1999).

lengths of the tubes, there is a relatively large difference in the number size distribution measured directly at the exit of the buffer volume and the PA detection cell, especially for particles with a characteristic diameter above 1 μm. On the other hand, the corrected size distribution data also show that almost all particles below 1.2 μm get into the PA detector. Furthermore, it is also important to note that neglecting the particle loss in the instrumentation modifies the calculated optical spectra quite drastically, as shown in Fig. 3. It affects not only their overall magnitude but also their shape as described in other works (Filep et al., 2013).

As it is expected, their MAC values are several orders of magnitude smaller than those of BC ($7.5\pm1.2\,m^2\,g^{-1}$ at 550 nm (Bond and Bergstrom, 2006)), $10\pm3.5\,m^2\,g^{-1}$ at 405 nm (Cross et al., 2010)) and BrC (0.5–$1.2\,m^2\,g^{-1}$ at 404 nm (Lack et al., 2012)). On the other hand, the PA signals were measured with a high signal to noise ratio even for the smallest MAC values; consequently all the presented data have high reliability.

However it is also worth noting that, for particulate matter with high scattering to absorption ratio, not only the signal to noise but also the signal to background ratio can also affect the reliability of the measured data. In our study, the ratio of the total photoacoustic signal to background signal was varied between 1.4 and 3.3 depending on the investigated components and the applied wavelengths. The signal to noise ratio was higher than 20 even in the less absorptive components such limestone and quartz; therefore, all the

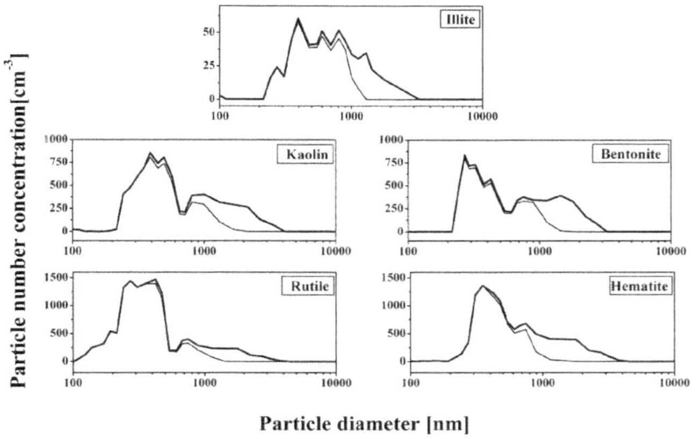

Figure 2. Number size distributions: in the buffer chamber measured by the SMPS and the OPC instruments (thick line) and in the PA cell calculated from the SMPS + OPC data by taking into account the calculated size transfer function of the PA system (thin line).

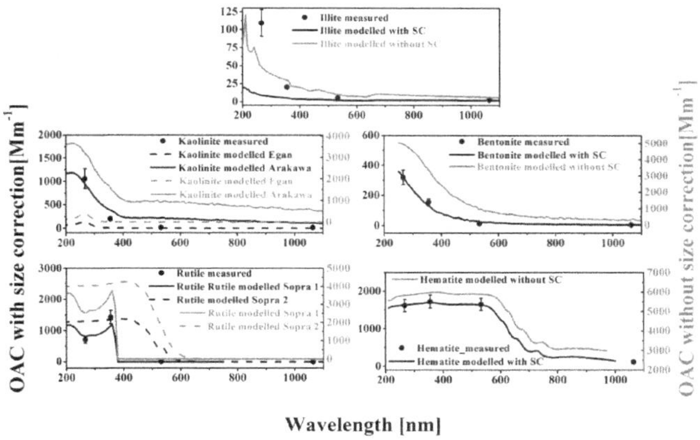

Figure 3. Aerosol optical absorption coefficients. The points represent OAC data calculated from PA signals with error bars representing the uncertainty of the OAC data (see text for details); the black and grey lines represent the Mie-theory-based calculations with or without size correction (SC – taking into account the size transfer function of the PA system), respectively. The dashed grey lines represent the results of calculations by using alternative RI parameters (see text for details).

presented data have high reliability. The κ values deduced from the presented measurements are compared to the values derived from the literature (Table 2). For some mineral dust components the differences between these values are within the combined uncertainty of the measurement and the calculation (e.g. in case of bentonite, hematite and rutile) while for other cases (such as for illite and kaolin) significant differences (above the uncertainty level) can be observed.

The uncertainties of the calculated parameters are affected by the fluctuations in the concentration of the generated aerosol, the noise of the PA measurements, the errors of the size distributions as well as the uncertainty of the literature values used in the retrieval procedure. In order to estimate the resulted uncertainty of the calculated parameters, a sensitivity study of the input parameters was performed. The uncertainty of the measured OAC values was determined directly by the uncertainty of the PA measurement including concentration instability. In case of MAC values the uncertainties of about ± 20 and ± 5 % were used for OAC and mass concentration measurements respectively, where the OAC uncertainty includes the concentration instability as well. The quadrature sum of these individual errors was found to be about 21 %. The uncertainty of the modelled OAC values is not determined here due to the lack of available information about the error of the literature *Ri* data. Error propagation of size distribution for the *Ri* retrieval was also investigated separately, and it was found that the uncertainty of the size distribution in between ± 15 % translates directly into the computed χ. Therefore, the resulting quadratic error of computed κ was found to be about 25 %.

As far as the measured and the calculated OAC spectra shown in Fig. 3 are concerned, it must be first of all emphasized that the presented research does not intend to provide a quantitative comparison of the theoretical and experimental results or to validate the optical models or the κ data used in the modelling because of the lack of detailed information on sample preparation and on measurement methods applied on bulk samples for the determination of κ. Rather, as the presented OAC values are based on a direct and aerosol-phase measurement of absorption using standardized measurement conditions and sample treatment, they can be used as reference values which reveal the deviation between the data presented earlier based on the differentiation of the applied methodology and the sample treatment. Moreover, they can contribute to the better understanding of the effect of bulk-aerosol transition on the optical properties.

Despite all the uncertainties in the bulk measurements there is an overall order of magnitude agreement between the measured and the calculated optical absorption spectra shown in Fig. 3. In some cases (i.e. for bentonite and for hematite) the agreement is remarkably good whenever the size transfer function is taken into account. For other MD components there are various possible explanations for the deviations. In case of kaolinite the two presented calculations give rather different results and neither of them has a good agreement with the results of the PA measurement. The publication of Arakawa et al. (1997) criticized the measurement method of Egan and Hilgeman (1979) by stating that their photon detector had insufficient collection efficiency below 400 nm. Indeed, the Arakawa et al. (1997) model shows much better agreement with the PA measurements in the short wavelength range than that of Egan and Hilgeman (1979). On the other hand, the calculation based on the data of Arakawa et al. (1997) overestimates the OAC compared to the PA results at longer wavelengths. As it was

Figure 4. SEM pictures of collected aerosol samples.

explained above, the low MAC values by the PA method still have high reliability; therefore, one can speculate that the overestimation by Egan and Hilgeman (1979) might be due to the fact that, in their transmission measurement made on a thin film, the wavelength of excitation in the long wavelength range became longer than the width of the film, and thus the light suffered multiple reflections, which resulted in the observed overestimation as described in detail elsewhere (McNeil and French, 2000; Arakawa et al., 1997). As far as illite is concerned, since it is measured with a similar method as kaolinite by Egan and Hilgeman (1979), the insufficient collection efficiency of their detector below 400 nm could again be a possible explanation for the discrepancies between the measurement and the calculation. However, the real differences between the measured and simulated values cannot be determined here due to the lack of revised data by Arakawa et al. (1997) in this case. Rutile provides a good example for the fact that even the same measurement method can result in strongly different spectra, most probably either due to differences in the sample preparation methods or due to differences in the samples themselves.

4 Summary and conclusions

The mass-specific optical absorption coefficient and the imaginary part of the complex refractive indices are determined for the most abundant MD components by using a recently developed multi-wavelength photoacoustic system. For modelling the OAC values by the Mie-theory-based simulation, we used the measured and particle-loss-corrected number size distribution, as well as tabulated *Ri* data based on bulk or thin film phase, an indirect measurement method taken from the published literature. Using the Mie-theory-based simulation and assuming spherical particles, we determined the refractive indices of the measured components and with the quoted limitations described in detail in Sect. 2.4.

We made a comparison between the measured and the simulated OAC data. We also experimentally demonstrated here that the reliability of both the measured and the simulated OAC values can be further increased by using particle-loss-corrected measurement data.

In some cases we have found good agreement between the photoacoustically measured and the simulated OAC data, while in other cases we could give plausible explanations for the differences. Nevertheless, due to the limited availability of information on the applied measurement techniques, on sample preparation and the detailed description of the measurement conditions in case of bulk samples, the scientific goal of this work was not the quantitative comparison of the measured and the simulated OAC values. Rather, we wanted to measure aerosol-phase optical absorption using a direct and light-scattering immune technique. Due to being free from all analytical and methodological artefacts (which may occur when using competitive techniques), the values measured and presented here can be used for comparison with the optical absorption measurements conducted on bulk-phase samples. Furthermore, the data deduced from the presented experiments open up novel possibilities to experimentally investigate the effect of the bulk-aerosol transition when the bulk-phase data are used in aerosol forcing calculations.

Acknowledgements. Financial support by the Hungarian Scientific Research Foundation (OTKA, project no. K101905) is gratefully acknowledged. The European Union and the European Social Fund have provided financial support to the project under the project no. TÁMOP-4.2.2.A-11/1/KONV-2012-0047 and TÁMOP FUTURICT 4.2.2.C-11/1/KONV-2012-0013. The research of Noémi Utry was supported by the European Union and the State of Hungary, co-financed by the European Social Fund in the framework of TÁMOP 4.2.4. A/2-11-1-2012-0001 "National Excellence Program".

Edited by: F. Pope

References

Ajtai, T., Filep, Á., Kecskeméti, G., Hopp, B., Bozóki, Z., and Szabó, G.: Wavelength dependent mass-specific optical absorption coefficients of laser generated coal aerosols determined from multi-wavelength photoacoustic measurements, App. Phys. A, 103, 1165–1172, 2010.

Andreae, M. O.: The dark side of aerosols, Nature, 409, 671–672, 2001.

Andreae, M. O. and Gelencsér, A.: Black carbon or brown carbon? The nature of light-absorbing carbonaceous aerosols, Atmos. Chem. Phys., 6, 3131–3148, doi:10.5194/acp-6-3131-2006, 2006.

Arakawa, E. T., Tuminello, P. S., Khare, B. N., Millham, M. E., Authier, S., and Pierce, J.: Measurement of optical properties of small particles, ORNL/CP–95872, Oak Ridge National Lab, United States, 1997.

Bohor, B. F. and Hughes, R. E.: Scanning electron microscopy of clays and clay minerals, Clay Clay Min., 9, 49–54, 1971.

Bond, T. C. and Bergstrom, R. W.: Light absorption by carbonaceous particles: An investigative review, Aerosol Sci. Technol., 40, 27–67, 2006.

Bozóki, Z., Pogány, A., and Szabó, G.: Photoacoustic instruments for practical applications: present, potentials, and future challenges, Appl. Spectr. Rev., 46, 1–37, 2011.

Cappa, C., Lack, D., Burkholder, J., and Ravishankara, A. R.: Bias in filter-based aerosol light absorption measurements due to organic aerosol loading: evidence from laboratory measurements, Aerosol Sci. Technol., 42, 1022–1032, 2008.

Chakrabarty, R. K., Moosmüller, H., Chen, L.-W. A., Lewis, K., Arnott, W. P., Mazzoleni, C., Dubey, M. K., Wold, C. E., Hao, W. M., and Kreidenweis, S. M.: Brown carbon in tar balls from smoldering biomass combustion, Atmos. Chem. Phys., 10, 6363–6370, doi:10.5194/acp-10-6363-2010, 2010.

Claquin, T., Schulz, M., Balkanski, Y., and Boucher, O.: Uncertainties in assessing radiative forcing by mineral dust, Tellus B, 50, 491–505, 1998.

Conant, W. C., Seinfeld, J. H., Wang, J., Carmichael, G. R., Tang, Y., Uno, I., Flatau, P. J., Markowicz, K. M., and Quinn, P. K.: A model for the radiative forcing during ACE-Asia derived from CIRPAS Twin Otter and R/V Ronald H. Brown data and comparison with observations, J. Geophys. Res., 108, 8661, doi:10.1029/2002JD003260, 2003.

Cross, E. S., Onasch, T. B., Ahern, A., Wrobel, W., Slowik, J. G., Olfert, J., Lack, D. A., Massoli, P., Cappa, C. D., Schwarz, J. P., Spackman, J. R., Fahey, D. W., Sedlacek, A., Trimborn, A., Jayne, J. T., Freedman, A., Williams, L. R., Ng, N. L., Mazzoleni, C., Dubey, M., Brem, B., Kok, G., Subramanian, R., Freitag, S., Clarke, A., Thornhill, D., Marr, L. C., Kolb, C. E., Worsnop, D. R., and Davidovits, P.: Soot particle studies – instrument intercomparison – project overview, Aerosol Sci. Technol., 44, 592–611, 2010.

DeSouza-Machado, S. G., Strow, L. L., Hannon, S. E., and Motteler, H. E.: Infrared dust spectral signatures from AIRS, Geophys. Res. Lett., 33, L03801, doi:10.1029/2005GL024364, 2006.

Dong, Z., Qin, D., Kang, S., Ren, J., Chen, J., Cui, X., and Qin, X.: Physicochemical characteristics and sources of atmospheric dust deposition in snow packs on the glaciers of western Qilian Mountains, China, Tellus B, 66, 79–95, 2014.

Egan, W. G. and Hilgeman T. W.: Optical Properties of Inhomogeneous Materials: Applications to Geology, Astronomy, Chemistry, and Engineering, Academic, San Diego, Calif., 1979.

Filep, Á., Ajtai, T., Utry, N., Pintér, M. D., Nyilas, T., Takács, S., and Szabó, G.: Absorption Spectrum of Ambient Aerosol and Its Correlation with Size Distribution in Specific Atmospheric Conditions after a Red Mud Accident, Aerosol Air Qual. Res., 13, 49–59, 2013.

Ghosh, G.: Dispersion-equation coefficients for the refractive index and birefringence of calcite and quartz crystals, Opt. Commu., 163, 95–102, 1999.

Guyon, P., Boucher, O., Graham, B., Beck, J., Mayol-Bracero, O. L., Roberts, G. C., Maenhaut, W., Artaxo, P., and Andreae, M. O.: Refractive index of aerosol particles over the Amazon tropical forest during LBA-EUSTACH 1999, J. Aerosol. Sci., 34, 883–907, 2003.

Hoffer, A., Gelencsér, A., Guyon, P., Kiss, G., Schmid, O., Frank, G. P., Artaxo, P., and Andreae, M. O.: Optical properties of humic-like substances (HULIS) in biomass-burning aerosols, Atmos. Chem. Phys., 6, 3563–3570, doi:10.5194/acp-6-3563-2006, 2006.

Kalashnikova, O. V. and Sokolik, I. N.: Modeling the radiative properties of nonspherical soil-derived mineral aerosols, J. Quant. Spectr. Rad. Trans., 87, 137–166, 2004.

Lack, D. A., Langridge, J. M., Bahreini, R., Cappa, C. D., Middlebrook, A. M., and Schwarz, J. P.: Brown carbon and internal mixing in biomass burning particles, Proc. Natl. Acad. Sci., 109, 4802–14807, 2012.

McConnell, C. L., Formenti, P., Highwood, E. J., and Harrison, M. A. J.: Using aircraft measurements to determine the refractive index of Saharan dust during the DODO Experiments, Atmos. Chem. Phys., 10, 3081–3098, doi:10.5194/acp-10-3081-2010, 2010.

McNeil, L. E. and French, R. H.: Multiple scattering from rutile TiO2 particles, Actamaterialia, 48, 4571–4576, 2000.

Moffet, R. C. and Prather, K. A.: Extending ATOFMS measurements to include refractive index and density, Anal. Chem., 77, 6535–6541, 2005.

Moosmüller, H., Chakrabarty, R. K., and Arnott, W. P.: Aerosol light absorption and its measurement: A review, J. Quant. Spectr. Rad. Trans., 110, 844–878, 2009.

Moosmüller, H., Engelbrecht, J. P., Skiba, M., Frey, G., Chakrabarty R. K., and Arnott, W. P.: Single Scattering Albedo of Fine Mineral Dust Aerosols Controlled by Iron Concentration, J. Geophys. Res., 117, D11210, doi:10.1029/2011JD016909, 2012.

Müller, T., Schladitz, A., Massling, A., Kaaden, N., Kandler, K., and Wiedensohler, A.: Spectral absorption coefficients and imaginary parts of refractive indices of Saharan dust during SAMUM-1. Tellus B, 61, 79–95, doi:10.3402/tellusb.v61i1.16816, 2011.

Nickovic, S., Vukovic, A., Vujadinovic, M., Djurdjevic, V., and Pejanovic, G.: Technical Note: High-resolution mineralogical database of dust-productive soils for atmospheric dust modeling, Atmos. Chem. Phys., 12, 845–855, doi:10.5194/acp-12-845-2012, 2012.

Otto, S., de Reus, M., Trautmann, T., Thomas, A., Wendisch, M., and Borrmann, S.: Atmospheric radiative effects of an in situ measured Saharan dust plume and the role of large particles, Atmos. Chem. Phys., 7, 4887–4903, doi:10.5194/acp-7-4887-2007, 2007.

Patashnick, H., Meyer, M. and Rogers, B.: Tapered element oscillating microbalance technology Proc N American/Ninth U.S. Mine Vent Symp, June 2002, 625–631, 2002.

Querry, M. R.: Optical constants, MISSOURI UNIV-KANSAS CITY, 1985.

Rao, K. N.: Influence of deposition parameters on optical properties of TiO2 films, Opt. Eng., 41, 2357–2364, 2002.

Sokolik, I., Andronova, A., and Johnson, T., Complex refractive index of atmospheric dust aerosols, Atmos. Environ., 27A, 2495–2502, 1993.

Sokolik, I. N. and Toon, O. B.: Direct radiative forcing by anthropogenic airborne mineral aerosols, Nature, 381, 681–683, 1996.

Sokolik, I. N. and Toon, O.: Incorporation of mineralogical composition into models of the radiative properties of mineral aerosol from UV to IR wavelengths, J. Geophys. Res., 104, 9423–9444, 1999.

Tegen, I. and Fung, I.: Contribution to the atmospheric mineral aerosol load from land surface modification, J. Geophys. Res.-Atmos., 100, 18707–18726, 1995.

Tegen, I. and Lacis, A. A.: Modeling of particle size distribution and its influence on the radiative properties of mineral dust aerosol, J. Geophys. Res.-Atmos., 101, 19237–19244, 1996.

Utry, N., Ajtai, T., Pintér, M., Bozóki, Z., and Szabó, G.: Wavelength-Dependent Optical Absorption Properties of Artificial and Atmospheric Aerosol Measured by a Multi-Wavelength Photoacoustic Spectrometer, Int. J. Thermophys., 35, 2246–2258, 2014.

von der Weiden, S.-L., Drewnick, F., and Borrmann, S.: Particle Loss Calculator – a new software tool for the assessment of the performance of aerosol inlet systems, Atmos. Meas. Tech., 2, 479–494, doi:10.5194/amt-2-479-2009, 2009.

Wang, J., Flagan, R. C., Seinfeld, J. H., Jonsson, H. H., Collins, D. R., Russell, P. B., Schmid, B., Redemann, J., Livingston, J. M., Gao, S., Hegg, D. A., Welton, E. J., and Bates, D.: Clear-column radiative closure during ACE-Asia: Comparison of multiwavelength extinction derived from particle size and composition with results from Sunphotometry, J. Geophys. Res., 107, 4688, doi:10.1029/2002JD002465, 2002.

Pressure-dependent calibration of the OH and HO_2 channels of a FAGE HO_x instrument using the Highly Instrumented Reactor for Atmospheric Chemistry (HIRAC)

F. A. F. Winiberg[1], S. C. Smith[1], I. Bejan[1], C. A. Brumby[1], T. Ingham[1,2], T. L. Malkin[1,*], S. C. Orr[1], D. E. Heard[1,2], and P. W. Seakins[1,2]

[1]School of Chemistry, University of Leeds, LS2 9JT, Leeds, UK
[2]National Centre for Atmospheric Science, University of Leeds, Leeds, LS2 9JT, UK
[*]now at: Institute for Climate and Atmospheric Science, School of Earth and Environment, University of Leeds, Woodhouse Lane, Leeds, LS2 9JT, UK

Correspondence to: P. W. Seakins (p.w.seakins@leeds.ac.uk)

Abstract. The calibration of field instruments used to measure concentrations of OH and HO_2 worldwide has traditionally relied on a single method utilising the photolysis of water vapour in air in a flow tube at atmospheric pressure. Here the calibration of two FAGE (fluorescence assay by gaseous expansion) apparatuses designed for HO_x (OH and HO_2) measurements have been investigated as a function of external pressure using two different laser systems. The conventional method of generating known concentrations of HO_x from H_2O vapour photolysis in a turbulent flow tube impinging just outside the FAGE sample inlet has been used to study instrument sensitivity as a function of internal fluorescence cell pressure (1.8–3.8 mbar). An increase in the calibration constants C_{OH} and C_{HO_2} with pressure was observed, and an empirical linear regression of the data was used to describe the trends, with $\Delta C_{OH} = (17 \pm 11)\,\%$ and $\Delta C_{HO_2} = (31.6 \pm 4.4)\,\%$ increase per millibar air (uncertainties quoted to 2σ). Presented here are the first direct measurements of the FAGE calibration constants as a function of external pressure (440–1000 mbar) in a controlled environment using the University of Leeds HIRAC chamber (Highly Instrumented Reactor for Atmospheric Chemistry). Two methods were used: the temporal decay of hydrocarbons for calibration of OH, and the kinetics of the second-order recombination of HO_2 for HO_2 calibrations. Over comparable conditions for the FAGE cell, the two alternative methods are in good agreement with the conventional method, with the average ratio of calibration factors (conventional : alternative) across the entire pressure range, $C_{OH(conv)}/C_{OH(alt)} = 1.19 \pm 0.26$ and $C_{HO_2(conv)}/C_{HO_2(alt)} = 0.96 \pm 0.18$ (2σ). These alternative calibration methods currently have comparable systematic uncertainties to the conventional method: $\sim 28\,\%$ and $\sim 41\,\%$ for the alternative OH and HO_2 calibration methods respectively compared to 35 % for the H_2O vapour photolysis method; ways in which these can be reduced in the future are discussed. The good agreement between the very different methods of calibration leads to increased confidence in HO_x field measurements and particularly in aircraft-based HO_x measurements, where there are substantial variations in external pressure, and assumptions are made regarding loss rates on inlets as a function of pressure.

1 Introduction

Short-lived free radicals play a crucial role in determining the composition of the atmosphere. The catalytic cycle of HO_x ($= OH + HO_2$) radicals is of central importance to tropospheric chemistry. OH acts as the primary daytime oxidant, initiating the degradation of most trace gases and thereby controlling their atmospheric concentrations and lifetimes. The short lifetime of the radicals generate HO_x concentrations which are uninfluenced by transport; therefore reproducing observed HO_x concentrations can be an excellent test of any chemical model (Heard and Pilling, 2003). However, it should be mentioned that agreement between measured and modelled $[HO_x]$ could be fortuitous as both sources and sinks of HO_x radicals could be simultaneously under- or overestimated. The development of detection techniques that permit the speciation of a wider range of atmospheric components (e.g. volatile organic compounds (VOCs) and aerosols), together with OH reactivity measurements, can help to further constrain modelling studies and reduce the potential for the coincidental agreement. OH has been detected by long-path differential absorption spectroscopy (DOAS) in the field (Brauers et al., 1996; Dorn et al., 1996) and in the SAPHIR chamber (Schlosser et al., 2009), and chemical ionisation mass spectrometric techniques (CIMSs) have also been used in field observations (Eisele and Tanner, 1991; Berresheim et al., 2002; Sjostedt et al., 2007; Kukui et al., 2008). However, a majority of field measurements have been made using laser-induced fluorescence (LIF) spectroscopy, and intercomparisons exist which have validated the technique against DOAS and CIMS in both chamber (Schlosser et al., 2007, 2009; Fuchs et al., 2012) and field environments, including aircraft-based measurements (Eisele et al., 2001, 2003). Low concentrations and potential interferences (Fuchs et al., 2011; Mao et al., 2012; Whalley et al., 2013) make HO_x measurements challenging. In addition, most HO_x detection methods are not absolute, and hence calibration is required.

Fluorescence assay by gas expansion, FAGE, is a low-pressure LIF technique commonly used for the detection of OH and HO_2 radicals (Heard, 2006, and references therein). The low concentrations of ambient OH ($\sim 10^6$ molecule cm^{-3}) require a viable measurement technique to discriminate between laser-scattered light and small signal levels. Originally this was attempted by exciting OH to the first vibrational level in the A state at 282 nm ($A^2\Sigma^+$ $(v'=1) \leftarrow X^2\Pi_i$ $(v''=0)$ OH transition) and observing off-resonant fluorescence at ~ 308 nm using an interference filter to help discriminate against scattered laser radiation (Davis et al., 1976). Although non-resonant LIF has been successful in stratospheric applications (Wennberg et al., 1994), in the troposphere 282 nm photolysis of ozone (and subsequent reaction of $O(^1D)$ with water vapour) generates an unacceptably high interfering OH signal, and on-resonant LIF with excitation at 308 nm is used instead. Expanding the sample through a pinhole to low pressure ($\sim$ 1–2 Torr) increases the fluorescence lifetime of the A state beyond the laser scatter pulse, allowing for temporal discrimination against the resonant 308 nm excitation pulse ($A^2\Sigma^+$ $(v'=0) \leftarrow X^2\Pi_i$ $(v''=0)$ OH transition). Injection of an OH scavenger (e.g. C_3F_6) allows quantification of any laser-generated OH interference (Mao et al., 2012; Novelli et al., 2014). HO_2 is converted into OH via reaction with added NO:

$$HO_2 + NO \rightarrow OH + NO_2 \tag{R1}$$

and the resultant OH is detected in the same way. Detection of OH and HO_2 either simultaneously or in series can be achieved using the same LIF detection axis (measurements in series; Creasey et al., 1997a), with two separate LIF axes within the same cell (simultaneous; Stevens et al., 1994) or with two separate detection cells (simultaneous; Whalley et al., 2010).

LIF is a very sensitive but non-absolute detection method, and therefore each channel of the instrument needs to be calibrated. The vacuum-ultraviolet (VUV) photolysis of H_2O vapour was originally developed for the calibration of HO_x measurement instruments in the 1990s (Aschmutat et al., 1994; Schultz et al., 1995; Heard and Pilling, 2003). Since then the methodology has become the HO_x measurement community standard. Upon the photolysis of a known H_2O vapour concentration (in synthetic air at atmospheric pressure) by a mercury (Hg) Pen-Ray lamp at 184.9 nm, OH and HO_2 are produced in unity ratio (Fuchs et al., 2011) via Reactions (R2) and (R3) (Schultz et al., 1995):

$$H_2O + h\nu \xrightarrow{\lambda=184.9\ \mathrm{nm}} OH + H, \tag{R2}$$

$$H + O_2 (+M) \xrightarrow{M=N_2,O_2} HO_2(+M). \tag{R3}$$

The radicals are then sampled by the HO_x instrument at atmospheric pressure; the concentrations of OH and HO_2 produced can be determined using Eq. (1):

$$[OH] = [HO_2] = [H_2O]\sigma_{H_2O,\ 184.9\ \mathrm{nm}}\Phi_{OH}F_{184.9\,\mathrm{nm}}\Delta t, \tag{1}$$

where $[H_2O]$ is the water vapour concentration; $\sigma_{H_2O,\ 184.9\,\mathrm{nm}}$ is the known absorption cross section of H_2O vapour at 184.9 nm ($(7.22 \pm 0.22) \times 10^{-20}$ cm^2 $molecule^{-1}$ (Cantrell et al., 1997; Creasey et al., 2000)); Φ_{OH} ($= \Phi_{HO_2} = 1$) is the photodissociation quantum yield of OH and HO_2 (Fuchs et al., 2011); $F_{184.9\,\mathrm{nm}}$ is the photon flux of 184.9 nm light; and Δt is the exposure time of the air to the Hg lamp output. There are two main methodologies used for obtaining the product $F_{184.9\,\mathrm{nm}}$, Δt in Eq. (1). In the first, the two parameters are measured separately, $F_{184.9\,\mathrm{nm}}$ using a calibrated phototube and Δt using knowledge of the volumetric flow rate and geometric parameters of the flow tube (Stevens et al., 1994). In the other, a chemical actinometer is used to obtain the product directly, with both O_2 and N_2O photolysis at 184.9 nm used to generate either O_3 or NO, which is subsequently detected using commercial analysers, with good sensitivity (Creasey et al., 1997a; Hofzumahaus et al., 1997; Heard and Pilling, 2003; Faloona et al., 2004). There are two main methods for delivery of the OH radicals to the FAGE inlet at atmospheric pressure, either using a laminar or turbulent flow tube. In the laminar flow regime there is a radial gradient in the OH concentration for which the so-called profile factor (P) has to be quantified (Holland et al., 1995; Creasey et al., 1997a), whereas in a turbulent flow system the radial OH concentration is constant except very close to the walls.

Alternative calibration methods have also been developed, but typically not deployed in the field, and examples of these will be employed in the current study. A detailed evaluation of calibration techniques has been presented by Dusanter et al. (2008). In some of the earliest field measurements, Hard et al. (1995) developed a calibration method based on hydrocarbon decays. The concentration of a hydrocarbon with a known and well-characterised rate coefficient for reaction with OH, k (in this case 1,3,5-trimethylbenzene), was measured as a function of time using gas chromatography, allowing determination of all the parameters in Eq. (2) with the exception of [OH]. The rate of loss of a hydrocarbon (HC) through reaction with OH is given by Eq. (2):

$$\frac{d[\mathrm{HC}]}{\mathrm{d}t} = k\,[\mathrm{OH}]\,[\mathrm{HC}]. \qquad (2)$$

This methodology has also been applied more recently to FAGE validation measurements in the EUPHORE chamber (Bloss et al., 2004).

For HO_2 the well-defined second-order recombination rate coefficient for Reaction (R4) can be used to determine $[HO_2]$, where for a second-order reaction the half-life of the decay is related to the initial starting concentration.

$$HO_2 + HO_2 \rightarrow H_2O_2 + O_2 \qquad (R4)$$

In a short set of experiments, Pilling et al. (2005) generated HO_2 from the photolysis of formaldehyde in the EUPHORE chamber and observed the second-order HO_2 decay with a FAGE instrument. The decays were in good agreement with the calibrated HO_2 measurements, but no systematic studies have been undertaken using this reaction as a calibration method.

The deployment of the FAGE technique for aircraft-based measurements (Faloona et al., 2000; Commane et al., 2010; Martinez et al., 2010) raises two issues. First, the need to sample air from outside of the boundary layer of the aircraft fuselage requires a significant length of flow tube before the gas sample is interrogated by the laser beam. Secondly, the pressure in the FAGE cell will vary as the aircraft changes altitude (e.g. 0–7 km, 1.3–2.8 mbar internal cell pressure range, from Commane et al., 2010), altering the instrumental sensitivity (Commane et al., 2010; Martinez et al., 2010) owing to changes, for example, in the nature of the initial expansion into the FAGE apparatus. The current design of the flow tube calibration method is limited to delivering the calibrated [OH] at atmospheric pressure; however, by using different nozzle pinhole diameters (typically 0.2–1.0 mm) it is possible to alter the pressure in the FAGE cell over the range typically encountered during a flight. Importantly, this method does not compensate for the changing pressure differential across the inlet nozzle experienced during a flight and what effect this might have on the expanding gas before it reaches the FAGE cell. The possible changes in radical surface losses due to the change in inlet pinhole diameter are also assumed to be negligible.

Potential systematic uncertainties around the application of calibrations performed at atmospheric pressure to HO_x data obtained whilst sampling from different pressures (e.g. in flight) highlight the need to obtain calibrations at relevant external pressures. Martinez et al. (2010) have investigated the effect of external pressure on instrument sensitivity by calibrating during flight, reporting an increase in the instrument sensitivity to OH in the free troposphere, compared to the boundary layer. It was not concluded whether this was an effect of the calibration source used (conventional H_2O vapour photolysis) or the instrument itself; however the increase was not characterised by the conventional calibrations performed on the ground before the flight.

We report here an intercomparison of HO_x calibrations based on the conventional flow tube methodology, using different inlet nozzle diameters to vary the internal fluorescence cell pressure, with two alternative calibration methods. Analysis of the decays of hydrocarbons was used to determine [OH], while analysis of the kinetics of HO_2 decay by self-reaction following the photolysis of formaldehyde was used to determine $[HO_2]$. The studies took place in the Highly Instrumented Reactor for Atmospheric Chemistry (HIRAC), which is a custom-built atmospheric simulation chamber providing the unique ability to simultaneously vary pressure and temperature whilst measuring the short-lived free radical species OH, HO_2 and NO_3 (Glowacki et al., 2007a; Malkin, 2010; Malkin et al., 2010). These features make HIRAC ideally suited to the study of the kinetics and mechanisms of atmospherically relevant reactions and the calibration, validation and development of atmospheric measurement instrumentation.

2 HIRAC and FAGE instrumentation

2.1 HIRAC

Experiments were conducted in HIRAC, a stainless-steel chamber with a total volume of $2.25\,m^3$ and total internal surfaces of $13\,m^2$ ($S/V \sim 5.8\,m^{-1}$). The chamber could operate over a wide range of pressures (10–1000 mbar), with multiple access ports used to connect an array of instrumentation and monitoring equipment (pressure gauges, thermocouples etc.). Further details on the construction can be found in Glowacki et al. (2007a) and Malkin et al. (2010).

The photolysis lamps, housed in eight quartz tubes mounted radially inside the reactive volume, were used to initiate photochemistry. The lamps were interchangeable depending on the target molecules; lamps, with primary emissions centred at 254 and 290 nm (GE Optica, GE55T8/HO and Philips, TL40W/12 RS), were used for the alternative OH and HO_2 calibration methods respectively (Sects. 3.2 and 3.3). The output of the lamps was temperature dependent outside of a narrow temperature range (~ 35–39 °C), and so the housings were flushed with N_2 to regulate the temperature and remove photolabile species. A photolysis-lamp-induced chamber temperature increase of ~ 2 K was seen over the course of a typical experiment (< 40 min), and was therefore considered negligible compared to the temperature of the chamber on any given day (293 ± 5 K).

Investigations into radical gradients across the HIRAC chamber have been conducted using direct FAGE measurements of OH produced from both photolytic (methyl nitrite) and non-photolytic (O_3+ *trans*-2-butene) sources using an extended inlet (800 mm) to probe across the chamber diameter. No significant OH radical gradient was observed until the FAGE sampling nozzle was ~ 200 mm from the wall and a maximum ~ 15 % decrease (compared to the centre of the chamber) was seen when the sampling inlet was flush with the chamber walls. Other than being close to the walls, the lack of gradient in OH radicals from both photolytic and non-photolytic sources provides direct evidence of the homogeneity of the lamp radiation profile and efficacy of mixing in the chamber, whilst showing that the standard FAGE inlet (280 mm, Sect. 2.2) samples well into the homogeneous area.

Ozone was monitored using a UV photometric O_3 analyser (Thermo Electron Corporation 49C, detection limit (d.l.) = 1.0 ppbv at 60 s averaging). The O_3 analyser had been calibrated using a commercial ozone primary standard (Thermo Electron Corporation 49i-PS), and an intercomparison with the Fourier transform infrared (FTIR) spectroscopy within HIRAC was linear (Glowacki et al., 2007a). A chemiluminescence NO_x analyser (TEC 42C, d.l. = 50 pptv at 60 s averaging) was used to determine that levels of NO_x (= $NO + NO_2$) were characteristically below the detection limit of the apparatus.

A calibrated gas chromatography instrument with flame ionisation detector (GC-FID, Agilent Technologies, 6890N) was used for the online detection of reactants (Sect. 3.2) using an evacuated sampling loop into which gas from the chamber was expanded. The GC was fitted with a CP-SIL-5 column (50 m, 0.32 mm, 5 μm) using He carrier gas and a constant oven temperature (40–75 °C depending on the hydrocarbon being detected) and was able to provide hydrocarbon measurements on a 2–6 min time resolution. Supporting measurements of *iso*-butene and $(CH_3)_3COOH$ were made via a long-path FTIR absorption facility. The FTIR spectrometer (Bruker, IFS/66) was coupled to a Chernin-type multipass cell (Glowacki et al., 2007b) and spectral resolution was maintained at 1 cm^{-1} across all experiments, using 32 co-added spectra for a 30 s time resolution.

Calibration experiments were conducted over a pressure range of 440–1000 mbar in an ultra-high-purity (UHP) 1 : 4 synthetic air mix of O_2 (BOC, zero-grade, > 99.999 %) and N_2 (BOC, zero-grade, > 99.998 %) to match the range of pressures from the pinhole calibration method (Sect. 3.1). The UHP gases help maintain low H_2O vapour (< 10 ppm, verified by dew-point hygrometer measurement), NO_x (< 1 ppbv) and non-methane hydrocarbons (< 1 ppbv) during experimental runs. Thorough mixing of reaction mixtures within HIRAC was achieved in ≤ 70 s by four vibrationally damped, variable speed circulation fans mounted in pairs at each end of the chamber. The chamber was evacuated to ~ 0.05 mbar for ~ 60 min following each experiment using a rotary-pump-backed roots blower (Leybold, trivac D40B and ruvac WAU251) to ensure removal of all reactants/products. Known concentrations of precursors were introduced to the chamber in the vapour phase through a 0.97 L stainless-steel delivery vessel. A combined sampling rate of ~ 9 sLm from the chamber required a counter flow of synthetic air maintaining the desired pressure and diluting the reactants ($(4.5 \pm 0.2) \times 10^{-5}\ s^{-1}$). This was regulated using two Brooks mass flow controllers (N_2 and O_2).

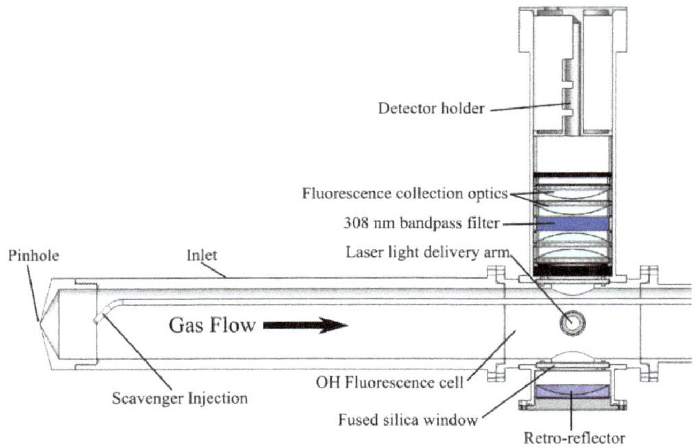

Figure 1. Schematic showing a side-on vertical cross section of the HIRAC FAGE OH fluorescence cell. The OH scavenger (*iso*-butane) was introduced ~ 40 mm from the inlet pinhole through an 1/8 in. internal-diameter stainless-steel tube mounted in between the OH and HO_2 cells (out of frame). The tube ran flush to the cell wall to reduce possible scattering of laser light, and the tip was angled slightly towards the centre of the main gas flow to improve mixing.

2.2 HO_x detection instrument

Calibrations were conducted using both the University of Leeds aircraft- and HIRAC-based FAGE instruments, brief operational details of which are shown in Table 1. The two FAGE systems were very similar in design except for the inlet length and pinhole size as highlighted in Table 1. The aircraft instrument was used as described in Commane et al. (2010) to validate the alternative HO_2 calibration technique only. The HIRAC-based FAGE instrument has also been described in the literature by Glowacki et al. (2007a), and hence only modifications since publication will be discussed here.

Figure 1 shows the cross-sectional schematic of the HIRAC FAGE instrument. Under typical operating conditions, air was sampled at ~ 6 sLm through a 1.0 mm diameter pinhole nozzle and passed down the inlet (length 280 mm,

Table 1. FAGE instruments used and their respective inlet designs, laser systems and calibration methods used.

FAGE instrument	Inlet design		Laser	Calibrations conducted
Aircraft	Pinhole diameter	0.6 mm	Photonics Industries (5 kHz PRF) Nd:YAG (DS-532-10) pumped Ti:Sa (TU-UV-308)	H_2O vapour photolysis, HCHO photolysis
	Length	420 mm		
HIRAC	Pinhole diameter	1.0 mm	Litron, NANO-TRL-250, (200 Hz PRF) Nd:YAG pumped dye laser (Lambda Physik, LPD3000)	H_2O vapour photolysis, HCHO photolysis, HC decay
	Length	280 mm		

50 mm diameter) into the OH detection axis maintained at low pressure (1.8–3.85 mbar) using a high capacity rotary-backed roots blower pumping system (Leybold, trivac D40B and ruvac WAU251). Using the same pump set, the aircraft instrument was operated with a 420 mm long inlet and a 0.6 mm pinhole. The long inlet was used to draw a sample away from the chamber walls where radical losses become significant (see Sect. 2.1). Both instruments were coupled to the HIRAC chamber using custom-made ISO-K160 flanges, ensuring the pinhole, in both cases, was kept ~225 mm from the chamber walls.

Concentrations of HO_2 were measured simultaneously in a second detection axis ~300 mm downstream of the OH detection axis. High-purity NO (BOC, N2.5 nitric oxide) was added ~20 mm before the HO_2 detection axis into the centre of the FAGE cell in the direction of gas flow through 1/8 in. stainless-steel tubing at a rate of 5 sccm (Brooks 5850S), converting HO_2 to OH.

Recently published material on the conversion of certain RO_2 radicals to OH upon reaction with NO in FAGE detections cells (Fuchs et al., 2011; Whalley et al., 2013) has shown a significant enhancement of the HO_2 signal in the presence of RO_2 derived from certain hydrocarbons. These effects have been thoroughly studied using a range of different hydrocarbons for the HIRAC FAGE apparatus and will be the subject of a further publication. The potential interferences associated with HO_2 measurements in the presence of certain hydrocarbons due to the presence of β-hydroxyperoxy radicals do not apply to either HO_2 calibration method. In addition, any interference from RO_2 radicals produced during the alternative calibration methods was experimentally demonstrated to be negligible under the conditions of these experiments (Winiberg, 2014).

Experiments with the HIRAC FAGE instrument used a new medium pulse repetition frequency (PRF) laser light source (= 200 Hz), with a different light delivery method to the detection cells, compared to that described by Glowacki et al. (2007a). The previously used JDSU Nd:YAG pumped Sirah Cobra Stretch system (PRF = 5 kHz) focussed the frequency-doubled 308 nm output into fibre optic cables

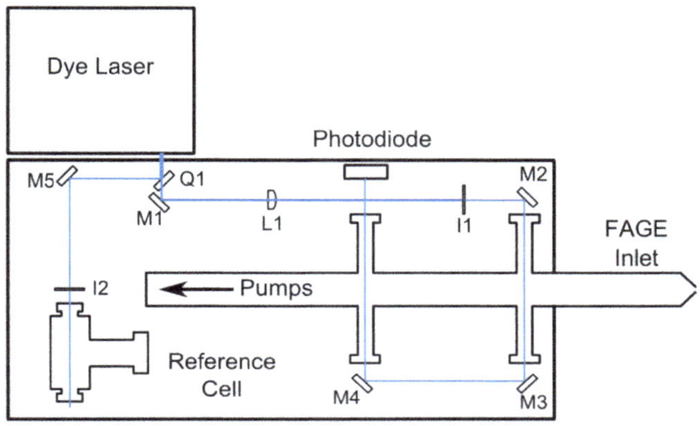

Figure 2. Top-down schematic of the FAGE instrument showing the laser beam path (blue line) through the OH and HO_2 detection cells, and the reference cell using the Litron/LPD3000, 200 Hz PRF laser source. Q – quartz flat; M – mirror; I – iris; and L – lens. The FAGE inlet is extended past the edge of the mounting table for insertion into the HIRAC chamber. The calibrated photodiode was used to normalise the fluorescence signals to fluctuations in laser power.

(10 m, Oz Optics), which were then attached directly to the FAGE cell arms via collimators (Oz Optics). Using the new Litron Nd:YAG (NANO-TRL-50-250) pumped Lambda Physik (LPD3000) dye laser system (PRF = 200 Hz), the high laser pulse energies were found to burn the ends of the fibre optic cables, and hence direct light delivery was applied using a combination of mirrors, lenses and irises to direct and shape the beam to the OH and HO_2 detection regions, as shown in the top-down schematic of the modified HIRAC FAGE instrument displayed in Fig. 2.

The UV light exiting the dye laser was split with a quartz flat (Fig. 2, Q1) to direct ~5 % of the laser light towards the reference cell (where OH was generated continuously from a hot wire filament in water-saturated air), which enabled precise tuning of the laser wavelength to the maxima of the OH Q1 (2) branch (within 98 %). The remaining light was aligned through the OH and HO_2 cells sequentially using a series of 308 nm centred turning optics (M1–M4, CVI Laser

Optics, Melles Griot). A lens was used (L1, $f = 100$ mm) in conjunction with an iris (I2), to help transmit the laser beam through both detection cells, avoiding collisions with any internal surfaces. Fluctuations in laser power were accounted for using a linear-response, UV-sensitive photodiode (UDT-555UV, Laser Components UK) at the exit arm of the HO_2 detection axis to normalise the LIF signal. Both laser systems provided 5–7 and 2–3 mW of 308 nm light to the OH and HO_2 detection axes respectively.

The OH fluorescence was collected orthogonal to the gas flow onto electronically gated Channeltron PhotoMultiplier tubes (CPM, Perkin Elmer, C943P) via a series of imaging lenses and a narrow-bandpass filter (Barr Associates, 308.8 ± 5.0 nm). A spherical concave back reflector was positioned underneath the cell, opposite the detection optics, to optimise light collection onto the CPM. To avoid detector saturation, the CPM was gated (i.e. switched off) for the duration of the laser pulse using a modified gating unit based on the original design by Creasey et al. (1997a). Signals from the CPM were analysed using photon-counting cards (Becker and Hickl PMS-400A).

A new OH scavenger system was installed to help discriminate between OH sampled from the chamber and laser-generated OH in the fluorescence cells due to the higher pulse energies associated with the 200 Hz PRF laser system (1×10^{14} compared to 5×10^{12} photons pulse^{-1} cm^{-2} at 5 kHz for laser power = 8 mW). A mixture of *iso*-butane (20 % in N_2) was injected $\sim$40 mm inside the inlet pinhole into the central flow (Fig. 1), through a 1/8 in. internal-diameter stainless-steel pipe at a rate of $\sim$20 sccm, reacting with the sampled OH before reaching the detection axis. The laser-generated OH was probed within the same laser pulse (12 ns) and hence was not suppressed by the scavenger injection. Multiple photolysis of the same gas sample was avoided as the residence time in the laser pulse cross section ($\sim$0.5 cm^2) was calculated at $\sim$0.4 ms, compared to a laser pulse every 5 ms at 200 Hz PRF (assuming plug flow at a 6 sLm ambient sampling rate). Neither a pressure increase nor attenuation of UV light was detected during the scavenger injection process at this flow rate and dilution.

3 Calibration procedures

3.1 Conventional H_2O vapour photolysis calibration

The requisite equation for calibration of FAGE by water vapour photolysis was given as

$$[OH] = [HO_2] = [H_2O]\sigma_{H_2O,\,184.9\,nm}\Phi_{OH}F_{184.9\,nm}\Delta t \quad (3)$$

and the principles were outlined above in Sect. 1. A schematic diagram of the H_2O vapour photolysis calibration source is presented in Fig. 3, consisting of a square cross-section flow tube ($12.7 \times 12.7 \times 300$ mm) through which 40 sLm of humidified air (BOC, BTCA 178) was

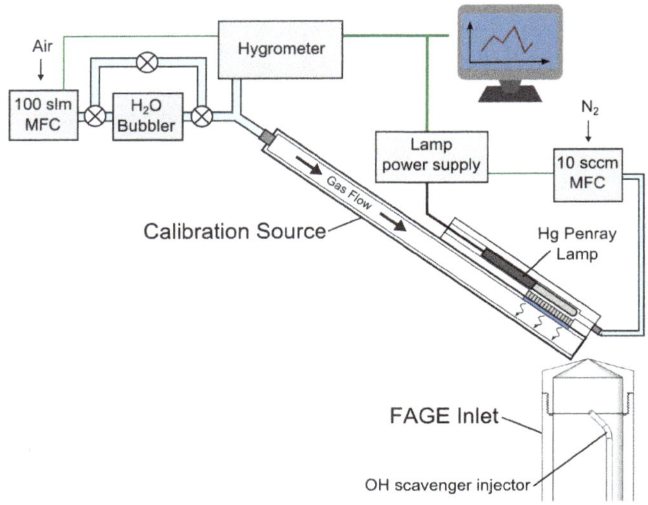

Figure 3. Schematic cross section of the H_2O vapour photolysis calibration source used in the calibration of the two FAGE instruments (inlet for HIRAC FAGE instrument shown here, for example). The $[H_2O]_{vapour}$ was measured prior to entering the square cross-section flow tube, and the concentration was controlled through a series of three taps around the bubbler. The Hg Pen-Ray lamp was housed in a second section of the wand, and the output was collimated through a Suprasil window using a honeycomb arrangement of $\o = 1$ mm aluminium tubes. The lamp was continuously flushed with N_2 to remove potential absorbers and photolabile species, and to help regulate temperature.

passed, resulting in a turbulent flow regime (Reynolds number ≥ 4000). The air was humidified by passing a fraction of the total air flow through a deionised water bubbler system, and $[H_2O]$ was measured using a dew-point hygrometer (CR4, Buck Research Instrument) prior to the flow tube. The collimated 184.9 nm output of a mercury Pen-Ray lamp (LOT-Oriel, Hg-Ar) was introduced to the end of the main flow tube, photolysing H_2O vapour (Reactions R2–R3). The gas output from the flow tube was directed towards the FAGE sampling inlet, where the overfill of the FAGE sample volume from the flow tube stopped the impingement of ambient air. A range of HO_x concentrations were produced by changing both the H_2O vapour concentration and the mercury lamp photon flux.

The flux of 184.9 nm light, $F_{184.9\,nm}$, was varied by altering the Hg lamp supply current and was dependent on the specific mercury lamp employed along with the lamp temperature and orientation (Hofzumahaus et al., 1997; Creasey et al., 2000; Dusanter et al., 2008). To this end, determinations of the flux from the specific mercury lamp used in the calibrations described in this work were made in situ for lamp supply currents between 0.2 and 3.0 mA using the N_2O actinometry method described in detail in a number of publications (Edwards et al., 2003; Heard and Pilling, 2003; Faloona et al., 2004; Glowacki et al., 2007a; Whalley et al., 2007).

The exposure time of the air to the 184.9 nm light, Δt, was calculated as a function of the known velocity of the air and the cross section of the photolysis region.

Various cell conditions and their effect on the sensitivity to OH and HO_2 have been reported in the literature (Faloona et al., 2004; Martinez et al., 2010; Regelin et al., 2013). Here, instrument sensitivity as a function of internal cell pressure has been determined for the HIRAC FAGE instrument using the 200 Hz PRF laser source only (Table 1). Different internal cell pressures (1.8–3.8 mbar) were achieved by changing the diameter of the FAGE inlet pinhole between 0.5 and 1.0 mm. For the aircraft FAGE instrument, inlet pinhole diameters between 0.3 and 0.6 mm were used giving internal cell pressures between 1.4 and 2.5 mbar.

3.2 Hydrocarbon decay method – OH calibration

Hydrocarbons (0.5–2.0×10^{13} molecule cm^{-3}) and the OH precursor, *tert*-Butyl hydroperoxide (TBHP, Sigma Aldrich ~40 % in H_2O, 2.0×10^{13} molecule cm^{-3}) were introduced to the chamber before the lamps were switched on, initiating the decay experiment. OH was produced directly from the photolysis of TBHP at $\lambda = 254$ nm and is, as far as we are aware, the first chamber experiment to use TBHP photolysis as a source of NO_x-free OH. Upon illumination of the chamber, rapid photolysis led to an instantaneous peak [OH] ~ 10^7 molecule cm^{-3} before OH decayed away over ~30 min as the TBHP was removed by photolysis, whilst OH was removed through reaction with TBHP ($k_{OH}(296\,K) = (3.58 \pm 0.54) \times 10^{-12}$ cm^3 molecule^{-1} s^{-1}; Baasandorj et al., 2010) and the selected hydrocarbon. The alternative OH calibrations presented here were conducted for the HIRAC-based FAGE instrument operating at 200 Hz PRF only.

Cyclohexane (> 99 %, Fisher Scientific), *n*-pentane (> 99 %, Fisher Scientific) and *iso*-butene (99 %, Sigma Aldrich) were employed as the hydrocarbons in this study due to their sufficiently fast and well-known rates of reaction with OH to provide a quantifiable decay compared to chamber dilution. The rate coefficient for OH with *iso*-butene has been evaluated by IUPAC as k_{OH} (298 K) = $(51 \pm 12) \times 10^{-12}$ cm^3 molecule^{-1} s^{-1} (IUPAC, 2007), and rate coefficients for the reaction of OH with cyclohexane and *n*-pentane have been reviewed by Calvert et al. (2008) as k_{OH} (298 K) = (6.97 ± 1.39) and $(3.96 \pm 0.76) \times 10^{-12}$ cm^3 molecule^{-1} s^{-1} respectively (all quoted to $\pm 2\sigma$). Whilst alkanes are known to have a pressure-independent rate coefficient for OH reactions, the reactions of OH with alkenes occur predominantly by addition, a process which is pressure dependent, with the rate coefficient increasing with pressure up to the high-pressure limit where the addition of OH is the rate-determining step (Pilling and Seakins, 1995). A study by Atkinson and Pitts (1975) into the reaction of various small-chain alkenes showed no pressure dependence for propene over 33–133 mbar of argon; therefore the reaction of OH with the larger *iso*-butene molecule is presumed to be pressure independent above 133 mbar (Atkinson, 1986; Atkinson et al., 2004). To confirm this, a relative rate study in air was conducted using isoprene as a reference (k_{OH} (298 K) = $(1.00 \pm 0.14) \times 10^{-10}$ cm^3 molecule^{-1} s^{-1}, IUPAC, 2007). Both direct and relative rate studies have shown that the reaction of isoprene and OH is at the high-pressure limit above 100 Torr (Campuzano-Jost et al., 2004; Park et al., 2004; Singh and Li, 2007). Figure 4 shows that there is no significant pressure dependence in k_{OH} for OH + *iso*-butene over the 250–1000 mbar pressure range within the uncertainty of the experiment (~25 %, $\pm 2\sigma$) and that the measured rate coefficient, k_{OH} (298 K = $(4.87 \pm 0.83) \times 10^{-11}$ cm^3 molecule^{-1} s^{-1}), is in good agreement with the literature values ($(5.07 \pm 0.51) \times 10^{-11}$ cm^3 molecule^{-1} s^{-1} for IUPAC; Atkinson, 2003; IUPAC, 2007).

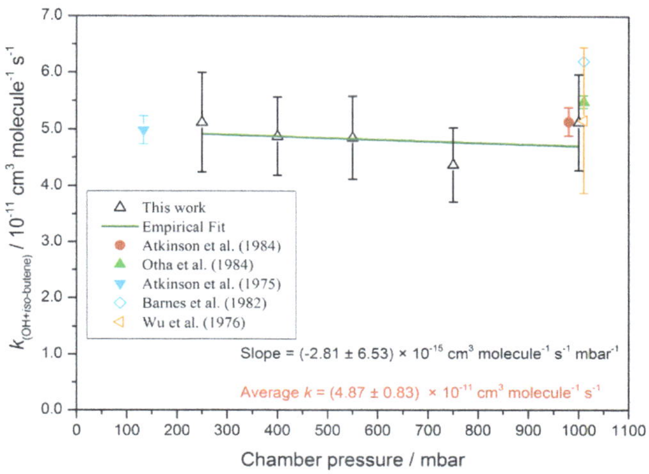

Figure 4. Rate constant, k_{OH}, for *iso*-butene + OH over the 250–1000 mbar pressure range measured relative to an isoprene reference in the HIRAC chamber. An empirical linear least-squares fit to the data is shown to emphasise lack of observed pressure dependence in the measured rate constant. Error bars represent the standard error ($\pm 1\sigma$) in the associated relative rate determination of k_{OH}, and linear regression is weighted to account for this.

The hydrocarbon decay method relies on the loss of hydrocarbon being solely due to reaction with OH, and hence the effects of O_3 and NO_3 as reagents must be considered as both are important in the oxidation of alkenes (Atkinson, 1994). Before photolysis, O_3 and NO_x were measured to be around the instrumental detection limits (0.5 and 0.050 ppb at 60 s averaging respectively) using commercial analysers (details given in Sect. 2.1). Upon photolysis a slow increase in O_3 and NO_2 was observed, to a maximum of ~40 and ~20 ppbv respectively. The [NO_3] upper limit was estimated at ~0.32 pptv using a simple steady-state approximation, where NO_3 production was controlled purely by

$O_3 + NO_2 \rightarrow NO_3$ (Atkinson et al., 2004) and loss by photolysis ($j(NO_2) = 1.93 \pm 0.10$; Glowacki et al., 2007a). Under these conditions it was estimated that $> 98\,\%$ of the loss of *iso*-butene would be due to OH and not O_3 or NO_3 where $k_{O_3} = (1.13 \pm 0.33) \times 10^{-17}$ cm^3 molecule^{-1} s^{-1} and $k_{NO_3} = (3.4 \pm 1.0) \times 10^{-13}$ cm^3 molecule^{-1} s^{-1} (Calvert et al., 2000).

3.3 Formaldehyde photolysis – HO_2 calibration

Formaldehyde was produced by direct heating of paraformaldehyde powder in a glass finger (Sigma Aldrich, 99 %) and was introduced in a flow of nitrogen into the chamber at concentrations $\sim 2 \times 10^{13}$ molecule cm^{-3} (determined manometrically). The chamber was irradiated (lamps: Philips TL40W/12 RS), resulting in an almost instantaneous HO_2 signal. Once an approximately steady-state HO_2 concentration was achieved, the photolysis lamps were turned off and the decay of HO_2 was monitored by FAGE for ~ 120 s until near-background signals levels were reached. The measurement of HO_2 decays was repeated up to five times before the laser wavelength was scanned to the offline position. Therefore five individual C_{HO_2} determinations could be achieved from one chamber fill, with the limiting factor being the increased complexity of the reaction mixture after repeated photolysis cycles. After five decays, the analysis often exhibited evidence of secondary chemistry starting to distort the HO_2 signal profiles, showing non-linearity in second-order plots. The absence of OH in these experiments was confirmed by simultaneous measurement of OH in the OH fluorescence cell, giving signals below the detection limit (1.6×10^6 molecule cm^{-3} for 60 s averaging for the 200 Hz PRF laser system).

Formaldehyde concentrations were kept low ($< 3 \times 10^{13}$ molecule cm^{-3}) to avoid removal of HO_2 via reaction with HCHO, ensuring that the loss of HO_2 occurs predominately via self-reaction and wall loss (Sect. 4.2). The HO_2 calibrations were conducted for the HIRAC-based FAGE instrument operating at 200 Hz PRF and the aircraft-based FAGE instrument operating at 5 kHz PRF. The chamber mixing fans were used for the majority of calibration data sets discussed here, representative of a typical experimental homogeneous gas mixture. A series of experiments were conducted without the mixing fans to probe the HO_2 recombination and wall-loss kinetics using the aircraft-based FAGE instrument, and these are discussed in greater detail in Sects. 4.2 and 5.3.

4 Data analysis

4.1 Hydrocarbon decay

Figure 5 shows the hydrocarbon decay for *iso*-butene at 750 mbar and 294 K measured by GC-FID and FTIR. Using the Guggenheim method (Guggenheim, 1926; Bloss et

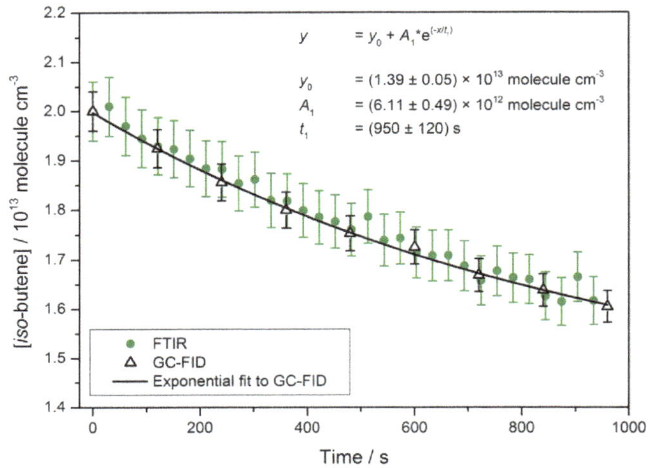

Figure 5. Decay of *iso*-butene as a function of time through reaction with OH in HIRAC (750 mbar, 294 K), measured using GC-FID on a 2 min time resolution. The data are fitted with a first-order exponential decay (purely empirical) to allow calculation of [HC] on the same timescale as the 60 s averaged FAGE data. Time = 0 s indicates photolysis lamp turn-on time and uncertainties are quoted to $\pm 1\sigma$. Error bars are representative of the precision in the GC-FID ($\sim 2\,\%$) and FTIR ($\sim 3\,\%$) measurements to 1σ.

al., 2004) the pseudo-first-order rate coefficient (k') for the hydrocarbon removal was calculated using Eq. (3):

$$k' = \frac{\ln([HC]_1/[HC]_2)}{(t_2 - t_1)}, \tag{4}$$

where $[HC]_1$ and $[HC]_2$ are the concentrations of the hydrocarbon at time t_1 and t_2 respectively. The mean [OH] between t_1 and t_2 was calculated using Eq. (5):

$$[OH] = \frac{(k' - k_{Dil})}{k_{OH}}, \tag{5}$$

where k_{Dil} is the dilution rate of the measured [HC] due to continuous sampling from instrumentation (e.g. FAGE). Bloss et al. (2004) found the Guggenheim method to be most effective when smoothing the inferred [OH] over five [HC] measurements (i.e. consider 10 measurements taken at times $t_1 - t_{10}$; [OH] at t_5 would take $[HC]_1$ and $[HC]_5$, [OH] at t_6 would take $[HC]_2$ and $[HC]_6$ etc.). Due to the short experiment time (20–30 min) and the 2–6 min time resolution on the GC measurements, this smoothing was not possible. For *iso*-butene, FTIR measurements were taken every 30 s, and these were typically found to be in excellent agreement with the GC-FID-measured HC decays, as shown in Fig. 5. However, measurement of small changes in the [HC], due to low steady-state [OH] in the chamber ($\sim 5 \times 10^6$ molecule cm^{-3}), led to large point-to-point variation in the inferred [OH], even after the smoothing was applied. A solution was found by fitting the hydrocarbon decay data with an empirical exponential function of the form

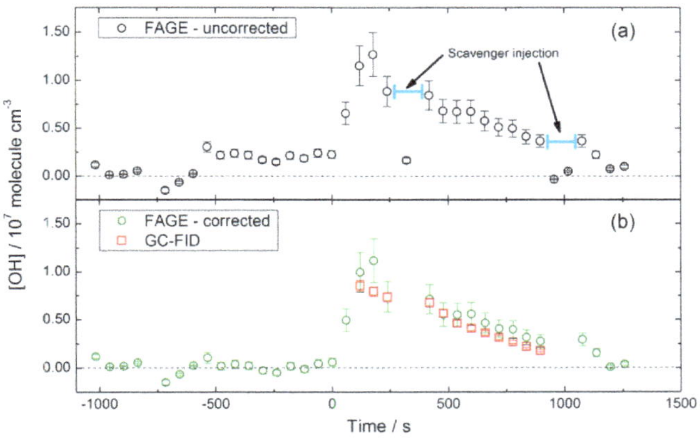

Figure 6. Comparison of [OH] traces measured using the HIRAC FAGE instrument (200 Hz PRF) during the photo-oxidation of *n*-pentane at 1000 mbar and 293 K before **(a)** and after **(b)**, correcting for laser-generated OH due to TBHP photolysis in the OH fluorescence cell. The uncorrected and corrected FAGE signal was converted to [OH] using $C_{OH} = 3.6 \times 10^{-8}$ counts cm^3 s^{-1} molecule^{-1} mW^{-1} determined using the conventional calibration method for comparison with GC-FID data. The TBHP (3.2×10^{13} molecule cm^{-3}) and *n*-pentane (2.1×10^{13} molecule cm^{-3}) were introduced into the chamber at $t \approx -500$ s, and the photolysis lamps were switched on at $t = 0$ s. The [OH] inferred from the HC decay method is also displayed in **(b)**. Dashed line at $y = 0$ given for clarity.

$y = A \times e^{(-x/t1)} + y_0$ as shown in Fig. 5, which allowed the accurate calculation of [HC] at the same time resolution as the FAGE instrument (20 s averaged). A negligible difference between inferred [OH] determined using the FTIR or GC-FID data was observed, and hence only GC-FID-measured hydrocarbon decays were used for direct comparison with *n*-pentane and cyclohexane.

Displayed in Fig. 6 is a typical [OH] profile for the photo-oxidation of *n*-pentane (2.1×10^{13} molecule cm^{-3}) in HIRAC at 1000 mbar and 293 K where photolysis of TBHP was used to produce $\sim 1.3 \times 10^7$ molecule cm^{-3} OH at $t = 0$. The OH was measured directly using the HIRAC FAGE instrument with the Litron Nd:YAG pumped dye laser light source, operating at 200 Hz PRF. Upon introduction of TBHP (3.2×10^{13} molecule cm^{-3}) to the dark chamber at $t \approx -500$ s, an OH signal equivalent to $\sim 2.5 \times 10^6$ molecule cm^{-3} was observed, and was typically $< 25\,\%$ of the total detected OH signal following lamp photolysis. The measured un-normalised OH fluorescence signal was observed to increase quadratically with laser power, suggesting a two-photon photolysis-probe process from the OH probe laser at 308 nm, as described by Reactions (R5)–(R7).

$$\text{TBHP} + h\nu \rightarrow \text{OH} + \text{products}, \quad \text{(R5)}$$

$$\text{OH} + h\nu \rightarrow \text{OH}(A), \quad \text{(R6)}$$

$$\text{OH}(A) \rightarrow \text{OH}(X) + h\nu(\text{LIF}). \quad \text{(R7)}$$

This phenomenon was not observed during a brief test of the HIRAC FAGE instrument with a 5 kHz PRF laser system (JDSU Nd:YAG pumped Sirah Cobra Stretch dye laser, as in Glowacki et al., 2007a, and Malkin et al., 2010), most likely due to much lower laser pulse energies for the 5 kHz system (1.6 μJ pulse^{-1} compared to 40 μJ pulse^{-1} at 200 Hz PRF). Using the scavenger injection system (Sect. 2.2), the decay of the TBHP could be accurately described (compared to simultaneous FTIR measurements) characterising the interference signal. At a time defined by the user, the *iso*-butane scavenger (20 % in N_2) was injected into the FAGE cell for ~ 90 s at ~ 20 sccm. Before the chamber photolysis lamps were initiated, the OH interference signal was measured with the scavenger off and on, and the difference in signal was observed to be negligible (within the uncertainty of the measurement). The OH interference profile during the hydrocarbon decay was characterised and accounted for using 3–4 scavenger injections per experiment. An empirical fit to the averaged signals was used to correct the measured OH signal from TBHP laser photolysis over time, shown here in Fig. 6b compared to the inferred [OH] from the GC-FID. The type of fitting parameter (e.g. linear or exponential) was judged depending on the quality of data.

The calibration procedure was completed by plotting the OH signals, normalised for laser power, measured by FAGE as a function of the calculated OH concentrations from the hydrocarbon decays producing a calibration plot with C_{OH}, in units of counts cm^3 s^{-1} mW^{-1} molecule^{-1}, as the gradient. A typical calibration plot is shown in Fig. 7 – produced using the decay of cyclohexane at 1000 mbar chamber pressure (see caption for detailed operating conditions). Uncertainties in C_{OH} are quoted to $\pm 2\sigma$, and error bars represent the standard error, and hence precision, of the measured S_{OH} to $\pm 1\sigma$. Error bars were kept at $\pm 1\sigma$ as this represented the precision used in the analysis procedure.

4.2 Formaldehyde photolysis

Calibration of the HO_2 detection cell required only the generation of HO_2 radicals in the HIRAC chamber, and a time-dependent measurement of their subsequent recombination using the FAGE instrument once the photolysis lamps were extinguished. Upon photolysis in air (lamps: Philips TL40W/12 RS), HCHO produced H + HCO and H_2+ CO (Reaction R9) in approximately a 60 : 40 ratio (Reactions R8, R9). Under the conditions in HIRAC, HCO reacted with O_2 to give HO_2+ CO (Reaction R10) and the H atom produced in Reaction (R8) reacted with O_2 to give HO_2 (Reaction R11). The loss of HO_2 was characterised by the competing bimolecular and termolecular self-reactions (Reactions R12 + R13) and a first-order wall-loss parameter (Reaction R14):

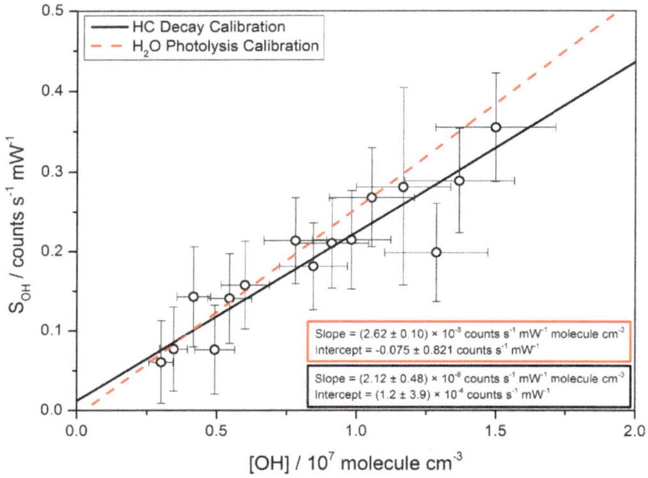

Figure 7. Calibration from the hydrocarbon decay method for cyclohexane at 1000 mbar and 293 K chamber pressure using the HIRAC FAGE instrument with the 200 Hz PRF laser system; inlet pressure = (3.81 ± 0.02) mbar; laser power = (7.0 ± 0.5) mW; $[H_2O]_{vapour} < 10$ ppmv. Extrapolated calibration from the H_2O photolysis calibration technique for inlet pressure = (3.79 ± 0.02) mbar, laser power = (6.0 ± 0.5) mW, $[H_2O]_{vapour} = (3900 \pm 20)$ ppmv and $[OH] = (0.5\text{–}1.5) \times 10^9$ molecule cm^{-3}. Both fits are weighted to errors in the x and y axes and error bars are representative of the standard error in the measurement to $\pm 2\sigma$. Uncertainties quoted for the slope and intercept represent the precision of the calibration processes to $\pm 2\sigma$.

$$HCHO + h\nu \rightarrow H + HCO \text{ or } H_2 + CO \quad (R8)$$

$$HCO + O_2 \rightarrow HO_2 + CO \quad (R9)$$

$$H + O_2 + M \rightarrow HO_2 + M \quad (R10)$$

$$HO_2 + HO_2 \rightarrow H_2O_2 + O_2 \quad (R11)$$

$$HO_2 + HO_2 + M \rightarrow H_2O_2 + O_2 + M \quad (R12)$$

$$HO_2 \rightarrow \text{Loss}. \quad (R13)$$

Therefore the rate of loss of HO_2 is given by:

$$\frac{d[HO_2]}{dt} = -\left(k_{loss}[HO_2] + +k_{HO_2} + HO_2[HO_2]^2\right), \quad (6)$$

where $k_{HO_2} + HO_2$is the HO_2 recombination rate coefficient; the sum of the pressure-independent (Reaction R12) and pressure-dependent (Reaction R13) rate coefficients as determined by IUPAC (2007). Solving analytically for $[HO_2]_t$ at a given time, t, integration of Eq. (5) becomes

$$\frac{1}{[HO_2]_t} = \left(\frac{1}{[HO_2]_0} + \frac{2 \cdot k_{HO_2+HO_2}}{k} \text{ loss}\right) \times e^{(k_{loss}t)} - \left(\frac{2 \cdot k_{HO_2+HO_2}}{k_{loss}}\right). \quad (7)$$

The $[HO_2]$ in Eq. (6) is unknown but is related to the normalised HO_2 signals measured by FAGE, S_{HO_2}, and the instrument sensitivity to HO_2, C_{HO_2}, and therefore

$$(S_{HO_2})_t = \left(\left(\left(\frac{1}{(S_{HO_2})_0} + \frac{2 \cdot k_{HO_2+HO_2}}{k_{loss} \cdot C_{HO_2}}\right) \times e^{(k_{loss}t)} - \left(\frac{2 \cdot k_{HO_2+HO_2}}{k_{loss} \cdot C_{HO_2}}\right)\right)\right)^{-1}, \quad (8)$$

where $(S_{HO_2})_t$ and $(S_{HO_2})_0$ are the HO_2 signal at time t and $t = 0$ respectively.

The measured decay of S_{HO_2} using FAGE and the fit described by Eq. (7) are displayed in Fig. 8a for a typical experiment (aircraft FAGE instrument (5 kHz PRF), 1000 mbar, 298 K, < 10 ppm $[H_2O]$, mixing fans on). Both k_{loss} and C_{HO_2} were determined by data fitting the S_{HO_2} decay using Eq. (7) with a Levenberg–Marquardt non-linear least-squares algorithm, fixing the initial signal and $k_{HO_2+HO_2}$. The first ~ 100 s of data were used, ensuring analysis after an almost complete decay of S_{HO_2}. Fitting was improved by the inclusion of upper and lower bounds of $\pm 10\,\%$ for the $(S_{HO_2})_0$ into the fitting routine, which accounted for the uncertainty in the determination of $(S_{HO_2})_0$ (see Sect. 5.4.3).

For the experimental 350–1000 mbar pressure range at 0 % H_2O vapour, $k_{HO_2+HO_2}$ was determined between (2.00–2.85) $\times 10^{-12}$ cm^3 molecule^{-1} s^{-1}, to the recommendation given by IUPAC (2007). A calibration was conducted at $[H_2O]_{vap} = 7500$ ppmv, to validate the calibration method at high water vapour concentrations, representative of the conventional H_2O vapour photolysis method. The $k_{HO_2+HO_2}$ therefore included a correction for the HO_2-H_2O vapour chaperone effect (Stone and Rowley, 2005) in accordance with the IUPAC recommendation (Atkinson et al., 2004). The wall-loss rate, k_{loss}, was dependent on daily chamber conditions and was therefore determined as part of the fitting procedure along with C_{HO_2}, typically between 0.032 and 0.073 s^{-1} with an uncertainty of $\pm 10\,\%$ (2σ). Variations in the wall-loss rates have implications for the uncertainty in C_{HO_2} derivation (Sect. 5.4).

5 Results and discussion

All results presented here were taken using the HIRAC FAGE instrument using Litron Nd:YAG pumped dye laser light source operating at 200 Hz PRF, unless otherwise stated. Tabulated data from the alternative calibration methods are displayed in the Supplement (Tables S1 and S2). All uncertainties displayed are quoted to $\pm 2\sigma$ and all regressions shown are empirical, unless otherwise stated.

5.1 Conventional H_2O vapour photolysis calibration

5.1.1 C_{OH} and C_{HO_2} as a function of internal cell pressure

The FAGE instrument sensitivity to OH (Fig. 9, HIRAC FAGE only) and HO_2 (Fig. 10, top) was determined as a

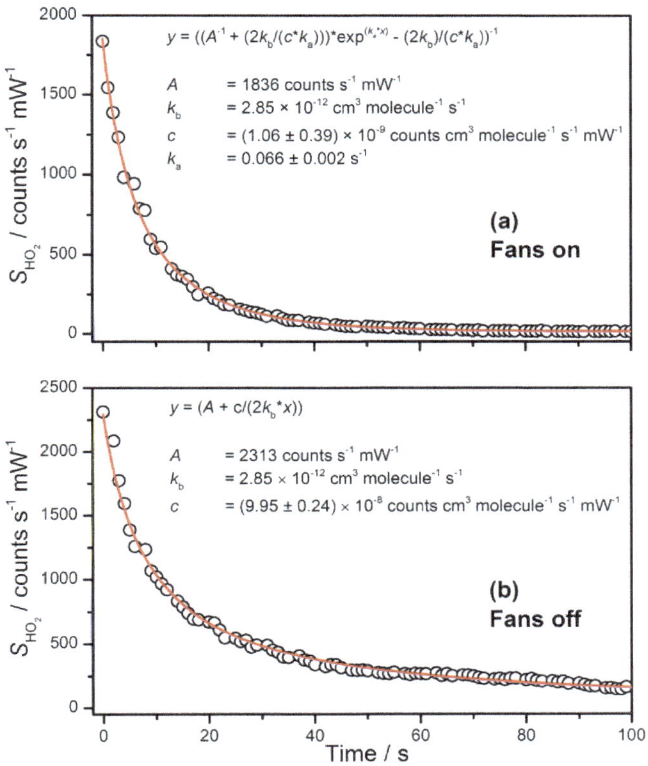

Figure 8. Normalised S_{HO_2} decay for the HCHO photolysis calibration method at 1000 mbar chamber pressure using the aircraft FAGE instrument (5 kHz PRF laser system) conducted with the HIRAC chamber mixing fans on **(a)** and off **(b)**; inlet pressure $= (2.53 \pm 0.02)$ mbar; laser power $= (8.25 \pm 0.25)$ mW. Data in **(a)** were fitted with Eq. (7) to give C_{HO_2} where $A = (S_{HO_2})_0$, $k_b = k_{HO_2} + HO_2$, $c = C_{HO_2}$, $k_a = k_{loss}$ and $z =$ offset, with uncertainties quoted to $\pm 1\sigma$. Parameters without quoted error were fixed.

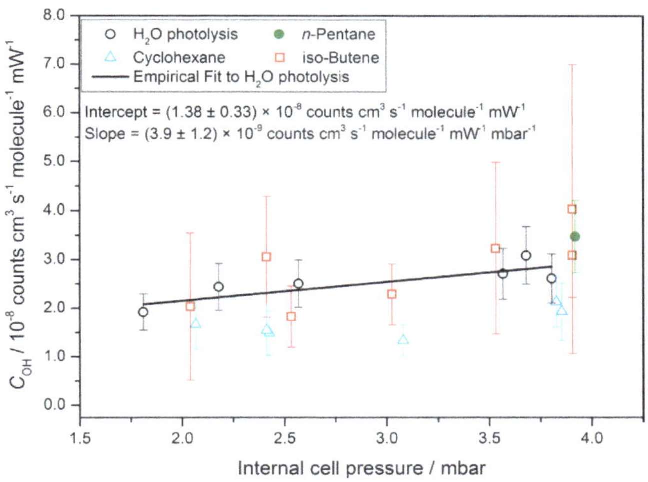

Figure 9. HIRAC FAGE instrument sensitivity to OH, C_{OH}, as a function of internal detection cell pressure as determined by the H_2O vapour photolysis and HC decay calibration techniques using the Litron Nd:YAG pumped dye laser operating at 200 Hz PRF. All calibrations were conducted at laser powers between 6.0 and 9.5 mW. Different internal cell pressures (1.8–3.8 mbar) were achieved by changing the diameter of the FAGE inlet pinhole between 0.5 and 1.0 mm. Conventional calibrations were conducted at constant $[H_2O]_{vap}$ (4500 ± 600 ppmv), whereas alternative calibrations were conducted in near-dry conditions (< 15 ppmv). HIRAC chamber pressures between 440 and 1000 mbar were used to induce internal cell pressures between 2.1 and 3.9 mbar. Error bars indicate the total uncertainty to $\pm 1\sigma$.

function of pressure using the H_2O vapour photolysis calibration method over the inlet pressure range of 1.8–3.8 mbar. Figures 9 and 10 display the H_2O calibration data compared to those from the respective alternative calibration methods, the results for which are discussed in Sects. 5.2 and 5.3; error bars in both figures are representative of the total uncertainty in the calibration (Sect. 5.4 for details). Constant laser power and $[H_2O]$ were maintained throughout the calibration process (8 ± 1 mW and 4500 ± 600 ppmv respectively).

The linear regressions were used to describe the sensitivity as a function of fluorescence cell pressure for experiments conducted in HIRAC, and are a valid description of the data inside the 1.8–3.8 mbar pressure range only. The C_{OH} and C_{HO_2} data sets shown here were not conducted at the same time, but 6 months apart. This was due to the chronological order of the development of the alternative calibration techniques, during which time the FAGE pump set was serviced, increasing the pumping capacity and generally lowering the internal cell pressures for each pinhole in the C_{OH} determination.

The fit displayed a greater increase in C_{HO_2} as a function of pressure compared to C_{OH}, where $\Delta C_{OH} = (17 \pm 11)\,\%$ and $\Delta C_{HO_2} = (31.6 \pm 4.4)\,\%$ increase between 1.3 and 3.8 mbar. Altering the pinhole diameter could change the flow dynamics inside the instrument, reducing NO mixing efficiency, and therefore HO_2 conversion efficiency, before the HO_2 cell. The decrease in C_{HO_2} at lower pressure has been reproduced in a more recent calibration of the HO_2 cell using the 5 kHz PRF laser source (slope $= (5.14 \pm 0.46) \times 10^{-9}$ counts cm^3 molecule^{-1} s^{-1} mW^{-1} mbar^{-1}), suggesting the process was not affected by changes in laser pulse energy.

The experimental parameters controlling the instrument sensitivity, C_{OH}, which are dependent upon pressure, are the OH concentration in the laser-excitation region, $[OH]_{cell}$; the fluorescence quantum yield following laser excitation to the OH $A^2\Sigma^+$ ($v' = 0$) excited state, φ_{fl}; and the fraction of the fluorescence decay which falls within the integrating gate of the photon counter, f_{gate} (Creasey et al., 1997b; Faloona et al., 2004). The OH concentration in the cell held at total density $[M]$ and the fluorescence quantum yield are given by

Eqs. (8) and (9):

$$[\mathrm{OH}]_{\mathrm{cell}} = \chi[M], \tag{9}$$

$$\varphi_{\mathrm{fl}} = \frac{A}{\left(A + k_{\mathrm{q}}[M]\right)}, \tag{10}$$

where χ is the mixing ratio of OH impinging at the pinhole (assuming no losses at the pinhole), A is the inverse of the radiative lifetime of OH and k_{q} is the rate coefficient for quenching of the excited $A^2\Sigma^+$ $(v' = 0)$ (averaged appropriately over all quenching species). Assuming that $f_{\mathrm{gate}} = 1$, then the overall pressure-dependent term for instrument sensitivity to OH can be described as the product of Eqs. (8) and (9), shown here in Eq. (10):

$$\varphi\chi[\mathrm{OH}]_{\mathrm{cell}} \times \Phi_{\mathrm{fl}} = \frac{[M]A}{\left(A + k_{\mathrm{q}}[M]\right)}. \tag{11}$$

When $[M] \rightarrow 0$, the product becomes $\chi[M]$, and C_{OH} is directly proportional to pressure ($[M]$). At higher pressures when $k_{\mathrm{q}}[M] \gg A$ (at 18 mbar the ratio is ~ 10), the product becomes $\sim \chi A/k_{\mathrm{q}}$ and C_{OH} is independent of $[M]$, and thus depends only on the mixing ratio of OH.

However, FAGE is an on-resonance technique, and therefore it is not possible to achieve the limit $f_{\mathrm{gate}} = 1$, because it is necessary to gate off the CPM during the laser pulse in order to avoid saturation of the detector. Thus, in these experiments f_{gate} was always < 1, and depended non-linearly on pressure because the photon-counting gate remained the same whilst the fraction of the total fluorescence collected within this gate (f_{gate}) changed as a result of changes in the total fluorescence lifetime of the excited-state OH radicals. Thus the effective area of integration under the fluorescence decay curve reduced with increasing pressure such that f_{gate} reduced non-linearly as pressure increased. For the conditions used in these experiments $f_{\mathrm{gate}} = 0.79$–0.63 (between 1.3 and 3.8 mbar). Hence the observation that C_{OH} increased linearly over pressures between 1.3 and 3.8 mbar in this study is consistent with the expected behaviour based purely on the balance between OH number density and the total fluorescence collected.

The inherent complexity that results from the multiple factors which control the sensitivity of FAGE instruments, and which also change with a variety of conditions (i.e. pressure, $[H_2O]$, laser power) and instrumental factors (e.g. time take for CPM to reach maximum gain), require that FAGE instruments are frequently and carefully calibrated.

Additional investigations into the FAGE instrument sensitivity to OH as a function of $[H_2O]_{\mathrm{vap}}$ and laser power are discussed in detail in the Supplement. The change in instrument sensitivity over $[H_2O]$ between 200 and 4500 ppmv was observed to be within the uncertainty of the calibration (35 % at 2σ) and was therefore considered negligible. For this reason no correction for sensitivity to $[H_2O]$ was applied to the date taken in the alternative calibration method where $[H_2O]$ < 10 ppmv.

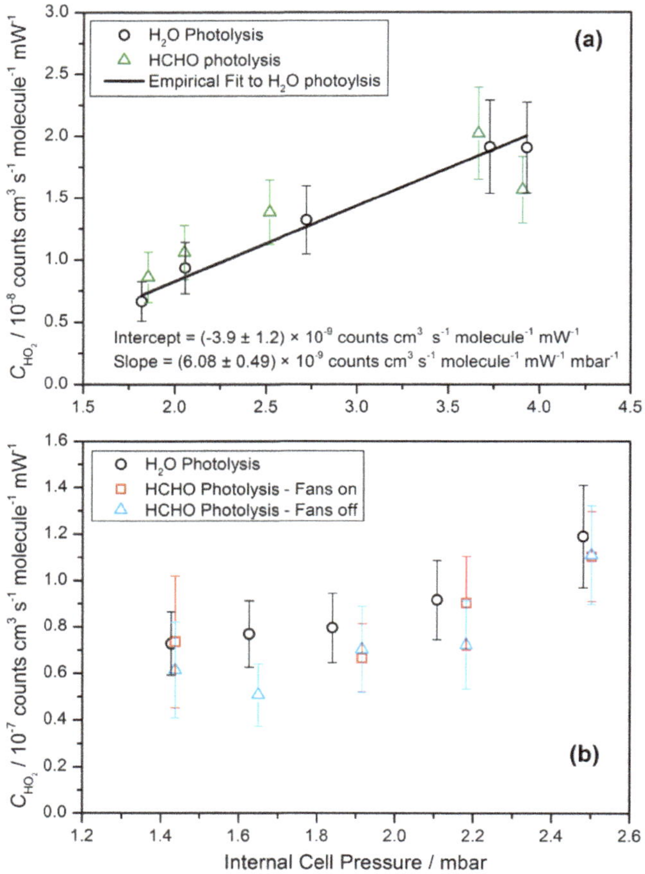

Figure 10. FAGE instrument sensitivity to HO_2, C_{HO_2}, as a function of internal detection cell pressure as determined by the H_2O vapour and HCHO photolysis calibration techniques using the HIRAC FAGE instrument operating at 200 Hz PRF **(a)** and the aircraft FAGE instrument operating at 5 kHz PRF **(b)**. Conventional calibrations were conducted at constant $[H_2O]_{\mathrm{vap}}$ (**(a)** 4500 ± 600 ppmv, **(b)** 6000 ± 600 ppmv), whereas alternative calibrations were conducted under low $[H_2O]_{\mathrm{vap}}$ (< 15 ppmv). HIRAC chamber pressures of 440–1000 mbar were used to induce internal cell pressures of **(a)** 1.8–3.8 mbar (pinhole diameter 0.5–1.0 mbar) and **(b)** 1.42–2.48 mbar (pinhole diameter 0.3–0.6 mbar). Error bars indicate the total uncertainty to $\pm 1\sigma$.

5.2 Hydrocarbon decay calibration

Figure 7 shows a direct comparison of analysed data from the decay of cyclohexane and H_2O vapour calibration method at ~ 3.80 mbar internal cell pressure (equivalent to 1000 mbar in HIRAC) using the 1.0 mm inlet pinhole and ~ 7 mW laser power. The C_{OH} was determined as $(2.13 \pm 0.52) \times 10^{-8}$ counts s^{-1} molecule^{-1} cm^3 mW^{-1}, within error of the traditional H_2O vapour photolysis calibration (2σ) at the same pressure ($(2.62 \pm 0.91) \times 10^{-8}$ counts s^{-1} molecule^{-1} cm^3 mW^{-1}). Error bars are representative of the total uncertainty at $\pm 1\sigma$. Additional example calibration plots for each hydrocarbon studied are included in the Supplement. Displayed in Fig. 9 is C_{OH} as a function of internal cell pressure using the HC decay calibration method, determined for *iso*-butene, cyclohexane and *n*-pentane. The HC decay calibration method was observed to be in agreement with the H_2O vapour photolysis calibration. The average of the ratio of calibration factors (conventional : alternative) was calculated for each alternative calibration point across the entire pressure range, $C_{OH(conv)}/C_{OH(alt)} = 1.19 \pm 0.26$, where $C_{OH(conv)}$ was determined from the fit to the H_2O photolysis data.

A large variability in the C_{OH} determined using the *iso*-butene decay was observed, with larger uncertainties associated with this calibration compared to cyclohexane and *n*-pentane, and the reason for this remains unclear. On average, the measured OH signals were closer to the detection limit of the FAGE instrument when using *iso*-butene. Initial concentrations of each of the hydrocarbons were 2.5×10^{13} molecule cm^{-3}, and hence a lower OH steady-state concentration is expected when *iso*-butene is present as the k_{OH} is an order of magnitude higher than those for *n*-pentane and cyclohexane. As S_{OH} approaches 0 counts s^{-1} mW^{-1}, the S_{OH} measurement becomes increasingly imprecise, and thus the uncertainty in the fitting of the calibration plot increases.

A general under-prediction of C_{OH}, compared to the H_2O vapour photolysis method, was observed when calculated using the decay of cyclohexane, $C_{OH(conv)}/C_{OH(Chex)} = 1.52 \pm 0.44$. The exact reason is unknown. Evaluation of the HC decay data with the k_{OH} adjusted at the upper limit of uncertainty recommended by Calvert et al. (2008) (25 % (2σ), $k_{OH} = 8.04 \times 10^{-12}$ cm^3 $molecule^{-1}$ s^{-1}) brings the two data sets into better agreement, $C_{OH(conv)}/C_{OH(Chex)} = 1.21 \pm 0.22$. The cyclohexane measurements were also affected to a greater extent by the chamber dilution due to the slower rate of reaction with OH, which contributed to 25–30 % of the total cyclohexane decay rate directly after the photolysis lamps were initiated, compared to 5–10 % for the *iso*-butene experiments. Correcting the cyclohexane data for a hypothetically enhanced chamber dilution could explain the lower sensitivity measurements (as the decay increases, $[OH]_{inf}$ increases), however the dilution rate was confirmed prior to photolysis of TBHP in each experiment.

5.3 Formaldehyde photolysis calibration

Figure 10a shows the instrument sensitivity to HO_2, C_{HO_2}, as a function of internal cell pressure for the newly developed formaldehyde photolysis calibration technique for the HIRAC FAGE instrument. Each data point corresponds to the average of up to five HO_2 decay traces (Fig. 8a), and the error bars are representative of the total calibration 1σ uncertainty (Sect. 5.4). All calibrations were completed over a 4–8 mW laser power range. The alternative calibration was observed to be in good agreement with the conventional H_2O vapour photolysis calibration technique over the operating internal cell pressure range of 1.8–3.8 mbar ($C_{HO_2(conv)}/C_{HO_2(alt)} = 0.96 \pm 0.09$) for the Litron-based FAGE system.

The kinetics of the HO_2 decay due to recombination and first-order wall loss (Eq. 7) were confirmed by studying the HO_2 decay profile with the chamber mixing fans on and off using the University of Leeds aircraft-based FAGE instrument. With the mixing fans off, the decay was accurately described by the recombination kinetics only (Fig. 8b), giving C_{HO_2} values within error of the fans on experiments, as shown in Fig. 10b. Good agreement between the conventional and alternative calibration methods was also observed across the 1.42–2.48 mbar internal cell pressure range, and the overall correlation between conventional and alternative calibration methods was calculated as $C_{HO_2(conv)}/C_{HO_2(alt)} = 1.07 \pm 0.09$ for the high-frequency aircraft-based FAGE instrument.

5.4 Calibration uncertainties

The overall uncertainty associated with the calibration methods presented here was calculated using the sum in quadrature of the accuracy and the precision terms of the calibration. The accuracy term accounted for any systematic uncertainty in the calculation of $[HO_x]$ for each calibration method, signal normalisation etc., and these are displayed in Table 2. The precision of the calibrations was defined as the random errors associated with each method. All uncertainties are quoted as 2σ.

5.4.1 H_2O photolysis calibration

The total uncertainty in the H_2O photolysis calibration method was estimated to be ~ 36 %. The accuracy was defined by the uncertainty associated with each term of Eq. (1) in the determination of $[HO_x]$ and was estimated to be ~ 35 %. The largest contribution to the accuracy of this calibration method came from the determination of the calibration source flux, $F_{184.9\,nm}$, with a total uncertainty of 32 %. The product of the flux and the irradiation time from Eq. (1), $F_{184.9\,nm} \times \Delta t$, was determined using N_2O actinometry which relied on the detection of trace levels of NO (0.5–3 ppbv, Sect. 2.1) followed by evaluation of the measurements using four rate constants each with ~ 20 % uncertainty. Although the actinometric method gives a direct determination of the product $F_{184.9\,nm} \times \Delta t$, in order to calculate [OH] from Eq. (1) any differences between the total volumetric flow rate during the actinometry experiment and the OH calibration needs to be accounted for as they change Δt. It is therefore necessary to account for the uncertainty in Δt, which was determined to be ~ 2 % using the uncertainty in the flow rates from the mass flow controllers (~ 1 %). For the remainder of the terms in Eq. (1) their contributions to the accuracy in the H_2O photolysis calibration method were

Table 2. The systematic uncertainties to 2σ in the various parameters that determine the accuracy in the OH and HO_2 calibration factors for all three calibration methods. Total accuracy is taken as the sum in quadrature of the individual uncertainties. The Online position error is the approximate error in the maximum line intensity that is achieved when positioning the laser wavelength at the centre of the OH transition.

$H_2O + h\nu$		Hydrocarbon decay		HCHO $+h\nu$	
Parameter	Uncertainty	Parameter	Uncertainty	Parameter	Uncertainty
$F_{184.9\,nm}$	32 %	k_{OH}	20–25 %	$k_{HO_2+HO_2}$	35 %
Δt	2 %	k_{Dil}	10 %	S_{HO_2} intial	20 %
$[H_2O]$	10 %	GC-FID	4 %	Laser power	6 %
$\sigma_{H_2O,184.9\,nm}$	6 %	Laser power	6 %	Online position	4 %
Laser power	6 %	Online position	4 %		
Online position	4 %				
Error	35 %	Error	24–28 %	Error	41 %

as follows: σ_{H_2O} was taken from Cantrell et al. (1997), with a reported total error of ±6 %; the error in $[H_2O]$ was taken from the hygrometer instrumental uncertainty (±10 %); and laser power was defined by the laser power meter (Molectron Powermax 500A, ±0.25 mW).

The precision was typically between 4 and 10 % for the flow tube calibration process and was taken from the standard error in the error-weighted fit of the calibration plot. The error bars were representative of the standard deviation in the S_{OH} and $[HO_x]$ for the x and y axes respectively. The flux output of the calibration source, hygrometer and CPM measurements were observed to have good point-to-point stability and therefore low standard deviations.

5.4.2 Hydrocarbon decay calibration

The accuracy of the hydrocarbon decay method was estimated to be better than that of the flow tube method (~ 28 % compared to 35 %). However, due to the large variation in the random errors that defined the precision of the experiment, the total uncertainty for the HC decay method was larger than the flow tube calibration method, with the total uncertainty estimated at ~ 45 %.

The accuracy in the calibration was intrinsic to the hydrocarbon used, being dependent on the uncertainty in k_{OH} and k_{Dil}. The largest uncertainty was in k_{OH}, taken from data reviews from the Calvert series or IUPAC recommendations: *n*-pentane, ±20 % (Calvert et al., 2008); cyclohexane, ±20 % (Calvert et al., 2008); and *iso*-butene, ±25 % (IUPAC, 2007). Uncertainty in k_{Dil} was calculated from repeated measurements of chamber dilution for each of the hydrocarbons, and induced errors in GC-FID calibration (4 %). The precision of the experiments for both *n*-pentane and cyclohexane was between 10 and 25 %, whereas *iso*-butene showed much greater variation of between 13 and 69 %; possible reasons for this have been discussed in Sect. 5.2.

5.4.3 Formaldehyde photolysis calibration

The total uncertainty for the HCHO photolysis calibration method has been estimated at ~ 46 %, which is 10 % greater than the conventional calibration method. The accuracy of the HCHO photolysis method was estimated as ~ 41 %; the largest contribution to this deriving from the uncertainty was in the HO_2 recombination rate constant (35 %), taken from the IUPAC recommendation (IUPAC, 2007). Determining the accurate initial S_{HO_2} (i.e. S_{HO_2} at t_0) is hard as HO_2 does not fully reach steady state before the photolysis lamps are switched off; therefore there is a certain amount of subjective choice in the value of $S(HO_2)_0$, and hence the uncertainty in the initial S_{HO_2} was based on the standard deviation of the mean "steady-state" HO_2 signal, which gives an estimation of the 1 s point-to-point variability for a chosen t_0 (~ 20 %).

The error associated with the precision of the experiment was taken from the error propagation of the standard error terms from the Levenberg–Marquardt iterative fitting procedure for Eq. (9) and Fig. 8. This includes both the error in the C_{HO_2} and k_{loss} parameters. The precision for this method was in line with the conventional flow tube calibration between 10 and 20 %.

6 Conclusions and outlook

The first pressure-dependent calibrations of a FAGE instrument for both OH and HO_2 have been successfully conducted using the HIRAC chamber. Previous pressure-dependent aircraft measurements had been extracted by assuming that the calibration factor could be determined by simply calibrating at the required internal FAGE cell pressure. Assumptions were therefore made that variations in radical losses on the inlet and the nature of the expansion caused by the varying pressure differential inside and outside the FAGE cell were insignificant. The results displayed in Figs. 9 and 10 validate the conventional calibration method with the alternative hydrocarbon decay and HCHO photolysis methods over a range of internal FAGE cell pressures. It should be emphasised that, strictly speaking, this validation applies only to these particular FAGE instruments, but this work suggests that the agreement between the different calibration techniques will translate to FAGE instruments of similar designs. As the calibration methods are quite different in principle, they are unlikely to be subject to the same systematic errors. The alternative calibration results presented here have been shown to be well within the combined uncertainty of their respective traditional calibration method, validating the pressure-dependent flow tube calibration technique and improving confidence in FAGE measurements both in the field and in kinetics experiments. Both alternative methods have also shown that calibrations conducted under high $[H_2O]_{vap}$ conditions (2000–4500 ppmv) can be applied to measurements at low $[H_2O]_{vap}$ (< 15 ppmv).

The hydrocarbon decay method has shown that the FAGE instrument can be calibrated over a range of external pressures using different hydrocarbons. Compared to the conventional calibration method, where $[HO_x]$ are generated typically at $> 10^8$ molecule cm^{-3}, the hydrocarbon decay method is conducted at a $[HO_x]$ relevant to chamber-based experimental measurements ($\sim 10^7$ molecule cm^{-3}) and much closer to typical ambient OH concentrations ($\sim 10^6$ molecule cm^{-3}).

Currently the total error associated with the hydrocarbon decay method is greater than that of the flow tube method (~ 45 % vs. 36 %). The accuracy or total systematic uncertainty associated with the alternative OH calibration method is lower than that of the flow tube calibration method (28 % vs. 35 %), and hence an improvement in the precision of the experiment could improve the overall uncertainty to be in line with the flow tube method. The primary source of random error arose in the detection of OH close to the detection limit. Increasing the steady-state OH concentration in the chamber would allow easier detection of the hydrocarbon decay compared to chamber dilution, as well as an OH measurement above the detection limit. The steady-state OH concentration could be increased by increasing the 254 nm intensity in the chamber; using new lamps or more lamps; altering the OH precursor, e.g. O_3+ alkenes or photolysis of methyl nitrite; or by lowering the initial [HC]. The latter would require a more sensitive hydrocarbon detection technique than GC-FID or FTIR, which are currently available in HIRAC. One such technique is proton transfer mass spectrometry (PTR-MS), which would reduce the uncertainty in the hydrocarbon decay measurements by providing higher-time-resolution measurements and allow for easier simultaneous measurement of multiple hydrocarbons at low concentrations, effectively providing multiple independent estimates of C_{OH} from a single experiment. Uncertainties in the rate coefficients could also be reduced by a concerted laboratory study including relative rate and direct flash photolysis methods, decreasing the systematic error. With careful experimental design, errors could potentially be reduced to closer to 10 % (Orkin et al., 2010; Carr et al., 2011; Glowacki et al., 2012).

A full range of pressure-dependent calibrations using this method would currently take ~ 2 days, compared to ~ 3 h for the flow-tube-based calibration. However, the timescale does not limit the suitability of the method for regular confirmation of C_{OH} obtained from the flow tube calibration method.

The total uncertainty in the HCHO photolysis method is ~ 46 % which is 10 % greater than that of the traditional H_2O photolysis method. The HCHO photolysis method is quick and reproducible. The time taken to complete the analysis and the errors is comparable with the flow tube technique. An advantage of the HCHO photolysis method is that several runs can be completed in one fill of the chamber, compared to the HC decay method, which requires one fill per experiment (although the proposed use of multiple HC decays will provide multiple estimates of C_{OH} from a single chamber fill).

Acknowledgements. This work was supported by funding from the Natural Environment Research Council (NERC grant NE/G523739/1), EUROCHAMP-2 and the National Centre for Atmospheric Science (NCAS). I. Bejan acknowledges a Marie Curie fellowship. We are grateful for helpful discussions with Andrew Goddard and Mark Blitz on technical aspects of HIRAC and lasers respectively.

Edited by: A. Hofzumahaus

References

Aschmutat, U., Hessling, M., Holland, F., and Hofzumahaus, A.: A tunable source of Hydroxyl (OH) and Hydroperoxy (HO_2) radicals: in the range between 10^6 and 10^9 cm^{-3}, in: Physico-Chemical Behaviour of Atmospheric Pollutants, edited by: Restelli, G. A. a. G., European Comission, Brussels, 811-816, 1994.

Atkinson, R.: Kinetics and mechanisms of the gas-phase reactions of the hydroxyl radical with organic compounds under atmospheric conditions, Chem. Rev., 86, 69–201, 1986.

Atkinson, R.: Gas-phase tropospheric chemistry of organic compounds, J. Phys. Chem. Ref. Data. Mono., 2, 1994.

Atkinson, R.: Kinetics of the gas-phase reactions of OH radicals with alkanes and cycloalkanes, Atmos. Chem. Phys., 3, 2233–2307, doi:10.5194/acp-3-2233-2003, 2003.

Atkinson, R. and Pitts, J. N. J.: Rate Constants for the Reaction of OH Radicals with Propylene and the Butenes over the Temperature Range 297–425 K, J. Chem. Phys., 63, 3591–3595, 1975.

Atkinson, R., Baulch, D. L., Cox, R. A., Crowley, J. N., Hampson, R. F., Hynes, R. G., Jenkin, M. E., Rossi, M. J., and Troe, J.:

Evaluated kinetic and photochemical data for atmospheric chemistry: Volume I – gas phase reactions of O_x, HO_x, NO_x and SO_x species, Atmos. Chem. Phys., 4, 1461–1738, doi:10.5194/acp-4-1461-2004, 2004.

Baasandorj, M., Papanastasiou, D. K., Talukdar, R. K., Hasson, A. S., and Burkholder, J. B.: $(CH_3)_3COOH$ (tert-butyl hydroperoxide): OH reaction rate coefficients between 206 and 375 K and the OH photolysis quantum yield at 248 nm, Phys. Chem. Chem. Phys., 12, 12101–12111, doi:10.1039/c0cp00463d, 2010.

Berresheim, H., Elste, T., Tremmel, H. G., Allen, A. G., Hansson, H. C., Rosman, K., Dal Maso, M., Makela, J. M., Kulmala, M., and O'Dowd, C. D.: Gas-aerosol relationships of H_2SO_4, MSA, and OH: Observations in the coastal marine boundary layer at Mace Head, Ireland, J. Geophys. Res.-Atmos., 107, 8100, doi:10.1029/2000JD000229, 2002.

Bloss, W. J., Lee, J. D., Bloss, C., Heard, D. E., Pilling, M. J., Wirtz, K., Martin-Reviejo, M., and Siese, M.: Validation of the calibration of a laser-induced fluorescence instrument for the measurement of OH radicals in the atmosphere, Atmos. Chem. Phys., 4, 571–583, doi:10.5194/acp-4-571-2004, 2004.

Brauers, T., Aschmutat, U., Brandenburger, U., Dorn, H. P., Hausmann, M., Hessling, M., Hofzumahaus, A., Holland, F., Plass-Dulmer, C., and Ehhalt, D. H.: Intercomparison of tropospheric OH radical measurements by multiple folded long-path laser absorption and laser induced fluorescence, Geophys. Res. Lett., 23, 2545–2548, 1996.

Calvert, J. G., Atkinson, R., Kerr, J. A., Madronich, S., Moortgat, G. K., Wallington, T. J., and Yarwood, G.: The Mechanism of Atmospheric Oxidation of the Alkenes, OUP, Oxford, 2000.

Calvert, J. G., Derwent, R. G., Orlando, J. J., Tyndall, G. S., and Wallington, T. J.: Mechanisms of Atmospheric Oxidation of the Alkanes, Oxford University Press, 2008.

Campuzano-Jost, P., Williams, M. B., D'Ottone, L., and Hynes, A. J.: Kinetics and mechanism of the reaction of the hydroxyl radical with h(8)-isoprene and d(8)-isoprene: Isoprene absorption cross sections, rate coefficients, and the mechanism of hydroperoxyl radical production, J. Phys. Chem. A, 108, 1537–1551, doi:10.1021/jp0363601, 2004.

Cantrell, C. A., Tyndall, G., and Zimmer, A.: Absorption cross sections for water vapour from 183 to 193 nm, Geophys. Res. Lett., 24, 2195–2198, 1997.

Carr, S. A., Blitz, M. A., and Seakins, P. W.: Site-Specific Rate Coefficients for Reaction of OH with Ethanol from 298 to 900 K, J. Phys. Chemi. A, 115, 3335–3345, doi:10.1021/jp200186t, 2011.

Commane, R., Floquet, C. F. A., Ingham, T., Stone, D., Evans, M. J., and Heard, D. E.: Observations of OH and HO_2 radicals over West Africa, Atmos. Chem. Phys., 10, 8783–8801, doi:10.5194/acp-10-8783-2010, 2010.

Creasey, D. J., Halford-Maw, P. A., Heard, D. E., Pilling, M. J., and Whitaker, B. J.: Implementation and initial deployment of a field instrument for measurement of OH and HO_2 in the troposphere by laser-induced fluorescence, J. Chem. Soc.-Faraday Trans., 93, 2907–2913, 1997a.

Creasey, D. J., Heard, D. E., Pilling, M. J., Whitaker, B. J., Berzins, M., and Fairlie, R.: Visualisation of a supersonic free-jet expansion using laser- induced fluorescence spectroscopy: Application to the measurement of rate constants at ultralow temperatures, Appl. Phys. B-Lasers Opt., 65, 375–391, 1997b.

Creasey, D. J., Heard, D. E., and Lee, J. D.: Absorption cross-section measurements of water vapour and oxygen at 185 nm. Implications for the calibration of field instruments to measure OH, HO_2 and RO_2 radicals, Geophys. Res. Lett., 27, 1651–1654, doi:10.1029/1999gl011014, 2000.

Davis, D. D., Heaps, W., and McGee, T.: Direct Measurements of Natural Tropospheric Levels of OH *via* an Aircraft Borne Tunable Dye-Laser, Geophys. Res. Lett., 3, 331–333, 1976.

Dorn, H. P., Brandenburger, U., Brauers, T., Hausmann, M., and Ehhalt, D. H.: In-situ detection of tropospheric OH radicals by folded long-path laser absorption, Results from the POPCORN field campaign in August 1994, Geophys. Res. Lett., 23, 2537–2540, doi:10.1029/96gl02206, 1996.

Dusanter, S., Vimal, D., and Stevens, P. S.: Technical note: Measuring tropospheric OH and HO_2 by laser-induced fluorescence at low pressure. A comparison of calibration techniques, Atmos. Chem. Phys., 8, 321–340, doi:10.5194/acp-8-321-2008, 2008.

Edwards, G. D., Cantrell, C., Stephens, S., Hill, B., Goyea, O., Shetter, R., Mauldin, R. L., Kosciuch, E., Tanner, D., and Eisele, F.: Chemical Ionization Mass Spectrometer Instrument for the Measurement of Tropospheric HO_2 and RO_2, Anal. Chem., 75, 5317–5327, 2003.

Eisele, F. L. and Tanner, D. J.: Ion-Assisted Tropospheric OH Measurements, J. Geophys. Res.-Atmos., 96, 9295–9308, 1991.

Eisele, F. L., Mauldin, L., Cantrell, C., Zondlo, M., Apel, E., Fried, A., Walega, J., Shetter, R., Lefer, B., Flocke, F., Weinheimer, A., Avery, M., Vay, S., G., S., Podolske, J., Diskin, G., Barrick, J. D., Singh, H. B., Brune, W., Harder, H., Martinez, M., Bandy, A., Thornton, D., Heikes, B., Kondo, Y., Riemer, D., Sandholm, S., Tan, D., Talbot, R., and Dibb, J.: Summary of measurement intercomparisons during TRACE-P, J. Geophys. Res., 108, 8791–8810, 2003.

Eisele, F. L., Mauldin, R. L., Tanner, D. J., Cantrell, C., Kosciuch, E., Nowak, J. B., Brune, B., Faloona, I., Tan, D., Davis, D. D., Wang, L., and Chen, G.: Relationship between OH measurements on two different NASA aircraft during PEM Tropics B, J. Geophys. Res.-Atmos., 106, 32683–32689, 2001.

Faloona, I., Tan, D., Brune, W. H., Jaegle, L., Jacob, D. J., Kondo, Y., Koike, M., Chatfield, R., Pueschel, R., Ferry, G., Sachse, G., Vay, S., Anderson, B., Hannon, J., and Fuelberg, H.: Observations of HOx and its relationship with NOx in the upper troposphere during SONEX, J. Geophys. Res.-Atmos., 105, 3771–3783, 2000.

Faloona, I. C., Tan, D., Lesher, R. L., Hazen, N. L., Frame, C. L., Simpas, J. B., Harder, H., Martinez, M., Di Carlo, P., Ren, X. R., and Brune, W. H.: A laser-induced fluorescence instrument for detecting tropospheric OH and HO_2: Characteristics and calibration, J. Atmos. Chem., 47, 139–167, doi:10.1023/B:JOCH.0000021036.53185.0e, 2004.

Fuchs, H., Bohn, B., Hofzumahaus, A., Holland, F., Lu, K. D., Nehr, S., Rohrer, F., and Wahner, A.: Detection of HO_2 by laser-induced fluorescence: calibration and interferences from RO_2 radicals, Atmos. Meas. Tech., 4, 1209–1225, doi:10.5194/amt-4-1209-2011, 2011.

Fuchs, H., Dorn, H.-P., Bachner, M., Bohn, B., Brauers, T., Gomm, S., Hofzumahaus, A., Holland, F., Nehr, S., Rohrer, F., Tillmann, R., and Wahner, A.: Comparison of OH concentration measurements by DOAS and LIF during SAPHIR chamber experiments at high OH reactivity and low NO concentration, Atmos. Meas. Tech., 5, 1611–1626, doi:10.5194/amt-5-1611-2012, 2012.

Glowacki, D. R., Goddard, A., Hemavibool, K., Malkin, T. L., Commane, R., Anderson, F., Bloss, W. J., Heard, D. E., Ingham, T., Pilling, M. J., and Seakins, P. W.: Design of and initial results from a Highly Instrumented Reactor for Atmospheric Chemistry (HIRAC), Atmos. Chem. Phys., 7, 5371–5390, doi:10.5194/acp-7-5371-2007, 2007a.

Glowacki, D. R., Goddard, A., and Seakins, P. W.: Design and performance of a throughput-matched, zero-geometric-loss, modified three objective multipass matrix system for FTIR spectrometry, Appl. Opt., 46, 7872–7883, doi:10.1364/ao.46.007872, 2007b.

Glowacki, D. R., Lockhart, J., Blitz, M. A., Klippenstein, S. J., Pilling, M. J., Robertson, S. H., and Seakins, P. W.: Interception of excited vibrational quantum states by O_2 in atmospheric association reactions, Science (Washington, D.C., 1883), 337, 1066–1067, doi:10.1126/science.1224106, 2012.

Guggenheim, E. A.: On the determination of the Velocity Constant of a Unimolecular Reaction, Philos. Mag., 2, 538–543, 1926.

Hard, T. M., George, L. A., and O'Brien, R. J.: FAGE Determination of Tropospheric HO and HO_2, J. Atmos. Sci., 52, 3354–3372, 1995.

Heard, D. E.: Atmospheric field measurements of the hydroxylradical using laser-induced fluorescence spectroscopy, Ann. Rev. Phys. Chem., 57, 191–216, doi:10.1146/annurev.physchem.57.032905.104516, 2006.

Heard, D. E. and Pilling, M. J.: Measurement of OH and HO_2 in the Troposphere, Chem. Rev., 103, 5163–5198, 2003.

Hofzumahaus, A., Brauers, T., Aschmutat, U., Brandenburger, U., Dorn, H. P., Hausmann, M., Heßling, M., Holland, F., Plass-Dulmer, C., Sedlacek, M., Weber, M., and Ehhalt, D. H.: The measurement of tropospheric OH radicals by laser-induced fluorescence spectroscopy during the POPCORN field campaign and Intercomparison of tropospheric OH radical measurements by multiple folded long-path laser absorption and laser induced fluorescence – Reply, Geophys. Res. Lett., 24, 3039–3040, 1997.

Holland, F., Hessling, M., and Hofzumahaus, A.: In-Situ Measurement of Tropospheric OH Radicals by Laser- Induced Fluorescence – a Description of the KFA Instrument, J. Atmos. Sci., 52, 3393–3401, 1995.

Kukui, A., Ancellet, G., and Le Bras, G.: Chemical ionisation mass spectrometer for measurements of OH and Peroxy radical concentrations in moderately polluted atmospheres, J. Atmos. Chem., 61, 133–154, doi:10.1007/s10874-009-9130-9, 2008.

Malkin, T. L.: Detection of free-radicals and other species to investigate atmospheric chemistry in the HIRAC chamber, University of Leeds (School of Chemistry), 2010., Leeds, xvii, 285, 215 pp., 2010.

Malkin, T. L., Goddard, A., Heard, D. E., and Seakins, P. W.: Measurements of OH and HO_2 yields from the gas phase ozonolysis of isoprene, Atmos. Chem. Phys., 10, 1441–1459, doi:10.5194/acp-10-1441-2010, 2010.

Mao, J., Ren, X., Zhang, L., Van Duin, D. M., Cohen, R. C., Park, J.-H., Goldstein, A. H., Paulot, F., Beaver, M. R., Crounse, J. D., Wennberg, P. O., DiGangi, J. P., Henry, S. B., Keutsch, F. N., Park, C., Schade, G. W., Wolfe, G. M., Thornton, J. A., and Brune, W. H.: Insights into hydroxyl measurements and atmospheric oxidation in a California forest, Atmos. Chem. Phys., 12, 8009–8020, doi:10.5194/acp-12-8009-2012, 2012.

Martinez, M., Harder, H., Kubistin, D., Rudolf, M., Bozem, H., Eerdekens, G., Fischer, H., Klüpfel, T., Gurk, C., Königstedt, R., Parchatka, U., Schiller, C. L., Stickler, A., Williams, J., and Lelieveld, J.: Hydroxyl radicals in the tropical troposphere over the Suriname rainforest: airborne measurements, Atmos. Chem. Phys., 10, 3759–3773, doi:10.5194/acp-10-3759-2010, 2010.

Novelli, A., Hens, K., Tatum Ernest, C., Kubistin, D., Regelin, E., Elste, T., Plass-Dülmer, C., Martinez, M., Lelieveld, J., and Harder, H.: Characterisation of an inlet pre-injector laser-induced fluorescence instrument for the measurement of atmospheric hydroxyl radicals, Atmos. Meas. Tech., 7, 3413–3430, doi:10.5194/amt-7-3413-2014, 2014.

Orkin, V. L., Martynova, L. E., and Ilichev, A. N.: High-Accuracy Measurements of OH Reaction Rate Constants and IR Absorption Spectra: CH_2 =CF-CF_3 and trans-CHF=CH-CF_3, J. Phys. Chem. A, 114, 5967–5979, doi:10.1021/jp9092817, 2010.

Park, J., Jongsma, C. G., Zhang, R. Y., and North, S. W.: OH/OD initiated oxidation of isoprene in the presence of O_2 and NO, J. Phys. Chem. A, 108, 10688–10697, doi:10.1021/jp040421t, 2004.

Pilling, M. J. and Seakins, P. W.: Reaction Kinetics, Oxford University Press, Oxford, 1995.

Pilling, M. J., Bloss, W. J., and Wirtz, K.: IALSI: Processes relevant to global change – Improvements and Access to a Large Simulation Chamber, Fundacion CEAM, Valencia, Spain, 2005.

Regelin, E., Harder, H., Martinez, M., Kubistin, D., Tatum Ernest, C., Bozem, H., Klippel, T., Hosaynali-Beygi, Z., Fischer, H., Sander, R., Jöckel, P., Königstedt, R., and Lelieveld, J.: HO_x measurements in the summertime upper troposphere over Europe: a comparison of observations to a box model and a 3-D model, Atmos. Chem. Phys., 13, 10703–10720, doi:10.5194/acp-13-10703-2013, 2013.

Schlosser, E., Brauers, T., Dorn, H.-P., Fuchs, H., Hofzumahaus, A., Holland, F., and Wahner, A.: Intercomparison of Two Hydroxyl Radical Measurement Techniques at the Atmosphere Simulation Chamber SAPHIR, J. Atmos. Chem., 56, 187–205, 2007.

Schlosser, E., Brauers, T., Dorn, H.-P., Fuchs, H., Häseler, R., Hofzumahaus, A., Holland, F., Wahner, A., Kanaya, Y., Kajii, Y., Miyamoto, K., Nishida, S., Watanabe, K., Yoshino, A., Kubistin, D., Martinez, M., Rudolf, M., Harder, H., Berresheim, H., Elste, T., Plass-Dülmer, C., Stange, G., and Schurath, U.: Technical Note: Formal blind intercomparison of OH measurements: results from the international campaign HOxComp, Atmos. Chem. Phys., 9, 7923–7948, doi:10.5194/acp-9-7923-2009, 2009.

Schultz, M., Heitlinger, M., Mihelcic, D., and Volzthomas, A.: Calibration source for peroxy-radicals with built-in actinometry using H_2O and O_2 photolysis at 185 nm, J. Geophys. Res.-Atmos., 100, 18811–18816, doi:10.1029/95jd01642, 1995.

Singh, S. and Li, Z. J.: Kinetics investigation of OH reaction with isoprene at 240-340 K and 1-3 Torr using the relative rate/discharge flow/mass spectrometry technique, J. Phys. Chem. A, 111, 11843–11851, doi:10.1021/jp074148h, 2007.

Sjostedt, S. J., Huey, L. G., Tanner, D. J., Peischl, J., Chen, G., Dibb, J. E., Lefer, B., Hutterli, M. A., Beyersdorf, A. J., Blake, N. J., Blake, D. R., Sueper, D., Ryerson, T., Burkhart, J., and Stohl, A.: Observations of hydroxyl and the sum of peroxy radicals at

Summit, Greenland during summer 2003, Atmos. Environ., 41, 5122–5137, 2007.

Stevens, P. S., Mather, J. H., and Brune, W. H.: Measurement of Tropospheric OH and HO_2 by Laser-Induced Fluorescence at Low-Pressure, J. Geophys. Res.-Atmos., 99, 3543–3557, 1994.

Stone, D. and Rowley, D. M.: Kinetics of the gas phase HO_2 self-reaction: Effects of temperature, pressure, water and methanol vapours, Phys. Chem. Chem. Phys., 7, 2156–2163, 2005.

Wennberg, P. O., Cohen, R. C., Hazen, N. L., Lapson, L. B., Allen, N. T., Hanisco, T. F., Oliver, J. F., Lanham, N. W., Demusz, J. N., and Anderson, J. G.: Aircraft-Borne, Laser-Induced Fluorescence Instrument for the in-Situ Detection of Hydroxyl and Hydroperoxyl Radicals, Rev. Sci. Instrum., 65, 1858–1876, 1994.

Whalley, L. K., Furneaux, K. L., Gravestock, T. J., Atkinson, H. M., Bale, C. S. E., Ingham, T., Bloss, W. J., and Heard, D. E.: Detection of iodine monoxide radicals in the marine boundary layer using laser induced fluorescence spectroscopy, J. Atmos. Chem., 58, 19–39, 2007.

Whalley, L. K., Furneaux, K. L., Goddard, A., Lee, J. D., Mahajan, A., Oetjen, H., Read, K. A., Kaaden, N., Carpenter, L. J., Lewis, A. C., Plane, J. M. C., Saltzman, E. S., Wiedensohler, A., and Heard, D. E.: The chemistry of OH and HO_2 radicals in the boundary layer over the tropical Atlantic Ocean, Atmos. Chem. Phys., 10, 1555–1576, doi:10.5194/acp-10-1555-2010, 2010.

Whalley, L. K., Blitz, M. A., Desservettaz, M., Seakins, P. W., and Heard, D. E.: Reporting the sensitivity of laser-induced fluorescence instruments used for HO_2 detection to an interference from RO_2 radicals and introducing a novel approach that enables HO_2 and certain RO_2 types to be selectively measured, Atmos. Meas. Tech., 6, 3425–3440, doi:10.5194/amt-6-3425-2013, 2013.

Winiberg, F. A. F.: Characterisation of FAGE apparatus for HOx detection and application in an environmental chamber, Ph.D, School of Chemistry, University of Leeds, Leeds, 233 pp., 2014.

Updated Smithsonian Astrophysical Observatory Ozone Monitoring Instrument (SAO OMI) formaldehyde retrieval

G. González Abad[1], X. Liu[1], K. Chance[1], H. Wang[1], T. P. Kurosu[2], and R. Suleiman[1]

[1]Harvard-Smithsonian Center for Astrophysics, Cambridge, MA, USA
[2]Jet Propulsion Laboratory, Pasadena, CA, USA

Correspondence to: G. González Abad (ggonzalezabad@cfa.harvard.edu)

Abstract. We present and discuss the Smithsonian Astrophysical Observatory (SAO) formaldehyde (H_2CO) retrieval algorithm for the Ozone Monitoring Instrument (OMI) which is the operational retrieval for NASA OMI H_2CO. The version of the algorithm described here includes relevant changes with respect to the operational one, including differences in the reference spectra for H_2CO, the fit of O_2–O_2 collisional complex, updates in the high-resolution solar reference spectrum, the use of a model reference sector over the remote Pacific Ocean to normalize the retrievals, an updated air mass factor (AMF) calculation scheme, and the inclusion of scattering weights and vertical H_2CO profile in the level 2 products. The setup of the retrieval is discussed in detail. We compare the results of the updated retrieval with the results from the previous SAO H_2CO retrieval. The improvement in the slant column fit increases the temporal stability of the retrieval and slightly reduces the noise. The change in the AMF calculation has increased the AMFs by 20 %, mainly due to the consideration of the radiative cloud fraction. Typical values for retrieved vertical columns are between 4×10^{15} and 4×10^{16} molecules cm^{-2}, with typical fitting uncertainties ranging between 45 and 100 %. In high-concentration regions the errors are usually reduced to 30 %. The detection limit is estimated at 1×10^{16} molecules cm^{-2}.

1 Introduction

H_2CO is one of the most abundant hydrocarbons in the troposphere and a key constituent, despite its short lifetime of 1.5 h, due to photolysis and oxidation by the hydroxyl radical (OH) (Anderson et al., 1996; Brune et al., 1999). Background levels of H_2CO (concentrations below 1 ppbv) are sustained by the oxidation of methane, but over continental regions the oxidation of short-lived, non-methane volatile organic compounds (NMVOCs) from anthropogenic, biogenic, and pyrogenic origins and direct emissions from fires and industrial activities can produce enhanced concentrations of H_2CO within the boundary layer of over 5 ppbv (Houweling et al., 1998; Kanakidou et al., 2005; Lowe and Schmidt, 1983; Parrish et al., 2012). Given its short lifetime, measurements of H_2CO from satellites can be used as a proxy for NMVOC emissions and can be used to estimate top-down emission inventories (Barkley et al., 2008; Fu et al., 2007; Marais et al., 2012; Stavrakou et al., 2009a, b).

Ground-based measurements of H_2CO total columns in the ultraviolet (UV) and infrared spectral ranges are well established in the scientific community. For example, over Lauder station in New Zealand, the mean column reported by Jones et al. (2009) using high-resolution Fourier-transform infrared spectroscopy is 4.9×10^{15} molecules cm^{-2}. The same technique was used by Vigouroux et al. (2009) to measure H_2CO over Réunion Island in the Indian Ocean. They also used the multi-axis differential optical absorption spectroscopy (MAX-DOAS) technique, obtaining good agreement between the two techniques with concentrations around 4.0×10^{15} molecules cm^{-2}. MAX-DOAS H_2CO measurements (Heckel et al., 2005; Peters et al., 2012; Wagner et al., 2011; Pinardi et al., 2013) will be of great value in validating our H_2CO measurements.

H_2CO has been measured from space using both UV and the infrared radiation. Infrared solar occultation measurements from the Atmospheric Chemistry Experiment–Fourier Transform Spectrometer (ACE-FTS) (Dufour et al.,

2009) and emission measurements from the Michelson Interferometer for Passive Atmospheric Sounding (MIPAS) (Steck et al., 2008) are able to measure H_2CO profiles from the stratosphere to the upper troposphere. UV nadir measurements from the Global Ozone Monitoring Experiment (GOME), GOME-2, the SCanning Imaging Absorption spectrometer for Atmospheric CHartographY (SCIAMACHY), and Ozone Monitoring Instrument (OMI) have all been used to retrieve H_2CO vertical columns from space (Chance et al., 2000; Palmer et al., 2001; Kurosu et al., 2004; Wittrock et al., 2006; De Smedt et al., 2008, 2012; Vrekoussis et al., 2010).

We consider the update of the operational Smithsonian Astrophysical Observatory (SAO) H_2CO retrieval necessary. Since 2009 there has been a significant increase of the retrieved background H_2CO values – e.g., Kim et al. (2011) – with data becoming noisier with time. There are new reference spectra available that need to be updated. We are also including for the first time scattering weight information in the level 2 files. We describe the theoretical basis of the SAO OMI H_2CO operational retrieval algorithm and the details of the setup we are using. This algorithm is used at SAO, with the necessary adjustments, for the retrieval of other trace gases (BrO (Raid Suleiman, personal communication), $C_2H_2O_2$ (Miller et al., 2014), H_2O (Wang et al., 2014)) from OMI.

This paper first focuses on explaining the column retrievals by describing the OMI instrument characteristics, the spectral fitting, the conversion of slant column density (SCD) to vertical column density (VCD), and the reference sector normalization. It then compares previous SAO H_2CO retrievals with these new ones and, finally, summarizes this study.

2 SAO OMI H_2CO observations

2.1 The OMI instrument

OMI (Levelt et al., 2006) was launched on 15 July 2004 onboard the NASA Earth Observing System (EOS) Aura platform into a Sun-synchronous polar orbit with an average altitude of 705 km and a period of 99 min, giving 14.7 orbits per day. The Equator-crossing time is 13:42 in the ascending node. OMI measures UV/visible (UV/Vis) solar backscattered radiation using two-dimensional charge-coupled devices (CCDs) from 270 to 500 nm in three channels: UV-1 from 264 to 311 nm, UV-2 from 307 to 383 nm, and Vis from 349 to 504 nm, with spectral resolutions of 0.63, 0.42, and 0.63 nm, respectively. We use UV-2 for H_2CO retrievals. In nominal operation mode, the information in the across-track direction is binned by a factor of 8, resulting in 60 binned ground pixels in the UV-2 channel. OMI has a cross-track field of view of 115°, resulting in a swath of 2600 km, with pixel size between $13 \times 24\,km^2$ at nadir and $26 \times 135\,km^2$ at the swath edges, providing daily global coverage.

2.2 SAO OMI H_2CO retrieval

We retrieve VCDs of H_2CO using a two-step approach. First, we spectrally fit OMI radiances to retrieve SCDs using the basic optical differential spectroscopy (BOAS) method (Chance, 1998). Second, we convert the SCDs to VCDs using air mass factors (AMFs) calculated offline with a radiative transfer model. Finally, we apply post-processing corrections based on a reference sector over the remote Pacific Ocean (Khokhar et al., 2005).

2.2.1 Spectral fitting

SCDs of H_2CO are determined by direct fitting of OMI radiances (Chance, 1998). First, irradiances and radiance wavelengths are calibrated using cross correlation with a high-resolution solar spectrum (Caspar and Chance, 1997). We use the high-resolution solar spectrum recently published by Chance and Kurucz (2010). Two variables may be fitted in the wavelength calibration: spectral shift and squeeze. As they are highly correlated except for very wide fitting regions, we only fit the spectral shift. Wavelength-calibrated radiances are fitted to model spectra. The construction of the model spectra starts with the radiances from the remote Pacific Ocean, which substitute for the daily solar irradiance, scaled by a factor plus the contributions from the molecular ring correction (Chance and Spurr, 1997) and a correction accounting for the aliasing introduced from spectral undersampling computed online (Chance, 1998; Chance et al., 2005). To this term we apply the Beer–Lambert law to account for the molecular absorption from O_3 (Malicet et al., 1995), NO_2 (Vandaele et al., 1998), BrO (Wilmouth et al., 1999), the O_2–O_2 collision complex (Thalman and Volkamer, 2013), and H_2CO (Chance and Orphal, 2011), all convolved with the OMI slit function measured prior to launch (Dirksen et al., 2006). After accounting for the molecular absorption, we include the common mode spectrum obtained by averaging the fitting residuals for each orbit. We finish the construction of the model spectra by considering two low-order closure polynomials that help to account for low-frequency features of the measured radiance in the fitting window (Eq. 1).

$$I = \left[\left(aI_0 + \sum_i \alpha_i X_i\right) e^{-\sum_j \alpha_j X_j} + \sum_k \alpha_k X_k\right] \sum_n \alpha_n X_n + \sum_m \alpha_m X_m \tag{1}$$

In Eq. (1), a is the factor applied to I_0, the solar irradiance, or in this case the radiance reference from the remote Pacific Ocean; $\alpha_i X_i$ are terms added before the trace gas contributions, including the ring spectrum and the undersampling correction; $\alpha_j X_j$ are the Beer–Lambert law contributions from the trace gases; $\alpha_k X_k$ are the terms added afterwards, which in this particular case is the common mode spectrum; and $\alpha_n X_n$ and $\alpha_m X_m$ are the scaling and baseline

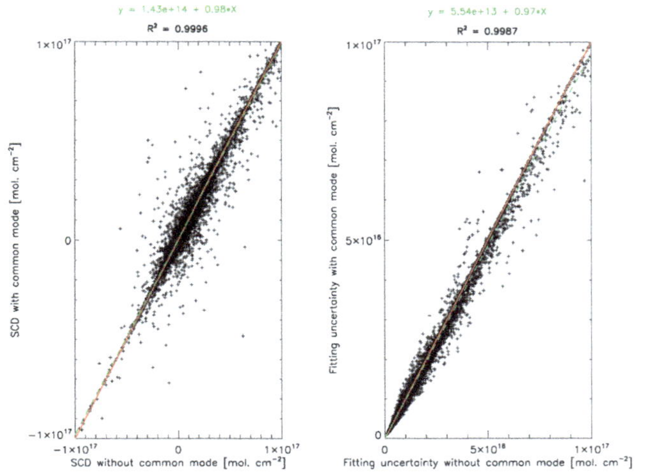

Figure 1. Fitting results of H_2CO SCDs for OMI Orbit 11118 with and without common mode (17 August 2006 overpass over South America).

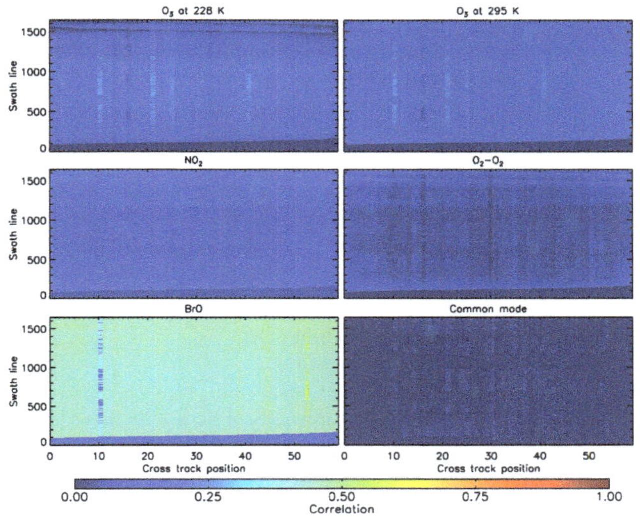

Figure 2. Correlations of H_2CO with the rest of molecules considered in the spectral fitting and the common mode spectrum for Orbit 8537 (21 February 2006 overpass over Africa and Europe).

closure polynomials. The inclusion of the common mode has the intention of reducing the fitting residuals without affecting the values of the retrieved slant columns since its correlation with the H_2CO cross sections is small. Figure 1 shows the scatterplots of the SCDs and the fitting uncertainties for Orbit 11118 comparing the retrieval when we use or do not use the common mode. The correlation values between the two setups is high (0.9996 for the SCDs and 0.9987 for the fitting uncertainties), with small changes in the SCD values. As said before, the main reason to include the common mode is to reduce the root mean square (rms) and therefore the fitting uncertainties, which, as shown in the right panel of Fig. 1, are reduced on average 2 %.

Figure 2 shows the correlations of the other molecules included in the fitting with H_2CO for Orbit 8537. We are also showing the correlation with the common mode. Correlations of H_2CO with O_3 (at 228 and 295 K), with NO_2, O_2–O_2 collision complex, and the common mode spectrum are all small. The orbit-averaged correlations are 0.15, 0.15, 0.13, 0.05, and 0.01, respectively. The common mode is weakly correlated with H_2CO. As expected from the correlation of cross sections between BrO and H_2CO, the correlation of the retrieved column amount between them is strong. The average correlation coefficient for the whole orbit is 0.44. As a result, H_2CO retrieval is affected if BrO is present in significant amounts in the atmosphere.

Instead of using the traditional solar irradiance as I_0 in Eq. (1), we use radiances over the remote Pacific Ocean obtained from an orbit as close to 165° W as possible, measured within 24 h of the orbit to be processed. We build the radiance reference by averaging the radiance spectra over the Pacific between 30° S and 30° N. Then we use this radiance reference as I_0. Due to the weak absorption of H_2CO, the use of the radiance reference over the Pacific helps to address some instrument issues observed in previous versions of the H_2CO retrieval, including the well-known striping problems of OMI (Veihelmann and Kleipool, 2006). The characteristics of the spectral fit, reference spectra, closure polynomials, and online computed corrections are summarized in Table 1. The SCD that minimizes the cost function χ^2 calculated from the difference between the measured and modeled radiance is retrieved using a nonlinear least-squares inversion method implemented in the ELSUNC software (Lindström and Wedin, 1988).

We have also implemented an iterative approach to remove pixels with outliers in the fitting residual. After performing an initial fit of the SCD, we check the value of the residuals for each spectral point. If the residual for a spectral pixel is outside 3σ of the mean fitting residual, then this point is rejected and the retrieval is performed again. This process is repeated for a maximum of two times. With this approach we can minimize the effect of the South Atlantic anomaly and keep the stability of the retrieval along the mission. This method has been used as well in GOME-2 NO_2 retrievals (Richter et al., 2011).

We use a fitting window from 328.5 to 356.5 nm, similar to other windows reported in the literature for GOME, GOME-2, and SCIAMACHY (De Smedt et al., 2008, 2012). It is a minor change compared with our previous fitting window, 327.5 to 356.5 nm, but its use coupled with the effects of the updates in the cross sections, the inclusion of the O_2–O_2 collision complex, and the iterative exclusion of spectral pixels with big fitting residuals helps to keep the fitting of SCDs stable over time, minimizing the effects of instrument degradation. We will further extend the analysis on the effect of each one of these changes in Sect. 3, where we compare the previous and the new H_2CO SAO product, the latter of which

Table 1. Fitting window and parameters for deriving H_2CO slant column density.

Fitting window	328.5–356.5 nm
Radiance reference spectrum	Computed online over the remote Pacific Ocean between 30° N and 30° S
Baseline polynomial	3rd order
Scaling polynomial	3rd order
Instrument slit function	Hyper-parameterization of pre-flight measurements Dirksen et al. (2006)
Solar reference spectrum	Chance and Kurucz (2010)
H_2CO cross sections	Chance and Orphal (2011), 300 K
O_3 cross sections	Malicet et al. (1995), 228 and 295 K
NO_2 cross sections	Vandaele et al. (1998), 220 K
BrO cross sections	Wilmouth et al. (1999), 228 K
O_2–O_2 collision complex cross sections	Thalman and Volkamer (2013), 293 K
Molecular ring cross sections	Chance and Spurr (1997)
Undersampling correction	Computed online Chance et al. (2005)
Residual common mode spectrum	Computed online between 30° N and 30° S

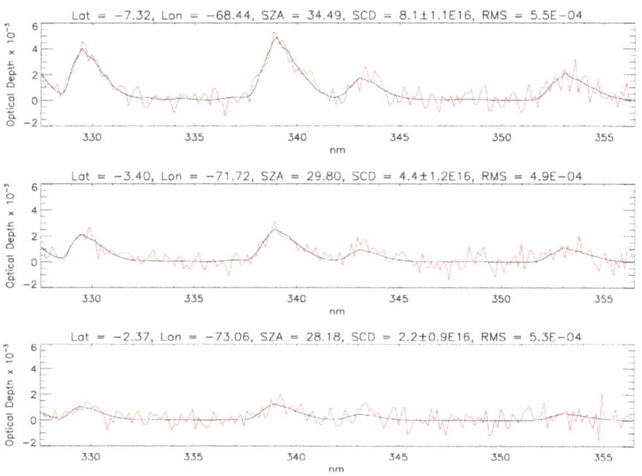

Figure 3. Fitting of H_2CO in OMI Orbit 11118 (17 August 2006 overpass over South America) for high (top), medium (middle), and low (bottom) amounts. The black line is the fitted H_2CO optical depth, and the red line is the fitted H_2CO optical depth plus the fitting residuals.

is presented here. Figure 3 shows the residual and the results of the fitting for three pixels of Orbit 11118 (17 August 2006) over South America. The thick black line is the fitted H_2CO optical depth, and the thin red line is the fitted H_2CO optical depth plus the residual, shown here to provide an idea of the fitting quality.

Typical values for the retrieved H_2CO SCDs are in the range from 4×10^{15} to 6×10^{16} molecules cm^{-2}. Fitting uncertainties for individual pixels range from 6×10^{15} molecules cm^{-2} ($\sim 30\,\%$) for pixels with high concentrations (above 2×10^{16} molecules cm^{-2}) to 100 % or larger for pixels with low concentrations. These fitting uncertainties are calculated for the variable j considering the diagonal term of the covariance matrix for the fitting variable $C_{j,j}$ and the fitting residual weighted by the number of degrees of freedom (Eq. 2). Other sources of error in the fitting of the slant columns such as forward model errors or errors associated with parameters not optimized during the fitting are not considered in the calculation of the fitting uncertainties. Fitting rms values vary between 0.4×10^{-3} and 2.0×10^{-3}, corresponding to detection limits between $\sim 6 \times 10^{15}$ and $\sim 3 \times 10^{16}$ molecules cm^{-2}.

$$\delta X_j^2 = \mathrm{RMS}^2 \left(\frac{m}{m-n}\right) (C_{(j,j)} C_{(j,j)}) \tag{2}$$

2.2.2 Vertical column density determination

The second step of our retrieval converts SCDs into VCDs. The photons reaching OMI are either reflected from the Earth's surface and transmitted or scattered to OMI or directly scattered to OMI without reaching the Earth's surface. The retrieved SCDs depend on the viewing geometry for each pixel (solar zenith angle, viewing zenith angle, and relative azimuth angle), the surface albedo, and the state of the atmosphere (clouds, trace gases concentrations, temperature, air density, pressure, etc.). The SCDs are related to the VCDs by AMFs. The relationship between SCDs and VCDs is

$$\mathrm{VCD} = \mathrm{SCD}/\mathrm{AMF}. \tag{3}$$

To determine AMFs, we follow the approach described in Palmer et al. (2001), summarized by Eq. (3).

$$\mathrm{AMF} = \int_{\mathrm{atm}} w(z, \theta_0, \theta, \varphi, a_s, h_s, C_f, C_h) S(z, \mathrm{lat}, \mathrm{lon}, \mathrm{month}) \mathrm{d}z \tag{4}$$

In this formulation, the scattering weights (w's) account for the sensitivity of the satellite measurements to the presence of H_2CO molecules at different altitudes. They are functions of altitude (z), solar zenith angle (θ_0), viewing zenith angle (θ), relative azimuth angle (φ), surface albedo (a_s), terrain height (h_s), effective cloud fraction (C_f), and cloud-top altitude (C_h). For an optically thin absorber such as H_2CO, the w's are independent of the column abundance of the species. The shape factor (S), dependent on the location and time of the measurement, represents the normalized vertical profile of the absorbing molecule and provides a priori estimation of the vertical distribution of H_2CO. The shape factor for a particular location and time is defined in Eq. (4).

$$S(z) = \frac{x_a(z)}{\int_{atm} x_a(z)dz} \qquad (5)$$

x_a is the a priori vertical profile of H_2CO partial columns. We have calculated vertical profiles using the GEOS-Chem (Bey et al., 2001) chemical transport model (Version v9-01-03), which is driven by Goddard Earth Observing System-5 (GEOS-5)-assimilated meteorological data products interpolated onto 47 vertical levels at a spatial resolution of 2° latitude × 2.5° longitude. We have developed a monthly mean climatology using GEOS-Chem simulations between 13:00 and 14:00 local time (LT) in 2007. Figure 4 shows a set of vertical profiles used in the AMF calculation for Orbit 11118 at different latitudes.

We have calculated the scattering weights using the radiative transfer model VLIDORT version 2.4RT (Spurr, 2006). $w(z)$ usually depends on the wavelength and the viewing geometry. However, for solar zenith angles below 70°, since the ozone absorption is not too strong in the 328.5–356.5 nm fitting window, the variation of $w(z)$ with wavelength is small (less than 5 %). Therefore, we have used the scattering weights at only one optimized wavelength, 340 nm. For solar zenith angles above 70°, the ozone absorption becomes large enough to make the assumption of an optically thin atmosphere not valid (we concentrate on SZA < 70°). We have prepared a set of weighting function look-up tables that are parameterized by the surface albedo and the relative azimuth angle, with entries for solar zenith angles (12), viewing zenith angles (8), altitudes (48), and cloud-top/surface pressures (6). These parameters are summarized in Table 2. OMCLDO2, the OMI cloud product we have used to obtain the cloud information (Acarreta et al., 2004), assumes the clouds to be opaque Lambertian surfaces at a single cloud height (C_h) with a cloud albedo (a_c) of 0.8–1.0. At the retrieval partly cloudy pixels are assumed to be a mixture of clear-sky scene and cloudy scene ($a_c = 0.8$) with an effective cloud fraction of C_f calculated using the independent pixel approximation (Martin et al., 2002). Under this approximation, the scattering weight for an inhomogeneous pixel with a given viewing geometry becomes

$$w = (1 - \Phi) \cdot w_{clear}(a_s, h_s) + \Phi \cdot w_{cloud}(a_c, C_h), \qquad (6)$$

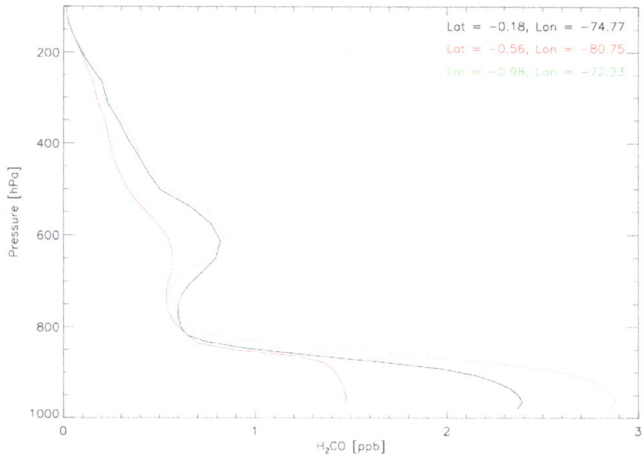

Figure 4. Three a priori H_2CO profiles used in the air mass factor calculations for Orbit 11118.

where w_{clear} is the scattering weight for the clear-sky part of the pixel, a_s is the surface albedo, h_s is the terrain height, w_{cloud} is the scattering weight of the cloudy part of the pixel, C_h is the cloud height, and Φ is the intensity-weighted cloud fraction or radiative cloud fraction calculated as

$$\Phi = \frac{C_f \cdot I_{cloud}}{(1 - C_f) \cdot I_{clear} + C_f \cdot I_{cloud}}. \qquad (7)$$

I_{clear} and I_{cloud} are the radiance intensities for clear-sky and cloudy scenes at the top of the atmosphere. The values of I_{clear} and I_{cloud} are calculated using VLIDORT and stored in a look-up table with the same parameter grid as the weighting function look-up table. The a_s used to evaluate w_{clear} and I_{clear} is obtained from the most frequent Lambertian equivalent reflectance (LER) field in version 3 of the OMI surface reflectance climatology (Kleipool et al., 2008). We calculate a_s by interpolating the values contained in the climatology to the wavelength where the AMFs are calculated, 340 nm. For the terrain height, h_s, we use the information provided with the OMI level 1B product (OML1BRUG). The description of clouds is consistent with the assumptions used in the cloud retrieval algorithm (Acarreta et al., 2004; Stammes et al., 2008) from which we obtain C_f and C_h distributed in the Aura OMI Cloud Data Product-OMCLDO2. Figure 5 shows the scattering weights for three pixels of Orbit 11118 with low, medium, and high effective cloud fractions. Once the scattering weights are calculated, the averaging kernels (AKs) (Rodgers and Connor, 2003) are straightforward to be approximated using the formulation shown below for DOAS retrievals (Eskes and Boersma, 2003):

$$AK(z) = \frac{w(z)}{AMF}. \qquad (8)$$

w_z is calculated as in Eq. (5), and AMFs are described by Eq. (3). Model comparisons with H_2CO VCDs, assimilation of OMI H_2CO products, and validation with H_2CO profiles

Table 2. Parameters of the weighting function look-up table.

Parameter	Grid values
Solar zenith angle	0, 15, 30, 45, 60, 70, 77, 81, 84, 86, 88, 89 [°]
Viewing zenith angle	0, 15, 30, 45, 60, 70, 75, 80 [°]
Altitude	0.06, 0.19, 0.32, 0.45, 0.59, 0.73, 0.86, 1.00, 1.15, 1.29, 1.44, 1.58, 1.76, 1.99, 2.25, 2.52, 2.79, 3.08, 3.44, 3.90, 4.38, 4.88, 5.42, 5.98, 6.59, 7.24, 7.95, 8.86, 9.96, 11.04, 12.11, 13.15, 14.19, 15.22, 16.24, 17.26, 18.81, 20.93, 23.12, 25.42, 29.16, 34.65, 40.93, 48.04, 55.86, 64.16, 74.49, 80.79 [km]
Cloud-top/surface pressure	1.00, 0.83, 0.63, 0.52, 0.37, 0.24 [atm]

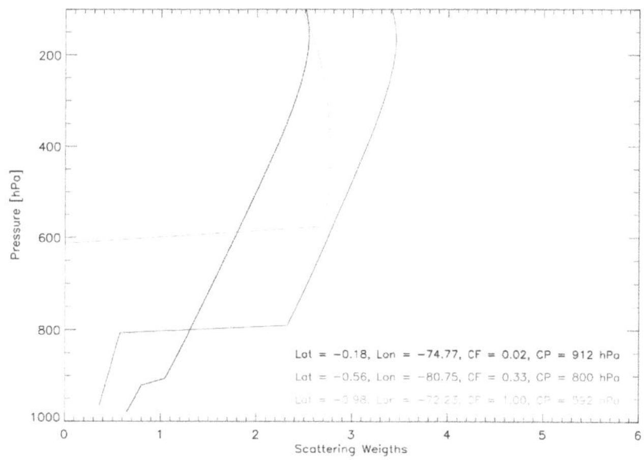

Figure 5. Example of calculated scattering weights for three pixels with different cloud fractions corresponding to the profiles plotted in Fig. 3. For cloud fractions equal to 1 the scattering weights below the cloud-top pressure become 0 (green line), while for other values there is a contribution below the cloud in the air mass factor calculation (red and black lines) proportional to the radiative cloud fraction.

should make use of the scattering weights and the used H_2CO vertical profiles, so we provide them along with the columns.

We have performed a sensitivity analysis to assess the behavior of the AMFs calculations with respect to the H_2CO vertical profile, the wavelength, the surface albedo, the viewing geometry, and the cloud parameters. We have defined a reference scenario for the AMF calculation considering a cloud-free pixel, with a surface albedo of 0.02, a solar zenith angle of 30°, and a viewing zenith angle of 0°. The wavelength considered for the standard calculation is 340 nm.

As mentioned above, the variation of the AMFs in function of the wavelength in the range of the fitting window is small, between 1 and 5 %. It is true and worth mentioning that at shorter wavelengths, below 330 nm, the AMFs have a stronger dependency on the wavelength. This is one of the reasons to keep the lower limit of the fitting window on 328.5 nm.

In the UV spectral range, most surfaces reflect between 1 and 10 % of the incoming light. Kleipool et al. (2008) estimated the errors for the LER in the UV to be around 40 %. We can then estimate the errors we are carrying over from the albedo data into the AMF calculation for different surfaces. For example, the AMF of a typical surface with an albedo of 0.03, such as a forest, will have a 6 % error associated with the OMLER uncertainty. But for brighter surfaces, the error will be around 14 % for the Sahara desert with an albedo of 0.1 and around 26 % for a snow surface with an albedo of 0.8. Therefore, we recommend the users to exercise caution when using retrievals over high albedo surfaces.

As mentioned before the dependence of the AMF in function of the SZA and VZA is strong. For example the AMF for our reference scenario can vary between 2 and 6. Fortunately the geometries of the satellite observations are determined precisely and the errors associated to the inaccuracy of the viewing geometry are neglected.

The cloud pressure has an important impact in the value of the computed AMFs. Considering the reference scenario and a pixel fully covered with clouds the AMFs can range from 0.6 to 1.2 when the cloud pressure increases from 800 to 900 hPa. The cloud product we use is derived using the O_2–O_2 absorption band at 477 nm (Acarreta et al., 2004). An alternative to this product is the OMCLDRR cloud product, derived from rotational Raman scattering around 350 nm (Joiner and Vassilkov, 2006). The difference between these two data sets was estimated to be between 2 and 45 hPa (Sneep et al., 2008). After propagating this difference in the cloud pressure to the calculations of the AMFs the values of the AMFs change around 20 %. The other cloud parameter we need to consider is the cloud fraction. Acarreta et al. (2004) estimated the systematic uncertainty of the cloud fraction to be ±0.1. In the case of considering a cloud pressure of 800 hPa a change in the cloud fraction of ±0.1 implies changes in the AMFs of around 25 %.

We have also tested the impact of the a priori profiles of H_2CO. Assuming a bias in the GEOS-Chem profiles of $\pm10\,\%$ (Millet et al., 2006; Barkley et al., 2011) we have found that the AMF may change for different regions up to 16 %.

To summarize, the errors associated with AMF calculations are estimated to be around 35 %, dominated by the cloud parameters, the surface albedo, and the H_2CO vertical profiles. The presence of aerosols in the atmosphere is an important factor, but we have not explicitly included them in our look-up tables to keep them consistent with the OMCLDO2 cloud product (Acarreta et al., 2004). The AMF uncertainties combined with the estimated errors on the SCDs (30–100 %) result in uncertainties associated with the VCDs between 45 and 105 %.

2.2.3 Post-processing normalization of retrieved columns

We derive a daily post-processing normalization for the retrieved columns according to the concentrations simulated by GEOS-Chem (Bey et al., 2001) over the remote Pacific, following an approach similar to Khokhar et al. (2005) and De Smedt et al. (2008). In this region, the H_2CO concentrations are assumed to be at background levels, their only source being CH_4 oxidation.

We define the reference sector as the region in-between 140 and 160° W, and 90° N and 90° S. Using 2.0° latitude by 2.5° longitude model results in 2007, we have built a monthly mean climatology of VCDs by longitudinally averaging concentrations for a given latitude and then latitudinally interpolating the results to a grid of 500 points extending from 90° N to 90° S with a resolution of 0.36°. With these climatological values, taking into consideration the viewing geometry and atmospheric conditions of each pixel (i.e., AMF), we correct the effects of the row anomaly and stripping and normalize the retrieved values to the reference sector to remove the artificially increasing background H_2CO values using the following approach.

We calculate a pixel-by-pixel correction (Eq. 8) over the remote Pacific Ocean (radiance reference granule) for the pixels whose SCD fitting has converged successfully above the noise level, and the retrieved SCD plus 2 times the fitting uncertainty are above 0. These pixels are identified in the file product as good pixels by having a main data quality flag of 0. Other possible values for the main quality flag are 1 for suspect pixels – with negative values for the retrieved column plus 2 times the fitting uncertainty but positive values for the column plus 3 times the fitting uncertainty, or with convergence of the fitting at the noise level – and value 2 for bad pixels, which are pixels where the fitting has not converged, or the retrieved values plus 3 times the fitting uncertainty remain below 0.

$$\begin{aligned}\text{Correction}(i,j) &= \text{OMI}_{\text{SCD\_Pacific}}(i,j)\\ &\quad - (\text{GEOS\_Chem}_{\text{ReferenceSector}}(\text{lat}) \times \text{AMF}(i,j)) \qquad (9)\end{aligned}$$

i is the a cross-track position (longitude) within an OMI orbit, j is the line position (latitude), and lat is the latitude at the center of pixel (i, j) of the orbit.

After working the pixel-by-pixel correction using the radiance reference orbit, we assume that the bias between the retrieved columns and the modeled columns over the remote Pacific Ocean (Correction(i, j)) is representative of a longitudinally invariant bias that affects the retrievals everywhere else on the globe. Therefore, we calculate a latitude-dependent correction of this bias for each cross-track position using the median values of the corrections calculated in Eq. (8) falling within two points of the reference sector latitude grid. Finally we apply this bias to the rest of the orbits by interpolating the values of the median bias to the corresponding latitude at the center of each pixel.

We attribute this bias mainly to instrumental issues whose effects are weaker than the row anomaly, and to correlations between BrO and H_2CO absorption cross sections, which are especially significant at high latitudes. One latitude-dependent bias correction is calculated for each radiance reference orbit. In fact this implies that the bias correction is calculated daily. We consider this bias correction to be orbit invariant for all the retrievals performed with the same radiance reference (within 24 h difference). Therefore we apply it to all OMI columns retrieved from different granules over the globe (typically between 7 and 17 granules using a particular radiance reference). The final corrected columns are calculated as

$$\text{VCD}_{\text{corrected}}(i,j) = \frac{\text{SCD}(i,j) - \text{Correction}(i,j)}{\text{AMF}(i,j)}. \qquad (10)$$

Figure 6 illustrates the GEOS-Chem results used in the calculation of the background levels for August 2007. Figure 7 shows the effect of the reference sector correction for the reference sector (140 to 160° W and 90° N to 90° S) for each month. We can see how at high latitudes in the winter months we consistently retrieve high values of H_2CO when compared with the GEOS-Chem reference sector slant columns (left panels). It is therefore no surprise that the reference sector correction is strongest for the winter months in the high latitudes. We consider the month-to-month variation of the GEOS-Chem climatology to be representative of the changes in the background values, in comparison to the rapid spatial and temporal variation in concentrations over the continental hot spots.

Even though the correction helps us to remove unrealistic biases associated with the instrument and correlations between the H_2CO and BrO cross sections, it introduces external a priori information subject to uncertainties and assumes that H_2CO over the reference sector does not change from

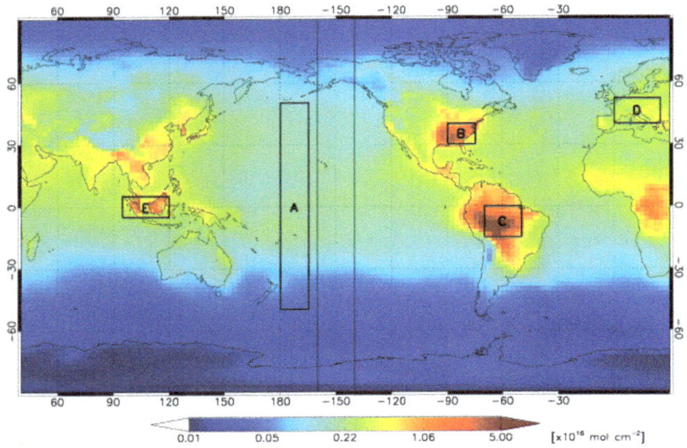

Figure 6. August 2007 GEOS-Chem monthly mean climatology used in the calculation of the reference sector background concentrations. The portion of the remote Pacific Ocean used is enclosed by the two black vertical lines. The region above Hawaii has been excluded from the longitudinally averaged concentrations. Note the logarithmic scale used for the colorbar. Overplotted are the five regions used in the time series shown in Fig. 9 as described in the text.

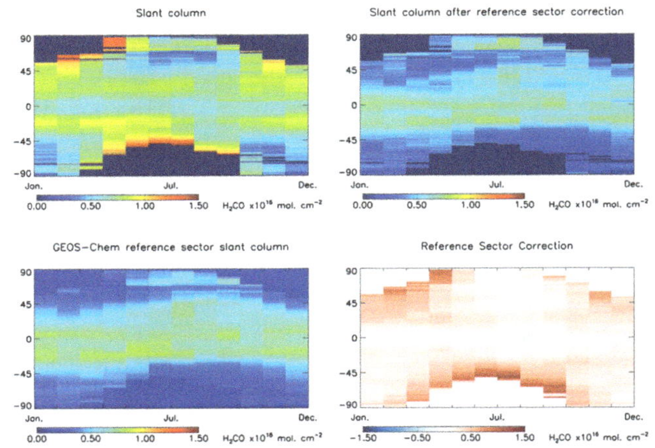

Figure 7. Analysis of the reference sector correction using monthly averages for the year 2007. Top left panel shows OMI-retrieved SCD for the reference sector (140 to 160° W), top right panel shows the OMI SCD after the reference sector correction is applied, bottom left panel the reference sector concentration from the GEOS-Chem climatology, and the bottom right panel the correction applied for each month and latitude bin.

year to year. This situation should be considered when using the reference-sector-corrected columns. We provide in our data files both the reference-sector-corrected columns and the non-corrected VCDs so each user can develop a suitable reference sector correction.

3 Comparison between previous and current H_2CO SAO product

We have changed the spectral fitting for the new version of the retrieval by updating some cross sections to newly available data, by considering the O_2–O_2 collision complex, and by changing the fitting window. Table 3 shows the mean SCD, mean fitting uncertainty, mean rms, and the percent of pixels flagged as good (main quality flag equal to 0) for OMI Orbit 11148 using different retrieval configurations. First we report the result from the old SAO retrieval and then the result after changing each one of the items listed below. The last line of Table 3 shows the results for the new retrieval. The mean SCD has changed from 7.52 to 11.02×10^{15} molecules cm^{-2}, the fitting uncertainty from 11 to 8.87×10^{15} molecules cm^{-2}, the rms from 1.31 to 1.22×10^{-3}, and the number of good pixels rises from 98.95 to 99.64 %.

The list of changes in the spectral fitting includes the high-resolution solar spectrum from Chance and Spurr (1997) to Chance and Kurucz (2010), the H_2CO cross sections from Cantrell et al. (1990) to Chance and Orphal (2011), and the use of two O_3 cross sections at 228 and 295 K instead of only one at 228 K (Malicet et al., 1995). Another change is that we are fitting the O_2–O_2 collisional complex cross section (Thalman and Volkamer, 2013), while in the old SAO retrieval the O_2–O_2 collisional complex was not considered. The O_2–O_2 collisional complex has two significative features at 342 and 360 nm that are relevant in the H_2CO fitting window. For that reason we consider it necessary to include it in the fitting. We have also modified the fitting window from 327.5–356.5 nm in the old SAO retrieval to 328.5–356.5 nm in the new retrieval.

To compare the long-term series of both retrievals, 3 years of data are used; years 2006 and 2007, near the beginning of the mission, are representative of a time when the previous SAO product is quite stable, while year 2012 is representative of a time when the previous SAO product is affected by problems, including the row anomaly and increasing background concentrations.

In general, the new product is less noisy, with smaller fitting uncertainties. This can be seen for the particular case of Orbit 43214 (August 2012) in Fig. 8. The decrease in uncertainty is evident with almost all pixels having an absolute uncertainty smaller than that in the older version of the SAO retrieval. There is also significant reduction of unrealistically high SCDs.

The general increase in the retrieved columns in the previous SAO retrieval is so large that it is difficult to use it after 2010 (Kim et al., 2011). In the new retrieval, the situation is greatly improved. In Fig. 9 (for Orbit 43214) we have plotted the retrieved VCDs for the old retrieval, the new one, and the new one with the reference sector correction applied to show how the situation has improved. The first panel, top left, gives VCDs retrieved with the old algorithm. Extremely high values (over 2.0×10^{16}) are common over the continents and

Table 3. Spectral fitting results for Orbit 11118. Results from the old SAO retrieval and from different new retrieval settings are included to illustrate the change induced by each one of them.

	Mean SCD molecules cm^{-2}	Mean uncertainty molecules cm^{-2}	Mean rms	% good pixels
Old SAO	7.52×10^{15}	1.10×10^{16}	1.31×10^{-3}	98.95
New fitting window	8.41×10^{15}	1.03×10^{16}	1.31×10^{-3}	99.14
New solar reference	5.56×10^{15}	1.02×10^{16}	1.34×10^{-3}	98.53
New H_2CO cross sections	4.48×10^{15}	9.14×10^{15}	1.33×10^{-3}	98.31
New O_3 cross sections	4.95×10^{15}	9.60×10^{15}	1.26×10^{-3}	97.67
O_2–O_2 cross sections	7.03×10^{15}	9.85×10^{15}	1.29×10^{-3}	98.77
New SAO	1.02×10^{16}	8.87×10^{15}	1.22×10^{-3}	99.64

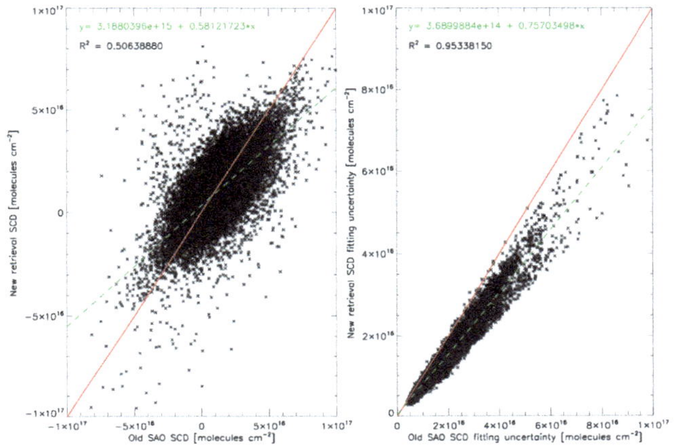

Figure 8. Scatterplots for Orbit 43214 comparing the old SAO retrieval and new SAO retrieval of SCDs and SCD fitting uncertainties.

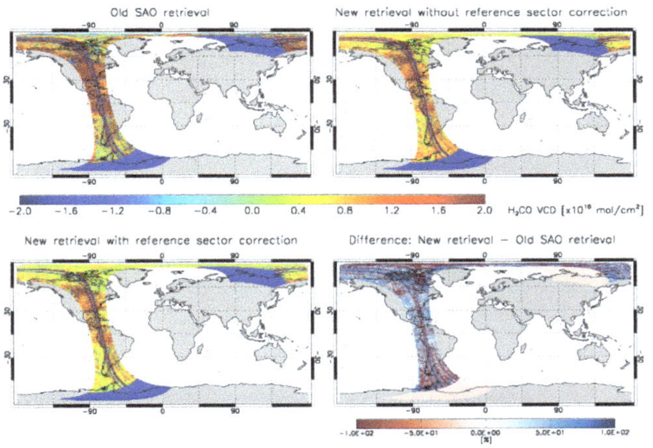

Figure 9. VCD retrieval for Orbit 43214, on 29 August 2012. The panels show the old SAO retrieval (top left), the new retrieval without reference sector correction applied to it (top right), the new retrieval with the reference sector correction applied (bottom left), and the difference between the new retrieval with the reference sector correction applied to it expressed as percent of the old operational retrieval (bottom right).

equatorial ocean areas, while there are areas of negative values at high latitudes.

With the new algorithm without the reference sector correction, top right panel, it is evident that there is a significant reduction in the concentrations almost everywhere. The contrast between high- and low-concentration regions increases. The new retrieval also has fewer pixels with negative values at high latitudes. After we have applied the reference sector correction (bottom left panel), the situation further improves. As mentioned earlier, the new retrieval still shows large negative concentrations – which, however, are smaller in the new version – at high latitudes due to correlations with BrO. Applying the reference sector correction helps to minimize this effect. Finally, we plot the difference between the VCDs retrieved with the old and the new products in the bottom right panel. The concentrations are reduced almost everywhere in the orbit (blue colors). We only see increases in concentrations where the old retrieval is consistently negative. There is also an improvement for pixels affected by the row anomaly in the new retrieval. However, we do not recommend using pixels affected by the row anomaly.

We now compare level 3 monthly averages to better assess changes in the concentrations and trends of the new retrieval with respect to the old SAO retrieval. To compute monthly averages, we only considered pixels with cloud fractions below 40 % and unaffected by the row anomaly. The level 3 data are calculated on a $0.2° \times 0.2°$ grid where values of the OMI pixel are weighted by the area falling within the grid box and by the fitting uncertainty.

We have defined six regions in this study. Background levels over the remote Pacific Ocean (A), southeastern USA (B), the Amazon Basin (C), Europe (D), and southeast Asia (E) plus an equatorial Pacific Ocean area which includes only the latitudes between $-20°$ S and $20°$ N to assess the evolution of the SCD standard deviation in a remote area. The areas used for each of these regions are shown in Fig. 6. We have focused our analysis on 2006, 2007, and 2012. While the old SAO retrieval is stable until the end of 2007, by 2012

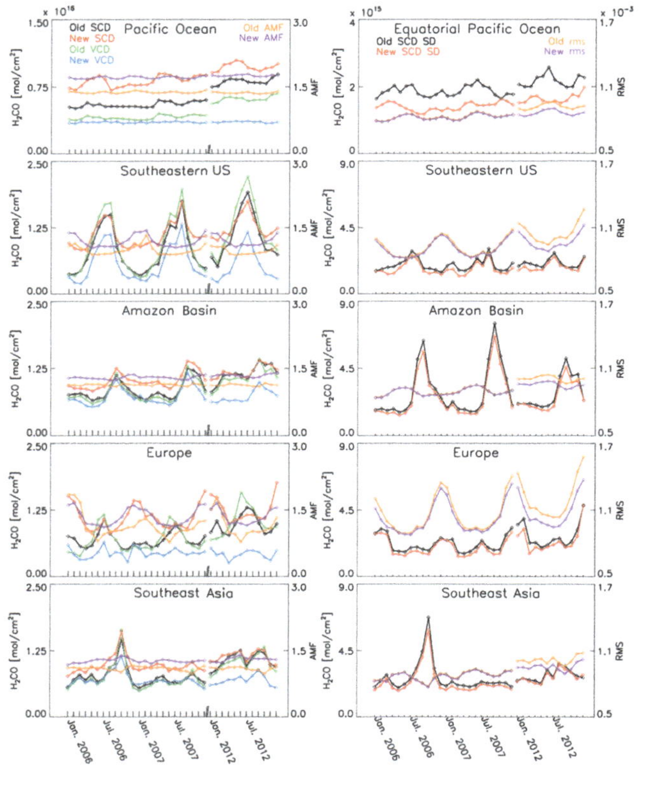

Figure 10. Year 2006, 2007, and 2012 monthly average of the $0.2° \times 0.2°$ gridded SCDs, VCDs (only reference sector corrected for the new retrieval), AMFs, rms's, and gridded SCD standard deviation time series for the regions depicted in Fig. 5 plus a sixth region named equatorial Pacific Ocean including the latitudes inside region A between $-20°$ S and $20°$ N added to study the evolution of the standard deviation of the gridded SCDs in a remote region. The break in the x axis indicates the transition from 2007 to 2012.

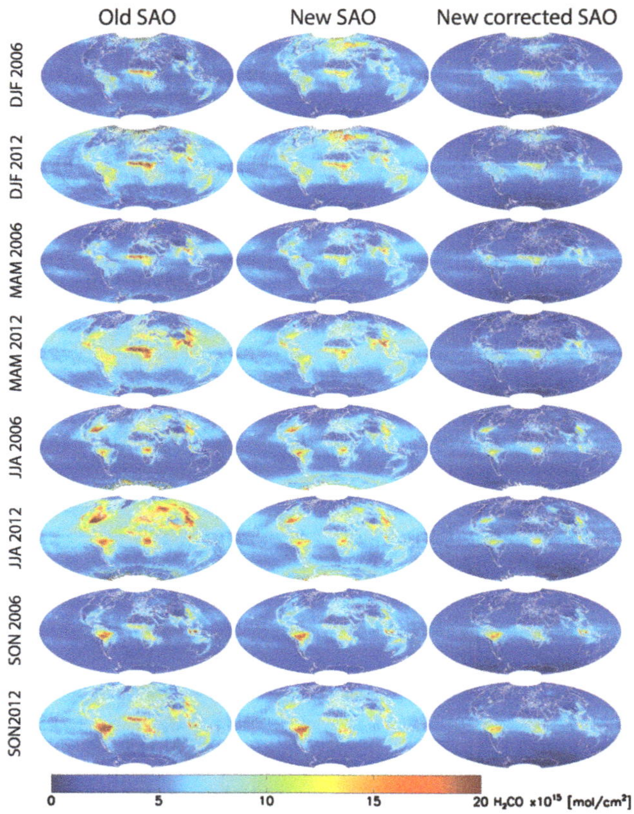

Figure 11. Seasonal averages for 2006 and 2012. Left column shows the old SAO retrieval, middle column shows the new retrieval without the reference sector correction, and right column shows new retrieval with the reference sector correction.

it is clearly producing unreasonable columns over the Pacific and an overall increase of the retrieved concentrations worldwide. The selected 3 years are thus appropriate for comparing the old and the new results.

Figure 10 shows the results of the time series comparison. Overall, the new retrieval (red and blue lines in the left panels) is more stable over time than the old retrieval (black and green lines in the left panels). The VCDs plotted for the new retrieval are the ones with the reference sector correction applied to them. In comparison with the old SAO retrieval the new one does not show a large increasing trend in the SCD or VCD concentrations. The background concentrations over the Pacific Ocean remain almost the same for the new retrieval, while the old SAO retrieval exhibits an increase of 0.3×10^{16} molecules cm^{-2} between the end of 2007 and the beginning of 2012. However there is still some increase, albeit greatly reduced. We have not be able to attribute it to any specific cause after analyzing possible anisotropies introduced by the row anomaly in the viewing geometry of the pixels. What we can say is that we do not observe a degradation of the fitting residuals (orange and purple lines of the right side panels in Fig. 10) or increases in the standard deviations (black and red lines in the right side panels) in the new retrieval in opposition to what was observed in the old retrieval. The other regions have similar seasonality in the old and new retrievals, but the sharp increase in the background levels observed in the old retrieval from 2007 to 2012 is not present in the new retrieval, especially in the southeastern USA and southeast Asia. The retrieved columns in the new version do not show significant jumps between 2007 and 2012 for the rest of the regions analyzed. The main remaining issue in the new retrieval is the too-high winter concentrations over Europe. We recommend at present the values for December, January, and February over northern Europe to be used with caution.

In the left side panels of Fig. 10 we have also included the AMF for the old and new retrieval (orange and purple lines). As mentioned above, the use of the radiative cloud fraction instead of the effective cloud fraction translates to a general increase of the AMFs of around 20%.

Figure 11 shows the results for the new retrieval without the reference sector correction applied in the middle panels,

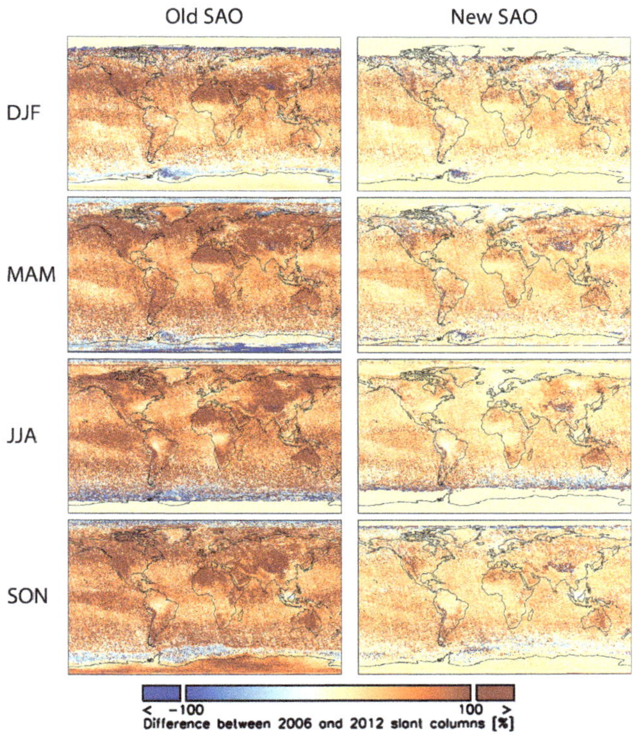

Figure 12. Difference between 2006 and 2012 slant column seasonal averages. Left column shows results for the old SAO retrieval, while the right column shows results for the new retrieval.

with the reference sector correction applied in the right panels, and the old SAO retrieval in the left panels. These plots are seasonal averages for DJF, MAM, JJA, and SON for the years 2006 and 2012. The reduction in the background values is evident in the new retrieval. Hot spots are more spatially resolved in the new retrieval, e.g., the hot spot over Central America in MAM. The high concentrations over the Mediterranean Sea in JJA are removed. To illustrate the reduction of the increasing trend in the new retrieval, Fig. 12 shows the difference of seasonally averaged slant columns between 2006 and 2012 (2012–2006) for the old (left column) and new (right column) retrievals. Apparently, the new retrieval has reduced the increasing trend significantly. The differences are well over 100 % in the old version for most continental and oceanic regions in all seasons. While there is still an increasing trend in the new retrieval, it has been decreased to between 0 and 50 % in most regions and seasons.

4 Conclusions

We have presented and discussed the new SAO OMI H_2CO retrieval, paying especially attention to the changes with respect to the old operational algorithm. We have updated the H_2CO cross sections and the high-resolution solar spectra. We have also included a second O_3 cross sections at 295 K and the O_2–O_2 collisional complex. We have implemented an iterative fitting residual analysis to remove spectral pixels with high fitting residuals. We have slightly modified the fitting window and used a reference sector over the remote Pacific Ocean to correct for possible biases and their temporal drift in the retrieved slant columns. We have also updated the AMF calculation to include the radiative cloud fraction. This change has increased the value of the AMFs by 20 % on average.

The columns retrieved by the new SAO OMI H_2CO algorithm show more stability over time. The increasing trend in concentrations is greatly reduced, as is the increase in the rms and therefore the fitting uncertainties and the standard deviations of the columns. This new product greatly extends the volume of useful H_2CO data from OMI beyond 2008 and also provides new data to facilitate its use. The scattering weights used in the SCD-to-VCD inversion and the associated H_2CO vertical profiles are now included in the level 2 data files for use in model comparisons. We now include a quality flag in our product for problems caused by instrument row anomaly. As a general rule, we recommend users not to use pixels affected by the row anomaly.

The new operational algorithm will be version 3.2.0.3 in the algorithm development sequence. Generically we will refer to it as version 3. A detailed error analysis is in preparation. We are also working on further validation of our new product.

Acknowledgements. This study is supported by NASA Atmospheric Composition Program/Aura Science Team (NNX11AE58G) and the Smithsonian Institution. The Dutch–Finnish OMI instrument is part of the NASA EOS Aura satellite payload. The OMI project is managed by NIVR and KNMI in the Netherlands. We acknowledge the OMI International Science Team for providing OMI data used in this study.

Edited by: F. Boersma

References

Acarreta, J. R., De Haan, J. F., and Stammes, P.: Cloud pressure retrieval using the O_2–O_2 absorption band at 477 nm, J. Geophys. Res.-Atmos., 109, D05204, doi:10.1029/2003JD003915, 2004.

Anderson, L. G., Lanning, J. A., Barrell, R., Miyagishima, J., Jones, R. H., and Wolfe, P.: Sources and sinks of formaldehyde and acetaldehyde: An analysis of Denver's ambient concentration data, Atmos. Environ., 30, 2113–2123, doi:10.1016/1352-2310(95)00175-1, 1996.

Barkley, M. P., Palmer, P. I., Kuhn, U., Kesselmeier, J., Chance, K., Kurosu, T. P., Martin, R. V., Helmig, D., and Guenther, A.: Net ecosystem fluxes of isoprene over tropical South America inferred from Global Ozone Monitoring Experiment (GOME) observations of HCHO columns, J. Geophys. Res.-Atmos., 113, D20304, doi:10.1029/2008JD009863, 2008.

Barkley, M. P., Palmer, P. I., Ganzeveld, L., Arneth, A., Hagberg, D., Karl, T., Guenther, A., Paulot, F., Wennberg, P. O., Mao, J., Kurosu, T. P., Chance, K., Müller, J.-F., De Smedt, I., Van Roozendael, M., Chen, D., Wang, Y., and Yantosca, R. M.: Can a "state of the art" chemistry transport model simulate Amazonian tropospheric chemistry?, J. Geophys. Res.-Atmos., 116, D16302, doi:10.1029/2011JD015893, 2011.

Bey, I., Jacob, D. J., Yantosca, R. M., Logan, J. A., Field, B. D., Fiore, A. M., Li, Q., Liu, H. Y., Mickley, L. J., and Schultz, M. G.: Global modeling of tropospheric chemistry with assimilated meteorology: Model description and evaluation, J. Geophys. Res.-Atmos., 106, 23073–23095, doi:10.1029/2001JD000807, 2001.

Brune, W. H., Tan, D., Faloona, I. F., Jaeglé, L., Jacob, D. J., Heikes, B. G., Snow, J., Kondo, Y., Shetter, R., Sachse, G. W., Anderson, B., Gregory, G. L., Vay, S., Singh, H. B., Davis, D. D., Crawford, J. H., and Blake, D. R.: OH and HO_2 chemistry in the North Atlantic free troposphere, Geophys. Res. Lett., 26, 3077–3080, doi:10.1029/1999GL900549, 1999.

Cantrell, C. A., Davidson, J. A., McDaniel, A. H., Shetter, R. E., and Calvert, J. G.: Temperature-dependent formaldehyde cross sections in the near-ultraviolet spectral region, J. Phys. Chem., 94, 3902–3908, doi:10.1021/j100373a008, 1990.

Caspar, C. and Chance, K.: GOME wavelength calibration using solar and atmospheric spectra, in: Third ERS Symposium on Space at the service of our Environment, edited by: Guyenne, T.-D. and Danesy, D., Vol. 414 of ESA Special Publication, p. 609, 1997.

Chance, K.: Analysis of BrO measurements from the Global Ozone Monitoring Experiment, Geophys. Res. Lett., 25, 3335–3338, doi:10.1029/98GL52359, 1998.

Chance, K. and Kurucz, R.: An improved high-resolution solar reference spectrum for earth's atmosphere measurements in the ultraviolet, visible, and near infrared, J. Quant. Spectrosc. Ra., 111, 1289–1295, doi:10.1016/j.jqsrt.2010.01.036, 2010.

Chance, K. and Orphal, J.: Revised ultraviolet absorption cross sections of H_2CO for the HITRAN database, J. Quant. Spectrosc. Ra., 112, 1509–1510, doi:10.1016/j.jqsrt.2011.02.002, 2011.

Chance, K., Palmer, P. I., Spurr, R. J. D., Martin, R. V., Kurosu, T. P., and Jacob, D. J.: Satellite observations of formaldehyde over North America from GOME, Geophys. Res. Lett., 27, 3461–3464, doi:10.1029/2000GL011857, 2000.

Chance, K., Kurosu, T. P., and Sioris, C. E.: Undersampling correction for array detector-based satellite spectrometers, Appl. Optics, 44, 1296–1304, doi:10.1364/AO.44.001296, 2005.

Chance, K. V. and Spurr, R. J. D.: Ring effect studies: Rayleigh scattering, including molecular parameters for rotational Raman scattering, and the Fraunhofer spectrum, Appl. Optics, 36, 5224–5230, doi:10.1364/AO.36.005224, 1997.

De Smedt, I., Müller, J.-F., Stavrakou, T., van der A, R., Eskes, H., and Van Roozendael, M.: Twelve years of global observations of formaldehyde in the troposphere using GOME and SCIAMACHY sensors, Atmos. Chem. Phys., 8, 4947–4963, doi:10.5194/acp-8-4947-2008, 2008.

De Smedt, I., Van Roozendael, M., Stavrakou, T., Müller, J.-F., Lerot, C., Theys, N., Valks, P., Hao, N., and van der A, R.: Improved retrieval of global tropospheric formaldehyde columns from GOME-2/MetOp-A addressing noise reduction and instrumental degradation issues, Atmos. Meas. Tech., 5, 2933–2949, doi:10.5194/amt-5-2933-2012, 2012.

Dirksen, R., Dobber, M., Voors, R., and Levelt, P.: Prelaunch characterization of the Ozone Monitoring Instrument transfer function in the spectral domain, Appl. Optics, 45, 3972–3981, doi:10.1364/AO.45.003972, 2006.

Dufour, G., Szopa, S., Barkley, M. P., Boone, C. D., Perrin, A., Palmer, P. I., and Bernath, P. F.: Global upper-tropospheric formaldehyde: seasonal cycles observed by the ACE-FTS satellite instrument, Atmos. Chem. Phys., 9, 3893–3910, doi:10.5194/acp-9-3893-2009, 2009.

Eskes, H. J. and Boersma, K. F.: Averaging kernels for DOAS total-column satellite retrievals, Atmos. Chem. Phys., 3, 1285–1291, doi:10.5194/acp-3-1285-2003, 2003.

Fu, T.-M., Jacob, D. J., Palmer, P. I., Chance, K., Wang, Y. X., Barletta, B., Blake, D. R., Stanton, J. C., and Pilling, M. J.: Space-based formaldehyde measurements as constraints on volatile organic compound emissions in east and south Asia and implications for ozone, J. Geophys. Res-Atmos., 112, D06312, doi:10.1029/2006JD007853, 2007.

Heckel, A., Richter, A., Tarsu, T., Wittrock, F., Hak, C., Pundt, I., Junkermann, W., and Burrows, J. P.: MAX-DOAS measurements of formaldehyde in the Po-Valley, Atmos. Chem. Phys., 5, 909–918, doi:10.5194/acp-5-909-2005, 2005.

Houweling, S., Dentener, F., and Lelieveld, J.: The impact of nonmethane hydrocarbon compounds on tropospheric photochemistry, J. Geophys. Res.-Atmos., 103, 10673–10696, doi:10.1029/97JD03582, 1998.

Joiner, J. and Vassilkov, A. P.: First results from the OMI rotational Raman scattering cloud pressure algorithm, IEEE T. Geosci. Remote Sens., 44, 1272–1282, doi:10.1109/TGRS.2005.861385, 2006.

Jones, N. B., Riedel, K., Allan, W., Wood, S., Palmer, P. I., Chance, K., and Notholt, J.: Long-term tropospheric formaldehyde concentrations deduced from ground-based fourier transform solar infrared measurements, Atmos. Chem. Phys., 9, 7131–7142, doi:10.5194/acp-9-7131-2009, 2009.

Kanakidou, M., Seinfeld, J. H., Pandis, S. N., Barnes, I., Dentener, F. J., Facchini, M. C., Van Dingenen, R., Ervens, B., Nenes, A., Nielsen, C. J., Swietlicki, E., Putaud, J. P., Balkanski, Y., Fuzzi, S., Horth, J., Moortgat, G. K., Winterhalter, R., Myhre, C. E. L., Tsigaridis, K., Vignati, E., Stephanou, E. G., and Wilson, J.: Organic aerosol and global climate modelling: a review, Atmos. Chem. Phys., 5, 1053–1123, doi:10.5194/acp-5-1053-2005, 2005.

Khokhar, M., Frankenberg, C., Roozendael, M. V., Beirle, S., Kuhl, S., Richter, A., Platt, U., and Wagner, T.: Satellite observations of atmospheric SO_2 from volcanic eruptions during the time-period of 19962002, Adv. Space Res., 36, 879–887, 2005.

Kim, J. H., Kim, S. M., Baek, K. H., Wang, L., Kurosu, T., De Smedt, I., Chance, K., and Newchurch, M. J.: Evaluation of satellite-derived HCHO using statistical methods, Atmos. Chem. Phys. Discuss., 11, 8003–8025, doi:10.5194/acpd-11-8003-2011, 2011.

Kleipool, Q. L., Dobber, M. R., de Haan, J. F., and Levelt, P. F.: Earth surface reflectance climatology from 3 years of OMI data, J. Geophys. Res.-Atmos., 113, D18308, doi:10.1029/2008JD010290, 2008.

Kurosu, T. P., Chance, K., and Sioris, C. E.: Preliminary results for HCHO and BrO from the EOS-Aura Ozone Monitoring Instrument, Proc. SPIE, 5652, 116–123, doi:10.1117/12.578606, 2004.

Levelt, P. F., van den Oord, G. H. J., Dobber, M. R., Mälkki, A., Visser, H., de Vries, J., Stammes, P., Lundell, J. O. V., and Saari, H.: The ozone monitoring instrument, IEEE T. Geosci. Remote., 44, 1093–1101, 2006.

Lindström, P. and Wedin, P.-A.: Methods and software for nonlinear least squares problems, Technical report UMINF-133.87, Inst. of Information Processing, University of Umeå, Umeå, Sweden, 1988.

Lowe, D. C. and Schmidt, U.: Formaldehyde (HCHO) measurements in the nonurban atmosphere, J. Geosphys. Res.-Oceans, 88, 10844–10858, doi:10.1029/JC088iC15p10844, 1983.

Malicet, J., Daumont, D., Charbonnier, J., Parisse, C., Chakir, A., and Brion, J.: Ozone UV spectroscopy. II. Absorption cross-sections and temperature dependence, J. Atmos. Chem., 21, 263–273, doi:10.1007/BF00696758, 1995.

Marais, E. A., Jacob, D. J., Kurosu, T. P., Chance, K., Murphy, J. G., Reeves, C., Mills, G., Casadio, S., Millet, D. B., Barkley, M. P., Paulot, F., and Mao, J.: Isoprene emissions in Africa inferred from OMI observations of formaldehyde columns, Atmos. Chem. Phys., 12, 6219–6235, doi:10.5194/acp-12-6219-2012, 2012.

Martin, R. V., Chance, K., Jacob, D. J., Kurosu, T. P., Spurr, R. J. D., Bucsela, E., Gleason, J. F., Palmer, P. I., Bey, I., Fiore, A. M., Li, Q., Yantosca, R. M., and Koelemeijer, R. B. A.: An improved retrieval of tropospheric nitrogen dioxide from GOME, J. Geophys. Res.-Atmos., 107, ACH9.1–ACH9.21, doi:10.1029/2001JD001027, 2002.

Miller, C. C., Abad, G. G., Wang, H., Liu, X., Kurosu, T., Jacob, D. J., and Chance, K.: Glyoxal retrieval from the Ozone Monitoring Instrument, Atmos. Meas. Tech. Discuss., 7, 6065–6112, doi:10.5194/amtd-7-6065-2014, 2014.

Millet, D. B., Jacob, D. J., Turquety, S., Hudman, R. C., Wu, S., Fried, A., Walega, J., Heikes, B. G., Blake, D. R., Singh, H. B., Anderson, B. E., and Clarke, A. D.: Formaldehyde distribution over North America: Implications for satellite retrievals of formaldehyde columns and isoprene emission, J. Geophys. Res.-Atmos., 111, D24S02, doi:10.1029/2005JD006853, 2006.

Palmer, P. I., Jacob, D. J., Chance, K., Martin, R. V., Spurr, R. J. D., Kurosu, T. P., Bey, I., Yantosca, R., Fiore, A., and Li, Q.: Air mass factor formulation for spectroscopic measurements from satellites: Application to formaldehyde retrievals from the Global Ozone Monitoring Experiment, J. Geophys. Res.-Atmos., 106, 14539–14550, doi:10.1029/2000JD900772, 2001.

Parrish, D. D., Ryerson, T. B., Mellqvist, J., Johansson, J., Fried, A., Richter, D., Walega, J. G., Washenfelder, R. A., de Gouw, J. A., Peischl, J., Aikin, K. C., McKeen, S. A., Frost, G. J., Fehsenfeld, F. C., and Herndon, S. C.: Primary and secondary sources of formaldehyde in urban atmospheres: Houston Texas region, Atmos. Chem. Phys., 12, 3273–3288, doi:10.5194/acp-12-3273-2012, 2012.

Peters, E., Wittrock, F., Großmann, K., Frieß, U., Richter, A., and Burrows, J. P.: Formaldehyde and nitrogen dioxide over the remote western Pacific Ocean: SCIAMACHY and GOME-2 validation using ship-based MAX-DOAS observations, Atmos. Chem. Phys., 12, 11179–11197, doi:10.5194/acp-12-11179-2012, 2012.

Pinardi, G., Van Roozendael, M., Abuhassan, N., Adams, C., Cede, A., Clémer, K., Fayt, C., Frieß, U., Gil, M., Herman, J., Hermans, C., Hendrick, F., Irie, H., Merlaud, A., Navarro Comas, M., Peters, E., Piters, A. J. M., Puentedura, O., Richter, A., Schönhardt, A., Shaiganfar, R., Spinei, E., Strong, K., Takashima, H., Vrekoussis, M., Wagner, T., Wittrock, F., and Yilmaz, S.: MAX-DOAS formaldehyde slant column measurements during CINDI: intercomparison and analysis improvement, Atmos. Meas. Tech., 6, 167–185, doi:10.5194/amt-6-167-2013, 2013.

Richter, A., Begoin, M., Hilboll, A., and Burrows, J. P.: An improved NO_2 retrieval for the GOME-2 satellite instrument, Atmos. Meas. Tech., 4, 1147–1159, doi:10.5194/amt-4-1147-2011, 2011.

Rodgers, C. D. and Connor, B. J.: Intercomparison of remote sounding instruments, J. Geophys. Res.-Atmos., 108, 4116, doi:10.1029/2002JD002299, 2003.

Sneep, M., de Haan, J. F., Stammes, P., Wang, P., Vanbauce, C., Joiner, J., Vasilkov, A. P., and Levelt, P. F.: Three-way comparison between OMI and PARASOL cloud pressure products, J. Geophys. Res.-Atmos., 113, doi:10.1029/2007JD008694, 2008.

Spurr, R. J.: VLIDORT: A linearized pseudo-spherical vector discrete ordinate radiative transfer code for forward model and retrieval studies in multilayer multiple scattering media, J. Quant. Spectrosc. Ra., 102, 316–342, 2006.

Stammes, P., Sneep, M., de Haan, J. F., Veefkind, J. P., Wang, P., and Levelt, P. F.: Effective cloud fractions from the Ozone Monitoring Instrument: Theoretical framework and validation, J. Geophys. Res.-Atmos., 113, D16S38, doi:10.1029/2007JD008820, 2008.

Stavrakou, T., Müller, J.-F., De Smedt, I., Van Roozendael, M., van der Werf, G. R., Giglio, L., and Guenther, A.: Evaluating the performance of pyrogenic and biogenic emission inventories against one decade of space-based formaldehyde columns, Atmos. Chem. Phys., 9, 1037–1060, doi:10.5194/acp-9-1037-2009, 2009a.

Stavrakou, T., Müller, J.-F., De Smedt, I., Van Roozendael, M., van der Werf, G. R., Giglio, L., and Guenther, A.: Global emissions of non-methane hydrocarbons deduced from SCIAMACHY formaldehyde columns through 2003–2006, Atmos. Chem. Phys., 9, 3663–3679, doi:10.5194/acp-9-3663-2009, 2009b.

Steck, T., Glatthor, N., von Clarmann, T., Fischer, H., Flaud, J. M., Funke, B., Grabowski, U., Höpfner, M., Kellmann, S., Linden, A., Perrin, A., and Stiller, G. P.: Retrieval of global upper tropospheric and stratospheric formaldehyde (H_2CO) distributions from high-resolution MIPAS-Envisat spectra, Atmos. Chem. Phys., 8, 463–470, doi:10.5194/acp-8-463-2008, 2008.

Thalman, R. and Volkamer, R.: Temperature dependent absorption cross-sections of O_2-O_2 collision pairs between 340 and 630 nm and at atmospherically relevant pressure, Phys. Chem. Chem. Phys., 15, 15371–15381, doi:10.1039/C3CP50968K, 2013.

Vandaele, A., Hermans, C., Simon, P., Carleer, M., Colin, R., Fally, S., Mérienne, M., Jenouvrier, A., and Coquart, B.: Measurements of the NO_2 absorption cross-section from 42 000 cm^{-1} to 10 000 cm^{-1} (238–1000 nm) at 220 K and 294 K, J. Quant. Spectrosc. Ra., 59, 171–184, 1998.

Veihelmann, B. and Kleipool, Q.: Reducing Along-Track Stripes in OMI-Level 2 Products, Technical report TN-OMIE-KNMI-785, KNMI, available at: http://disc.sci.gsfc.nasa.gov/Aura/data-holdings/OMI/documents/v003/RD08_TN785_i1_Reducing_AlongTrack_Stripes.pdf (last access: 13 March 2006), 2006.

Vigouroux, C., Hendrick, F., Stavrakou, T., Dils, B., De Smedt, I., Hermans, C., Merlaud, A., Scolas, F., Senten, C., Vanhaelewyn, G., Fally, S., Carleer, M., Metzger, J.-M., Müller, J.-F., Van Roozendael, M., and De MaziÈre, M.: Ground-based FTIR and MAX-DOAS observations of formaldehyde at Réunion Island and comparisons with satellite and model data, Atmos. Chem. Phys., 9, 9523–9544, doi:10.5194/acp-9-9523-2009, 2009.

Vrekoussis, M., Wittrock, F., Richter, A., and Burrows, J. P.: GOME-2 observations of oxygenated VOCs: what can we learn from the ratio glyoxal to formaldehyde on a global scale?, Atmos. Chem. Phys., 10, 10145–10160, doi:10.5194/acp-10-10145-2010, 2010.

Wagner, T., Beirle, S., Brauers, T., Deutschmann, T., Frieß, U., Hak, C., Halla, J. D., Heue, K. P., Junkermann, W., Li, X., Platt, U., and Pundt-Gruber, I.: Inversion of tropospheric profiles of aerosol extinction and HCHO and NO_2 mixing ratios from MAX-DOAS observations in Milano during the summer of 2003 and comparison with independent data sets, Atmos. Meas. Tech., 4, 2685–2715, doi:10.5194/amt-4-2685-2011, 2011.

Wang, H., Liu, X., Chance, K., Gonzalez Abad, G., and Chan Miller, C.: Water vapor retrieval from OMI visible spectra, Atmos. Meas. Tech., 7, 1901–1913, 2014, http://www.atmos-meas-tech.net/7/1901/2014/.

Wilmouth, D. M., Hanisco, T. F., Donahue, N. M., and Anderson, J. G.: Fourier Transform Ultraviolet Spectroscopy of the A 2Π3/2 ← X 2Π3/2 Transition of BrO†, J. Phys. Chem. A, 103, 8935–8945, doi:10.1021/jp991651o, 1999.

Wittrock, F., Richter, A., Oetjen, H., Burrows, J. P., Kanakidou, M., Myriokefalitakis, S., Volkamer, R., Beirle, S., Platt, U., and Wagner, T.: Simultaneous global observations of glyoxal and formaldehyde from space, Geophys. Res. Lett., 33, L16804, doi:10.1029/2006GL026310, 2006.

First continuous measurements of $\delta^{18}O$-CO_2 in air with a Fourier transform infrared spectrometer

S. N. Vardag[1], S. Hammer[1], M. Sabasch[1], D. W. T. Griffith[2], and I. Levin[1]

[1]Institut für Umweltphysik, Heidelberg University, Heidelberg, Germany
[2]Department of Chemistry, University of Wollongong, Wollongong, Australia

Correspondence to: S. N. Vardag (svardag@iup.uni-heidelberg.de)

Abstract. The continuous in situ measurement of $\delta^{18}O$ in atmospheric CO_2 opens a new door to differentiating between CO_2 source and sink components with high temporal resolution. Continuous ^{13}C–CO_2 measurement systems have already been commercially available for some time, but until now, only few instruments have been able to provide a continuous measurement of the oxygen isotope ratio in CO_2. Besides precise $^{13}C/^{12}C$ observations, the Fourier transform infrared (FTIR) spectrometer is also able to measure the $^{18}O/^{16}O$ ratio in CO_2, but the precision and accuracy of the measurements have not yet been evaluated. Here we present a first analysis of $\delta^{18}O$-CO_2 (and $\delta^{13}C$-CO_2) measurements with the FTIR analyser in Heidelberg. We used Allan deviation to determine the repeatability of $\delta^{18}O$-CO_2 measurements and found that it decreases from 0.25 ‰ for 10 min averages to about 0.1 ‰ after 2 h and remains at that value up to 24 h. We evaluated the measurement precision over a 10-month period (intermediate measurement precision) using daily working gas measurements and found that our spectrometer measured $\delta^{18}O$-CO_2 to better than 0.3 ‰ at a temporal resolution of less than 10 min. The compatibility of our FTIR-spectrometric measurements to isotope-ratio mass-spectrometric (IRMS) measurements was determined by comparing FTIR measurements of cylinder gases and ambient air with IRMS measurements of flask samples, filled with gases of the same cylinders or collected from the same ambient air intake. Two-sample t tests revealed that, at the 0.01 significance level, the FTIR and the IRMS measurements do not differ significantly from each other and are thus compatible. We describe two weekly episodes of ambient air measurements, one in winter and one in summer, and discuss what potential insights and new challenges combined highly resolved CO_2, $\delta^{13}C$-CO_2 and $\delta^{18}O$-CO_2 records may provide in terms of better understanding regional scale continental carbon exchange processes.

1 Introduction

Quantitative understanding of the processes governing the carbon cycle is vital in order to assess the impact and fate of increasing anthropogenic CO_2 emissions to the atmosphere. The stable isotopes in CO_2 can provide information about the fluxes between the different carbon reservoirs, such as the atmosphere, the biosphere and the oceans. $^{13}CO_2$ measurements can be used to distinguish between terrestrial biosphere and marine fluxes (Keeling et al., 1989; Ciais et al., 1995), and are also used as a tracer for anthropogenic emissions, as most fossil fuel CO_2 emissions are depleted in ^{13}C relative to those of the biosphere (Tans, 1981). The interpretation of atmospheric $\delta^{18}O$-CO_2 is more complex, since ^{18}O in CO_2 is strongly coupled to the water cycle (e.g. Francey and Tans, 1987; Farquhar et al., 1993; Cuntz et al., 2003a; 2003b, Buenning et al., 2014). During CO_2 exchange with soil and leaves, the ^{18}O isotopes of CO_2 are exchanged with those of H_2O (Hesterberg and Siegenthaler, 1991). Carbonic anhydrase facilitates the equilibration with leaf water (Gillon and Yakir, 2001; Farquhar et al., 1993). The isotopic composition of soil water is determined by the isotopic composition of precipitation, which itself has strong spatial variations (IAEA/WMO GNIP database available at http://isohis.iaea.org). Since precipitation at higher latitudes is depleted in ^{18}O (Dansgaard, 1964), the soil water and consequently the CO_2 from root respiration and heterotrophic respiration is also depleted in ^{18}O at higher latitudes (Farquhar et al., 1993). The soil invasion flux will further influence the apparent soil respiration signature as the CO_2 diffuses into the soil, partially equilibrates with soil water and retro-diffuses out of the soil with a new isotopic composition (Tans, 1998; Miller et al., 1999). Isotopic exchange during soil invasion might even be enhanced due to carbonic anhydrase in soils (Wingate et al., 2009). Miller et al. (1999) reported that in most settings and especially in dry ground and for short residence times of air close to the soil surface (corresponding to high boundary layer mixing heights), the effect will be smaller than 5 ‰. Due to ^{18}O enrichment during evapotranspiration, the plant leaf water is enriched in ^{18}O relative to the soil water (Farquhar et al., 1993). During photosynthesis, CO_2 equilibrates with leaf water and about two-thirds of the CO_2 retro-diffuses into the atmosphere without being assimilated (Tans, 1998). The retro-diffused CO_2 changes the atmospheric $\delta^{18}O\text{-}CO_2$ value, depending on the isotopic signature of the leaf water. In central Europe, we expect the discrimination against ^{18}O during net CO_2 assimilation to be positive (Farquhar et al., 1993; Cuntz et al. 2003b; Wingate et al., 2009). Still et al. (2009), Welp et al. (2011) and Buenning et al. (2014) have studied the susceptibility of atmospheric $\delta^{18}O\text{-}CO_2$ to environmental parameters, such as precipitation, relative humidity, temperature, solar radiation and cloud cover, and estimated the influences of these parameters on the atmospheric $\delta^{18}O\text{-}CO_2$ using regional and global scale models. They also assessed the effect of the isotopic composition of precipitation and water vapour. They found that many of these parameters should not be neglected when quantifying biospheric gross $^{18}O\text{-}CO_2$ fluxes. They also highlight the complexity and the large uncertainties of the processes and sensitivities influencing atmospheric $\delta^{18}O\text{-}CO_2$. Thus, in order to understand atmospheric $\delta^{18}O\text{-}CO_2$ measurements in all their complexity, information about the regional isotopic composition of precipitation, environmental parameters such as temperature and water vapour deficit and a comprehensive land-surface model are necessary (Yakir and Wang, 1996; Ciais et al., 1997; Langendörfer et al., 2002; Cuntz et al., 2003a; Buenning et al., 2014).

The first step to understanding the $^{18}O\text{-}CO_2$ fluxes to and from the terrestrial biosphere is to make reliable and comparable measurements at high temporal resolution. However, measurements via isotope-ratio mass-spectrometry (IRMS) are elaborate and time-consuming, limiting the number of continuous records of ^{18}O in CO_2 that exist to date (Flanagan et al., 1997; Langendörfer et al., 2002; Bowling et al., 2003; Pataki et al., 2003). A quantum cascade laser-based absorption spectrometer measuring $^{12}C^{16}O^{16}O$,$^{13}C^{16}O^{16}O$ and $^{12}C^{16}O^{18}O$ with a high temporal resolution provided first continuous records (Tuzson et al., 2011; Sturm et al., 2013). $^{12}C^{16}O^{16}O$ and $^{13}C^{16}O^{16}O$ have also been determined by Fourier transform infrared (FTIR) spectroscopy in several previous studies (e.g. Esler et al., 2000; Mohn et al., 2008; Cambaliza, 2010; Griffith et al., 2012). In principle, FTIR spectroscopy can also provide continuous measurements of $^{12}C^{16}O^{18}O$. However, in their original study, Esler et al. (2000) remarked that the degree of precision is too poor for a useful determination in natural abundances using a 1 cm^{-1} resolution spectrometer. Given improvements in the instrumentation and spectral analysis methods since that time, we have revisited the practicality of continuous measurements of $\delta^{18}O$ in CO_2 using FTIR spectroscopy.

The scope of this manuscript is to answer two important questions: first, is it possible to measure $\delta^{18}O\text{-}CO_2$ using FTIR spectroscopy, and if yes, how well can we measure it in terms of precision, accuracy and compatibility to conventional IRMS observations? Second, what insight into regional scale carbon exchange processes can one gain from a highly resolved $\delta^{18}O\text{-}CO_2$ record (along with the continuous CO_2, CO and $\delta^{13}C\text{-}CO_2$ records) in the catchment area of our measurement site?

2 FTIR measurement principle and calibration procedure

The in situ FTIR analyser used in Heidelberg was developed and built at the University in Wollongong, Australia and is described in detail by Griffith et al. (2012) and Hammer et al. (2013a). It was used during two travelling instrument campaigns by Hammer et al. (2013b) and Vardag et al. (2014) for CO_2, CH_4 and N_2O measurements. Briefly, the FTIR spectrometer obtains a broadband transmittance spectrum of the sample air as the ratio of the infrared spectra measured with and without a sample in the optical cell. The measured transmittance spectrum is fitted by non-linear least squares using the program MALT (Multi-Layer Absorption Transmittance) to model the spectrum (Griffith, 1996; Griffith et al., 2012). The model adjusts sample composition and instrument parameters to obtain the best fit to the measured spectrum, and the best-fit sample concentrations are taken as the retrieved values.

The analyser is a prototype of the now commercially available Spectronus FTIR trace gas analyser (Ecotech, Knoxfield, Australia). While functionally equivalent, there are some component differences. The FTIR spectrometer is an IRcube (Bruker Optics, Ettlingen, Germany) coupled to a glass multipass White cell (model 24 PA, IRanalysis Inc., Anaheim, CA) with 3.5 L volume and 24 m optical path. Spectra were recorded at 1 cm^{-1} resolution and typically co-added to 3 min averages (approximately one cell exchange time at the typical flow rate). The sample handling system comprises four selectable inlets, an optional dryer (Nafion, Permapure PD-100T-24SS), followed by a granulated magnesium perchlorate trap), two mass flow controllers (model D-5111, Bronkhorst, Germany) and a four-head diaphragm vacuum pump (model MV2, Vacuubrand, Germany). One mass flow controller upstream of the cell controls sample air-

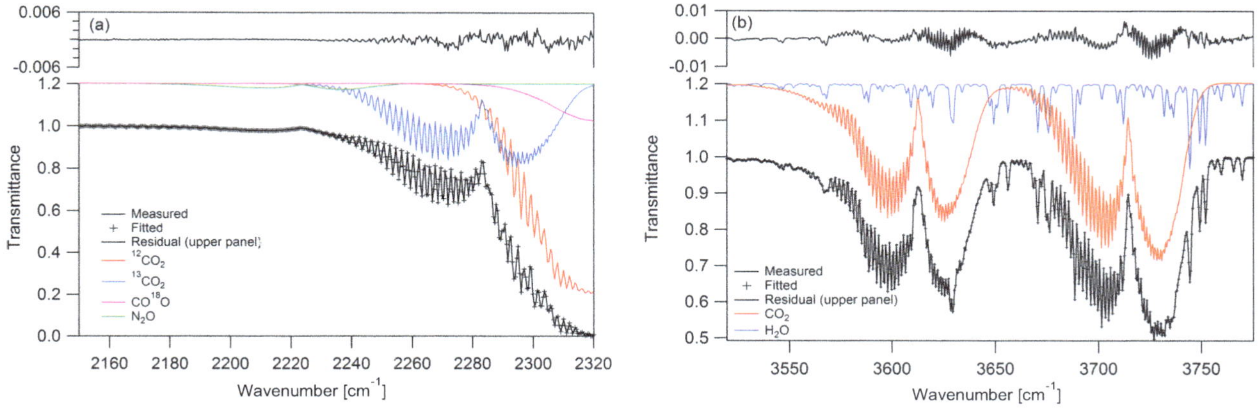

Figure 1. (a) Spectra of CO_2 isotopologues and N_2O in the 2150–2320 cm^{-1} region. The coloured traces show the individual isotopologues, the black spectra are a measured air spectrum (black line), calculated best-fit spectrum (+ symbols) and the fitting residual (black, upper panel). **(b)** Spectrum including CO_2 and H_2O near 3600 cm^{-1}. The individual trace gas and isotopologue spectra are shifted by 0.2 upwards for clarity.

flow, while the other downstream of the cell actively controls pressure via a proportional–integral software control loop to better than ±0.1 hPa. The FTIR housing and the cell are both thermostated and stable within 0.01 °C (1σ). A Windows PC controls sample flow, spectrum collection and online analysis. Sample air is delivered to the analyser at 1500–1800 hPa pressure through a clean diaphragm pump (model N86K.18, KNF Neuberger, Freiburg, Germany). In this work all measurements of both air samples and tank gases were dried ($< 10\,\mu mol\,mol^{-1}$ water vapour) and made at 1100 hPa pressure, 30 °C and a flow of 1 SLPM (standard litre per minute). The measurements were performed in the laboratory under stable temperature conditions (±1 °C).

Figure 1a shows the CO_2 isotopologue components (coloured traces) of the infrared absorption spectrum of air in the 2150–2320 cm^{-1} spectral region routinely used for CO_2 FTIR analysis by the analyser (Griffith et al., 2012). The black traces show a measured spectrum and typical fit to the composite air spectrum including $^{12}C^{16}O_2$, $^{13}C^{16}O_2$ and $^{12}C^{18}O^{16}O$. Although the $^{12}C^{18}O^{16}O$ isotopologue is heavily overlapped by the parent and $^{13}C^{16}O_2$ isotopologues, its contribution to the total absorption is significant and repeatable and provides the basis for quantification of this isotopologue. The upper panel of Fig. 1a shows a typical spectral residual which is well above the detector noise level above 2240 cm^{-1}. This residual is systematic and constant in shape from spectral fit to fit. The MALT spectrum calculation model is not able to improve this fit, which may be due to either (or both) an imperfect instrument line shape (ILS) or actual line shapes, which are not Voigt shaped as assumed in the model. To investigate the ILS contribution further, we have recorded spectra of air under the same conditions (temperature, pressure, resolution, cell path length) in a Bruker IFS 125/HR spectrometer at the University of Wollongong. This high resolution spectrometer is maintained in a well-aligned condition as part of the Total Carbon Column Observing Network (TCCON, Wunch et al., 2011) and its ILS is well characterised through high resolution test cell measurements (Hase et al., 2013) to be very close to the theoretically ideal shape calculated in the MALT model (modulation efficiency > 0.99, phase error $< 0.5°$). Fitting these IFS 125/HR spectra resulted in residuals very similar in shape and magnitude to that in Fig. 1a (upper panel), which indicates that imperfect ILS is not the primary cause of the lack of fit. To investigate the possible effects of non-Voigt molecular line shapes, typical FTIR analyser spectra were fitted with two independent spectrum fitting models, GFIT (Geoff Toon, Jet Propulsion Laboratory) and PROFITT (Frank Hase, Karlsruhe Institute of Technology). These spectrum models optionally extend to non-Voigt line shapes including effects of speed dependent cross sections, line narrowing and line mixing (e.g. Ngo et al., 2013). Using several different line shape models did not remove the spectral residuals – in some cases they were slightly reduced or of different shape, but total residuals were reduced by at most 25 %.

Thus, from these two tests we conclude that the residuals are not primarily due to an imperfect instrument line shape, but rather due to the inadequacy of currently available line shape models for the calculation. The imperfect fit is exacerbated by the fact that the residuals are dominated by absorption in the line wings of strongly absorbed lines, which are the least accurately modelled. We must therefore accept the imperfect fits as unavoidable until further advances in line shape models become available. If $^{12}C^{18}O^{16}O$ is removed from the fit, the residuals are two to three times larger; in this case, the least squares fit routine adjusts the amounts of $^{12}C^{16}O_2$ and $^{13}C^{16}O_2$ in the fit by unrealistic amounts to attempt to minimise the residual. The results in this paper show that the $^{12}C^{18}O^{16}O$ amount retrieved from fitting these

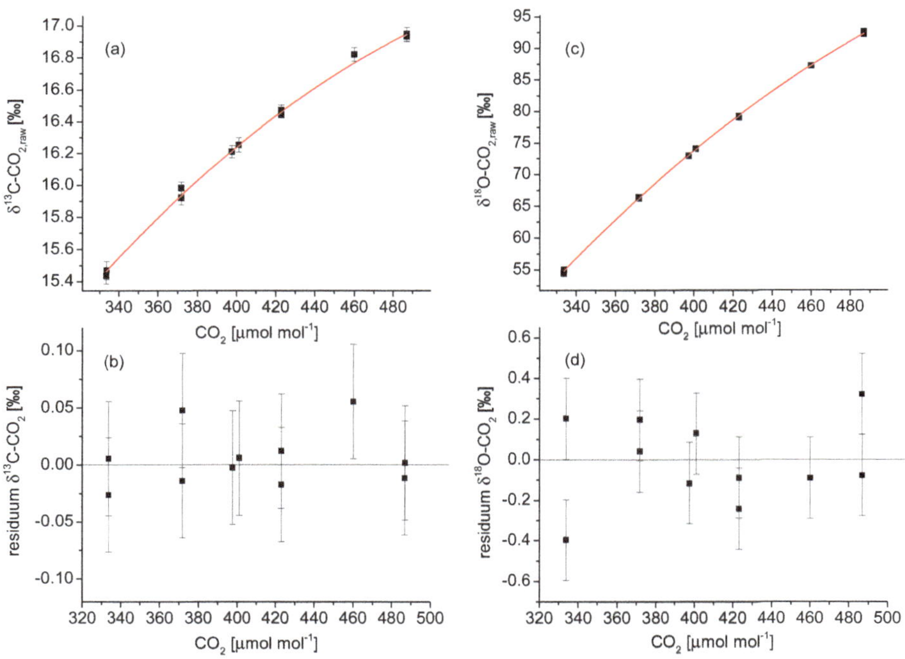

Figure 2. CO_2 dependence of raw $\delta^{13}C\text{-}CO_2$ and $\delta^{18}O\text{-}CO_2$ (**a** and **b**) and their residuals from the cubic fit (**b** and **d**). The experimental results shown here were obtained in August 2012; the same experiment was repeated in March 2014 and showed no significant difference to the earlier measurements.

spectra, despite the residuals, provides a consistent basis for quantification of this isotopologue.

Total CO_2 can also be retrieved from the region around $3600\,cm^{-1}$ without isotopic discrimination. The fit to this region is shown in Fig. 1b, from Griffith et al. (2012). Retrieval of CO_2 from this region is more precise (i.e. lower noise, better repeatability) than that of $^{12}C^{16}O_2$ near $2300\,cm^{-1}$, firstly because the bands are not saturated and are of near optimum absorption (50 %), and secondly, because the whole bands have lower temperature sensitivity due to the inclusion of both high and low-J lines with both positive and negative temperature sensitivity. As detailed below, the total CO_2 retrieval, scaled if required, can be used as a proxy for $^{12}C^{16}O_2$ in isotopic calculations with acceptable accuracy.

2.1 Data evaluation and calibration

In the following, we describe the data evaluation and calibration procedure for the isotopologue ratio $\delta^{18}O\text{-}CO_2$, but the procedure is analogous for $\delta^{13}C\text{-}CO_2$.

Step 1: Calculate the raw δ^{18}O-CO_2 value from FTIR measurements

The FTIR computes the raw $\delta^{18}O\text{-}CO_2$ value using the ratio of the raw value of the rare isotopologue and the raw value of the common isotopologue:

$$\delta^{18}O\text{-}CO_{2\,\mathrm{HITRAN}} = \left[\frac{\left(\frac{^{12}C^{18}O^{16}O_{\mathrm{raw}}}{^{12}C^{16}O^{16}O_{\mathrm{raw}}}\right)_{\mathrm{sample}}}{R_{\mathrm{HITRAN}}} - 1 \right] \cdot 1000\,‰ \quad (1)$$

with $R_{\mathrm{HITRAN}} = 0.0040104$ (Rothman et al., 2005). For ^{13}C, the equivalent value of R_{HITRAN} is 0.0112372.

Following Coplen (2011) and common usage, we use the terminology $\delta^{18}O\text{-}CO_2$, even though the δ-notation is originally defined with the isotope ratio (in contrast to isotopologue ratio). The FTIR analysis implicitly uses the HITRAN scale (Rothman et al., 2005), which is referred to Vienna Pee Dee Belemnite (VPDB) for $\delta^{13}C$ and to Vienna Standard Mean Ocean Water (VSMOW) for $\delta^{18}O$; during the calibration (step 3) the final reference scale of the calibrated data can be changed to any other scale. We chose the VPDB scale for $\delta^{13}C\text{-}CO_2$ $((^{13}C^{16}O_2/^{12}C^{16}O_2)_{\mathrm{VPDB}} = 0.0112372)$ and VPDB-CO_2 scale for $\delta^{18}O\text{-}CO_2$ $((^{12}C^{18}O^{16}O/^{12}C^{16}O_2)_{\mathrm{VPDB-CO_2}} = 0.0041767)$ following Allison et al. (1995). We abbreviate $\delta^{13}C\text{-}CO_2$ and $\delta^{18}O\text{-}CO_2$ on the VPDB-CO_2 scale with $\delta^{13}C\text{-}CO_{2,\mathrm{VPDB}}$ and $\delta^{18}O\text{-}CO_{2,\mathrm{VPDB}}$ respectively.

Step 2: Cross-sensitivity and interspecies interference corrections

To first order, the fitting software MALT takes into account pressure, temperature and interspecies overlapping absorp-

Table 1. Interspecies interference and cross-sensitivity correction factors for $\delta^{13}C\text{-}CO_2$ and $\delta^{18}O\text{-}CO_2$ used in Eqs. (2) and (3). Reference values were $T_{ref} = 31.8\,°C$, $F_{ref} = 1.0$ SLPM, $P_{ref} = 1100$ hPa and $H_2O_{ref} = 0\,\mu$mole mole^{-1}, for temperature, flow, pressure and water vapour content, respectively.

	$\delta^{13}C\text{-}CO_2$	$\delta^{18}O\text{-}CO_2$
dC_{raw}/dT [‰ °C^{-1}]	0.127	4.256
dC_{raw}/dF [‰ SLPM^{-1}]	−0.91424	−2.92166
dC_{raw}/dP [‰ hPa^{-1}]	0.00249	−0.18694
dC_{raw}/dH_2O [‰ (μmole mole^{-1})$^{-1}$]	0	0
a [‰]	−10.344	−252.786
b [‰ (μmole mole^{-1})$^{-1}$]	0.0461902	1.162269
c [‰ (μmole mole^{-1})$^{-2}$]	−0.0000658108	−0.00179787
d [‰ (μmole mole^{-1})$^{-3}$]	0.000000034299	0.000001093919

tion bands in the fit. However, small second order effects remain due to real imperfections in temperature and pressure measurements, spectrometer instrumental line shape and the assumption of the MALT models (such as Voigt line shapes, see above), necessitating small empirical corrections to the raw measured mole fraction (C_{raw}) (Griffith et al., 2012; Hammer et al., 2013a). A cross-sensitivity correction for sample temperature (T) and pressure (P), H_2O amount and flow rate (F), as well as an interspecies-sensitivity correction for CO_2 mole fraction (corr(CO_2)) is applied for every measurement following Eq. (2):

$$C_{corr} = \frac{dC_{raw}}{dT}\cdot(T - T_{ref}) - \frac{dC_{raw}}{dF}\cdot(F - F_{ref}) - \frac{dC_{raw}}{dP}\cdot(P - P_{ref}) - \frac{dC_{raw}}{dH_2O}\cdot(H_2O - H_2O_{ref}) - \mathrm{corr}(CO_2) \quad (2)$$

Where P_{ref}, T_{ref}, etc. are the reference values to which pressure, temperature etc. are corrected, and the CO_2 correction follows:

$$\mathrm{corr}(CO_2) = a + b\cdot C_{raw} + c\cdot {C_{raw}}^2 + \mathrm{d}\cdot {C_{raw}}^3 \quad (3)$$

Table 1 lists all cross-sensitivity parameters and CO_2-interspecies interference corrections.

Hammer et al. (2013a) describe in detail the set-up of the experiment to determine the CO_2 sensitivity. We use a cubic fit to describe the CO_2 interspecies correction (Fig. 2a and b; coefficient of determination $R^2 = 0.99$ for $\delta^{13}C\text{-}CO_2$ and for $\delta^{18}O\text{-}CO_2$), with residuals showing no further concentration dependence (Fig. 2b and d).

Step 3: Calibration

The cross-sensitivity corrected data are calibrated on the VPDB gas scale using a linear instrument response function (typically linear to the degree of $R^2 = 0.9998$). We derive the calibration response function weekly from three reference tanks with known values for CO_2, $\delta^{13}C\text{-}CO_2$ and $\delta^{18}O\text{-}CO_2$. Our reference standards span ranges from about 370 to 470 μmol mol^{-1} for CO_2 mole fraction, a $\delta^{13}C\text{-}CO_{2,VPDB}$ range from −8.7 to −12.8 ‰ and a $\delta^{18}O\text{-}CO_{2,VPDB}$ range from −1.9 to −5.0 ‰ as determined by the Heidelberg IRMS (Neubert, 1998).

Step 4: Smoothed working standard correction

We have found that regular measurements of different cylinder gases on the FTIR analyser show small but correlated sub-weekly variations of $\delta^{18}O\text{-}CO_2$. One can thus use a smoothed working standard correction in order to account for these small instrumental variations on a sub-weekly time scale. For this purpose, we smooth daily working gas measurements using a 10-point moving average and interpolate the residual variation to the date of sample measurement using a cubic spline interpolation. We then subtract the smoothed residual variations from the long-term mean value of this gas vs. the reference standards from all sample measurements. By performing this correction, typically less than 0.2 ‰, the standard deviation of a weekly measured target or surveillance gas reaches about 0.2 ‰ for $\delta^{18}O\text{-}CO_2$. Step 4 is not obligatory, but further increases the precision of the measurement. In the data presented in Sect. 4, we have applied this smoothed working standard correction.

Figure 3 illustrates the application for the entire calibration procedure. For $\delta^{13}C\text{-}CO_2$, Fig. 3 shows raw (a), corrected (b) and calibrated (c) FTIR measurements against IRMS reference values of $\delta^{13}C\text{-}CO_2$, and Fig. 3d shows the difference between calibrated FTIR measurements and IRMS values (FTIR-IRMS) against CO_2 mole fractions. Figure 3e–h show corresponding data for $\delta^{18}O\text{-}CO_2$. The cross-sensitivity correction forces $\delta^{13}C\text{-}CO_2$ and $\delta^{18}O\text{-}CO_2$ onto a linear regression line (Fig. 3b and f), so that we can then apply a linear calibration. The large correction for $\delta^{18}O\text{-}CO_2$ is most likely related to the systematic residual in the fitting of the spectra (Fig. 1a).

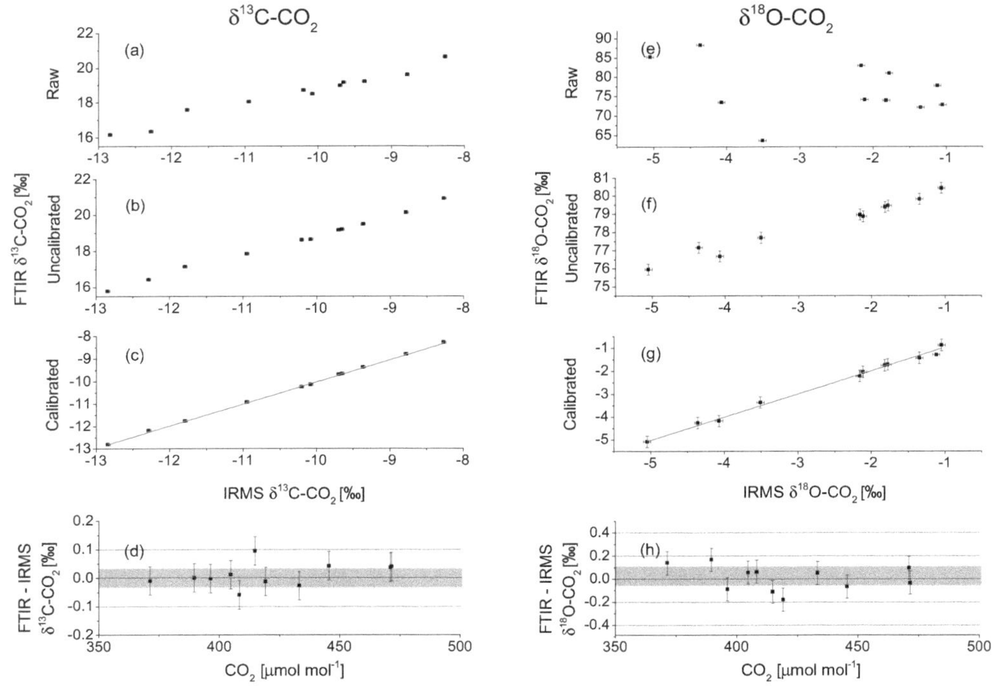

Figure 3. **(a)** Raw, **(b)** cross- and interspecies corrected (but still un-calibrated) and **(c)** calibrated δ^{13}C-CO_2 measurements and **(e)** raw, **(f)** cross- and interspecies corrected (but still un-calibrated); **(g)** calibrated δ^{18}O-CO_2 measurements of different target cylinders against the IRMS measurement of the same cylinders. Lowest panels: **(d)** calibrated FTIR δ^{13}C-CO_2 value minus reference value measured by the Heidelberg IRMS, **(h)** same as **(d)** for δ^{18}O-CO_2, both plotted versus the CO_2 mole fraction of the samples. The red lines in the lowest panels give the mean difference between the FTIR and the IRMS measurements. Grey areas illustrate the standard deviation of the differences.

2.2 Remarks on the calibration procedure

2.2.1 Using total CO_2 instead of $^{12}C^{16}O_2$ to calculate δ^{13}C-CO_2 and δ^{18}O-CO_2

As pointed out above, the precision of total CO_2 measurement in the 3600 cm^{-1} range is significantly higher ($\sim 50\,\%$) than that of $^{12}C^{16}O_2$ in the region of 2300 cm^{-1}, due to an optimum absorption strength and a lower temperature sensitivity. $^{12}C^{16}O_2$, $^{13}C^{16}O_2$ as well as $^{12}C^{16}O^{18}O$ absorb in this region, but the minor isotopologue absorptions are weak and are barely distinguishable. Thus, we calculate the raw δ^{13}C-CO_2 and δ^{18}O-CO_2 values using total CO_2 from the 3600 cm^{-1} region instead of $^{12}C^{16}O_2$. There is a small bias between measurements of CO_2 and $^{12}C^{16}O_2$, but as long as the isotopic composition of the sample is close to the isotopic composition of the reference standards, the bias in δ^{13}C-CO_2 and δ^{18}O-CO_2 is negligible ($< 0.03\,‰$ for δ^{13}C-CO_2 and $< 0.05\,‰$ for δ^{18}O-CO_2) after calibration (step 3). However, for strongly depleted cylinder gases, as may be the case for synthetic gas mixtures, the biases may become as large as 0.2 ‰. If necessary, the bias introduced by total CO_2 can be corrected iteratively using Eqs. (8) and (9) of Griffith et al. (2012):

$$^{12}C^{16}O_2 = \frac{CO_2}{X} \tag{4}$$

where X is an isotopic partition sum with a value very close to unity.

2.2.2 Direct isotopologue calibration

Griffith et al. (2012) described two methods for calibration of isotopic fractionations, either

a. the isotopologue amounts are calibrated independently and the isotopologue δ values calculated directly from the calibrated isotopologue amounts, or

b. the isotopologue δ values are calculated from raw measurements of the isotopologues and the calibration is carried through on the δ-values.

These methods were referred to as "absolute" and "empirical" calibration respectively by Griffith et al. (2012), but to avoid ambiguity we will refer to them here as (a) isotopologue calibration and (b) ratio or δ-calibration. The correction and calibration method described above and used in this work is the ratio calibration, (b). In principle, it is equally valid to use (direct) isotopologue calibration. In this case, we correct the isotopologue amounts $^{16}O^{12}C^{16}O$, $^{16}O^{13}C^{16}O$ and $^{16}O^{12}C^{18}O$ (step 2), calibrate them individually (step 3) and finally compute $\delta^{18}O$-CO_2 and $\delta^{13}C$-CO_2 from the calibrated amounts, i.e.

$$\delta^{18}O - CO_{2,\mathrm{VPDB}} = \left[\frac{\left(\frac{^{12}C^{18}O^{16}O}{^{12}C^{16}O^{16}O}\right)_{\mathrm{sample}}}{\left(\frac{^{12}C^{18}O^{16}O}{^{12}C^{16}O^{16}O}\right)_{\mathrm{VPDB-CO_2}}} - 1 \right] \cdot 1000\,‰ \quad (5)$$

with $(^{12}C^{18}O^{16}O/^{12}C^{16}O_2)_{\mathrm{VPDB-CO_2}} = 0.0041767$ (Allison et al., 1995), which takes into account that CO_2 contains two oxygen atoms.

In principle, both methods should lead to the same results, but they are sensitive to errors in different ways (Griffith et al., 2012). In practice, we find they differ by about 0.11 ± 0.03 ‰ for $\delta^{13}C$-CO_2 and by 0.08 ± 0.15 ‰ for $\delta^{18}O$-CO_2 (mean $\pm$ standard deviation for a 2-month period in 2014). The discrepancy between both calibration methods is most likely due to small inaccuracies in interspecies interference corrections. The ratio calibration requires a large CO_2-interspecies interference correction over a large CO_2 range (see Fig. 2c). Only if the CO_2 interspecies interference correction is well determined can we obtain a reliable $\delta^{18}O$-CO_2 value from the ratio method. For the independent isotopologue calibration, no explicit interspecies CO_2 correction is required, but a very accurate determination of all CO_2 isotopologue calibration equations is vital. The decision on which method to use should thus be based on which correction can be performed with higher accuracy. In this work, we have found for the Heidelberg spectrometer that the empirical calibration method better fits the Heidelberg IRMS values.

2.3 Direct cylinder comparison to mass spectrometric values

In order to check the FTIR calibration as well as the compatibility of the FTIR and the Heidelberg IRMS Finnigan MAT 252, we analysed measurements of different test cylinders in March and April 2014 on both instruments. The IRMS values are linked to the VPDB scale via three pure CO_2 reference gases (RM8562, RM8563 and RM8564). The FTIR reference cylinders were calibrated by the IRMS and thus the FTIR and the IRMS are on the same scale. For all cylinder measurements with the IRMS, we filled cylinder air into evacuated flasks from an intermediate transfer volume; we then analysed these flasks by both techniques like regular flask samples, since pressure regulator effects have often disturbed the IRMS analyses. The precision of the IRMS is about 0.02–0.03 ‰ for $\delta^{13}C$-CO_2 and 0.05–0.1 ‰ for $\delta^{18}O$-CO_2 (standard deviation of repeated flask measurements). Further, Wendeberg et al. (2013) have shown that the Heidelberg IRMS scale does not exhibit any significant scale contraction errors or errors through cross contamination between sample and standard measurements in the IRMS. For more details on the IRMS, see Neubert (1998). A two-sample t test reveals that, at the 0.01 significance level, the means of the FTIR and the IRMS measurements (Fig. 3d and h) for $\delta^{13}C$-CO_2 and for $\delta^{18}O$-CO_2 do not differ significantly and thus, are compatible.

3 Characterisation of $\delta^{18}O$-CO_2 and $\delta^{13}C$-CO_2 measurements with the Heidelberg FTIR

3.1 Allan deviation

We performed an Allan deviation repeatability test (Werle et al., 1993; Werle et al., 2011) on the FTIR system over 6 days from 17 September 2011 to 23 September 2011, with flowing sample supplied from a reference gas cylinder with a $\delta^{13}C$-CO_2 value of about -10.1 ‰ and a $\delta^{18}O$-CO_2 value of about -3.7 ‰. We used the Allan deviation as a measure for the repeatability (following JCGM, 2008) as shown in Fig. 4. Allan deviation is the standard deviation of the pairwise differences between adjacent measurements averaged over different averaging periods. In the absence of drift and with only white (random) noise, the Allan deviation will decrease with the square root of the averaging time. We found that the Allan deviations after 10 min were $\delta^{13}C$-$CO_2 = \pm 0.03$ ‰ and $\delta^{18}O$-$CO_2 = \pm 0.25$ ‰ (Fig. 4a and b). After 30 min, the Allan deviations decrease to $\delta^{13}C$-$CO_2 = \pm 0.02$ ‰ and $\delta^{18}O$-$CO_2 = \pm 0.15$ ‰. From 2 hours to up to 1 day, the Allan deviations stayed below about $\delta^{13}C$-$CO_2 = \pm 0.02$ ‰ and $\delta^{18}O$-$CO_2 = \pm 0.10$ ‰. No significant increase in Allan deviation could be observed within 1 day, since drifts on this time scale are small compared to the noise. Further, we can confirm that the frequency of smoothed working standard correction is adequate, since between daily working standard gas measurements the system remains stable within 0.02 ‰ for $\delta^{13}C$-CO_2 and 0.10 ‰ for $\delta^{18}O$-CO_2. In Heidelberg, a typical diurnal variation of $\delta^{18}O$-CO_2 is of the order of 1 ‰ (see Sect. 4). Thus, the system is stable enough to resolve diurnal ambient $\delta^{18}O$-CO_2 variations (see Sect. 4).

3.2 Intermediate measurement precision

We monitored the intermediate measurement precision (following JCGM, 2008) by measuring standard gases every day or week under reproducible conditions. The averaging time for each cylinder measurement was 9 min. We used the standard deviation of the 9 min cylinder gas averages to estimate the intermediate measurement precision of our instrumental set-up. For $\delta^{13}C$-CO_2 and $\delta^{18}O$-CO_2, we found that the intermediate measurement precision was 0.04 ‰ and 0.27 ‰, respectively for the period from December 2012 to October 2013 (see Fig. 5a and b). The Allan deviation at 9 min is very close to the standard deviation of the daily working cylinder measurements, which shows that for our system and laboratory conditions the repeatability dominates the intermediate measurement precision.

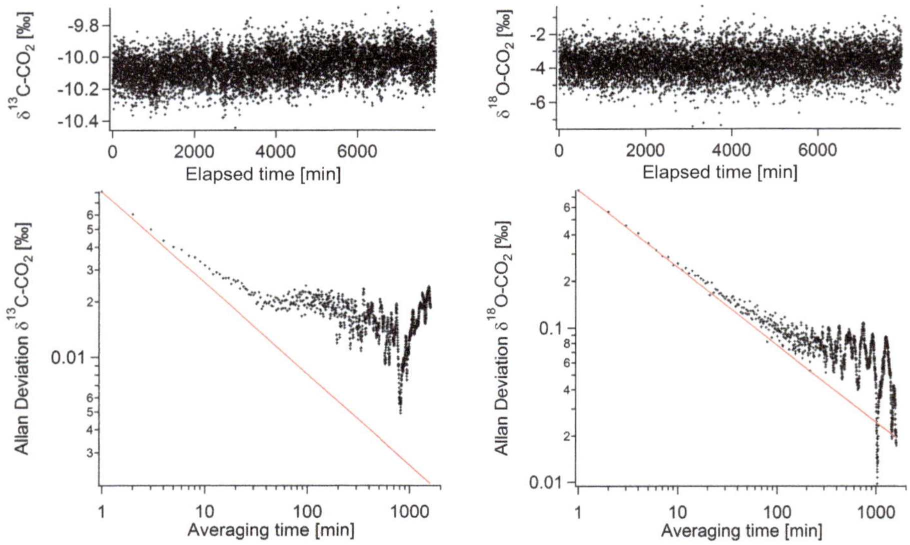

Figure 4. Allan deviation of $\delta^{13}C\text{-}CO_2$ (left) and $\delta^{18}O\text{-}CO_2$ (right) measured over the course of 6 days in September 2011 with the FTIR.

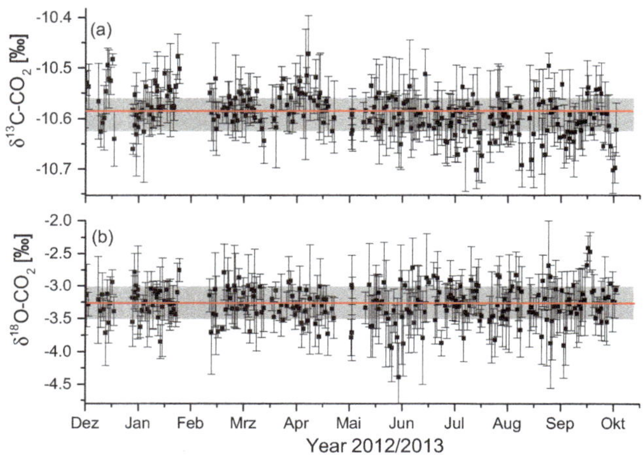

Figure 5. Repeated daily working gas measurements (9 min averages) depict an intermediate measurement precision of **(a)** ± 0.04 ‰ for $\delta^{13}C\text{-}CO_2$ and of **(b)** ± 0.27 ‰ for $\delta^{18}O\text{-}CO_2$ **(b)** for the period from December 2012 to October 2013. Red lines: mean values, grey areas: standard deviation.

Note that in our calibration procedure we now use the daily measured cylinder (working standard gas) in a final correction step (step 4) to account for sub-weekly variations of the instrument response. Since we only recognised the need to correct for this variability well after commencement of the measurements, we do not yet have a long-term record for a real surveillance cylinder. Therefore, Fig. 5 displays the working standard measurements without any sub-weekly smoothing applied, and thus gives an upper estimate of the intermediate measurement precision of real measurements where we apply step 4 of our calibration procedure in addition.

3.3 Compatibility of ambient air measurements

In the previous sections, we have evaluated the repeatability, as well as the intermediate measurement precision of the FTIR measurements. The results make us confident that the FTIR spectrometer is of sufficient precision and stability to resolve atmospheric signals, such as the diurnal variation of $\delta^{13}C\text{-}CO_2$ and $\delta^{18}O\text{-}CO_2$. Further, we have shown in Sect. 2.3 that the FTIR cylinder gas measurements are compatible to those of the Heidelberg IRMS. In order to show that not only the direct cylinder measurements, but also the ambient air measurements are compatible with the IRMS analyses, we compared in situ ambient air samples, which we measured with both instruments.

For this purpose, an automated flask sampler (Neubert et al., 2004) collected dried (dew point $-40\,°C$) ambient air from the same intake line as the FTIR into 2.5 litre glass flasks. Every flask was flushed with a flow rate of about 1.1 SLPM for 2 h and then pressurised to 2000 hPa absolute pressure and closed. Then the automated flask sampler opened, flushed and filled the next flask to 2000 hPa. Pressurising the flasks took about 5 min. With this procedure, we were able to capture a diurnal isotopic profile with a 2-hourly resolution in the flasks, which could be analysed by mass spectrometry. We then compared these values to the continuous values measured by the FTIR spectrometer; the results are shown in Fig. 6. We used 9 min averaged values from the FTIR spectrometer to compare them to the flask results to account for atmospheric variability and to minimise differences due to lack of temporal synchronisation between the event sampler and the FTIR, and to reduce the noise on the FTIR measurement. We found that the mean residual and standard error is 0.01 ± 0.02 ‰ for $\delta^{13}C\text{-}CO_2$ and 0.08 ± 0.14 ‰ for $\delta^{18}O$-

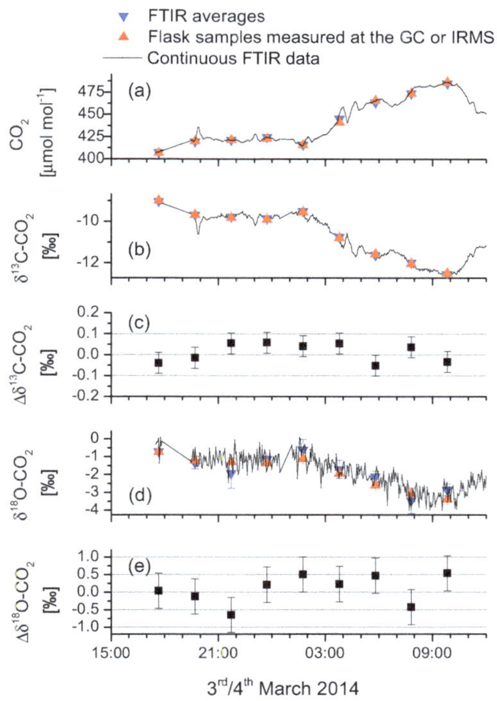

Figure 6. Diurnal cycle event sampled on the 3–4 March 2014 at the Institut für Umweltphysik in Heidelberg. Red: GC concentration (in case of CO_2) or IRMS isotopologue value (in case of isotopologues) of flasks samples; blue: 9 min averaged values from FTIR; black: continuous 3 min values from the FTIR. **(a)** CO_2 mole fraction; **(b)** $\delta^{13}C$-CO_2 value; **(c)** residual of 9 min average $\delta^{13}C$-CO_2 FTIR and IRMS measurement (FTIR - IRMS); **(d)** $\delta^{18}O$-CO_2 value; **(e)** residual of 9 min averaged $\delta^{18}O$-CO_2 FTIR and IRMS measurement (FTIR – IRMS). All error bars on the (blue) averaged FTIR data are the standard deviation during the 9 min of averaging time. The error bars on the (red) IRMS values show the typical intermediate measurement precision of our IRMS measurements. The residual (FTIR-IRMS) has an error bar, which combines the IRMS uncertainty and the FTIR uncertainty and the variability of atmospheric signal during the flask filling time.

CO_2 (FTIR – IRMS). We tested the compatibility between the FTIR and the IRMS ambient air measurements with a two-sample t test and found that at the 0.01 significance level, the means of the FTIR and the IRMS measurements in ambient air do not differ from each other for $\delta^{13}C$-CO_2 or for $\delta^{18}O$-CO_2. Note, that the standard deviation of the differences between the FTIR and the IRMS is 0.05 ‰ for $\delta^{13}C$-CO_2 and 0.42 ‰ for $\delta^{18}O$-CO_2 and with that the standard deviation for $\delta^{18}O$-CO_2 differences is higher than expected from the combined Allan deviation (0.25 ‰ for 9 min averages) and the uncertainty of the IRMS measurement (ca. 0.05–0.1 ‰).

The slightly larger variability in $\delta^{18}O$-CO_2 ambient air comparison than in cylinder gas comparisons (Sect. 2.3) reflects the fact that there are more contributions to the difference between the FTIR and the IRMS flask measurement. There are the storage effect of the flasks themselves, which could be slightly wet and thus alter the $\delta^{18}O$ value of the CO_2 in the flask, or some other possible interference of the automated flask sampler (i.e. varying integration time due to flow and pressure variations).

4 Example period of continuous trace gas and stable isotopologue measurements in Heidelberg

In this section, we illustrate how we might potentially use a highly resolved $\delta^{18}O$-CO_2 record at a typical European monitoring station, such as Heidelberg, in order to disentangle regional scale carbon exchange processes. Note, however, that for a quantitative evaluation, we would require explicit information on local CO_2 source signatures and on the exchanging water reservoirs. We look here at two very different periods in which the FTIR measured $\delta^{18}O$-CO_2 along with $\delta^{13}C$-CO_2, total CO_2 and CO in Heidelberg (see Fig. 7).

In order to interpret the atmospheric $\delta^{18}O$-CO_2 variation, we must estimate the isotopic signature or discrimination of the processes influencing the isotopic content. The Heidelberg catchment area is typical of many European urban areas with the most important CO_2 fluxes associated with plant photosynthesis, leaf and soil respiration, as well as fossil fuel burning. In the greater catchment area, discrimination during photosynthesis tends to enrich atmospheric CO_2 with respect to ^{13}C and ^{18}O (Cuntz et al., 2003b). Typical mean $\delta^{13}C$ fractionation relative to the atmosphere during photosynthesis is about – (2–8) ‰ for C4 plants and about – (12–20) ‰ for C3 plants (Mook, 1994). As a first approximation, the $^{13}CO_2$ / $^{12}CO_2$ ratio captured during photosynthesis is released during respiration, which leads to an overall depletion of the atmospheric $^{13}CO_2$ / $^{12}CO_2$ ratio. In addition, ^{18}O discrimination during respiration tends to deplete the atmosphere in its $\delta^{18}O$-CO_2 value. Neubert (1998) measured the isotopic composition of soil-respired CO_2 in the surroundings of Heidelberg and found values of $\delta^{18}O$-$CO_{2,VPDB} \approx -10$ ‰ with a tendency of slightly more depleted values in winter (−15 ‰) than in summer (−5 ‰) and $\delta^{13}C$-$CO_{2,VPDB} \approx -25$ ‰. For the discrimination during photosynthesis, typical mean values for the central European continent are between 0 and +20 ‰ for ^{18}O (Farquhar et al., 1993; Cuntz et al., 2003b). Further, the invasion flux will influence the apparent soil respiration signature (Tans, 1998; Miller et al., 1999), but we cannot quantify the magnitude of this effect for our catchment area without intensive sampling and isotopic soil flux modelling. Therefore, we only consider the invasion flux in a sense that a larger range must be attributed to the signature of the apparent soil respiration flux when qualitatively discussing our atmospheric $\delta^{18}O$-CO_2 records here.

For the isotopic signature of fossil fuels, most studies assume a common $\delta^{18}O$ fossil fuel signature of $\delta^{18}O$-$CO_{2,VPDB} \approx -17$ ‰, corresponding to the ambient oxy-

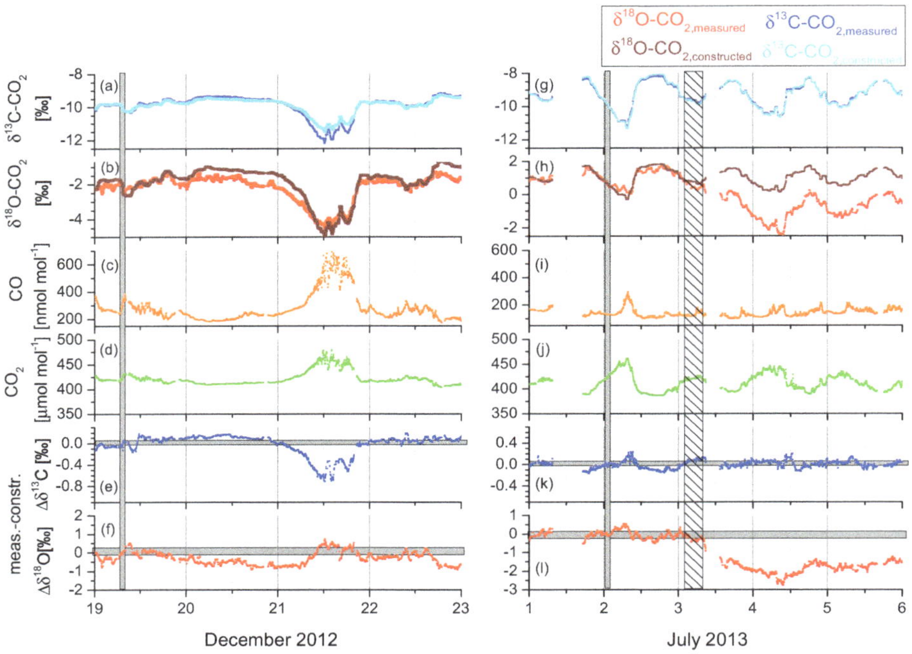

Figure 7. Trace gas records in winter (left panel) and summer (right panel) in Heidelberg. **(a)** and **(g)** show the measured (dark blue) and artificially constructed (light blue) $\delta^{13}C\text{-}CO_2$ value, **(b)** and **(h)** the measured (red) and artificially constructed (burgundy) $\delta^{18}O\text{-}CO_2$ value, **(c)** and **(i)** the measured CO value, **(d)** and **(j)** the measured CO_2 value. Panels **(e)** and **(k)** give the difference between the measured and constructed $\delta^{13}C\text{-}CO_2$ value with a mean isotopic source signature of $\delta^{13}C\text{-}CO_{2,VPDB} \approx -25$ ‰ in the wintertime and $\delta^{13}C\text{-}CO_{2,VPDB} \approx -27$ ‰ in the summertime. Panels **(f)** and **(l)** give the difference between the measured and constructed $\delta^{18}O\text{-}CO_2$ value with a mean isotopic signature of $\delta^{18}O\text{-}CO_{2,VPDB} \approx -28$ ‰ in the wintertime and $\delta^{18}O\text{-}CO_{2,VPDB} \approx -12$ ‰ in the summertime. Grey vertical bars indicate the "reference periods", in which the isotopic source signature for artificially constructed $\delta^{13}C\text{-}CO_2$ and $\delta^{18}O\text{-}CO_2$ was determined from Keeling plots of about 20 individual atmospheric 3 min average measurements. The dashed vertical bar in the right panel shows a period of high precipitation. Grey horizontal bars in **(f)** and **(l)** mark the 1σ-uncertainty of the isotope measurements.

gen isotopic signature, but incomplete combustion can lead to a range of different isotopic signatures. The ^{18}O signature of fossil fuel emissions varies from about $\delta^{18}O\text{-}CO_{2,VPDB} \approx -11$ to -40 ‰ (Schumacher et al., 2011). Traffic exhausts tend to be less depleted in ^{18}O relative to other fossil fuel CO_2 emissions ($\delta^{18}O\text{-}CO_{2,VPDB} \approx -15$ ‰), followed by natural gas burning ($\delta^{18}O\text{-}CO_{2,VPDB} \approx -28$ ‰). Combustion of coal, on the other hand, leads to a $\delta^{18}O\text{-}CO_2$ value of about -38 ‰ (Schumacher et al., 2011). To our knowledge, the potential range of these values is not well known. For $\delta^{13}C$, typical signatures are $\delta^{13}C\text{-}CO_{2,VPDB} \approx -29$ ‰ for traffic exhausts, -25 ‰ for coal combustion and -39 ‰ for natural gas emissions (Widory and Javoy, 2003; Kaul, 2007). With these examples of isotopic signatures, we can now look at our atmospheric CO_2 records that show values of $\delta^{13}C\text{-}CO_2$ between $\delta^{13}C\text{-}CO_{2,VPDB} \approx -8$ and -12 ‰, while $\delta^{18}O\text{-}CO_2$ varies between $\delta^{18}O\text{-}CO_{2,VPDB} \approx -2$ to -4 ‰ in winter and 0 to -2 ‰ in summer (Fig. 7).

Since all CO_2 sources with a negative isotopic signature relative to atmospheric CO_2 lead to $\delta^{13}C\text{-}CO_2$ or $\delta^{18}O\text{-}CO_2$ depletion, a differentiation between different depleting sources is difficult. Therefore we used the following approach: We first constructed an artificial $\delta^{13}C\text{-}CO_2$ and $\delta^{18}O\text{-}CO_2$ record using the slope (a_{ref}) and offset (b_{ref}) of so-called "Keeling plots" (Keeling, 1958), determined from measured atmospheric δ- and CO_2 concentration values in an exemplary and short nighttime reference period (grey bars in Fig. 7) according to:

$$\delta_{meas} = a_{ref} \cdot \frac{1}{CO_2} + b_{ref} \tag{6}$$

Note that in the nighttime reference periods, for which the reference slope and offsets were calculated, we can neglect photosynthetic sinks. Therefore, we can interpret the $\delta^{13}C$ source signature of the reference period as the flux-weighted average of all sources (Miller and Tans, 2003). We then applied the parameters (a_{ref} and b_{ref}) from the reference period to the entire CO_2 record to calculate artificially constructed $\delta^{13}C_{constr}$ and $\delta^{18}O_{constr}$ values:

$$\delta_{constr(t)} = a_{ref} \cdot \frac{1}{CO_2(t)} + b_{ref}, \tag{7}$$

Fig. 7a, b, g and h show the constructed $\delta^{13}C\text{-}CO_2$ and $\delta^{18}O\text{-}CO_2$ records in burgundy and light blue. During the reference period in which the Keeling plot slopes and offsets were derived, the Keeling plot had a high correlation coefficient ($r^2 > 0.85$) and showed an isotopic ^{13}C and ^{18}O source signature that was typical for the respective season (source $\delta^{13}C$-$CO_2 \approx -25$ ‰ in the winter and -27 ‰ in the summer period, $\delta^{18}O$-$CO_2 \approx -28$ ‰ in the winter and -12 ‰ in the summer period). To identify influences from enriching or depleting sources and sinks relative to those in the reference period, we then calculated the difference between the measured and the artificially constructed (Eq. 6) $\delta^{13}C$-CO_2 and $\delta^{18}O$-CO_2 record (Fig. 7e, k, f and l):

$$\Delta\delta(t) = \delta_{meas}(t) - \delta_{constr}(t). \quad (8)$$

Negative $\Delta\delta$ values occur in periods when the apparent sources are more depleted than in the reference period and positive values occur when apparent sources are more enriched than in the reference period. During photosynthetic CO_2 uptake, the equilibration of back-diffusing CO_2 with enriched leaf water leads to an enrichment of atmospheric $\delta^{18}O$-CO_2 and thus to positive $\Delta\delta^{18}O$ values. We now have a tool that allows differentiation between more and less depleted fluxes relative to the reference period.

In the wintertime, relative fossil fuel contributions in the Heidelberg catchment area are higher than in the summertime (Levin et al., 2003). Fossil fuel CO_2 emissions lead to high concentration of CO_2 (Fig. 7d) and deplete atmospheric CO_2 in its heavy isotopes ^{13}C and ^{18}O (original measurements: dark blue and red in Fig. 7a, b). During incomplete combustion of fossil fuels, CO (Fig. 7c) is often emitted as well. A typical example of a pollution event is shown in Fig. 7 (left panel) on 21 December 2012. The difference between the measured and artificially constructed $\delta^{13}C$-CO_2 (Fig. 7e) decreases rapidly on 21 December. Environmental parameters such as relative humidity, global radiation and temperature (not shown here) remain constant during the event, but low wind speed leads to an atmospheric inversion and, accompanied with a slight change of wind direction, to a more local source (mix), which is more depleted in ^{13}C than during the reference period ($\delta^{13}C_{ref} = -25$ ‰). The strong influence of a more ^{13}C depleted source mix points towards a high contribution from fossil fuel sources, including domestic heating (natural gas). At the same time, the isotopic signature of $\delta^{18}O$-CO_2 is very close to the isotopic signature during the reference period (-28 ‰) and increases during the pollution event. The different behaviour of $\delta^{13}C$ and $\delta^{18}O$ in CO_2 points towards a larger influence from traffic or natural gas combustion, as both sources are slightly more enriched in ^{18}O, but less enriched in ^{13}C with respect to coal-fired combustion (Schumacher et al., 2011). One can see that the fact that different fossil fuel types influence both stable isotopes ^{13}C and ^{18}O in CO_2 in a different way can potentially be used to differentiate between different emission groups in situations when biogenic fluxes are low (i.e. in winter). However, for a quantitative analysis we must know the exact isotopic signatures of all fluxes in the area of influences.

In the summertime, we expect biosphere fluxes to be much larger than during winter and at the same time fossil fuel (especially residential heating) emissions to be smaller than in winter. In fact, we do not find large deviations in $\delta^{13}CO_2$ from those determined in the reference period (-27 ‰), pointing towards a relatively constant mixture of biogenic and fossil fuel emissions. On the other hand, the measured $\delta^{18}O$-CO_2 decreased rapidly on 3 July, compared to the reference period with a source isotopic signature of ≈ -12 ‰. This decrease is not accompanied by changes of any other tracer, such as CO, $\delta^{13}CO_2$ or CO_2, and also not by drastic changes of environmental parameters such as relative humidity, temperature or wind speed (not shown here). A possible explanation for the decrease is a change in the hydrological conditions. After 4 dry days, a sudden heavy rain occurred in Heidelberg on 3 July (see dashed bar in Fig. 7, right panel). The rainfall replenished the water reservoirs with an ^{18}O-depleted signature (Daansgard et al., 1964) and equilibration between the soil and leaf water reservoirs and CO_2 most probably caused the atmospheric $\delta^{18}O$-CO_2 to become depleted relative to the reference period. This example illustrates the close coupling between $\delta^{18}O$ in the water and carbon cycle. It is thus crucial to study also the hydrological conditions, such as precipitation and its isotopic signature, in order to quantitatively use the $\delta^{18}O$-CO_2 records for carbon cycle research.

5 Discussion

5.1 Instrumental performance

The main scope of this work was to ascertain whether the FTIR analyser is capable of measuring $\delta^{18}O$-CO_2 in the atmosphere and if so, to assess how well it performs. We have seen that the FTIR succeeds in measuring atmospheric $\delta^{18}O$-CO_2 with a high repeatability (Allan deviation after 1 day: 0.1 ‰) and good intermediate measurement precision ($\delta^{18}O$-$CO_2 = 0.27$ ‰ for daily repeated working standard gas measurements on 9 min averages over 10 months). We were also able to confirm a good compatibility to the IRMS. Some optimisation, concerning the calibration, the fitted spectral windows and the theoretical spectrum modelling could lead to improved results. However, the current performance of the spectrometer suffices to quantify typical diurnal and synoptic variations at an urban site, which is an important step towards quantification of gross biospheric fluxes using FTIR-based $\delta^{18}O$-CO_2 measurements.

5.2 Quantitative interpretation of continuous $\delta^{13}C$-CO_2 and $\delta^{18}O$-CO_2 record

We further investigated which insight may be gained from continuous isotopologue records at an urban site. For $\delta^{13}C$, the different carbon sources and sinks are relatively well understood, but for $\delta^{18}O$, high temporal variability of the respiratory and photosynthetic fluxes (due to a strong variation of environmental parameters such as precipitation, temperature and humidity) makes it difficult to separate the different CO_2 fluxes. For our qualitative study, we could use observations from Neubert (1998) in the catchment area of Heidelberg, as well as globally resolved model data for assimilation isofluxes from Cuntz et al. (2003b). However, for a quantitative apportionment of the CO_2 fluxes at a high temporal resolution, sampling of the isotopic content of precipitation, soil respiration and foliage gas exchange in the catchment area will be necessary with similarly high temporal resolution (Stern et al., 1999; Langendörfer et al., 2002). Further, isotope soil-atmosphere flux models are required to quantify the effect of this process at the measurement site. All of these unknowns largely limit current applicability of our new continuous isotope measurements. Future sophisticated regional models of the water and the carbon cycle may, however, be able to fully exploit the wealth of new information now available.

6 Summary and conclusion

The analysis of $\delta^{18}O$ in CO_2 using FTIR spectroscopy is novel. We evaluated the measurements of ^{18}O in CO_2 using the FTIR with respect to repeatability, intermediate measurement precision and compatibility. The Allan deviation test showed that the instrument measures $\delta^{18}O$-CO_2 with good stability over the course of a day (the frequency of the working standard measurement) to within 0.1 ‰. Averages of 9 min show a standard deviation of about 0.25 ‰, which is in agreement with the intermediate measurement precision based on daily working standard gas measurements.

Evaluation of diurnal ambient air variations is therefore possible using, for example, 30 min averages. The high temporal resolution of the FTIR measurement is a major advantage over the IRMS analyses. Even though the FTIR precision does not reach the WMO inter-laboratory compatibility targets (WMO, 2012), a number of interesting scientific applications seem possible using FTIR spectroscopy. In particular, investigation of the processes that govern the $\delta^{18}O$-CO_2 variability of atmospheric CO_2 on the regional scale seem very promising if comprehensive knowledge on the isotopic signature of different CO_2 sources and sinks, as well as of the influencing water reservoirs, is available.

Acknowledgements. We thank Geoff Toon, Frank Hase and Joseph Mendonca for valuable discussions on molecular line shapes spectra and for fitting our spectra with different line shape models. Four anonymous reviewers are gratefully acknowledged for their valuable comments, which helped to improve our manuscript. This work has been funded by the InGOS EU project (284274). We further acknowledge the financial support given by Deutsche Forschungsgemeinschaft and Ruprecht-Karls-Universität Heidelberg within the funding program Open Access Publishing.

Edited by: G. Phillips

References

Allison, C., Francey, R., and Meijer, H.: Recommendations for the reporting of stable isotope measurements for carbon and oxygen in CO_2 gas, Reference and intercomparison materials for stable isotopes of light element, IAEA-TECDO, 155–162, 1995.

Bowling, D. R., McDowell, N. G., Welker, J. M., Bind, B. J., Law, B.E., and Ehleringer, J. R.: Oxygen isotope content of CO_2 in nocturnal ecosystem respiration: 1. Observations in forests along a precipitation transect in Oregon, USA, Global Biogeochem. Cy., 17, 1120, doi:10.1029/2003GB002081, 2003.

Buenning, N., Noone, D., Randerson, J., Riley, W. J., and Still, C.: The response of the $^{18}O/^{16}O$ composition of atmospheric CO_2 to changes in environmental conditions, J. Geophys. Res.-Biogeosci., 119, 55–79, doi:10.1002/2013JG002312, 2014.

Cambaliza, M. O. L.: Measurement of forest ecosystem-atmosphere exchange of $\delta^{13}C$-CO_2 using Fourier Transform Infrared spectroscopy and disjunct eddy covariance, Dissertation, Washington State University, Department of Civil and Environmental Engineering, USA, 2010.

Ciais, P., Tans, P. P., White, J. W. C., Trolier, M., Francey, R. J., Berry, J. A., Randall, D. R., Sellers, P. J., Collatz, J. G., and Schimel, D. S.: Partitioning of ocean and land uptake of CO_2 as inferred by $\delta^{13}C$ measurements from the NOAA Climate Monitoring and Diagnostics Laboratory Global Air Sampling Network, J. Geophys. Res., 100, 5051–5070, doi:10.1029/94JD02847, 1995.

Ciais, P., Denning, A. S., Tans, P. P., and Berry, J. A.: A three-dimensional synthesis study of $\delta^{18}O$ in atmospheric CO_2 1.surface fluxes, J. Geophys. Res., 102, 5857–5872, doi:10.1029/96JD02360, 1997.

Coplen, T. B.: Guidelines and recommended terms for expression of stable-isotope-ratio and gas-ratio measurement results. Rapid Commun. Mass Spectrom., 25, 2538–2560, doi:10.1002/rcm.5129, 2011.

Cuntz, M., Ciais, P., Hoffmann, G., and Knorr, W.: A comprehensive global three-dimensional model of $\delta^{18}O$ in atmospheric CO_2: 1. Validation of surface processes, J. Geophys. Res., 108, 4527, doi:10.1029/2002jd003153, 2003a.

Cuntz, M., Ciais, P., Hoffmann, G., Allison, C. E., Francey, R. J., Knorr, W., Tans, P. P., White, J. W. C., and Levin, I.: A comprehensive global three-dimensional model of $\delta^{18}O$ in atmospheric CO_2: 2. Mapping the atmospheric signal, J. Geophys. Res., 108, 4528, doi:10.1029/2002JD003154, 2003b.

Daansgard, W.: Stable isotopes in precipitation, Tellus, 16, 436–468, doi:10.1111/j.2153-3490.1964.tb00181.x, 1964.

Esler, M. B., Griffith, D. W. T, Wilson, S. R., and Steele, L. P.: Precision trace gas analysis by FT-IR Spectroscopy. 2. The $^{13}C/^{12}C$ Isotope Ratio of CO_2, Anal. Chem, 72, 216–221, 2000.

Farquhar, G. D., Lloyd, J., Taylor, J. A., Flanagan, L. B., Syvertsen, J. P., Hubick, K. T., Wong, S. C., and Ehleringer, J. R.: Vegetation effects on the isotope composition of oxygen in atmospheric CO_2, Nature, 363, 439–443, doi:10.1038/363439a0, 1993.

Flanagan, L. B., Brooks, J. R., Varney, G. T., and Ehleringer, J. R.: Discrimination against $C^{18}O^{16}O$ during photosynthesis and the oxygen isotope ratio respired CO_2 in boreal forest ecosystems, Global Biogeochem. Cy., 11, 83–98, 1997.

Francey, R. J. and Tans, P. P.: Latitudinal variation in oxygen-18 of atmospheric CO_2, Nature, 327, 495–497, doi:10.1038/327495a0, 1987.

Gillon, J. and Yakir, D.: Influence of carbonic anhydrase activity in terrestrial vegetation on the ^{18}O content of atmospheric CO_2, Science, 291, 2584–2587, doi:10.1126/science.1056374, 2001.

Griffith, D. W. T.: Synthetic calibration and quantitative analysis of gas phase infrared spectra, Appl. Spectrosc., 50, 59–70, 1996.

Griffith, D. W. T., Jamie, I. M., and Meyers, R. A. (Eds.): Fourier Transform Infrared Spectrometry in Atmospheric and Trace Gas Analysis, Encyclopedia of Analytical Chemistry, Wiley, Chichester, 2000.

Griffith, D. W. T., Deutscher, N. M., Caldow, C., Kettlewell, G., Riggenbach, M., and Hammer, S.: A Fourier transform infrared trace gas and isotope analyser for atmospheric applications, Atmos. Meas. Tech., 5, 2481–2498, doi:10.5194/amt-5-2481-2012, 2012.

Hammer, S., Griffith, D. W. T., Konrad, G., Vardag, S., Caldow, C., and Levin, I.: Assessment of a multi-species in situ FTIR for precise atmospheric greenhouse gas observations, Atmos. Meas. Tech., 6, 1153-1170, doi:10.5194/amt-6-1153-2013, 2013a.

Hammer, S., Konrad, G., Vermeulen, A. T., Laurent, O., Delmotte, M., Jordan, A., Hazan, L., Conil, S., and Levin, I.: Feasibility study of using a "travelling" CO_2 and CH_4 instrument to validate continuous in situ measurement stations, Atmos. Meas. Tech., 6, 1201–1216, doi:10.5194/amt-6-1201-2013, 2013b.

Hase, F., Drouin, B. J., Roehl, C. M., Toon, G. C., Wennberg, P. O., Wunch, D., Blumenstock, T., Desmet, F., Feist, D. G., Heikkinen, P., De Mazière, M., Rettinger, M., Robinson, J., Schneider, M., Sherlock, V., Sussmann, R., Té, Y., Warneke, T., and Weinzierl, C.: Calibration of sealed HCl cells used for TCCON instrumental line shape monitoring, Atmos. Meas. Tech., 6, 3527–3537, doi:10.5194/amt-6-3527-2013, 2013.

Hesterberg, R. and Siegenthaler, U.: Production and stable isotopic composition of CO_2 in a soil near Bern, Switzerland. Tellus B, 43, 197–205, doi:10.1034/j.1600-0889.1991.00013.x, 1991.

JCGM: Joint Committee for Guides in Metrology: International vocabulary of metrology – basic and general concepts and associated terms (VIM), 3rd Edn., 2008.

Kaul, M.: Isotopenverhältnisse im atmosphärischem Kohlendioxid und seine Quellen im Raum Heidelberg, Staatsexamen thesis, Universität Heidelberg, Heidelberg, Germany, 2007.

Keeling, C. D.: The concentration and isotopic abundance of atmospheric carbon dioxide in rural areas, Geochim. Cosmochim. Ac. 13, 322–334, 1958.

Keeling, C. D., Bacastow, R. B., Carter, A. F., Piper, S. C., Whorf, T. P., Heimann, M., Mook, W. G. and Roeloffzen, H.: A three-dimensional model of atmospheric CO_2 transport based on observed winds: 1. Analysis of observational data, Aspects of Climate Variability in the Pacific and the Western Americas, edited by: Peterson, D. H., Vol. 55, American Geophysical Union, Washington, D. C., 165–236, 1989.

Langendörfer, U., Cuntz, M., Ciais, P., Peylin, P., Bariac, T., Milyukova, I., Kolle, O., Naegler, T., and Levin, I.: Modelling of biospheric CO_2 gross fluxes via oxygen isotopes in a spruce forest canopy: a ^{222}Rn calibrated box model approach. Tellus B, 54, 476–496, doi:10.1034/j.1600-0889.2002.01345.x, 2002.

Levin, I., Kromer, B., Schmidt, M., and Sartorius, H.: A novel approach for independent budgeting of fossil fuel CO_2 over Europe by $^{14}CO_2$ observations, Geophys. Res. Lett., 30, 2194, doi:10.1029/2003GL018477, 2003.

Miller, J. B. and Tans, P. P.: Calculating isotopic fractionation from atmospheric measurements at various scales, Tellus B, 55, 207–214. doi:10.1034/j.1600-0889.2003.00020.x, 2003.

Miller, J. B., Yakir, D., White, J. W. C., and Tans, P. P.: Measurement of $^{18}O/^{16}O$ in the soil-atmosphere CO_2 flux, Global Biogeochem. Cy., 13, 761–774, doi:10.1029/1999GB900028, 1999.

Mohn, J., Zeeman, M. J., Werner, R. A., Eugster, and W., Emmenegger, L.: Continuous field measurements of $\delta^{13}C - CO_2$ and trace gases by FTIR spectroscopy, Isotopes Environ. Health Stud., 44, 241–251, 2008.

Mook, W.: Environmental isotopes in the hydrological cycle – Principles and applications, Introductory course on Isotope Hydrology, Department of Hydrogeology and Geographical Hyrdology, 1994.

Neubert, R.: Measurement of stable isotopomers of atmospheric carbon dioxide, PhD thesis, University of Heidelberg, Heidelberg, Germany, 1998.

Neubert, R. E. M., Spijkervet, L. L., Schut, J. K., Been, H. A., and Meijer, H. A. J.: A computer-controlled continuous air drying and flask sampling system, J. Atmos. Ocean. Technol., 21, 651–659, 2004.

Ngo, N. H., Lisak, D., Tran, H., and Hartmann, J.-M.: An isolated lineshape model to go beyond the Voigt profile in spectroscopic databases and radiative transfer codes. J. Quant. Spectrosc. Ra., 129, 89–100, 2013.

Pataki, D. E., Bowling, D. R., and Ehleringer, J. R.: Seasonal cycle of carbon dioxide and its isotopic composition in an urban atmosphere: Anthropogenic and biogenic effects, J. Geophys. Res., 108, 4735, doi:10.1029/2003JD003865, 2003.

Rothman, L. S., Jacquemart, D., Barbe, A., Benner, D. C., Birk, M., Brown, L. R., Carleer, M. R., C. Chaucerian, J., Chance, K., Dana, V., Devi, V. M., Flaud, J.-M., Gamache, R. R., Goldman, A., Hartmann, J.-M., Jucks, K. W., Maki, A. G., Mandin, J.-Y., Massie, S. T., Orphali, J., Perrin, A., Rinsland, C. P., Smith, M. A. H., Tennyson, J., Tolchenov, R. N., Toth, R. A., Auwera, J. V., Varanasi, P., and Wagner, G.: The HITRAN 2004 molecular spectroscopic database, J. Quant. Spectrosc. Ra., 96, 139–204, doi:10.1016/j.jqsrt.2004.10.008, 2005.

Schumacher, M., Werner, R. A., Meijer, H. A. J., Jansen, H. G., Brand, W. A., Geilmann, H., and Neubert, R. E. M.: Oxygen isotopic signature of CO_2 from combustion processes, Atmos. Chem. Phys., 11, 1473–1490, doi:10.5194/acp-11-1473-2011, 2011.

Seibt, U., Wingate, L., Lloyd, J., and Berry, J. A.: Diurnally variable $\delta^{18}O$ signatures of soil CO_2 fluxes indicate carbonic anhydrase activity in a forest soil, J. Geophys. Res., 111, G04005, doi:10.1029/2006JG000177, 2006.

Stern, L., Baisden, W. T., and Amundson, R.: Processes controlling the oxygen isotopic ratio of soil CO_2: analytical and numerical modelling, Geochim Cosmochim. Ac. 63, 799–814, 1999.

Still, C. J., Riley, W. J., Biraud, S. C., Noone, D. C., Buenning, N. H., Randerson, J. T., Tom, M. S., Welker, J., White, J .W. C., Vachon, R., Farquhar, G. D., and Berry, J. A.: Influence of clouds and diffuse radiation on ecosystem-atmosphere

CO_2 and $CO^{18}O$ exchanges, J. Geophys. Res., 114, G01018, doi:10.1029/2007JG000675, 2009.

Sturm, P., Tuzson, B., Henne, S., and Emmenegger, L.: Tracking isotopic signatures of CO_2 at the high altitude site Jungfraujoch with laser spectroscopy: analytical improvements and representative results, Atmos. Meas. Tech., 6, 1659–1671, doi:10.5194/amt-6-1659-2013, 2013.

Tans, P. P.: Oxygen isotopic equilibrium between carbon dioxide and water in soils. Tellus B, 50, 163–178, doi:10.1034/j.1600-0889.1998.t01-1-00004.x, 1998.

Tans, P. P. and Bolin, B. (Eds.): $^{13}C/^{12}C$ of industrial CO_2, Carbon Cycle Modelling, Scope Ser., 16, 127–129, 1981.

Tuzson, B., Henne, S., Brunner, D., Steinbacher, M., Mohn, J., Buchmann, B., and Emmenegger, L.: Continuous isotopic composition measurements of tropospheric CO_2 at Jungfraujoch (3580 m a.s.l.), Switzerland: real-time observation of regional pollution events, Atmos. Chem. Phys., 11, 1685–1696, doi:10.5194/acp-11-1685-2011, 2011.

Vardag, S. N., Hammer, S., O'Doherty, S., Spain, T. G., Wastine, B., Jordan, A., and Levin, I.: Comparisons of continuous atmospheric CH_4, CO_2 and N_2O measurements – results from a travelling instrument campaign at Mace Head, Atmos. Chem. Phys., 14, 8403–8418, doi:10.5194/acp-14-8403-2014, 2014.

Welp, L. R., Keeling, R. F., Meijer, H. A., Bollenbacher, A. F., Piper, S. C., Yoshimura, K., Francey, R. J., Allison, C. E., and Wahlen, M.: Interannual variability in the oxygen isotopes of atmospheric CO_2 driven by El Niño, Nature, 477, 579–582, 2011.

Wendeberg, M., Richter, J. M., Rothe, M., and Brand, W. A.: Jena Reference Air Set (JRAS): a multi-point scale anchor for isotope measurements of CO_2 in air, Atmos. Meas. Tech., 6, 817–822, doi:10.5194/amt-6-817-2013, 2013.

Werle, P. O., Mücke, R., and Slemr, F.: The limits of signal averaging in atmospheric trace-gas monitoring by tunable diode-laser absorption spectroscopy (TDLAS), Appl. Phys. B, 57.2, 131–139, 1993.

Werle, P.: Accuracy and precision of laser spectrometers for trace gas sensing in the presence of optical fringes and atmospheric turbulence, Appl. Phys. B, 10, 251–253, 2011.

Widory, D. and Javoy, M.: The carbon isotope composition of atmospheric CO_2 in Paris, Earth Planet. Sci. Lett., 215, 289–298, 2003.

Wingate, L., Seibt, U., Maseyk, K., Ogée, J., Almeida, P., Yakir, D., Pereira, J. S., and Menunccini, M.: Evaporation and carbonic anhydrase activity recorded in oxygen isotope signatures of net CO_2 fluxes from a Mediterranean soil, Global Change Biol., 14, 2178–2193, doi:10.1111/j.1365-2486.2008.01635.x, 2008.

Wingate, L., Ogée, J., Cuntz, M., Genty, B., Reiter, I., Seibt, U., Yakir,D., Maesyk, K., Pendall, E. G., Barbour, M. M., Mortazavi, B., Burlett, R., Peylin, P., Miller, J., Mencuccini, M., Shim, J. H., Hunt, J. Grace, J.: The impact of soil microorganisms on the global budget of $\delta^{18}O$ in atmospheric CO_2 Proc. Natl. Acad. Sci., 106, 22411–22415, doi:10.1073/pnas.0905210106, 2009.

WMO: Report of the 16th WMO/IAEA Meeting of Experts on Carbon Dioxide, Other Greenhouse Gases and Related Tracers Measurement Techniques (GGMT-2011): Wellington, New Zealand, 25–28 October 2011, GAW Report No. 206, available at: http://www.wmo.int/pages/prog/arep/gaw/gaw-reports.html (last access: 23 October 2014), Welligton, New Zealand, 2012.

Wunch, D., Toon, G. C., Blavier, J.-F., Washenfelder, R., Notholt, J., Connor, B., Griffith, D. W. T.. and Wennberg, P. O.: The Total Carbon Column Observing Network (TCCON). Phil. Trans. Roy. Soc. A, 369, 2087–2112, 2011.

Yakir, D. and Wang, X.-F.: Fluxes of CO_2 and water between terrestrial vegetation and the atmosphere estimated from isotope measurements, Nature, 380, 515–517, doi:10.1038/380515a0, 1996.

18

Next-generation angular distribution models for top-of-atmosphere radiative flux calculation from CERES instruments: methodology

W. Su[1], J. Corbett[2], Z. Eitzen[2], and L. Liang[2]

[1]MS420, NASA Langley Research Center, Hampton, Virginia, USA
[2]Science Systems & Applications, Inc., Hampton, Virginia, USA

Correspondence to: W. Su (wenying.su-1@nasa.gov)

Abstract. The top-of-atmosphere (TOA) radiative fluxes are critical components to advancing our understanding of the Earth's radiative energy balance, radiative effects of clouds and aerosols, and climate feedback. The Clouds and the Earth's Radiant Energy System (CERES) instruments provide broadband shortwave and longwave radiance measurements. These radiances are converted to fluxes by using scene-type-dependent angular distribution models (ADMs). This paper describes the next-generation ADMs that are developed for Terra and Aqua using all available CERES rotating azimuth plane radiance measurements. Coincident cloud and aerosol retrievals, and radiance measurements from the Moderate Resolution Imaging Spectroradiometer (MODIS), and meteorological parameters from Goddard Earth Observing System (GEOS) data assimilation version 5.4.1 are used to define scene type. CERES radiance measurements are stratified by scene type and by other parameters that are important for determining the anisotropy of the given scene type. Anisotropic factors are then defined either for discrete intervals of relevant parameters or as a continuous functions of combined parameters, depending on the scene type. Significant differences between the ADMs described in this paper and the existing ADMs are over clear-sky scene types and polar scene types. Over clear ocean, we developed a set of shortwave (SW) ADMs that explicitly account for aerosols. Over clear land, the SW ADMs are developed for every 1° latitude × 1° longitude region for every calendar month using a kernel-based bidirectional reflectance model. Over clear Antarctic scenes, SW ADMs are developed by accounting the effects of sastrugi on anisotropy. Over sea ice, a sea-ice brightness index is used to classify the scene type. Under cloudy conditions over all surface types, the longwave (LW) and window (WN) ADMs are developed by combining surface and cloud-top temperature, surface and cloud emissivity, cloud fraction, and precipitable water. Compared to the existing ADMs, the new ADMs change the monthly mean instantaneous fluxes by up to 5 W m^{-2} on a regional scale of 1° latitude × 1° longitude, but the flux changes are less than 0.5 W m^{-2} on a global scale.

1 Introduction

The Clouds and the Earth's Radiant Energy System (CERES) project has been providing data products critical to advancing our understanding of the effects of clouds and aerosols on radiative energy within the Earth–atmosphere system. CERES data are used by the science community to study the Earth's energy balance (e.g., Trenberth et al., 2009; Kato et al., 2011; Loeb et al., 2012; Stephens et al., 2012), aerosol direct radiative effects (e.g., Satheesh and Ramanathan, 2000; Zhang et al., 2005; Loeb and Manalo-Smith, 2005; Su et al., 2013), aerosol–cloud interactions (Loeb and Schuster, 2008; Quaas et al., 2008; Su et al., 2010b), and to evaluate global general circulation models (e.g., Pincus et al., 2008; Su et al., 2010a; Wang and Su, 2013; Wild et al., 2013).

The CERES instrument consists of a three-channel broadband scanning radiometer (Wielicki et al., 1996). The scanning radiometer measures radiances in shortwave (SW, 0.3–5 μm), window (WN, 8–12 μm), and total (0.3–200 μm) channels at a spatial resolution of ~ 20 km at nadir. The longwave (LW) component is derived as the difference between total and SW channels. These measured radiances at a given Sun–Earth–satellite geometry are converted to outgoing reflected solar and emitted thermal top-of-atmosphere (TOA) radiative fluxes. To do so, we must account for the angular distribution of the radiance field, which is scene type dependent. Here scene type is a combination of variables (e.g., surface type, cloud fraction, cloud optical depth, cloud phase, aerosol optical depth, precipitable water, lapse rate) that we use to group the data to develop distinct angular distribution models (ADMs). To facilitate the construction of ADMs, there are pairs of identical CERES instruments on both Terra and Aqua spacecrafts. At the beginning of these missions one of the instruments on each spacecraft was always placed in a rotating azimuth plane (RAP) scan mode. In this mode, the instrument scans in elevation as it rotates in azimuth, thus acquiring radiance measurements from a wide range of viewing combinations. To provide accurate scene type information within CERES footprints, imager (Moderate Resolution Imaging Spectroradiometer (MODIS) on Terra and Aqua) cloud and aerosol retrievals are averaged over CERES footprints by accounting for the CERES point spread function (PSF; Smith, 1994) and are used for scene type classification.

The first set of CERES ADMs was developed using 9 months of CERES and Visible Infrared Scanner (VIRS) data from the Tropical Rainfall Measuring Mission (TRMM) satellite (Loeb et al., 2003). This set of ADMs represents a more improved anisotropy characterization than the ADMs used for Earth Radiation Budget Experiment (ERBE), which only provided anisotropy for 12 scene types (Smith et al., 1986; Suttles et al., 1988). Because TRMM is in a 350 km circular precessing orbit with a 35° inclination angle, CERES TRMM sampled the full range of solar zenith angle over regions between 38° S and 38° N every 46 days.

The second set of CERES ADMs was developed using 2 years of CERES and MODIS data on Terra (Loeb et al., 2005). The same methodology is also applied to CERES measurements on Aqua. The scene identification used to develop these ADMs is based upon the cloud algorithms described in Minnis et al. (2011). TOA fluxes inverted from these ADMs and cloud properties used for scene identification are in the Edition 2 Single Scanner Footprint TOA/Surface Fluxes and Clouds (Ed2SSF) product. The overall bias in global monthly mean TOA flux from ADM uncertainties is less than $0.2\,\mathrm{W\,m^{-2}}$ in the SW and 0.2–$0.4\,\mathrm{W\,m^{-2}}$ in the LW (Loeb et al., 2007). However, validation studies also reveal that the uncertainties of fluxes over snow/ice are larger than those over ocean/land, and uncertainties of fluxes over clear ocean are dependent on the MODIS fine-mode fraction (Loeb et al., 2007). These findings led to this investigation to further improve the CERES ADMs.

ADMs are scene type dependent; thus accuracy in scene identification affects the characterization of anisotropy. The CERES team has improved upon the cloud algorithms used for Ed2SSF and recently delivered the improved cloud algorithms for Edition 4 SSF (Ed4SSF). The improvements in cloud algorithms include using regional mean boundary apparent lapse rates developed using collocated CALIPSO and MODIS data to determine low cloud-top height (Sun-Mack et al., 2014); using a CO_2-slicing method to retrieve high cloud over low-lying clouds (Chang et al., 2010); using a rough ice crystal model (Yang et al., 2008) to improve ice cloud retrieval; using the $1.24\,\mu$m channel for cloud optical depth retrieval over snow, and other changes as discussed by Minnis et al. (2010). We utilize the improved cloud algorithm (scene identification) and all available RAP data to develop the improved next-generation CERES ADMs. This paper documents the methodology used to develop ADMs for the SW and LW (WN) channels over different scene types. The accuracy of the TOA SW, LW, and WN fluxes derived from these new ADMs will be assessed in a future publication.

2 Observations

CERES instruments on Terra and Aqua are flying with MODIS instruments so that the higher-resolution imager can provide cloud conditions for every CERES field of view (FOV). The CERES/MODIS cloud algorithms retrieve cloud fraction (f, in %), cloud optical depth, cloud phase (liquid = 1, and ice = 2), cloud top and effective temperature/pressure (among other variables) based on MODIS pixel-level measurements. These pixel-level cloud properties are spatially and temporally matched with the CERES FOV. If the cloud top pressure within a CERES FOV is significantly different, up to two non-overlapping cloud layers are defined (Loeb et al., 2003). For each cloud layer, pixel-level cloud properties are weighted by the CERES PSF to provide layer-mean cloud properties. Layer-mean cloud properties are then weighted by the layer cloud fractions to derive the cloud properties over the CERES FOV.

Other pixel-level retrievals such as aerosol optical depth and Ångström exponent from MODIS collection 5.1 (Hsu et al., 2004; Remer et al., 2008; Levy et al., 2010) are also averaged over the CERES footprints using the PSF. Similarly, spectral radiances from MODIS observations are averaged over CERES FOV weighted by the CERES PSF. These radiances are used to develop clear ocean ADMs (Sect. 4.1.1), clear land ADMs (Sect. 4.2.1), and sea ice ADMs (Sect. 4.5).

Surface elevation data are from digital elevation model GTOPO30 and resampled at a resolution of 1/6° equal angle grid box. The elevation variability (EV) for a given grid box is calculated as the standard deviation of the elevations of this grid box and the eight surrounding grid boxes. Meteorological fields are from the Global Modeling and Assimilation Office's Goddard Earth Observing System (GEOS) version 5.4.1 data assimilation system for CERES. This version provides consistent analysis over the entire CERES data record period. Surface types are obtained from the International Geosphere Biosphere Program (IGBP) global land cover data set. Fresh snow and sea ice surface types are derived from a combination of the National Snow and Ice Data Center (NSIDC) microwave snow/ice map and the National Environmental Satellite, Data and Information Service (NESDIS) snow/ice map. NESDIS uses imager data to identify snow and sea ice and provide snow and sea ice information near the coast, whereas NSIDC does not provide microwave retrievals within 50 km of the coast. Another snow and ice data source is from the CERES team's snow and sea ice fraction over the clear portions of CERES footprints using MODIS reflectances. The presence of snow and sea ice is determined using thresholds of the reflectance at 1.6 and 0.65 μm, and the brightness temperature differences at 3.7 and 11 μm (Minnis et al., 2008). Hereafter, we refer to this snow and sea ice fraction as cloud mask snow/ice fraction.

These cloud properties, aerosol properties, spectral radiances, meteorological data, and CERES broadband radiances are all included in the CERES Ed4SSF product. We use all available RAP data on Terra ($\sim$ 60 months) and Aqua ($\sim$ 32 months) to construct their individual ADMs over most scene types. However, over certain scene types (i.e., SW over permanent snow and sea ice, and LW over clear permanent snow, fresh snow, and sea ice) when we do not have enough samples from a single spacecraft, combined Terra and Aqua ADMs are developed.

The general strategy of constructing ADMs is to sort the measured radiances into angular bins over different scene types. Over a given scene type (j), the measured radiances are sorted into discrete angular bins. Averaged radiances in all angular bins ($\hat{I}$) are calculated and all radiances in the upwelling directions are integrated to provide the ADM flux ($\hat{F}$). The ADM anisotropic factors (R) for scene type j are calculated as

$$R_j(\theta_0,\theta,\phi) = \frac{\pi \hat{I}_j(\theta_0,\theta,\phi)}{\int_0^{2\pi}\int_0^{\frac{\pi}{2}} \hat{I}_j(\theta_0,\theta,\phi)\cos\theta\sin\theta \mathrm{d}\theta \mathrm{d}\phi} = \frac{\pi \hat{I}_j(\theta_0,\theta,\phi)}{\hat{F}_j(\theta_0)}, \quad (1)$$

where θ_0 is the solar zenith angle, θ is the CERES viewing zenith angle, and ϕ is the relative azimuth angle between CERES and the solar plane. For an observed radiance (I_o) under the same scene type, it is then converted to flux by using the anisotropic factor that we derived:

$$F(\theta_0) = \frac{\pi I_o(\theta_0,\theta,\phi)}{R_j(\theta_0,\theta,\phi)}. \quad (2)$$

Besides the sort into angular bin method for developing ADMs, we also employ analytical functions when appropriate to characterize the anisotropies over some scene types (see Sects. 4 and 5).

3 Angular distribution model evaluation

ADMs are sets of anisotropic factors used to convert the observed TOA radiances to fluxes. Because we do not know the instantaneous radiance angular distributions for different scene types, it is a challenging task to evaluate the ADMs that we developed. To overcome this challenge, we design an evaluation method by combining Eq. (1) with Eq. (2) to rewrite the TOA flux as

$$F(\theta_0) = \frac{I_o(\theta_0,\theta,\phi)}{\hat{I}_j(\theta_0,\theta,\phi)} \hat{F}_j(\theta_0). \quad (3)$$

Note that the TOA flux is sensitive to the ratio of $\hat{F}_j$ to $\hat{I}_j$, which implies that it is more important to get the shape of $\hat{I}_j$ correct rather than its magnitude. To facilitate the comparison between the shape of observed and ADM-predicted radiances, we first normalize the observed and predicted radiances with their own means for each 1° latitude × 1° longitude region. We then calculate the root-mean-square (rms) error between the normalized predicted radiance and the normalized observed radiance as

$$\mathrm{rms} = \sqrt{\frac{1}{n}\sum_{j=1}^{n}\left(\frac{\hat{I}_j}{\bar{\hat{I}}} - \frac{I_j^{\mathrm{o}}}{\bar{I^{\mathrm{o}}}}\right)^2}. \quad (4)$$

The smaller the rms error, the closer the shape of the predicted radiance to the shape of the observed radiance, which indicates the ADMs are more accurate. We use the rms error for every scene type to assess the improvement of the new ADMs we describe below. Note that this rms error is not the same as the flux rms error.

4 Shortwave angular distribution models

4.1 Ocean

4.1.1 Clear sky

Over ocean, clear footprints are defined as having $f < 0.1$ %. Previously, ADMs over clear ocean were developed as a function of wind speed with an angular resolution of 2° in θ_0, θ, and ϕ. Aerosol effects on ADMs were not explicitly considered; dependence on aerosol optical depth (AOD) is accounted for by theoretical adjustment (Loeb et al., 2005). Subsequent validation indicated that the TOA flux uncertainty depends on the MODIS AOD and fine-mode fraction (Loeb et al., 2007).

To improve the characterization of anisotropy over clear ocean, we develop new clear ocean ADMs which explicitly account for aerosols using the same angular resolution and the same wind speed bins (0–2, 2–4, 4–6, 6–8, 8–10, and $> 10\,\mathrm{m\,s^{-1}}$). Over clear ocean, about 48 % of the CERES cross-track measurements are taken when the sun glint angles are less than 40° (glint region). The reflectance distributions in the glint region differ sharply from those outside

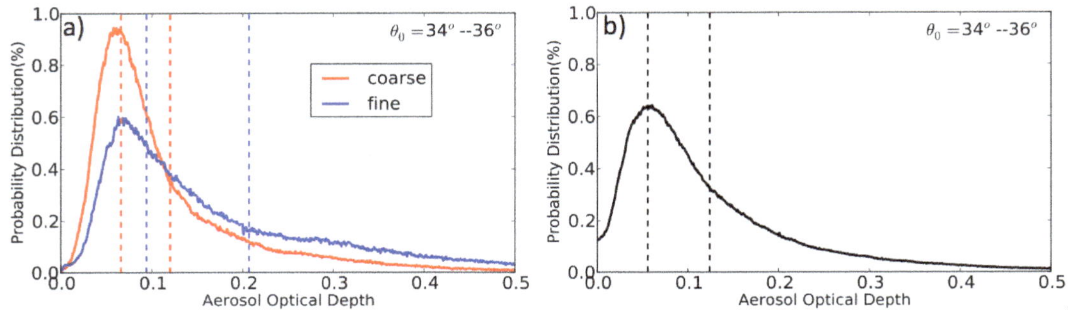

Figure 1. Probability distributions of aerosol optical depth retrieved from Ed4SSF using Terra measurements from 2000 to 2005 for $\theta_0 = 34$–$36°$. **(a)** For glint angle $> 40°$, the AOD threshold values are 0.066 and 0.12 for coarse-mode-like aerosols and are 0.094 and 0.207 for fine-mode-like aerosols. **(b)** For glint angle $\leq 40°$, the AOD threshold values are 0.056 and 0.124.

the glint region; we thus treat glint and non-glint regions separately. For the non-glint region, we develop AOD- and aerosol-type-dependent ADMs; for the glint region, we develop AOD-dependent ADMs only.

Non-glint regions

For clear footprints with glint angles greater than 40°, AOD and fine-mode fraction are available from the MODIS team (Remer et al., 2008). However, these retrievals are produced using a cloud mask algorithm developed by Martins et al. (2002), which is different from the cloud mask algorithm used by the CERES cloud group. Thus, not every clear CERES FOV has a MODIS aerosol retrieval (about 8.3 % of the clear CERES footprints do not have valid MODIS aerosol retrievals). To ensure self-consistency, we develop an aerosol retrieval algorithm using PSF-weighted mean MODIS radiances over clear CERES footprints. This algorithm retrieves AOD based upon five-band lookup tables (LUTs) calculated using maritime tropical and urban aerosol models from Hess et al. (1998). These two aerosol models are used to represent coarse-mode-like and fine-mode-like aerosols. The LUTs are calculated for 29 AOD values ranging from 0 to 2, and for six wind speeds (midpoints of the wind speed bins). The five MODIS bands used for aerosol retrieval are 0.47, 0.55, 0.66, 0.87, and 1.24 μm on Terra (2.13 μm on Aqua). For a given clear CERES footprint, the AOD for each aerosol model is first retrieved by minimizing the error (ϵ), which is defined as

$$\epsilon_c = \sum_{i=1}^{5} \left(\frac{\rho_i^{obs} - \rho_i^{LUT}(\mathrm{AOD^c}, |\boldsymbol{u}|)}{\rho_i^{obs} + 0.01} \right)^2,$$

$$\epsilon_f = \sum_{i=1}^{5} \left(\frac{\rho_i^{obs} - \rho_i^{LUT}(\mathrm{AOD^f}, |\boldsymbol{u}|)}{\rho_i^{obs} + 0.01} \right)^2, \quad (5)$$

where ρ_i^{obs} is the MODIS observed reflectance for the ith band, and ρ_i^{LUT} is the LUT reflectance calculated for different AOD and wind speed ($|\boldsymbol{u}|$). For the coarse-mode-like LUT, $\mathrm{AOD^c}$ yields the smallest error ϵ_c; while for the fine-mode-like LUT, $\mathrm{AOD^f}$ yields the smallest error ϵ_f. If $\epsilon_c < \epsilon_f$, then the aerosols of this footprint are coarse-mode-like with optical depth of $\mathrm{AOD^c}$; otherwise they are fine-mode-like aerosols with optical depth of $\mathrm{AOD^f}$. Our AOD retrievals exhibit very similar global distributions as those from MODIS, but are higher than MODIS AOD (not shown). For every θ_0 bin, the distributions of fine- and coarse-mode AODs are calculated. We then divide the AODs into three groups by using AOD percentiles: less than 33 % is the low-AOD bin, between 33 and 66 % is the mid-AOD bin, and greater than 66 % is the high-AOD bin. Figure 1a shows an example of fine- and coarse-mode AOD distributions for $\theta_0 = 34$–$36°$, where the 33 and 66 % AODs are 0.094 and 0.207 for fine-mode aerosols and 0.066 and 0.120 for coarse-model aerosols. The percentile classification approach we adopted here to construct ADMs relaxes the requirement on absolute aerosol retrieval accuracy.

For each angular bin, the observed radiances are first sorted by wind speed, then separated by aerosol type and within each type further separated into three AOD percentiles. Thus, it is possible to construct a total number of 36 (6 wind speeds, 2 aerosol types, and 3 AOD bins) ADMs over non-glint clear oceans if the observations cover all variable space. For each case, the radiances are integrated in all upwelling directions to derive the ADM fluxes, and the missing angular bins are filled with radiative transfer calculations using the same fine- and coarse-mode aerosol models mentioned above. The anisotropic factors are then derived following Eq. (1).

Glint regions

Aerosol retrievals over glint regions are difficult because of the rapid changes in reflectance and wave-slope distribution around the specular point. Thus, we expect large uncertainties in aerosol retrieval over glint region and do not think we can discriminate aerosol types. We adopt the same retrieval method as for non-glint region but only use one

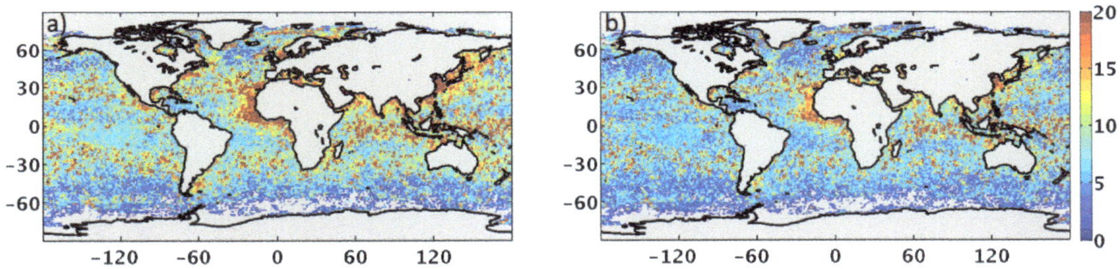

Figure 2. The rms error between normalized measured radiances and normalized ADM-predicted radiances, **(a)** using ADMs from Loeb et al. (2005), and **(b)** using the new ADMs. All clear-sky footprints from Terra RAP measurements in 2002 were used.

aerosol type (maritime tropical) and stratify the AOD distributions in each θ_0 bin into three percentile bins (0–33, 33–66, and 66–100 %). Figure 1b shows the AOD distribution for $\theta_0 = 34$–$36°$ bin retrieved within the glint region. Radiances taken with AOD < 0.056 are classified as the low-AOD bin, radiances taken with $0.056 \leq$ AOD < 0.124 are classified as the mid-AOD bin, and radiances taken with AOD ≥ 0.124 are classified as the high-AOD bin. We then construct ADMs for these three AOD bins. However, reflectances over glint regions do not always monotonically increase with AOD, resulting in ambiguous retrievals. This often happens at small glint angles where specular reflectances are large. At small AOD the aerosol scattering reduces the specular reflectance from the ocean surface; then as AOD becomes large enough, the reflectance from aerosol layers starts to increase with increasing AOD. About 9 % of the glint footprints produce ambiguous AOD retrievals (two AOD solutions). For these footprints, we construct ADMs without accounting for aerosols but rely on the theoretical adjustment as in Loeb et al. (2005). When a FOV is in close proximity to the specular reflection direction, we also perform a glint check by calculating the variability of the clear ocean anisotropy:

$$\sigma_R = (1 - f - f_{\text{ice}})\sigma_{R_{\text{clr}}}, \quad (6)$$

where f and f_{ice} are the fraction of the FOV covered by cloud and sea ice, and $\sigma_{R_{\text{clr}}}$ is the standard deviation of clear ocean R in the angular bins adjacent to the observation angle. If $\sigma_{R_{\text{clr}}} \geq 0.05$ for a given FOV, the ADM flux ($\hat{F}$) corresponding to the FOV's scene type is used.

The aerosol-dependent ADMs characterize the clear ocean anisotropy more accurately than the ADMs developed by Loeb et al. (2005). Figure 2 shows the rms error between normalized measured radiances and ADM-predicted radiances (Eq. 4) using all clear-sky measurements of 2002 from CERES Terra RAP measurements. The global area-weighted mean rms error is 9.2 % using the aerosol-dependent ADMs, whereas it is 10.9 % using ADMs from Loeb et al. (2005). Significant improvements are noted over coastal regions where large amounts of dust and pollution are expected.

4.1.2 Cloudy sky

Over cloudy ocean, we adopt the same method as Loeb et al. (2005) to construct ADMs using an analytical function that relates CERES radiances and cloud properties retrieved from MODIS. Under cloudy conditions, the magnitude of CERES radiance in a given angular bin is most sensitive to cloud fraction (f, in %), cloud optical depth ($\tilde{\tau}$), and effective cloud phase (ECP) over the non-glint regions (glint angle $> 20°$) or over glint regions with sufficiently thick clouds and/or extensive cloud coverage ($\ln(f\,\tilde{\tau}) > 6$). Over the glint regions with $\ln(f\,\tilde{\tau}) < 6$, ADMs are constructed using mean radiances for 6 wind speed bins (same as clear ocean) and four $\ln(f\,\tilde{\tau})$ bins ($<$ 3.5, 3.5–4.5, 4.5–5.5, 5.5–6). Figure 3 shows the relationship between CERES-measured SW radiance and $\ln(f\,\tilde{\tau})$ for liquid clouds in the angular bin of $\theta_0 = 44$–$46°$, $\theta = 18$–$20°$, $\phi = 88$–$90°$ and for ice clouds in the angular bin of $\theta_0 = 54$–$56°$, $\theta = 14$–$16°$, $\phi = 176$–$178°$ using CERES RAP measurements on Terra. The instantaneous radiances are separated into five equal sample number bins based upon f. The minimum and maximum cloud fractions for each bin are listed in the legend. The mean radiances for each 0.02 interval of $\ln(f\,\tilde{\tau})$ are shown as black dots. We then fit the mean radiances with a five-parameter sigmoidal function:

$$I = I_0 + \frac{a}{\left[1 + e^{-(x - x_0)/b}\right]^c}, \quad (7)$$

where x_0, I_0, a, b, and c are coefficients of the sigmoidal fit, and $x = \ln(f\,\tilde{\tau})$. The sigmoidal fits over cloudy ocean are derived with an angular resolution of 2° in θ_0, θ, ϕ and separately for liquid ($1.00 \leq$ ECP < 1.01), mixed ($1.01 \leq$ ECP ≤ 1.75), and ice ($1.75 <$ ECP ≤ 2.00) clouds.

The relative rms error between the mean radiances and the fitted radiances is 3.0 % for the water cloud case and 3.4 % for the ice cloud case. Figure 4 shows the cumulative probability distribution of the relative rms errors between mean radiances and fitted radiances for all angular bins with θ_0 between 10 and 80° for liquid, mixed, and ice clouds. There are 86.6, 74.6, and 67.7 % of the angular bins with relative rms errors less than 5 %, and 98.4, 95.3, and 89.3 % of the angular bins with relative rms errors less than 10 % for liquid,

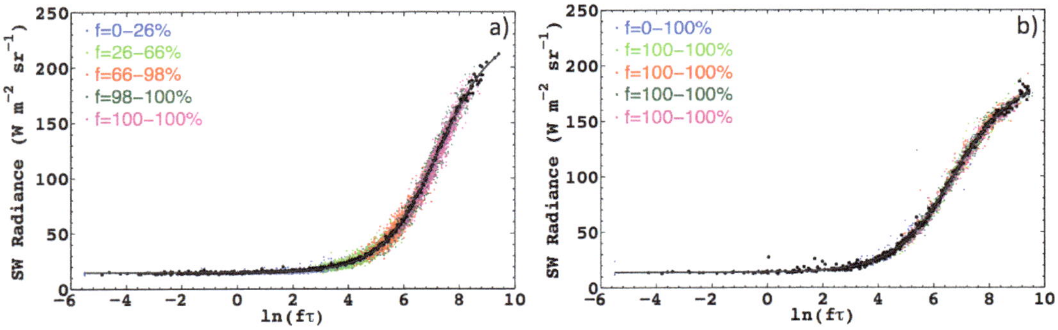

Figure 3. Relationship between SW radiances and $\ln(f\,\widetilde{\tau})$ over ocean using CERES Terra measurements. **(a)** for liquid cloud in the angular bin of $\theta_0 = 44$–$46°$, $\theta = 18$–$20°$, and $\phi = 88$–$90°$ and **(b)** for ice cloud in the angular bin of $\theta_0 = 54$–$56°$, $\theta = 14$–$16°$, and $\phi = 176$–$178°$. Colored dots are instantaneous SW radiances; black dots are mean SW radiances calculated for each 0.02 intervals of $\ln(f\,\widetilde{\tau})$. Black lines are the sigmoidal fits.

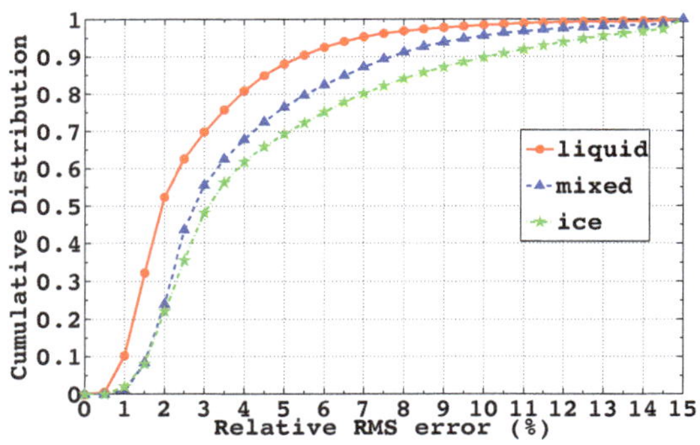

Figure 4. Cumulative distributions of the relative rms errors for the sigmoidal fits over cloudy ocean using CERES Terra measurements.

mixed, and ice clouds. The mean relative rms errors for all these angles are 3.0, 4.1, and 4.9 % for liquid, mixed, and ice clouds, which are smaller than those (between 5 and 10 %) derived by Loeb et al. (2005). The reduction in relative rms errors is mostly due to the improved cloud algorithms which provide more reliable f, $\widetilde{\tau}$, and cloud phase retrievals.

The TOA ADM flux in each $\ln(f\,\widetilde{\tau})$ interval is obtained by integrating SW radiances inferred from the sigmoidal fits in all upwelling directions. Anisotropic factors for different $\ln(f\,\widetilde{\tau})$ values are derived for each angular bin following Eq. (1). Figure 5 shows the anisotropy factors at $\theta_0 = 44$–$46°$ in the principal plane for liquid clouds with three $\ln(f\,\widetilde{\tau})$ values and for different cloud phases with $\ln(f\,\widetilde{\tau}) = 6$. Anisotropy depends on the magnitude of $\ln(f\,\widetilde{\tau})$ and cloud phase. The liquid and mixed clouds exhibit well-defined peaks in anisotropy due to cloud glory and rainbow features, while ice clouds exhibit peaks in anisotropy in the specular reflectance direction. Cloudy scenes tend to be more isotropic as $\ln(f\,\widetilde{\tau})$ increases and as clouds change from liquid to ice phase.

4.2 Land

4.2.1 Clear sky

Over land, clear footprints are defined as having $f < 0.1\,\%$. Loeb et al. (2005) used a eight-parameter fit from Ahmad and Deering (1992) to characterize the bidirectional reflectance distribution function (BRDF) over clear land/desert. However, a kernel-based bidirectional model has proven to be effective in modeling the surface BRDF (Lucht et al., 2000; Chopping et al., 2008). Here we adopt the reciprocal RossThick–LiSparse model (Ross–Li; Roujean et al., 1992; Li and Strahler, 1992) to develop the reflectance over clear land/desert:

$$\rho(\mu_0, \mu, \phi) = k_0 + k_1 \cdot B_1(\mu_0, \mu, \phi) + k_2 \cdot B_2(\mu_0, \mu, \phi). \quad (8)$$

The first term quantifies the isotropic contribution to BRDF; the second term estimates the directional reflectance of a flat surface with randomly distributed and oriented protrusions; and the third term approximates the radiative transfer within a vegetation canopy, modified by Maignan et al. (2004) to account for the hot spot effect.

We derive the kernel weights (k_0, k_1, $k2$) using CERES clear-sky RAP measurements for every 1° latitude by 1° longitude region for each calendar month to quantify the TOA BRDF. The measured radiance in each CERES FOV is converted to a reflectance as

$$\rho(\theta_0, \theta, \phi) = \frac{\pi I(\theta_0, \theta, \phi)}{\mu_0 E_0}(1 + e_{se})^2, \quad (9)$$

where $\mu_0 = \cos\theta_0$, E_0 is the TOA solar incoming flux (1361 W m^{-2}), and $(1 + e_{se})$ is the Earth–Sun distance in astronomical units. To account for the effects of solar zenith angle, vegetation coverage, and surface roughness on BRDF, the weights are derived separately for different intervals of μ_0 (0.2), TOA normalized difference vegetation index (NDVI, 0.1), and for two categories of elevation variability (EV) over rough terrain. A 1° × 1° region is regarded

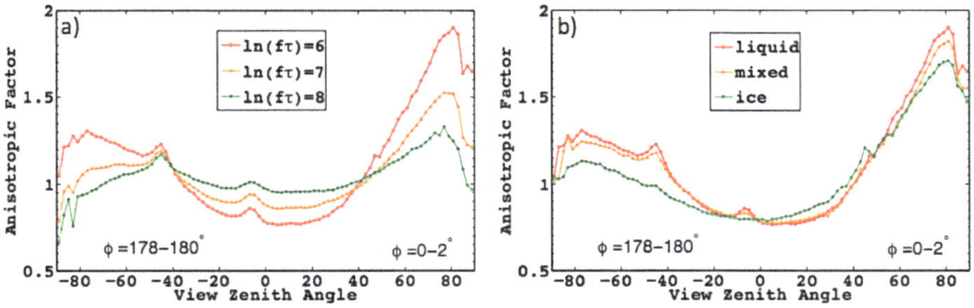

Figure 5. CERES SW anisotropic factors over ocean in the principal plane for **(a)** liquid clouds with different $\ln(f\,\widetilde{\tau})$ values, **(b)** clouds of different phases with $\ln(f\,\widetilde{\tau})=6$. Anisotropic factors are derived for $\theta_0 = 44$–$46°$ based upon 60 months of CERES Terra measurements.

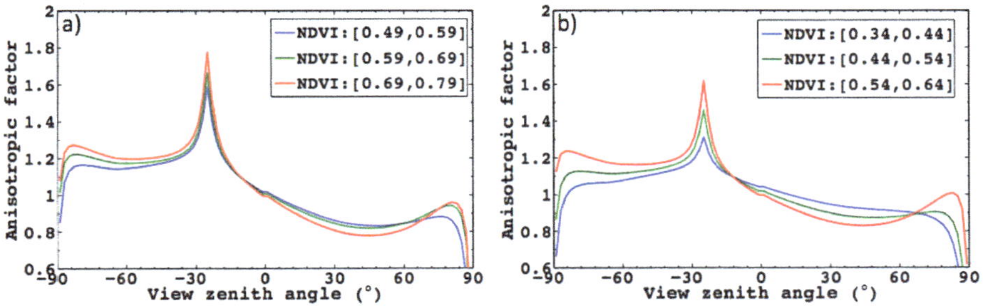

Figure 6. CERES SW anisotropic factors in the principal plane for a region centered at 6.5° S and −59.5° W for **(a)** July and **(b)** September for $\theta_0 = 24°$ for all available NDVI bins. Anisotropic factors are derived based upon 60 months of CERES Terra measurements over clear CERES footprints.

as rough terrain if its median EV is greater than 25 m, and this region is then separated into two groups using the median EV. The TOA NDVI over clear land/desert/fresh-snow CERES FOV is determined using PSF-weighted mean MODIS 0.64 μm ($\widetilde{I}_{0.64}$) and 0.86 μm ($\widetilde{I}_{0.86}$) radiances:

$$\mathrm{NDVI} = \frac{\widetilde{I}_{0.86} - \widetilde{I}_{0.64}}{\widetilde{I}_{0.86} + \widetilde{I}_{0.64}}. \tag{10}$$

The TOA NDVI is used to separate subregions within a 1° × 1° region that have different surface characteristics (surface type or vegetation coverage) and/or different aerosol properties. Note that the TOA NDVI is not corrected for atmospheric effect; hence it tends to decrease as aerosol optical depth increases as the aerosols are more effective scatters at 0.64 μm than at 0.86 μm.

To construct ADMs from the BRDF fits, albedos for each μ_0 interval in which the BRDF fit was derived are calculated by directly integrating the BRDFs. Next, a fit from Rahman et al. (1993) is used to account for the albedo dependence on solar zenith angle in each μ_0 interval. The instantaneous anisotropic factor at a given location is derived from the ratio of reflectance to albedo for a combination of μ_0, TOA NDVI, and EV.

Figure 6 shows the clear-sky anisotropic factors derived from the Ross–Li model for a 1° × 1° region in the Amazon (6–7° S, 59–60° W) for July and September, the dry season for this location. The TOA NDVI ranges from 0.49 to 0.79 for July, but it is much lower (from 0.34 to 0.64) for September. The anisotropic factors are more isotropic as TOA NDVI decreases and the hot spots become less prominent. Because this region has a weak seasonality in phenology (Bradley et al., 2011), the sharp decline in TOA NDVI from July to September can not be explained solely by changes in vegetation. However, biomass burning reaches its peak in September and the aerosol optical depth is 3–4 times larger than that of July (Bevan et al., 2009), which leads to the decrease in NDVI. Figure 6 demonstrates that using NDVI to stratify the CERES measurements can provide distinctly different ADMs under different aerosol conditions, as NDVI contains combined spectral information of land surface and aerosols.

The Ross–Li model produces a more accurate characterization of anisotropy when compared to the model used by Loeb et al. (2005). Figure 7a and b show the rms error between the normalized measured radiances and the normalized predicted radiances (Eq. 4) from the two versions of ADMs using all clear-sky measurements of July 2002 from CERES Terra RAP mode. The mean rms error using the ADMs from Loeb et al. (2005) is 7.0 %, and it is reduced to 5.7 % using the new ADMs. Significant improvements are noted over high-latitude regions. Table 1 lists the mean rms error for January, April, July, and October for 2002 CERES Terra and 2004 CERES Aqua. The new ADMs reduce the

Table 1. The rms error (%) between normalized measured radiances and ADM-predicted radiances over clear-sky land. Two versions of CERES ADMs are used (Loeb and New). The new ADM results are stratified into two populations by using the median aerosol optical depth (AOD_m) of each 1° by 1° region.

	Terra 2002				Aqua 2004			
ADMs AOD	Loeb all	New all	New < AOD_m	New > AOD_m	Loeb all	New all	New < AOD_m	New > AOD_m
Jan	6.7	6.6	5.8	5.8	7.2	5.6	6.5	6.5
Apr	6.4	5.7	5.3	5.4	6.8	6.0	5.8	5.7
Jul	7.0	5.7	5.0	5.2	7.1	5.8	5.3	5.3
Oct	6.8	6.0	5.4	5.5	7.5	6.4	6.0	5.9

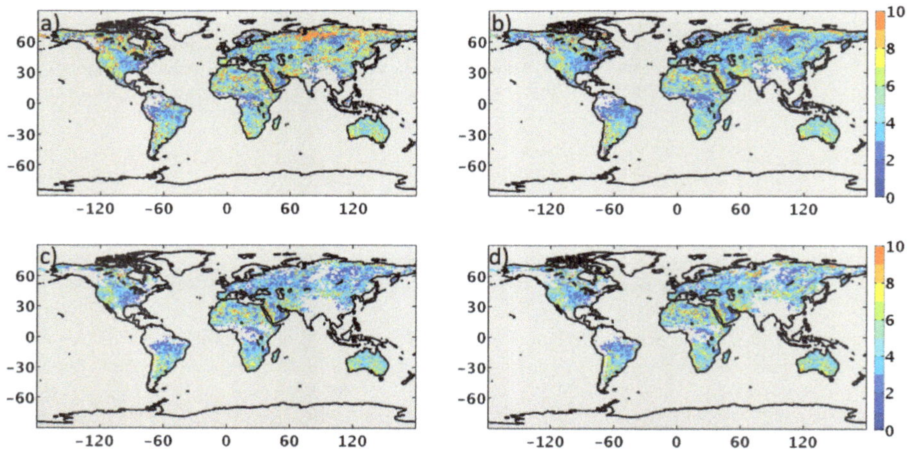

Figure 7. The rms error between normalized measured radiances and normalized ADM-predicted radiances for clear-sky land, **(a)** using ADMs from Loeb et al. (2005), **(b)** using the new ADMs, **(c)** using the new ADMs only for footprints with AODs less than the median AOD, and **(d)** using the new ADMs only for footprints with AODs greater than the median AOD.

clear-sky rms error for all seasons and large reductions occur in boreal summer and fall.

To investigate if the rms error depends on aerosol loading and type, we stratify the data in each 1° × 1° region into two populations by using the median AOD of the region. For a given footprint, if AOD from the MODIS dark target retrieval (Levy et al., 2010) is not available we use that from the Deep Blue retrieval (Hsu et al., 2004). Note that MODIS aerosol retrievals are not available for every clear CERES footprint because different cloud mask algorithms were used. These footprints are not used in calculating the rms errors for the two aerosol populations. The rms errors for the population with AOD less than and greater than the median are shown in Fig. 7c and d, and they are almost identical. Over northwestern Africa, the median AOD exceeds 0.5 (not shown) and is predominantly dust aerosols. Over eastern China, the median AOD exceeds 0.5 and is predominantly polluted aerosols. Over Australia, the median AOD is less than 0.1. Although these regions have very different aerosol loadings and are of different types, the rms errors over these regions do not vary with aerosols. Table 1 also lists the rms errors for these two aerosol populations and they differ by less than 0.2 %.

4.2.2 Cloudy sky

Over cloudy land/desert, we adopt the same methodology as Loeb et al. (2005), which we briefly review here. The anisotropy over cloudy land/desert depends not only on cloud properties but also on the underlying surface type, especially for thin and broken cloud conditions. A similar approach as clouds over ocean (Sect. 4.1.2) is used to develop the analytical ADMs over cloudy land/desert by accounting for the contribution from the surface. The observed radiance is split into four terms:

$$I(\mu_0,\mu,\phi) = (1-f)\frac{\mu_0 E_0}{\pi}\rho^{\mathrm{clr}}(\mu_0,\mu,\phi) + f I^{\mathrm{cld}}(\mu_0,\mu,\phi) + f\frac{\mu_0 E_0}{\pi}\left[\rho^{\mathrm{clr}}(\mu_0,\mu,\phi)e^{\frac{-\pi}{\mu_0}}e^{\frac{-\pi}{\mu}} + \overline{\alpha}^{\mathrm{clr}}\frac{t^{\mathrm{cld}}(\widetilde{\tau},\mu_0)\,t^{\mathrm{cld}}(\widetilde{\tau},\mu)}{1-\overline{\alpha}^{\mathrm{clr}}\overline{\alpha}^{\mathrm{cld}}(\widetilde{\tau})}\right], \quad (11)$$

where the first and second terms represent reflection from the cloud-free and cloudy areas, and the third and fourth terms correspond to scattering by the surface and atmosphere transmitted through the clouds. Here ρ^{clr} is the clear-sky

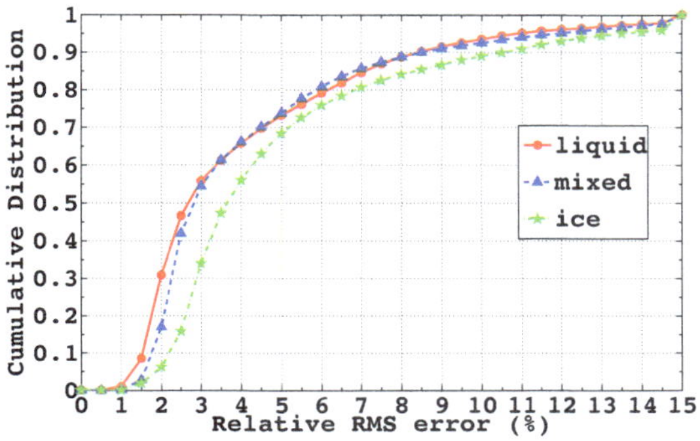

Figure 8. Cumulative distributions of the relative rms errors for the sigmoidal fits over cloudy land using CERES Terra measurements.

BRDF and $\overline{\alpha}^{\mathrm{clr}}$ is the clear-sky spherical albedo (Thomas and Stamnes, 1999), and they are derived from $1° \times 1°$ BRDF fits (Sect. 4.2.1); I^{cld} is the radiance from the cloudy layer; $\overline{\alpha}^{\mathrm{cld}}$ is the cloudy-sky spherical albedo and t^{cld} is the diffuse transmittance of the cloud. The lookup tables of $\overline{\alpha}^{\mathrm{cld}}$ and t^{cld} are generated using the broadband Langley Fu–Liou radiative transfer model (Fu and Liou, 1993; Rose et al., 2013).

Cloudy area contribution ($f\,I^{\mathrm{cld}}$) is derived from Eq. (11). The mean $f\,I^{\mathrm{cld}}$ is calculated for each 0.04 interval of $\ln(f\,\widetilde{\tau})$ and the relationship between them is quantified using sigmoidal fit (Eq. 7) separately for liquid, mixed, and ice clouds. An angular resolution of 5° in θ_0, θ, and ϕ is used for clouds over land/desert. Figure 8 shows the cumulative probability distribution of the relative rms errors between mean $f\,I^{\mathrm{cld}}$ and fitted radiances for all angular bins with θ_0 between 10 and 80° for liquid, mixed, and ice clouds. There are 71.2, 72.0, and 65.8 % of the angular bins with relative rms errors less than 5, and 93.0, 92.1, and 88.6 % of the angular bins with relative rms errors less than 10 % for liquid, mixed, and ice clouds. The mean relative rms errors for all these angles are 4.4, 4.5, and 5.4 % for the three cloud phases, which are slightly larger than the rms errors over cloudy ocean.

The TOA ADM flux is computed by integrating Eq. (11) in all upwelling directions for each $\ln(f\,\widetilde{\tau})$ interval using SW radiances inferred from the sigmoidal fits for cloudy area and BRDF fits for clear area. Anisotropic factors for different $\ln(f\,\widetilde{\tau})$ values are derived for each angular bin following Eq. (1). Figure 9 shows the anisotropic factors in the principal plane for $\theta_0 = 40$–45° for liquid clouds with three $\ln(f\,\widetilde{\tau})$ values and for different cloud phases with $\ln(f\,\widetilde{\tau})=6$. Cloudy scenes tend to be more isotropic as $\ln(f\,\widetilde{\tau})$ increases and as clouds change from liquid to ice phase.

4.3 Fresh snow

4.3.1 Clear sky

Over fresh snow, clear sky is defined as footprints with $f < 1$ %. Previously clear-sky fresh-snow ADMs were developed for discrete intervals of snow fractions (from the cloud mask). Because the anisotropy of fresh snow is also affected by the underlying surface type, we adopt the method that we used for clear land to account for the effect of surface type on fresh-snow anisotropy. The clear land method can be extended to fresh snow because we derive the kernel weights for each 0.1 NDVI interval. For NDVI values around 0.1, it is likely there is snow on the ground (Hall et al., 2002). Furthermore, recent airborne measurements also indicate that the Ross–Li BRDF model is suitable over snow surfaces (Lyapustin et al., 2010).

Figure 10 shows the anisotropic factor in the principal plane for grid box centered at 49.5° N and −119.5° W derived using all clear footprints from January measurements on Terra. There are two NDVI bins. The surface types that correspond to the low and high NDVI bins are fresh snow and evergreen needleleaf forest, respectively. The anisotropic factor for the snow-covered surface is more isotropic than snow-free surface.

4.3.2 Cloudy sky

The cloudy-sky fresh-snow ADMs are derived for discrete intervals of cloud fraction and snow fraction (Loeb et al., 2005; Kato and Loeb, 2005). For overcast conditions ($f > 99$ %), the ADMs are derived separately for bright and dark surfaces and for thin ($\widetilde{\tau} \leq 10$) and thick ($\widetilde{\tau} > 10$) clouds. A CERES FOV is determined to be bright or dark by comparing to the monthly climatological fresh-snow map. The snow map is calculated as follows: (1) determine mean MODIS 0.64 µm near-nadir ($\theta < 25°$) reflectances as a function of solar zenith angle using all clear fresh-snow CERES FOVs; (2) within a 1° latitude × 1° longitude region, if the MODIS radiances of more than 50 % of CERES FOVs are greater than the mean radiances, this region is classified as bright. Otherwise, this region is classified as dark. The same method is also used to classify sea ice and permanent snow into bright and dark categories by Loeb et al. (2005) and Kato and Loeb (2005).

4.4 Permanent snow

4.4.1 Clear sky

Over permanent snow, footprints with $f < 0.1\%$ are defined as clear. Loeb et al. (2005) divided clear permanent snow footprints into dark and bright categories using the same method outlined in Sect. 4.3.2 and built corresponding ADMs. However, the surface conditions are very different between Antarctica and Greenland. The snow surface over Antarctica is covered by sastrugi because of the strong and consistent katabatic wind flow there. Sastrugi are dune-like features that are generally aligned parallel to the prevailing wind direction. They can be over a meter high (Warren et al., 1998) and thus have significant effect on snow surface BRDF

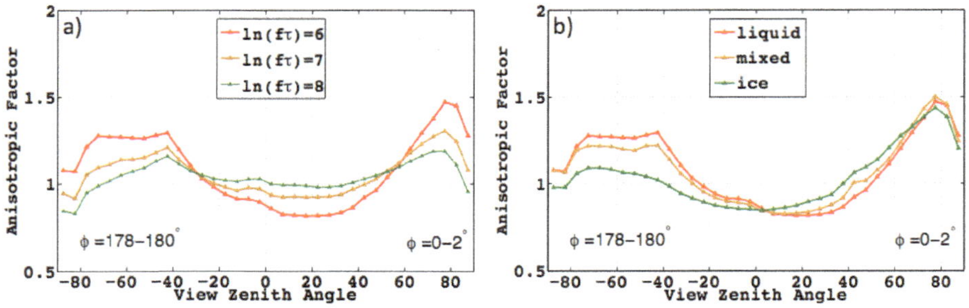

Figure 9. CERES SW anisotropic factors over land in the principal plane for **(a)** liquid clouds with different $\ln(f\,\widetilde{\tau})$ values, **(b)** clouds of different phases with $\ln(f\,\widetilde{\tau}) = 6$. Anisotropic factors are derived for $\theta_0 = 40$–$45°$ based upon 60 months of CERES Terra measurements.

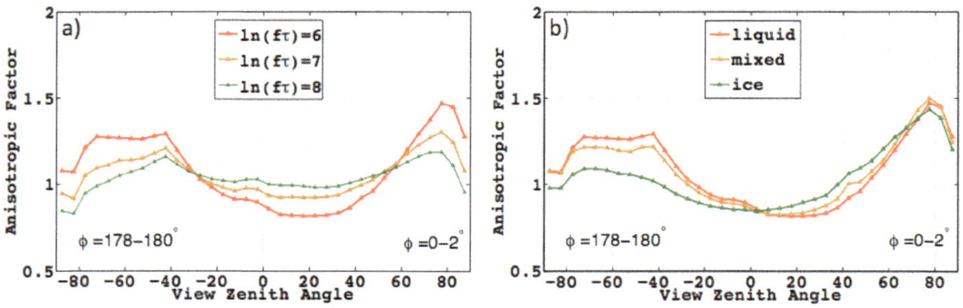

Figure 10. CERES clear-sky SW anisotropic factors in the principal plane for a region centered at 49.5° N and −119.5° W for January for $\theta_0 = 70°$ for all available NDVI bins. The IGBP surface type for NDVI between 0.1 and 0.2 is fresh snow and the IGBP surface type for NDVI between 0.2 and 0.3 is evergreen needleleaf forest. Anisotropic factors are derived based upon 60 months of CERES Terra measurements over clear CERES footprints.

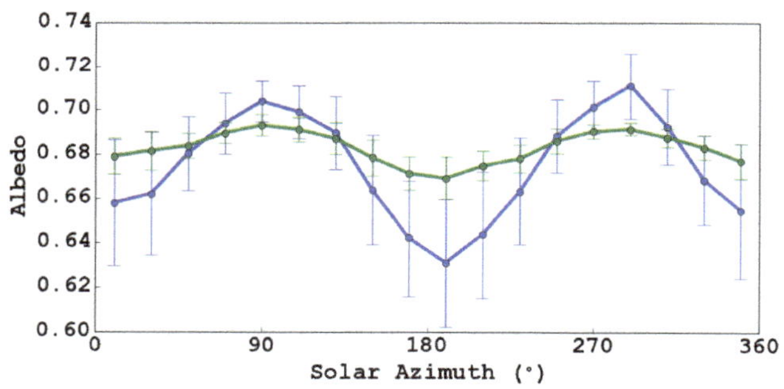

Figure 11. Mean TOA albedo and standard deviation as a function of solar azimuth angle using clear footprints in a South Pole region (88–89° S, and 76–83° W) during all Decembers of 2000–2004, for $\theta_0 = 65$–$70°$. The blue line shows albedos calculated using the ADMs from Loeb et al. (2005) and the green line shows albedos calculated with the new ADMs.

(e.g., Warren et al., 1998; Kuchiki et al., 2011). However, sastrugi are not found to be a noticeable feature on Greenland. To account for the differences in surface conditions, we develop separate ADMs for Antarctica and Greenland.

Antarctica

Corbett et al. (2012) demonstrated that the instantaneous albedo over clear permanent snow calculated from Ed2SSF shows a sine wave shape as solar azimuth angle changes (while ϕ remains within a 5° bin, blue line of Fig. 11). This is because the effect that sastrugi have on the snow BRDF depends on their orientation relative to the sun. When they are aligned parallel to the sun the reflectance in the forward scattering direction is relatively high, and the back scattering reflectance is relatively low in comparison to when the sastrugi are aligned perpendicular to the sun.

To understand how sastrugi affect snow BRDF, we use measurements from the Multi-angle Imaging Spectroradiometer (MISR; Diner et al., 1998), which provides near-simultaneous measurements of a target from nine cameras (four forward, four aftward and one nadir). These multi-angle measurements are used to quantify the effect of sastrugi on reflectances at different viewing angles, and a set of adjustment factors are developed to account for the effect of sastrugi on ADMs based upon co-located MISR and CERES measurements (Loeb et al., 2006).

We can demonstrate the effect of sastrugi over clear-sky Antarctica by examining the joint distributions of the standard scores of 0.86 μm reflectances (ρ_m) from any two pairs of MISR cameras. The standard score is defined as

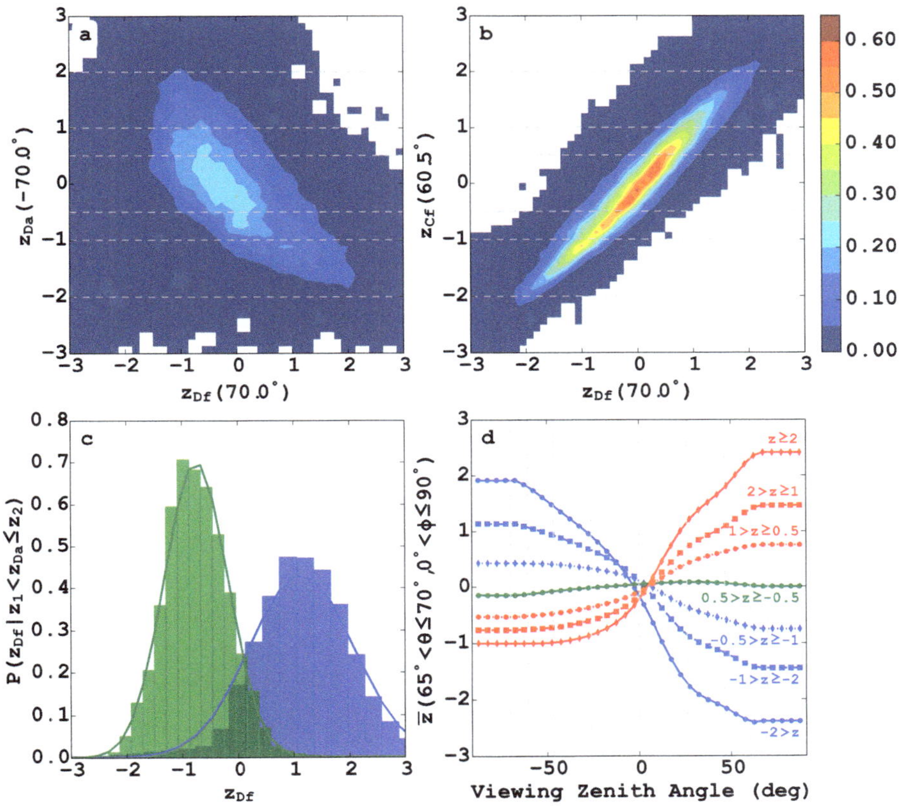

Figure 12. (a) Joint probability distribution of 0.86 µm reflectance standard scores between MISR's Df and Da cameras. The dashed lines indicate the standard score ranges used to develop adjustment factors. **(b)** As in **(a)** but for MISR's Df and Cf cameras. **(c)** The Df conditional probabilities of standard score when the Da standard score is between −1 and −2 (blue histogram) and between 1 and 2 (green histogram). **(d)** Adjustment factor against viewing zenith angle for the different standard score ranges for $\theta = 65$–$70°$ in the forward direction. Dashed lines indicate the nine MISR camera angles. These values are used to adjust the mean CERES reflectances in order to create the ADMs.

$$z(\theta_0,\theta_m,\phi) = \frac{\rho_m(\theta_0,\theta_m,\phi) - \overline{\rho_m}(\theta_0,\theta_m,\phi)}{\sigma_{\rho_m}(\theta_0,\theta_m,\phi)}, \quad (12)$$

where the averages ($\overline{\rho_m}$) and standard deviation (σ_{ρ_m}) are calculated in 5° bins for all available θ_0 and ϕ at each MISR viewing angle θ_m. The absolute value of z represents the departure of a measurement from the mean in units of the standard deviation, with negative z indicating the measurement is below the mean and positive z is above the mean.

Figure 12 shows the joint distributions of z between forward viewing Df ($\theta_m = 70.5°$) camera and aftward viewing Da ($\theta_m = -70.5°$) camera, and between Df camera and forward viewing Cf ($\theta_m = 60°$) camera using all available clear-sky measurements (including multiple θ_0 and ϕ bins). The two forward viewing cameras (Fig. 12b) show a strong positive correlation, but the forward and aftward cameras (Fig. 12a) show a distinct negative correlation. This indicates that the sastrugi are increasing the reflectance in one direction while at the same time decreasing the reflectance in the opposite direction. Thus, if the sastrugi cause the reflectance at a given angle to deviate from the mean by a certain amount, the joint distributions characterize how the sastrugi will affect the reflectances at other angles. We then use these to get the conditional probabilities for seven discrete intervals of z (dashed lines in Fig. 12a and b), allowing us to estimate the most likely z of one camera, given another camera's z. As z does not change with θ_0 and is only sensitive to ϕ being in the forward or backward direction, we average z over all θ_0 and forward and backward directions ($z(\theta_m, \phi_{f/b})$). For example, the most likely z value for the Df camera is 1.1 when z is between −2 and −1 for the Da camera (blue histogram in Fig. 12c), and the most likely z value for the Df camera is −0.8 when z is between 1 and 2 for the Da camera (green histogram in Fig. 12c). Thus, for each MISR camera, depending on the magnitude of z at this angle, we can estimate the most likely z at the eight other camera angles.

These standard score relationships established using nine MISR camera angles (vertical dashed lines in Fig. 12d) are interpolated onto 5° θ_c bins (symbols in Fig. 12d) for use with CERES measurements. For a reflectance measurement at a given θ and ϕ, depending on how much it deviates from the mean, we can adjust the reflectances at other θ_c bins accordingly. The adjustment factors also depend on ϕ; hence they are derived separately for the forward and backward directions and are denoted as $\beta(\theta_c, \phi_{f/b}, z)_{\theta,\phi}$. Figure 12d shows an example of the adjustment factors for $\theta = 65$–$70°$ and for the forward scattering direction. The seven curves are for the seven standard score groups, and the shape of these curves indicate how much the reflectance at other viewing angles should be adjusted. If the standard score of a reflectance taken at $\theta = 65$–$70°$ is greater than 2, the shape of the red solid line shows how much the reflectances at other viewing angles should be adjusted. For example, the reflectance at $\theta_c = 20°$ should be 0.7 standard deviations above the mean, and the reflectance at $\theta_c = -50°$ should be 1.0 standard deviations below the mean. Note that, regardless of the standard score range, the most likely value at nadir is always close to zero (no deviation from the mean reflectance). This is expected as sastrugi only have a large effect on the BRDF at oblique angles.

To construct ADMs, we first calculate the mean CERES reflectance ($\overline{\rho}$) and standard deviations (σ_ρ) in 2° θ_0 bins and 5° θ and ϕ bins using clear footprints over Antarctica. For each CERES θ bin, we apply the adjustment factors to $\overline{\rho}$:

$$\rho_a(\theta_0, \theta, \phi, \theta_c, \phi_{f/b}, z) = \overline{\rho}(\theta_0, \theta, \phi) + \beta(\theta_c, \phi_{f/b}, z)_{\theta,\phi} \times \sigma_\rho(\theta_0, \theta, \phi), \quad (13)$$

where $\rho_a(\theta_0, \theta, \phi, \theta_c, \phi_{f/b}, z)$ is the adjusted reflectance for a given z range; if $\phi < 90°$, we pick the forward adjustment factor $\beta(\theta_c, \phi_f, z)_{\theta,\phi}$; otherwise we use $\beta(\theta, \phi_b, z)_{\theta,\phi}$. These adjusted reflectances are integrated over all upwelling directions and the ADMs are derived from Eq. (1). Thus for each θ bin, we develop 14 ADMs representing how anisotropy changes as the reflectance deviates from the mean. To improve sampling, combined Terra and Aqua ADMs were developed over Antarctica as no significant differences were found over polar regions between measurements on these two spacecrafts.

For every instantaneous measurement, we need to know the standard score to selected the correct ADMs. The standard score is determined by using the mean reflectance and standard deviation calculated using all (RAP and cross-track) CERES measurements from Terra and Aqua for each sun-satellite viewing geometry. A linear interpolation between standard score bin midpoints is used to interpolate the anisotropic factors. The new ADMs are successful in removing the albedo dependence on solar azimuth angle (green line in Fig. 11).

Greenland

Over Greenland, we create only one set of ADMs for clear sky, as we do not have enough samples to develop separate ADMs for bright and dark surfaces (Sect. 4.3.2). For the same reason, we also combined data from Terra and Aqua to develop combined ADMs. These ADMs are also applied to all non-Antarctic permanent snow scenes.

4.4.2 Cloudy sky

For partly cloudy scenes ($0.1\,\% < f < 99.9\,\%$) over permanent snow, we use the same method as Loeb et al. (2005) and Kato and Loeb (2005) to develop ADMs for discrete intervals of f and adding linear interpolation between midpoints of f bins. For overcast scenes ($f \geq 99.9\,\%$), we separate the footprints by ECP and $\widetilde{\tau}$. We use two ECP bins (ECP < 1.5 and ECP ≥ 1.5) and four $\widetilde{\tau}$ bins ($\ln\widetilde{\tau} < 1$, $1 \leq \ln\widetilde{\tau} < 2$, $2 \leq \ln\widetilde{\tau} < 3$, and $\ln\widetilde{\tau} \geq 3$) to construct ADMs. These $\widetilde{\tau}$ bin values roughly correspond to quartiles of the overcast cloud optical depths over permanent snow. Surface brightness is not considered here. Additionally, no distinction of cloud phase is made for $\theta_0 < 60°$ to preserve sampling. Similar to the anisotropy of clouds over ocean/land, the ice phase ADMs are more isotropic than the liquid phase ADMs over permanent snow.

4.5 Sea ice

The surface of sea ice is highly variable in space and time, ranging from snow-covered ice to bare ice to melt ponds. This variability in surface type affects the characterization of SW sea ice anisotropy. As there are very few data sets available that provide information about the surface conditions of the Arctic/Antarctic sea ice, one way to obtain information is through the use of spectral channels (e.g., Rösel et al., 2012; Tschudi et al., 2008). The reflectances of 0.47 and 0.86 μm channels from MODIS provide large contrast for different sea ice surface types (Rösel et al., 2012), and we define the sea ice brightness index as

$$\eta = 1 - \frac{\rho_{0.47} - \rho_{0.86}}{\rho_{0.47} + \rho_{0.86}}. \quad (14)$$

For bright surfaces, such as snow over sea ice, η is close to 1 as the spectrum is relatively flat between the two channels. As the snow/ice melts and melt ponds begin to form, the 0.86 μm reflectance decreases faster than the 0.47 μm reflectance, causing η to decrease, with the lowest values occurring for open ocean. This is confirmed by the evolution of monthly η maps in the Arctic region (not shown). In April, η is high (above 0.9 over most of the Arctic oceans), as the sea ice is mostly covered by snow after the winter. As the year progresses into summer and snow/ice melts, η starts to decrease.

4.5.1 Clear sky

Over sea ice, clear sky is defined as footprints with $f < 1\,\%$. The ADMs for clear sky are developed for six sea ice fraction (from cloud mask algorithms) bins with angular resolutions of 5° for all θ_0, θ, and ϕ. For scenes with sea ice fraction greater than 99 %, they are further divided into three η bins (Table 2). The BRDFs of the three η classes for $\theta_0 = 65$–$70°$ are shown in Fig. 13. The lowest η bin has the lowest albedo and highest anisotropy in its BRDF. This is what we would

Table 2. Sea ice SW ADM scene type definitions by cloud fraction (%), sea ice fraction (%), sea ice brightness index (η), cloud optical depth ($\tilde{\tau}$), and effective cloud phase (ECP).

Cloud fraction (%)	Sea ice fraction (%)	Sea ice brightness	Cloud optical depth	Cloud phase
0–1	0–1	all	–	–
(clear)	1–25	all	–	–
	25–50	all	–	–
	50–75	all	–	–
	75–99	all	–	–
	99–100	$\eta \leq 0.85, 0.85 < \eta \leq 0.935, \eta > 0.935$	–	–
1–25	0–1	all	all	all
25–50	1–25	all	all	all
50–75	25–50	all	all	all
	50–75	all	all	all
	75–99	all	all	all
	99–100	$\eta \leq 0.85, 0.85 < \eta \leq 0.935, \eta > 0.935$	all	all
75–99	0–1	all	$\ln\tilde{\tau} < 1, \ln\tilde{\tau} \geq 1$	all
	1–25	all	$\ln\tilde{\tau} < 1, \ln\tilde{\tau} \geq 1$	all
	25–50	all	$\ln\tilde{\tau} < 1, \ln\tilde{\tau} \geq 1$	all
	50–75	all	$\ln\tilde{\tau} < 1, \ln\tilde{\tau} \geq 1$	all
	75–99	all	$\ln\tilde{\tau} < 1, \ln\tilde{\tau} \geq 1$	all
	99–100	$\eta \leq 0.85, 0.85 < \eta \leq 0.935, \eta > 0.935$	$\ln\tilde{\tau} < 1, \ln\tilde{\tau} \geq 1$	all
99–100	all	0–0.6	continuous in $\ln\tilde{\tau}$	ECP < 1.5, ECP ≥ 1.5
	all	0.6–0.7	continuous in $\ln\tilde{\tau}$	ECP < 1.5, ECP ≥ 1.5
	all	0.7–0.8	continuous in $\ln\tilde{\tau}$	ECP < 1.5, ECP ≥ 1.5
	all	0.8–0.9	continuous in $\ln\tilde{\tau}$	ECP < 1.5, ECP ≥ 1.5
	all	0.9–1.0	continuous in $\ln\tilde{\tau}$	ECP < 1.5, ECP ≥ 1.5

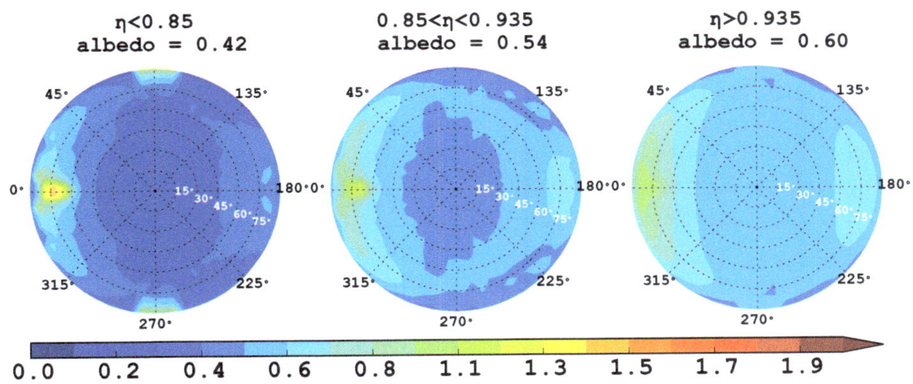

Figure 13. Clear-sky sea ice BRDFs for three η classes for $\theta_0 = 65$–$70°$.

expect, as the surface types that are being represented in this range are known to be more anisotropic than the surface represented by the higher η bins, such as snow.

The ADMs that we developed for the high η bin correspond well to the "bright" ADMs from Loeb et al. (2005) and Kato and Loeb (2005), while the ADMs for the low η bin correspond well to the "dark" ADMs. The ADMs for the mid-η bin sit between the high and low ones. The use of η provides a measure to objectively separate the scenes into different brightness categories.

4.5.2 Cloudy sky

For partly cloudy scenes ($1 \leq f < 99$ %), we use a similar approach to that of Loeb et al. (2005) and Kato and Loeb (2005). However, for partly cloudy scenes we redefine the sea ice fraction to be the fraction of the clear portion of the footprint covered by sea ice (namely, normalize the sea ice fraction by the clear-area fraction). Previously the sea ice fraction used was the fraction of the footprint covered by sea ice, which, as the sea ice under the clouds is unknown, could not be greater than the clear-area fraction. This redefinition increases the number of ADMs for partly cloudy scenes (Table 2) and allows us to be consistent in how we treat the clear-sky sea ice, regardless of the cloud fraction. For scenes with sea ice fraction greater than 99 %, we also split the scenes into three η bins (same as clear sky). We use the PSF-weighted MODIS reflectance over clear portion of the footprint to calculate η. For scenes with $75 \le f < 99\,\%$, we also construct ADMs for thin ($\ln \tilde{\tau} < 1$) and thick ($\ln \tilde{\tau} \ge 1$) clouds.

For overcast conditions ($f \ge 99\,\%$), the new ADMs are based on a linear regression between reflectance and $\ln\tilde{\tau}$. Separate regressions are derived for liquid (ECP < 1.5) and ice (ECP ≥ 1.5) clouds, and for five η bins (Table 2). As η is calculated using clear MODIS pixels, we create maps of η for each month for each 1° equal angle region from clear-sky and partly cloudy footprints. Mean reflectances are calculated for each θ_0, θ, ϕ, ECP, η, and $\ln \tilde{\tau}$ bins. The angular resolution is 5° for all three angles and the width of the $\ln \tilde{\tau}$ bin is 1. Linear fits are then performed between these mean reflectance values and the $\ln \tilde{\tau}$ bin midpoints. Figure 14a shows an example of the linear regression between mean reflectance and $\ln \tilde{\tau}$ for $\theta_0 = 55$–$60°$, $\theta = 35$–$40°$, and $\phi = 45$–$50°$. Reflectances are larger for higher η bins than for lower η bins. However, once the $\ln \tilde{\tau}$ is high enough, the reflectances intersect. This suggests there is a point where the optical thickness overcomes the effect of the surface brightness.

Reflectances from linear regressions are integrated to derive albedo for different values of $\ln \tilde{\tau}$. Albedo increases linearly with $\ln \tilde{\tau}$, and a linear regression is used to represent the relationship between the albedo and $\ln \tilde{\tau}$ (Fig. 14b). For the same $\ln \tilde{\tau}$ value, albedos are higher for ice clouds than for liquid clouds, and they are also higher for larger η. Albedos and reflectances from the linear regressions are used to derive anisotropic factors.

4.6 Mixed-scene fields of view

For CERES FOVs that consist of multiple surface types, the SW anisotropic factors are determined by accounting for the area coverage by each surface type:

$$R(\theta_0, \theta, \phi) = \frac{\pi\left(a_W \hat{I}_W + a_L \hat{I}_L + a_S \hat{I}_S\right)}{a_W \hat{F}_W + a_L \hat{F}_L + a_S \hat{F}_S}, \tag{15}$$

where a_W, a_L, and a_S are the area fraction over a FOV covered by water, land, and snow/ice; $\hat{I}_W$, $\hat{I}_L$, and $\hat{I}_S$ are the corresponding mean radiances from ADMs for the three surface types; and $\hat{F}_W$, $\hat{F}_L$, and $\hat{F}_S$ are the corresponding ADM fluxes.

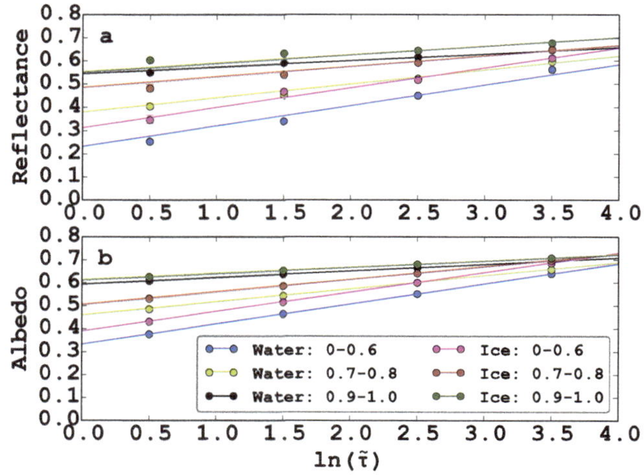

Figure 14. (a) Mean CERES reflectance values against $\ln \tilde{\tau}$ for different cloud phases and mean η values for $\theta_0 = 55$–$60°$, $\theta = 35$–$40°$, and $\phi = 45$–$50°$. The fitted linear regression lines are also shown. **(b)** The albedo values against $\ln \tilde{\tau}$ for different cloud phases and mean η values and the fitted regressions.

5 Longwave and window angular distribution models

The LW/WN anisotropy is a weak function of θ_0 and ϕ; therefore, the LW/WN ADMs are developed only as a function of θ (with a 2° angular bin resolution). Day and night ADMs are developed separately, because different cloud retrieval algorithms are used for daytime and nighttime scenes.

5.1 Clear sky over ocean, land, and desert

The clear-sky LW and WN ADMs are defined in terms of surface and meteorological properties. As in Loeb et al. (2005), there are six surface types (ocean, forests, savannas, grassland/cropland, dark deserts, and bright deserts). For footprints with mixed surface types, the anisotropy is calculated by averaging the ADM radiance and flux among the different types, weighted by the surface area of each type (Eq. 15). The meteorological variables used for the LW/WN ADMs are precipitable water (w), lapse rate (ΔT), and surface skin temperature (T_s). The lapse rate is defined as the difference between T_s and the temperature at a level 300 hPa below the surface pressure. The intervals for w, ΔT, and T_s are listed in Table 3. Note there are now additional bins used for T_s.

For clear-sky scenes, the most important variable for LW/WN anisotropy is T_s, as the ADMs become more anisotropic when T_s increases. The 20 K interval used for T_s by Loeb et al. (2005) meant that a scene with $T_s = 291$ K would be assumed to have the same anisotropy as one with $T_s = 309$ K. This leads to errors in predicted radiance in areas that have T_s close to the bin boundaries. To correct this, the skin temperature bin size was reduced from 20 to 10 K, and interpolation between the T_s bins is performed when possible. The interpolation is performed in radiance space,

Table 3. Precipitable water (w), lapse rate (ΔT), and surface skin temperature (T_S) intervals used to determine LW and WN ADMs under clear-sky conditions over the ocean, land, and desert. There are 4 w bins, 4 ΔT bins, and 10 T_S bins.

w (cm)	ΔT (K)	T_S (K)
0–1	< 15	< 260
1–3	15–30	260–340 every 10 K
3–5	30–45	> 340
> 5	> 45	

where the new anisotropic factor is calculated from a mean radiance $\hat{I}(w, \Delta T, T_s)$ linearly interpolated between $\hat{I}(w, \Delta T, T_{s_1})$ and $\hat{I}(w, \Delta T, T_{s_2})$, where $T_{s_1} < T_s < T_{s_2}$. Note that, because skin temperatures are not necessarily evenly distributed within a T_s bin, the interpolation is between bin-averaged values of T_s. There are cases where interpolation is impossible because there is not a valid ADM above or below that of the scene's ADM. In these cases, the bin's mean ADM is used as in Loeb et al. (2005). The use of smaller T_s bins and interpolation reduces the rms error (Eq. 4) over the oceans near 30° S, where the T_s is around 290 K, and the global daytime monthly mean rms error is reduced from 1.7–1.8 to 1.2–1.4 %.

We also consider additional w bins, but this has very little effect on reducing the rms error. Elevation variability is also tested as an additional LW ADM variable for daytime clear land scenes, in an attempt to reduce the rms error over rough terrain. However, adding elevation variability has very little impact on the rms error.

5.2 Clear sky over permanent snow, sea ice, and fresh snow

The ADMs over clear permanent snow, sea ice, and fresh snow are constructed for different T_s bins (Table 4), as T_s is the most important variable for LW/WN anisotropy. The Ed4SSF produces relatively few clear-sky footprints over snow and ice surfaces, particularly at the limb. For this reason, the ADMs for these surfaces are constructed using RAP measurements from both the Terra and Aqua satellites. In cases where there were no clear footprints for a given combination of surface type, θ, and T_s, the mean radiance at the given θ and T_s over all snow and ice surface types (from both day and night footprints) is used. This procedure is only used for 51 out of 1485 (or 3.4 %) total θ bins, and the bins affected tend to be at the limb, where $\hat{F}$ has low sensitivity to uncertainties in radiance. Note that for mixed snow and ice surfaces, the predominant surface type is used to determine the anisotropy.

One additional factor for clear-sky anisotropy over the poles is stratospheric temperature. Over Antarctica, the surface temperature falls suddenly to approximately 215 K with the onset of the polar night in April, but the temperature at 20 km falls from approximately 210 K in April to 185 K in July (Hudson and Brandt, 2005). Since there is very little emission from water vapor bands for such cold atmospheres, the emission from CO_2 bands in the stratosphere is the most important departure from blackbody emission, and changes in the stratospheric temperature have some impact on the anisotropy. Experiments with the Principal Component-based Radiative Transfer Model (Liu et al., 2006) using the temperature profiles from Hudson and Brandt (2005) indicate that the anisotropic factor associated with a nadir viewing zenith angle would rise from 0.995 in April to 1.002 in July. However, there is currently no stratospheric temperature information available on the SSF data product; the effect of stratospheric temperature on anisotropy will be addressed in the future.

5.3 Cloudy sky over ocean, land, and desert

When clouds are present over ocean, land, and desert surfaces, we adopt the same method as Loeb et al. (2005) to construct ADMs as a function of "pseudoradiance" (ψ), which characterizes scenes in terms of f, w, T_s, and surface–cloud temperature difference (ΔT_{sc}):

$$\psi(w, T_s, T_c, f, \epsilon_s, \epsilon_c) = (1-f)\epsilon_s B(T_s) + \sum_{j=1}^{2}\left[\epsilon_s B(T_s)\left(1-\epsilon_{c_j}\right) + \epsilon_{c_j} B\left(T_{c_j}\right)\right] f_j, \quad (16)$$

where ϵ_s is the surface infrared emissivity, ϵ_c is the cloud infrared emissivity, T_c is the cloud-top temperature, and $B(T) = \pi^{-1}\sigma T^4$ is the blackbody radiance, where σ is the Stefan–Boltzmann constant. Subscript j indicates the jth cloud layer within a CERES footprint. The quantity ϵ_{c_j} is determined according to

$$\epsilon_{c_j} = 1 - e^{-\tau_{a_j}}, \quad (17)$$

where the infrared absorption cloud optical depth of the jth layer is denoted by τ_{a_j}, and is calculated according to the approach given in Minnis et al. (1998) using the visible optical depth and ice/liquid radius retrievals from the SSF product. The pseudoradiance is meant to characterize emission from the surface and cloud tops. For a given surface type, and intervals of w, T_s, ΔT_{sc}, and f (Table 5), the average radiance is calculated for every 1 W m^{-2} sr^{-1} ψ bin. Separate ADMs are calculated for day and night, and for the Terra and Aqua satellites using RAP measurements. Here we increase the number of f bins compared to that used by Loeb et al. (2005). This additional f bin had little impact on the ADMs, but it was implemented to be consistent with the f bins used for other surface types.

A more substantial change in the cloudy-sky ADMs is in the derivation of estimated radiance from pseudoradiance.

Table 4. Surface skin temperature (T_S) intervals used to determine LW and WN ADMs under clear-sky conditions over permanent snow, sea ice, and fresh snow.

Surface type	T_S (K)	Number of T_S bins
Permanent snow (day)	< 210; 210–270 every 10 K; > 270	7
Permanent snow (night)	< 210; 210–250 every 10 K; > 250	6
Sea ice (day)	< 240; 240–270 every 10 K; > 270	5
Sea ice (night)	< 230; 230–250 every 10 K; > 250	4
Fresh snow (day)	< 240; 240–270 every 10 K; > 270	5
Fresh snow (night)	< 230; 230–270 every 10 K; > 270	6

Table 5. Surface type, precipitable water (w), cloud fraction (f), surface–cloud temperature difference (ΔT_{SC}), and surface skin temperature (T_S) intervals used to determine LW and WN ADMs under cloudy conditions over the ocean, land, and desert surface. There are 4 w bins, 5 f bins, 22 ΔT_{SC} bins, and 11 T_S bins.

Surface type	w (cm)	f (%)	ΔT_{SC} (K)	T_S (K)
Ocean	0–1	0.1–25	< −15	< 275
Land	1–3	25–50	−15 to 85 every 5 K	275 to 320 every 5 K
Desert	3–5	50–75	> 85	> 320
	> 5	75–99.9		
		99.9–100		

Loeb et al. (2005) used a third-order polynomial to characterize the relationship between radiance and ψ. The polynomial fit can be problematic for clouds with $\Delta T_{SC} > 85$ K (Fig. 15). The problems appear to arise because the polynomial is fitted to the entire range of ψ for a given interval of w, T_S, ΔT_{SC}, f, and θ. Most of the observed clouds are at low ψ values (where the observed radiance increases rapidly with ψ), while the upper part of the ψ range is for optically thin clouds (where the observed radiance increases slowly with ψ). Instead of a polynomial fit, the new ADMs use the mean observed radiance ($\hat{I}$) for each ψ bin, when possible. If the mean observed radiance at a given ψ bin is not available, we use the polynomial fit as a backup. Examples of the mean radiance at each ψ bin for three θ bins are shown in Fig. 16a for daytime ocean scenes with $w > 5$ cm, $T_S = 300$–305 K, $\Delta T_{SC} = 65$–70 K, and $f = 100$ %. We see that there is a larger decrease in $\hat{I}$ with θ for higher values of ψ, which correspond to optically thinner clouds for a fixed w, T_S, ΔT_{SC}, and f. The mean radiances $\hat{I}$ are calculated for each θ bin for different surface types for the intervals of w, f, T_S, and ΔT_{SC}. These radiances are then integrated over θ to produce $\hat{F}$, and then R is calculated from Eq. (1).

The anisotropic factors for $\psi = 51.5$ and 99.5 W m^{-2} sr^{-1} are shown in Fig. 16b for clouds with the same characteristics as those shown in Fig. 16a. We see that, for low ψ values (thick clouds), R does not change much with θ because there is little contribution from the surface, even at nadir. For high ψ values (thin clouds), a significant amount of radiance from the warm ocean surface is transmitted through the cloud at nadir, but the cold cloud top dominates as θ increases. In addition to the changes listed above, additional bins at higher values of ΔT_{SC} were tested, but this did not have a significant effect on the retrieved fluxes.

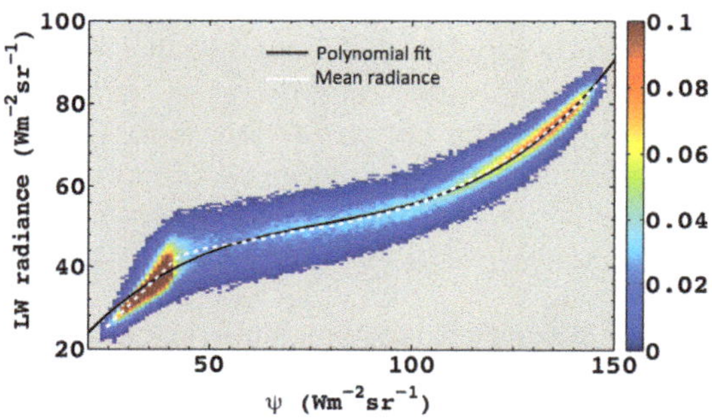

Figure 15. Observed LW radiances (contoured by percentage) for nighttime overcast clouds over ocean surfaces with $\Delta T_{SC} > 85$ K, $300 < T_S < 305$ K, $w > 5$ cm, and $60° < \theta < 62°$ plotted against pseudoradiance. Third-order polynomial fit is shown with solid black line, and mean radiance for each 1 Wm^{-2} sr^{-1} bin is shown with broken white line.

5.4 Cloudy sky over permanent snow, sea ice, and fresh snow

Loeb et al. (2005) developed ADMs over snow and ice surfaces for discrete intervals of f, T_S and ΔT_{SC}. For the new cloudy-sky ADMs over snow and ice, we use the pseudoradiance approach that is used for cloudy-sky land, ocean, and desert (Sect. 5.3). The bin intervals of f, T_S and ΔT_{SC}

Table 6. Surface type, cloud fraction (f), surface–cloud temperature difference (ΔT_{sc}), and surface skin temperature (T_s) intervals used to determine LW and WN ADMs over cloudy permanent snow, sea ice, and fresh-snow surfaces. There are 5 f bins, 10 ΔT_{sc} bins, and 8 T_s bins.

Surface type	f (%)	ΔT_{sc} (K)	T_s (K)
Permanent snow	0.1–25	< -10	< 210
Fresh snow	25–50	-10 to 70 every 10 K	210 to 270 every 10 K
Sea ice	50–75	> 70	> 270
	75–99.9		
	99.9–100		

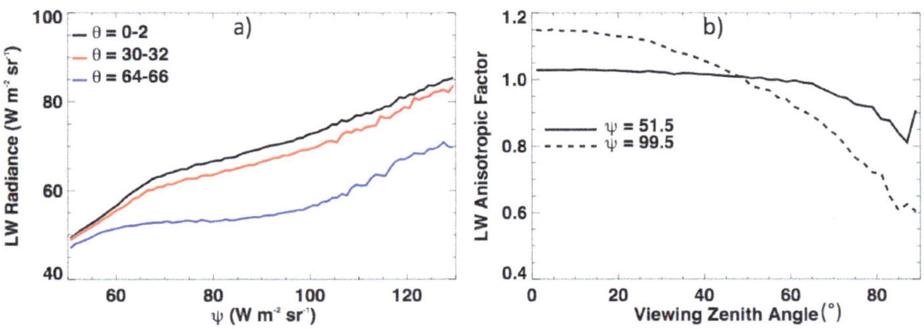

Figure 16. (a) LW radiance against ψ for three θ bins over ocean for $w > 5$ cm, $65 < \Delta T_{sc} < 70$ K, $300 < T_s < 305$ K, and $f = 100$ %. **(b)** LW anisotropic factors for $\psi = 51.5$ Wm^{-2} sr^{-1} and $\psi = 99.5$ Wm^{-2} sr^{-1}.

used for snow and ice are listed in Table 6. The pseudoradiance formulation allows the LW/WN ADMs to take cloud emissivity into account (via the cloud optical depth) over permanent snow, sea ice, and fresh-snow surfaces. Variations in cloud emissivity were not considered previously, since the only cloud property used was cloud fraction. The ADMs are constructed separately for day and night, and separately for the Terra and Aqua satellites using RAP measurements. For mixed cloudy snow and ice surface types, the predominant surface type is used to determine the anisotropy.

An example of the variation of LW radiance with ψ for three θ bins is shown in Fig. 17a for daytime permanent snow scenes with $T_s = 240$–250 K, $\Delta T_{sc} = 30$–40 K, and $f = 100$ %. The anisotropic factors for the same scenes are shown in Fig. 17b for $\psi = 39.5$ and 51.5 W m^{-2} sr^{-1}. Here the anisotropy increases with ψ, although the increment with ψ is smaller than that over warm surfaces (Fig. 16b). This is because the atmospheric (and associated cloud top) temperatures over polar regions tend to be closer to those at the surface than over warm oceans.

6 Footprints with insufficient imager information

For some CERES footprints, there is insufficient imager coverage because of missing MODIS data and/or the lack of scene identifications due to cloud property retrievals. When the total fraction of unknown cloud properties is over 35 %, we rely on artificial neural network (ANN) simulations to determine the anisotropic factors (Loukachine and Loeb, 2004). The percentage of footprints that are categorized as having insufficient imager information is greatly reduced in Ed4SSF (less than 2.0 % compared to 5.6 % for Ed2SSF) because of the improvement in cloud algorithms. We follow the methodology of Loukachine and Loeb (2004) and use 1 year of Terra and 1 year of Aqua RAP measurements to train the respective ANN ADMs. The instantaneous fluxes from the new ANN ADMs are validated against the fluxes inverted from the original ADMs. For SW, the rms difference between these two sets of fluxes is 5.3 %. For daytime LW (WN), the rms difference is about 1.0 % (1.3 %), and for nighttime LW (WN), the rms difference is 1.5 % (1.6 %). These rms differences are about half of those found in Loeb et al. (2005).

7 Effects of scene identifications and ADMs on TOA fluxes

TOA fluxes in Ed4SSF are different from those in Ed2SSF because of the changes in cloud algorithms and ADMs. To separate the changes in TOA fluxes caused by cloud algorithms from ADMs, we apply ADMs from Loeb et al. (2005) to the new cloud algorithms used for Ed4SSF (hereafter refer to as Ed4SSFβ). Thus, the differences between Ed4SSF and Ed4SSFβ are due to changes in the ADMs, and differences between Ed4SSFβ and Ed2SSF are due to changes in the cloud algorithms. A more stringent test for the ADM effects on flux would be updating the ADMs developed by Loeb

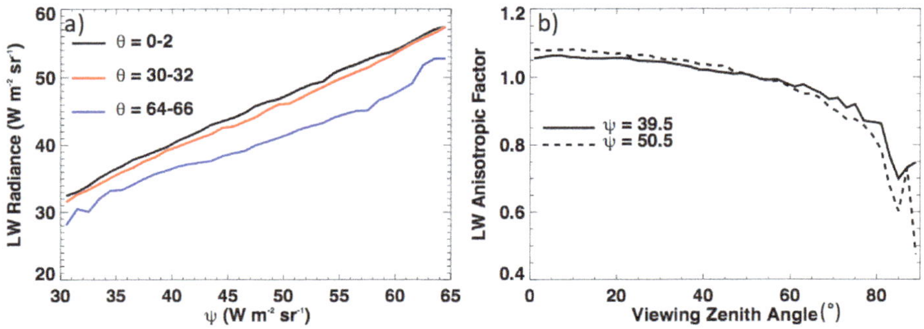

Figure 17. **(a)** LW radiance against ψ for three θ bins over permanent snow for $30 < \Delta T_{sc} < 40$ K, $240 < T_s < 250$ K, and $f = 100\,\%$. **(b)** LW anisotropic factors for $\psi = 39.5\ \mathrm{Wm^{-2}\,sr^{-1}}$ and $\psi = 50.5\ \mathrm{Wm^{-2}\,sr^{-1}}$.

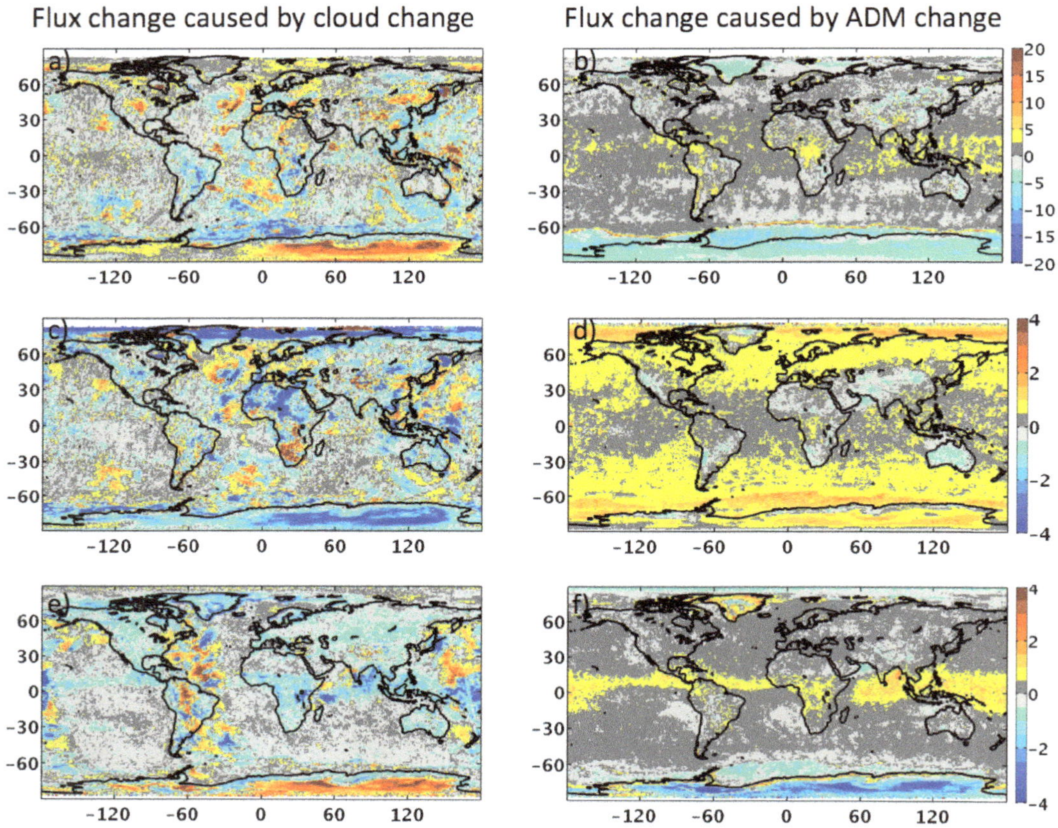

Figure 18. Instantaneous flux differences caused by changes in cloud algorithms (left panels, Ed4SSFβ-Ed2SSF) and by changes in ADMs (right panels, Ed4SSF-Ed4SSFβ): **(a)** and **(b)** for SW, **(c)** and **(d)** for daytime LW, and **(e)** and **(f)** for nighttime LW, using Terra measurements of October 2002.

et al. (2005) with the new cloud algorithms. However, the updated ADMs were not available. The ADM effects on flux would have been expected to be slightly smaller if we had used the updated ADMs than what we present below.

Figure 18 shows the effects of cloud algorithms (left panels) and ADMs (right panels) on instantaneous monthly gridded ($1° \times 1°$) TOA SW, daytime LW, and nighttime LW fluxes for October 2002 using Terra measurements. Large regional flux differences are due to changes in cloud algorithms, as the new cloud algorithms detect more clouds during both daytime and nighttime (except over Antarctica), and the clouds are also thinner (except over the polar regions during daytime) than the cloud algorithms used for Ed2SSF (Fig. 19). Additionally, because more MODIS data are available for the Ed4SSF processing and the improvement in the new cloud algorithms, there are more FOVs that provide

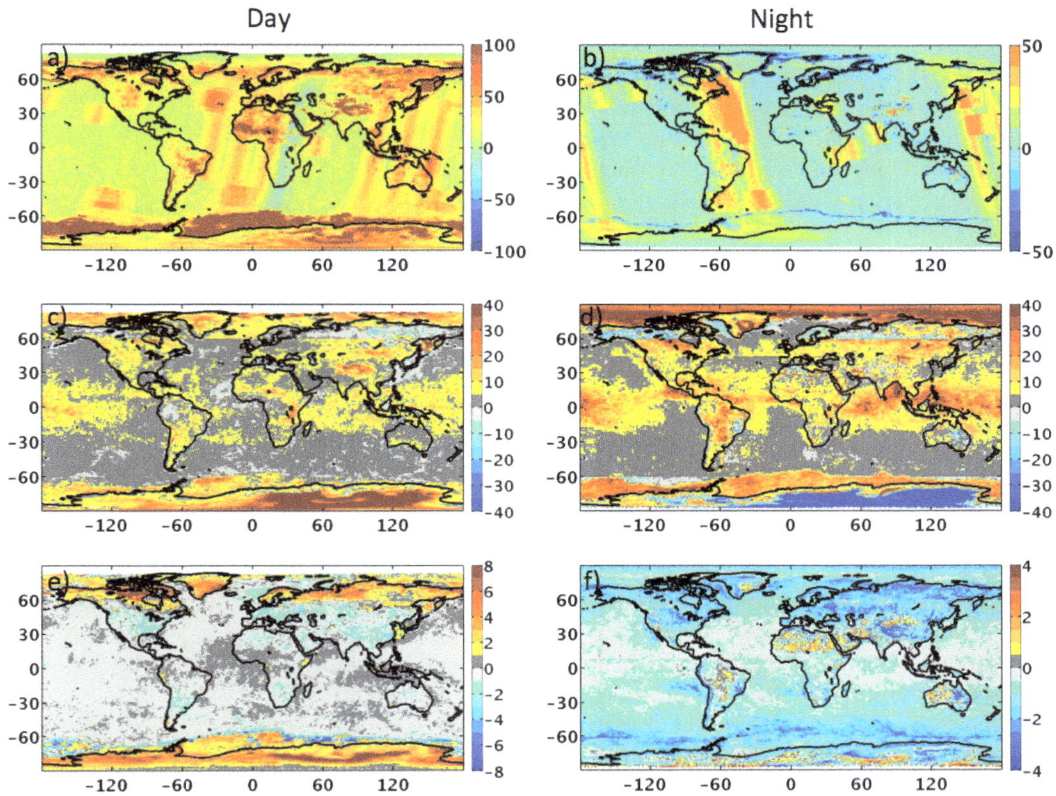

Figure 19. Daytime (right panels) and nighttime (left panels) sample number changes **(a, b)**, cloud fraction changes (%, **c** and **d**), and cloud optical depth changes **(e, f)** between Ed2SSF and Ed4SSF for each 1° latitude × 1° longitude region, using Terra measurements of October 2002.

valid flux inversions (Fig. 19a and b). Globally, the monthly mean instantaneous TOA fluxes decrease by 0.3, 0.4, and 0.3 W m^{-2} for SW, daytime LW, and nighttime LW between the new and old cloud algorithms. The new ADMs that we describe in this paper tend to increase the TOA SW fluxes (except over the polar regions), the daytime LW fluxes (except over some portion of the land regions), and the nighttime LW fluxes (except over the Antarctic). The global monthly mean instantaneous TOA fluxes increased by 0.6, 0.4, and 0.2 W m^{-2} for SW, daytime LW, and nighttime LW.

For monthly mean instantaneous TOA fluxes, we expect to see large differences (> 15 W m^{-2} for SW, > 3 W m^{-2} for LW) on regional scales due to changes in cloud algorithms and ADMs (difference between Ed2SSF and Ed4SSF). However, differences in monthly global mean instantaneous fluxes are much smaller. For example, the monthly global mean differences for Terra 2002 range from −0.2 to 0.6 W m^{-2} for SW, from −0.8 to 0.2 W m^{-2} for daytime LW, and from −0.1 to 0.0 W m^{-2} for nighttime LW. Thus, differences in 24 h averaged fluxes would be even smaller than the instantaneous flux differences, especially for SW fluxes.

8 Conclusions

To study the Earth's radiative energy budget, the CERES-measured SW and LW radiances must be converted to outgoing reflected SW and emitted LW fluxes at the TOA. To accomplish this conversion, we developed scene-type-dependent CERES SW, LW, and WN ADMs. Scene types are determined by using surface types, cloud properties, aerosol properties, and meteorological data from GEOS-5. ADMs are derived separately for Terra and Aqua using all available CERES RAP measurements on each spacecraft for most scene types. Over some snow/ice scenes, combined Terra and Aqua ADMs are developed to improve sampling.

Over clear ocean, CERES SW ADMs over the non-glint regions are developed for six wind speed bins, three aerosol optical depth bins, and two aerosol types; and SW ADMs over the glint regions are developed without considering the aerosol types. Over clear land/desert/fresh snow, SW ADMs are developed for every 1° latitude × 1° longitude region for every calendar month using Ross–Li BRDF model. The kernels for the BRDF model are derived for different intervals of μ_0, NDVI, and two categories of EV over rough terrain. We construct clear permanent snow ADMs separately for Antarctica and Greenland. Over Antarctica, we use MISR measurements to construct ADMs that account for the effects of sastrugi on anisotropy (but we do not need MISR measurements to apply the ADMs); over Greenland, we develop only one set of ADMs as sastrugi are not a noticeable feature there. Over clear sea ice scenes, the SW ADMs are developed for six sea ice fraction bins, and when sea ice fraction is greater than 99 %, the scenes are further divided into three sea ice brightness index bins.

Over cloudy ocean, SW ADMs are defined as continuous functions of $\ln(f\widetilde{\tau})$, where f and $\widetilde{\tau}$ are cloud percentage coverage and cloud optical depth derived from collocated imager pixels. The relationship between CERES-measured SW radiance and $\ln(f\widetilde{\tau})$ is quantified using a sigmoidal fit to develop ADMs, and the ADMs are derived separately for liquid, mixed, and ice phase clouds. Over cloudy land/desert, the sigmoidal fit is also used to relate the CERES SW radiance from clouds with $\ln(f\widetilde{\tau})$, by accounting for the surface anisotropy using the clear-sky ADMs. Over cloudy fresh-snow surfaces, the SW ADMs are constructed for discrete intervals of cloud fraction and snow fraction; under overcast conditions, ADMs are also derived by accounting for surface brightness and cloud optical depth. Over cloudy permanent snow, the SW ADMs are constructed for discrete intervals of cloud fraction; under overcast conditions, cloud optical depth and cloud phase are also included in ADM development. Over cloudy sea ice, ADMs are developed accounting for cloud fraction, sea ice fraction and brightness; for scenes with cloud fraction greater than 75 %, ADMs are developed for two categories of cloud optical depth; for overcast conditions, linear regression between reflectance and $\ln\widetilde{\tau}$ are used to develop ADMs for liquid and ice phase clouds under different sea ice brightness bins.

Under clear-sky conditions, the CERES LW/WN ADMs are developed for different surface types. Over land and ocean, ADMs are derived for discrete intervals of precipitable water, surface skin temperature, and lapse rate; over snow and ice surface, ADMs are derived only for discrete intervals of surface skin temperature. For cloudy conditions, the LW/WN ADMs are developed by combining surface and cloud-top temperature, surface and cloud emissivity, cloud fraction, and precipitable water (only over ocean, land, desert) into a "pseudoradiance" parameter.

The new ADMs described in this paper change the monthly mean $1^\circ \times 1^\circ$ instantaneous fluxes by less than $5\,\mathrm{W\,m^{-2}}$, compared to the ADMs developed by Loeb et al. (2005). However, the changes in global monthly mean instantaneous TOA fluxes are much smaller, generally less than $0.5\,\mathrm{W\,m^{-2}}$.

Acknowledgements. The authors thank Norman Loeb, Patrick Minnis, Sunny Sun-Mack, Walter Miller, Natividad Monalo-Smith, Constantine Lukashin, Seiji Kato, and Qing Trepte for helpful discussions.

Edited by: S. Schmid

References

Ahmad, S. P. and Deering, D. W.: A simple analytical function for bidirectional reflectance, J. Geophys. Res., 97, 18867–18886, 1992.

Bevan, S. L., North, P. R. J., Grey, W. M. F., Los, S. O., and Plummer, S. E.: Impact of atmospheric aerosol from biomass burning on Amazon dry-season drought, J. Geophys. Res., 114, D09204, doi:10.1029/2008JD011112, 2009.

Bradley, A. V., Gerard, F. F., Barbier, N., Weedon, G. P., Anderson, L. O., Huntingford, C., Aragao, L. E. O. C., Zelazowski, P., and Arai, E.: Relationships between phenology, radiation, and precipitation in the Amazon region, Global Change Biol., 17, 2245–2260, 2011.

Chang, F.-L., Minnis, P., Ayers, J. K., McGill, M. J., Palikonda, R., Spangenberg, D. A., Smith Jr., W. L., and Yost, C. R.: Evaluation of satellite-based upper troposphere cloud top height retrieval in multilayer cloud conditions during TC4, J. Geophys. Res., 115, D00J05, doi:10.1029/2009JD013305, 2010.

Chopping, M., Su, L., Rango, A., Martonchik, J. V., Peters, D. P. C., and Laliberte, A.: Remote sensing of woody shrub cover in desert grasslands using MISR with a geometric-optical canopy reflectance model, Remote Sens. Environ., 112, 19–34, doi:10.1016/j.rse.2006.04.023, 2008.

Corbett, J. G., Su, W., and Loeb, N. G.: Observed effects of sastrugi on CERES top-of-atmosphere clear-sky reflected shortwave flux over Antarctica, J. Geophys. Res., 117, D18104, doi:10.1029/2012JD017529, 2012.

Diner, D. J., Beckert, J. C., H.Reilly, T., Bruegge, C. J., Conel, J. E., Kahn, R. A., Martonchik, J. V., Ackerman, T. P., Davies, R., Gerstl, S. A. W., and Gordon, H. R.: Multi-angle Imaging SpectroRadiometer (MISR) instrument description an dexperiment overview, IEEE T. Geosci. Remote, 36, 1072–1087, 1998.

Fu, Q. and Liou, K.-N.: Parameterization of the radiative properties of cirrus clouds, J. Atmos. Sci., 50, 2008–2025, 1993.

Hall, D. K., Riggs, G. A., Salomonson, V. V., DiGirolamo, N. E., and Bayr, K. J.: MODIS snow-cover products, Remote Sens. Environ., 83, 181–194, 2002.

Hess, M., Koepke, P., and Schult, I.: Optical properties of aerosols and clouds: The software package OPAC, B. Am. Meteorol. Soc., 79, 831–844, 1998.

Hsu, C., Tsay, S., King, M. D., and Herman, M.: Aerosol properties over bright-reflecting source regions, IEEE T. Geosci. Remote, 42, 557–569, 2004.

Hudson, S. R. and Brandt, R. E.: A look at the surface-based temperature inversion on the Antarctic Plateau, J. Climate, 18, 1673–1696., 2005.

Kato, S. and Loeb, N. G.: Top-of-atmosphere shortwave broadband observed radiance and estimated irradiance over polar regions from Clouds and the Earth's Radiant Energy System (CERES) instruments on Terra, J. Geophys. Res., 110, D07202, doi:10.1029/2004JD005308, 2005.

Kato, S., Rose, F. G., Sun-Mack, S., Miller, W. F., Chen, Y., Rutan, D. A., Stephens, G. L., Loeb, N. G., Minnis, P., Wielicki, B. A., Winker, D. M., Charlock, T. P., Stackhouse Jr., P. W., Xu, K.-M., and Collins, W. D.: Improvements of top-of-atmosphere and surface irradiance computation with CALIPSO-, and MODIS-derived cloud and aerosol properties, J. Geophys. Res., 116, D19209, doi:10.1029/2011JD016050, 2011.

Kuchiki, K., Aoki, T., Niwano, M., and Motoyoshi, H.: Effect of sastrugi on snow bidirectional reflectance and its application to MODIS data, J. Geophys. Res., 116, D18110, doi:10.1029/2011JD016070, 2011.

Levy, R. C., Remer, L. A., Kleidman, R. G., Mattoo, S., Ichoku, C., Kahn, R., and Eck, T. F.: Global evaluation of the Collection 5 MODIS dark-target aerosol products over land, Atmos. Chem. Phys., 10, 10399–10420, doi:10.5194/acp-10-10399-2010, 2010.

Li, X. and Strahler, A. H.: Geometric-optical bidirectional reflectance modeling of the discrete crown vegetation canopy: effect of crown shape and mutual shadowing, IEEE T. Geosci. Remote, 30, 276–292, 1992.

Liu, X., Smith, W. L., Zhou, D. K., and Larar, A.: Principal component-based radiative transfer model for hyperspectral sensors: theoretical concept, Appl. Optics, 45, 201–209, 2006.

Loeb, N. G. and Manalo-Smith, N.: Top-of-atmosphere direct radiative effect of aerosols over global oceans from merged CERES and MODIS observations, J. Climate, 18, 3506–3526, 2005.

Loeb, N. G. and Schuster, G. L.: An observational study of the relationship between cloud, aerosol and meteorology in broken low-level cloud conditions, J. Geophys. Res., 113, D14214, doi:10.1029/2007JD009763, 2008.

Loeb, N. G., Manalo-Smith, N., Kato, S., Miller, W. F., Gupta, S. K., Minnis, P., and Wielicki, B. A.: Angular Distribution Models for Top-of-Atmosphere Radiative Flux Estimation from the Clouds and the Earth's Radiant Energy System Instrument on the Tropical Rainfall Measuring Mission Satellite, Part I: Methodology, J. Appl. Meteorol., 42, 240–265, 2003.

Loeb, N. G., Kato, S., Loukachine, K., and Manalo-Smith, N.: Angular Distribution Models for Top-of-Atmosphere Radiative Flux Estimation from the Clouds and the Earth's Radiant Energy System Instrument on the Terra Satellite, Part I: Methodology, J. Atmos. Ocean. Tech., 22, 338–351, 2005.

Loeb, N. G., Sun, M., Miller, W. F., Loukachine, K., and Davies, R.: Fusion of CERES, MISR, and MODIS measurements for top-of-atmosphere radiative flux validation, J. Geophys. Res., 111, D18209, doi:10.1029/2006JD007146, 2006.

Loeb, N. G., Kato, S., Loukachine, K., Manalo-Smith, N., and Doelling, D. R.: Angular Distribution Models for Top-of-Atmosphere Radiative Flux Estimation from the Clouds and the Earth's Radiant Energy System Instrument on the Terra Satellite, Part II: Validation, J. Atmos. Ocean. Tech., 24, 564–584, 2007.

Loeb, N. G., Lyman, J. M., Johnson, G. C., Allan, R. P., Doelling, D. R., Wong, T., Soden, B. J., and Stephens, G. L.: Observed changes in top-of-the-atmosphere radiation and upper-ocean heating consistent within uncertainty, Nat. Geosci., 5, 110–113, doi:10.1038/NGEO1375, 2012.

Loukachine, K. and Loeb, N. G.: Top-of-atmosphere flux retrievals from CERES using artifical neural networks, Remote Sens. Environ., 93, 381–390, 2004.

Lucht, W., Schaaf, C. B., and Strahler, A. H.: An anogorithm for the retrieval of albedo from space using semiempirical BRDF models, IEEE T. Geosci. Remote, 38, 977–998, 2000.

Lyapustin, A., Gatebe, C. K., Kahn, R., Brandt, R., Redemann, J., Russell, P., King, M. D., Pedersen, C. A., Gerland, S., Poudyal, R., Marshak, A., Wang, Y., Schaaf, C., Hall, D., and Kokhanovsky, A.: Analysis of snow bidirectional reflectance from ARCTAS Spring-2008 Campaign, Atmos. Chem. Phys., 10, 4359–4375, doi:10.5194/acp-10-4359-2010, 2010.

Maignan, F., Breon, F.-M., and Lacaze, R.: Bidirectional reflectance of Earth targets: Evaluation of analytical models using a large set of spaceborne measurements with emphasis on the hot spot, Remote Sens. Environ., 90, 210–220, 2004.

Martins, J. V., Tanre, D., Remer, L. A., Kaufman, J. Y., Mattoo, S., and Levy, R.: MODIS cloud screening for remote sensing of aerosols over oceans using spatial variability, Geophys. Res. Lett., 29, 1619–1622, 2002.

Minnis, P., Garber, D. P., Young, D. F., Arduini, R. F., and Takano, Y.: Parameterizations of reflectance and effective emittance for satellite remote sensing of cloud properties, J. Atmos. Sci., 55, 3313–3339, 1998.

Minnis, P., Trepte, C. R., Sun-Mack, S., Chen, Y., Doelling, D. R., Young, D. F., Spangenberg, D. A., Miller, W. F., Wielicki, B. A., Brown, R. R., Gibson, S. C., and Geier, E. B.: Cloud detection in nonpolar regions for CERES using TRMM VIRS and TERRA and AQUA MODIS data, IEEE T. Geosci. Remote, 46, 3857–3884, 2008.

Minnis, P., Sun-Mack, S., Trepte, Q. Z., Chang, F.-L., Heck, P. W., Chen, Y., Yi, Y., Arduini, R. F., Ayers, K., Bedka, K., Bedka, S., and Brown, R.: CERES Edition 3 Cloud Retrievals, in: 13th Conference on Atmospheric Radiation, Am. Meteorol. Soc., Oregon, Portland, 2010.

Minnis, P., Sun-Mack, S., Young, D. F., Heck, P. W., Garber, D. P., Chen, Y., Spangenberg, D. A., Arduini, R. F., Trepte, Q. Z., Smith, W. L. J., Ayers, J. K., Gibson, S. C., Miller, W. F., Chakrapani, V., Takano, Y., Liou, K., and Xie, Y.: CERES Edition-2 cloud property retrievals using TRMM VIRS and TERRA and AQUA MODIS data, Part I: Algorithms, IEEE T. Geosci. Remote, 49, 4374–4400, doi:10.1109/TGRS.2011.2144601, 2011.

Pincus, R., Batstone, C. P., Hofmann, R. J. P., Taylor, K. E., and Glecker, P. J.: Evaluating the present-day simulation of clouds, precipitation, and radiation in climate models, J. Geophys. Res., 113, D14209, doi:10.1029/2007JD009334, 2008.

Quaas, J., Boucher, O., Bellouin, N., and Kinne, S.: Satellite-based estimate of the direct and indirect aerosol climate forcing, J. Geophys. Res., 113, D05204, doi:10.1029/2007JD008962, 2008.

Rahman, H., Pinty, B., and Verstraete, M. M.: Coupled surface-atmosphere reflectance (CSAR) model 2. Semiempirical surface model usable with NOAA Advanced Very High Resolution Radiometer data, J. Geophys. Res., 98, 20,791–20,801, 1993.

Remer, L. A., Kleidman, R. G., Levy, R. C., Kaufman, Y. J., Tanre, D., Mattoo, S., Martins, J. V., Ichoku, C., Koren, I., Yu, H., and Holben, B.: Global aerosol climatology from the MODIS satellite sensors, J. Geophys. Res., 113, D14S07, doi:10.1029/2007JD009661, 2008.

Rose, F. G., Rutan, D. A., Charlock, T. P., Smith, G. L., and Kato, S.: An algorithm for the constraining of radiative transfer calculations to CERES-observed broadband top-of-atmosphere irradiance, J. Atmos. Ocean. Tech., 30, 1091–1106, doi:10.1175/JTECH-D-12-00058.1, 2013.

Rösel, A., Kaleschke, L., and Birnbaum, G.: Melt ponds on Arctic sea ice determined from MODIS satellite data using an artificial neural network, The Cryosphere, 6, 431–446, doi:10.5194/tc-6-431-2012, 2012.

Roujean, J.-L., Leroy, M., and Deschamps, P.-Y.: A bidirectional reflectance model of the Earth's surface for the correction of remote sensing data, J. Geophys. Res., 97, 20455–20468, 1992.

Satheesh, S. K. and Ramanathan, V.: Large differences in tropcial aerosol forcing at the top of the atmosphere and Earth's surface, Nature, 405, 60–63, 2000.

Smith, G. L.: Effects of time response on the point spread function of a scanning radiometer, Appl. Optics, 33, 7031–7037, 1994.

Smith, G. L., Green, R. N., Raschke, E., Avis, L. M., Suttles, J. T., Wielicki, B. A., and Davies, R.: Inversion methods for satellite studies of the Earth's radiation budget: development of algorithms for the ERBE mission, Rev. Geophys., 24, 407–421, 1986.

Stephens, G. L., Li, J.-L., Wild, M., Clayson, C. A., Loeb, N. G., Kato, S., L'Ecuyer, T., Stackhouse., P. W., Lebsock, M., and Andrews, T.: An update on Earth's energy balance in light of the latest global observations, Nat. Geosci., 5, 691–696, doi:10.1038/NGEO1580, 2012.

Su, W., Bodas-Salcedo, A., Xu, K.-M., and Charlock, T. P.: Comparison of the tropical radiative flux and cloud radiative effect profiles in a climate model with Clouds and the Earth's Radiant Energy System (CERES) data, J. Geophys. Res., 115, D01105, doi:10.1029/2009JD012490, 2010a.

Su, W., Loeb, N. G., Xu, K., Schuster, G. L., and Eitzen, Z. A.: An estimate of aerosol indirect effect from satellite measurements with concurrent meteorological analysis, J. Geophys. Res., 115, D18219, doi:10.1029/2010JD013948, 2010b.

Su, W., Loeb, N. G., Schuster, G. L., Chin, M., and Rose, F. G.: Global all-sky shortwave direct radiative forcing of anthropogenic aerosols from combined satellite observations and GOCART simulations, J. Geophys. Res., 118, 1–15, doi:10.1029/2012JD018294, 2013.

Sun-Mack, S., Minnis, P., Chen, Y., Kato, S., Yi, Y., Gibson, S. C., Heck, P. W., and Winker, D. M.: Regional apparent boundary layer lapse rates determined from CALIPSO and MODIS data fro cloud height determination, J. Appl. Meteorol. Clim., 53, 990–1011, doi:10.1175/JAMC-D-13-081.1, 2014.

Suttles, J. T., Green, R. N., Minnis, P., Smith, G. L., Staylor, W. F., Wielicki, B. A., Walker, I. J., Young, D. F., Taylor, V. R., and Stowe, L. L.: Angular radiation models for Earth-atmosphere system, Vol. I, Shortwave radiation, NASA RP-1184, Tech. rep., NASA Langley Research Center, Hampton, Virginia, 1988.

Thomas, G. E. and Stamnes, K.: Radiative Transfer in the Atmosphere and Ocean, Cambridge University Press, 1999.

Trenberth, K. E., Fasullo, J. T., and Kiehl, J.: Earth's global energy budget, B. Am. Meteorol. Soc., 90, 311–323, doi:10.1175/2008BAMS2634.1, 2009.

Tschudi, M., Maslanik, J., and Perovich, D.: Derivation of melt pond coverage on Arctic sea ice using MODIS observations, Remote Sens. Environ., 112, 2605–2614, 2008.

Wang, H. and Su, W.: Evaluating and understanding top of the atmosphere cloud radiative effects in Intergovernmental Panel on Climate Change (IPCC) fifth assessment report (AR5) coupled model intercomparison project phase 5 (CMIP5) models using satellite observations, J. Geophys. Res., 118, 1–17, doi:10.1029/2012JD018619, 2013.

Warren, S., Brandt, R., and Hinton, P.: Effect of surface roughness on bidirectional reflectance of Antarctic snow, J. Geophys. Res., 103, 25789–25807, 1998.

Wielicki, B. A., Barkstrom, B. R., Harrison, E. F., Lee, R. B., Smith, G. L., and Cooper, J. E.: Clouds and the Earth's Radiant Energy System (CERES): An Earth Observing System Experiment, B. Am. Meteorol. Soc., 77, 853–868, 1996.

Wild, M., Folini, D., Schar, C., Loeb, N. G., Dutton, E. G., and Konig-Langlo, G.: The global energy balance from a surface perspective, Clim. Dynam., 40, 3107–3134, doi:10.1007/s00382-012-1569-8, 2013.

Yang, P., Kattawar, G. W., Hong, G., Minnis, P., and Hu, Y.: Uncertainties associated with the surface texture of ice particles in satellite-based retrieval of cirrus clouds – Part I: single-scattering properties of ice crystals with surface roughness, IEEE T. Geosci. Remote, 46, 1940–1947, 2008.

Zhang, J., Christopher, S. A., Remer, L. A., and Kaufman, Y. J.: Shortwave aerosol radiative forcing over cloud-free oceans from Terra: 2. Seasonal and global distributions, J. Geophys. Res., 110, D10S24, doi:10.1029/2004JD005009, 2005.

Permissions

All chapters in this book were first published in AMT, by Copernicus Publications; hereby published with permission under the Creative Commons Attribution License or equivalent. Every chapter published in this book has been scrutinized by our experts. Their significance has been extensively debated. The topics covered herein carry significant findings which will fuel the growth of the discipline. They may even be implemented as practical applications or may be referred to as a beginning point for another development.

The contributors of this book come from diverse backgrounds, making this book a truly international effort. This book will bring forth new frontiers with its revolutionizing research information and detailed analysis of the nascent developments around the world.

We would like to thank all the contributing authors for lending their expertise to make the book truly unique. They have played a crucial role in the development of this book. Without their invaluable contributions this book wouldn't have been possible. They have made vital efforts to compile up to date information on the varied aspects of this subject to make this book a valuable addition to the collection of many professionals and students.

This book was conceptualized with the vision of imparting up-to-date information and advanced data in this field. To ensure the same, a matchless editorial board was set up. Every individual on the board went through rigorous rounds of assessment to prove their worth. After which they invested a large part of their time researching and compiling the most relevant data for our readers.

The editorial board has been involved in producing this book since its inception. They have spent rigorous hours researching and exploring the diverse topics which have resulted in the successful publishing of this book. They have passed on their knowledge of decades through this book. To expedite this challenging task, the publisher supported the team at every step. A small team of assistant editors was also appointed to further simplify the editing procedure and attain best results for the readers.

Apart from the editorial board, the designing team has also invested a significant amount of their time in understanding the subject and creating the most relevant covers. They scrutinized every image to scout for the most suitable representation of the subject and create an appropriate cover for the book.

The publishing team has been an ardent support to the editorial, designing and production team. Their endless efforts to recruit the best for this project, has resulted in the accomplishment of this book. They are a veteran in the field of academics and their pool of knowledge is as vast as their experience in printing. Their expertise and guidance has proved useful at every step. Their uncompromising quality standards have made this book an exceptional effort. Their encouragement from time to time has been an inspiration for everyone.

The publisher and the editorial board hope that this book will prove to be a valuable piece of knowledge for researchers, students, practitioners and scholars across the globe.

List of Contributors

P. S. Chhabra
Aerodyne Research, Inc. Billerica, Massachusetts, USA
Department of Chemical Engineering, University of Texas, Austin, Texas, USA

A. T. Lambe
Aerodyne Research, Inc. Billerica, Massachusetts, USA
Chemistry Department, Boston College, Chestnut Hill, Massachusetts, USA

M. R. Canagaratna
Aerodyne Research, Inc. Billerica, Massachusetts, USA

H. Stark
Aerodyne Research, Inc. Billerica, Massachusetts, USA
Cooperative Institute for Research in Environmental Sciences (CIRES), University of Colorado, Boulder, Colorado, USA

J. T. Jayne
Aerodyne Research, Inc. Billerica, Massachusetts, USA

T. B. Onasch
Aerodyne Research, Inc. Billerica, Massachusetts, USA
Chemistry Department, Boston College, Chestnut Hill, Massachusetts, USA

P. Davidovits
Chemistry Department, Boston College, Chestnut Hill, Massachusetts, USA

J. R. Kimme
Aerodyne Research, Inc. Billerica, Massachusetts, USA
Cooperative Institute for Research in Environmental Sciences (CIRES), University of Colorado, Boulder, Colorado, USA
TOFWERK AG, Thun, Switzerland

D. R. Worsnop
Aerodyne Research, Inc. Billerica, Massachusetts, USA

M. F. Schibig
Climate and Environmental Physics, Physics Institute and Oeschger Centre for Climate Change Research, University of Bern, Bern, Switzerland

M. Steinbacher
Swiss Federal Laboratories for Materials Science and Technology (Empa), Dübendorf, Switzerland

B. Buchmann
Swiss Federal Laboratories for Materials Science and Technology (Empa), Dübendorf, Switzerland

I. T. van der Laan-Luijkx
Climate and Environmental Physics, Physics Institute and Oeschger Centre for Climate Change Research, University of Bern, Bern, Switzerland
Meteorology and Air Quality, Wageningen University, Wageningen, the Netherlands

S. van der Laan
Climate and Environmental Physics, Physics Institute and Oeschger Centre for Climate Change Research, University of Bern, Bern, Switzerland
Centre for Ocean and Atmospheric Sciences, School of Environmental Sciences, University of East Anglia, Norwich, UK

S. Ranjan
Climate and Environmental Physics, Physics Institute and Oeschger Centre for Climate Change Research, University of Bern, Bern, Switzerland

M. C. Leuenberger
Climate and Environmental Physics, Physics Institute and Oeschger Centre for Climate Change Research, University of Bern, Bern, Switzerland

M. Kaufmann
Institute of Energy and Climate Research, Stratosphere (IEK-7), Research Centre Jülich, 52425 Jülich, Germany

J. Blank
Institute of Energy and Climate Research, Stratosphere (IEK-7), Research Centre Jülich, 52425 Jülich, Germany

T. Guggenmoser
Institute of Energy and Climate Research, Stratosphere (IEK-7), Research Centre Jülich, 52425 Jülich, Germany

J. Ungermann
Institute of Energy and Climate Research, Stratosphere (IEK-7), Research Centre Jülich, 52425 Jülich, Germany

A. Engel
Institute for Atmospheric and Environmental Sciences, Goethe University Frankfurt, 60438 Frankfurt am Main, Germany

M. Ern
Institute of Energy and Climate Research, Stratosphere (IEK-7), Research Centre Jülich, 52425 Jülich, Germany

F. Friedl-Vallon
Institute for Meteorology and Climate Research, Karlsruhe Institute of Technology, Karlsruhe, Germany

D. Gerber
RAL Space, STFC Rutherford Appleton Laboratory, Harwell Oxford, Didcot, OX11 0QX, UK

J. U. Grooß
Institute of Energy and Climate Research, Stratosphere (IEK-7), Research Centre Jülich, 52425 Jülich, Germany

G. Guenther
Institute of Energy and Climate Research, Stratosphere (IEK-7), Research Centre Jülich, 52425 Jülich, Germany

M. Höpfner
Institute for Meteorology and Climate Research, Karlsruhe Institute of Technology, Karlsruhe, Germany

A. Kleinert
Institute for Meteorology and Climate Research, Karlsruhe Institute of Technology, Karlsruhe, Germany

E. Kretschmer
Institute for Meteorology and Climate Research, Karlsruhe Institute of Technology, Karlsruhe, Germany

Th. Latzko
Institute for Meteorology and Climate Research, Karlsruhe Institute of Technology, Karlsruhe, Germany

G. Maucher
Institute for Meteorology and Climate Research, Karlsruhe Institute of Technology, Karlsruhe, Germany

T. Neubert
Central Institute of Engineering, Electronics and Analytics (ZEA-2) - Electronic Systems, Research Centre Jülich, Jülich, Germany

H. Nordmeyer
Institute for Meteorology and Climate Research, Karlsruhe Institute of Technology, Karlsruhe, Germany

H. Oelha
Institute for Meteorology and Climate Research, Karlsruhe Institute of Technology, Karlsruhe, Germany

F. Olschewski
Physics Department, University of Wuppertal, 42097 Wuppertal, Germany

J. Orphal
Institute for Meteorology and Climate Research, Karlsruhe Institute of Technology, Karlsruhe, Germany

P. Preusse
Institute of Energy and Climate Research, Stratosphere (IEK-7), Research Centre Jülich, 52425 Jülich, Germany

H. Schlager
German Aerospace Center, Institute of Atmospheric Physics, Atmospheric Trace Species, 82234 Oberpfaffenhofen-Wessling, Germany

H. Schneider
Central Institute of Engineering, Electronics and Analytics (ZEA-2) - Electronic Systems, Research Centre Jülich, Jülich, Germany

D. Schuettemeyer
European Space Agency, Mission Science Division, 2200AG Noordwijk, the Netherlands

F. Stroh
Institute of Energy and Climate Research, Stratosphere (IEK-7), Research Centre Jülich, 52425 Jülich, Germany

O. Suminska-Ebersoldt
Institute for Meteorology and Climate Research, Karlsruhe Institute of Technology, Karlsruhe, Germany

B. Vogel
Institute of Energy and Climate Research, Stratosphere (IEK-7), Research Centre Jülich, 52425 Jülich, Germany

C. M. Volk
Physics Department, University of Wuppertal, 42097 Wuppertal, Germany

W. Woiwode
Institute for Meteorology and Climate Research, Karlsruhe Institute of Technology, Karlsruhe, Germany

M. Riese
Institute of Energy and Climate Research, Stratosphere (IEK-7), Research Centre Jülich, 52425 Jülich, Germany
Google Inc., Dublin, Ireland

D. Tátrai
University of Szeged, Department of Optics and Quantum Electronics, Szeged, Hungary
SZTE-MTA Research Group on Photoacoustic Spectroscopy, Szeged, Hungary

Z. Bozóki
University of Szeged, Department of Optics and Quantum Electronics, Szeged, Hungary
SZTE-MTA Research Group on Photoacoustic Spectroscopy, Szeged, Hungary

H. Smit
Forschungszentrum Jülich, Institute for Energy and Climate Research Troposphere (IEK-8), Jülich, Germany

C. Rolf
Forschungszentrum Jülich, Institute for Energy and Climate Research Stratosphere (IEK-7), Jülich, Germany

N. Spelten
Forschungszentrum Jülich, Institute for Energy and Climate Research Stratosphere (IEK-7), Jülich, Germany

M. Krämer
Forschungszentrum Jülich, Institute for Energy and Climate Research Stratosphere (IEK-7), Jülich, Germany

A. Filges
Max Planck Institute for Biogeochemistry, Department of Biogeochemical Systems, Jena, Germany

C. Gerbig
Max Planck Institute for Biogeochemistry, Department of Biogeochemical Systems, Jena, Germany

G. Gulyás
Hilase Ltd., Szeged, Hungary

G. Szabó
University of Szeged, Department of Optics and Quantum Electronics, Szeged, Hungary
SZTE-MTA Research Group on Photoacoustic Spectroscopy, Szeged, Hungary

D. Cimini
IMAA-CNR, Potenza, Italy
CETEMPS, University of L'Aquila, L'Aquila, Italy

M. Nelson
Radiometrics, Boulder, Colorado, USA

J. Güldner
DWD, Meteorological Observatory Lindenberg – Richard Aßmann Observatory, Lindenberg, Germany

R. Ware
Radiometrics, Boulder, Colorado, USA
UCAR, Boulder, Colorado, USA

T. D. Thornberry
NOAA ESRL Chemical Sciences Division, Boulder, Colorado, USA
Cooperative Institute for Research in Environmental Sciences, University of Colorado, Boulder, Colorado, USA

A.W. Rollins
NOAA ESRL Chemical Sciences Division, Boulder, Colorado, USA
Cooperative Institute for Research in Environmental Sciences, University of Colorado, Boulder, Colorado, USA

R. S. Gao
NOAA ESRL Chemical Sciences Division, Boulder, Colorado, USA

L. A. Watts
NOAA ESRL Chemical Sciences Division, Boulder, Colorado, USA
Cooperative Institute for Research in Environmental Sciences, University of Colorado, Boulder, Colorado, USA

S. J. Ciciora
NOAA ESRL Chemical Sciences Division, Boulder, Colorado, USA

R. J. McLaughlin
NOAA ESRL Chemical Sciences Division, Boulder, Colorado, USA
Cooperative Institute for Research in Environmental Sciences, University of Colorado, Boulder, Colorado, USA

D. W. Fahey
NOAA ESRL Chemical Sciences Division, Boulder, Colorado, USA
Cooperative Institute for Research in Environmental Sciences, University of Colorado, Boulder, Colorado, USA

H. C. Pumphrey
School of GeoSciences, The University of Edinburgh, Edinburgh, UK

W. G. Read
Jet Propulsion Laboratory, California Institute of Technology, Pasadena, CA, USA

N. J. Livesey
Jet Propulsion Laboratory, California Institute of Technology, Pasadena, CA, USA

K. Yang
Department of Atmospheric and Oceanic Science, University of Maryland, College Park, MD, USA

A. Sanchez-Romero
Department of Physics, University of Girona, Girona, Spain

J. A. González
Department of Physics, University of Girona, Girona, Spain

J. Calbó
Department of Physics, University of Girona, Girona, Spain

A. Sanchez-Lorenzo
Department of Physics, University of Girona, Girona, Spain
Pyrenean Institute of Ecology, Spanish National Research Council (CSIC), Zaragoza, Spain

T. H. Raupach
Environmental Remote Sensing Laboratory, School of Architecture, Civil, and Environmental Engineering, École Polytechnique Fédérale de Lausanne (EPFL), 1015 Lausanne, Switzerland

A. Berne
Environmental Remote Sensing Laboratory, School of Architecture, Civil, and Environmental Engineering, École Polytechnique Fédérale de Lausanne (EPFL), 1015 Lausanne, Switzerland

C. Frankenberg
Jet Propulsion Laboratory, California Institute of Technology, Pasadena, CA, USA

R. Pollock
Jet Propulsion Laboratory, California Institute of Technology, Pasadena, CA, USA

R. A. M. Lee
Jet Propulsion Laboratory, California Institute of Technology, Pasadena, CA, USA

R. Rosenberg
Jet Propulsion Laboratory, California Institute of Technology, Pasadena, CA, USA

J.-F. Blavier
Jet Propulsion Laboratory, California Institute of Technology, Pasadena, CA, USA

D. Crisp
Jet Propulsion Laboratory, California Institute of Technology, Pasadena, CA, USA

C. W. O'Dell
Colorado State University, Fort Collins, CO, USA

G. B. Osterman
Jet Propulsion Laboratory, California Institute of Technology, Pasadena, CA, USA

C. Roehl
California Institute of Technology, Pasadena, CA, USA

P. O. Wennberg
California Institute of Technology, Pasadena, CA, USA

D. Wunch
California Institute of Technology, Pasadena, CA, USA

Y. Ren
Deutsches Zentrum für Luft- und Raumfahrt (DLR), Institut für Physik der Atmosphäre, Oberpfaffenhofen, Germany

R. Baumann
Deutsches Zentrum für Luft- und Raumfahrt (DLR), Institut für Physik der Atmosphäre, Oberpfaffenhofen, Germany

H. Schlager
Deutsches Zentrum für Luft- und Raumfahrt (DLR), Institut für Physik der Atmosphäre, Oberpfaffenhofen, Germany

M. Cazorla
Atmospheric Chemistry and Dynamics Lab, NASA Goddard Space Flight Center, Greenbelt, MD, USA
Universidad San Francisco de Quito, Instituto de Investigaciones Atmosféricas, Colegio de Ciencias e Ingeniería, Diego de Robles y Vía Interoceánica, Quito, Ecuador

G. M. Wolfe
Atmospheric Chemistry and Dynamics Lab, NASA Goddard Space Flight Center, Greenbelt, MD, USA
Joint Center for Earth Systems Technology, University of Maryland Baltimore County, Baltimore, MD, USA

S. A. Bailey
Atmospheric Chemistry and Dynamics Lab, NASA Goddard Space Flight Center, Greenbelt, MD, USA

A. K. Swanson
Atmospheric Chemistry and Dynamics Lab, NASA Goddard Space Flight Center, Greenbelt, MD, USA
Goddard Earth Sciences Technology and Research, University of Maryland Baltimore County, Baltimore, MD, USA

H. L. Arkinson
Department of Atmospheric and Oceanic Science, University of Maryland, College Park, MD, USA

T. F. Hanisco
Atmospheric Chemistry and Dynamics Lab, NASA Goddard Space Flight Center, Greenbelt, MD, USA

J. R. Campbell
Naval Research Laboratory, Monterey, California, USA

M. A. Vaughan
NASA Langley Research Center, Hampton, Virginia, USA

M. Oo
Space Sciences and Engineering Center, University of Wisconsin, Madison, Wisconsin, USA

R. E. Holz
Space Sciences and Engineering Center, University of Wisconsin, Madison, Wisconsin, USA

J. R. Lewis
Joint Center for Earth Systems Technology, University of Maryland Baltimore County, Baltimore, Maryland, USA

E. J. Welton
NASA/Goddard Space Flight Center, Greenbelt, Maryland, USA

N. Utry
Department of Optics and Quantum Electronics, University of Szeged, Szeged, Hungary

T. Ajtai
MTA-SZTE Research Group on Photoacoustic Spectroscopy, Szeged, Hungary

M. Pintér
Department of Optics and Quantum Electronics, University of Szeged, Szeged, Hungary

E. Tombácz
Department of Physical Chemistry and Materials Science, University of Szeged, Szeged, Hungary

E. Illés
Department of Physical Chemistry and Materials Science, University of Szeged, Szeged, Hungary

Z. Bozóki
MTA-SZTE Research Group on Photoacoustic Spectroscopy, Szeged, Hungary

G. Szabó
MTA-SZTE Research Group on Photoacoustic Spectroscopy, Szeged, Hungary

F. A. F. Winiberg
School of Chemistry, University of Leeds, LS2 9JT, Leeds, UK

S. C. Smith
School of Chemistry, University of Leeds, LS2 9JT, Leeds, UK

I. Bejan
School of Chemistry, University of Leeds, LS2 9JT, Leeds, UK

C. A. Brumby
School of Chemistry, University of Leeds, LS2 9JT, Leeds, UK

T. Ingham
School of Chemistry, University of Leeds, LS2 9JT, Leeds, UK
National Centre for Atmospheric Science, University of Leeds, Leeds, LS2 9JT, UK

T. L. Malkin
School of Chemistry, University of Leeds, LS2 9JT, Leeds, UK
Institute for Climate and Atmospheric Science, School of Earth and Environment, University of Leeds, Woodhouse Lane, Leeds, LS2 9JT, UK

S. C. Orr
School of Chemistry, University of Leeds, LS2 9JT, Leeds, UK

D. E. Heard
School of Chemistry, University of Leeds, LS2 9JT, Leeds, UK
National Centre for Atmospheric Science, University of Leeds, Leeds, LS2 9JT, UK

P. W. Seakins
School of Chemistry, University of Leeds, LS2 9JT, Leeds, UK
National Centre for Atmospheric Science, University of Leeds, Leeds, LS2 9JT, UK

G. González Abad
Harvard-Smithsonian Center for Astrophysics, Cambridge, MA, USA

X. Liu
Harvard-Smithsonian Center for Astrophysics, Cambridge, MA, USA

K. Chance
Harvard-Smithsonian Center for Astrophysics, Cambridge, MA, USA

H. Wang
Harvard-Smithsonian Center for Astrophysics, Cambridge, MA, USA

T. P. Kurosu
Jet Propulsion Laboratory, Pasadena, CA, USA

R. Suleiman
Harvard-Smithsonian Center for Astrophysics, Cambridge, MA, USA

S. N. Vardag
Institut für Umweltphysik, Heidelberg University, Heidelberg, Germany

S. Hammer
Institut für Umweltphysik, Heidelberg University, Heidelberg, Germany

M. Sabasch
Institut für Umweltphysik, Heidelberg University, Heidelberg, Germany

D. W. T. Griffith
Department of Chemistry, University of Wollongong, Wollongong, Australia

I. Levin
Institut für Umweltphysik, Heidelberg University, Heidelberg, Germany
Department of Chemistry, University of Wollongong, Wollongong, Australia

W. Su
MS420, NASA Langley Research Center, Hampton, Virginia, USA

J. Corbett
Science Systems & Applications, Inc., Hampton, Virginia, USA

Z. Eitzen
Science Systems & Applications, Inc., Hampton, Virginia, USA

L. Liang
Science Systems & Applications, Inc., Hampton, Virginia, USA